Progress in Mathematical Physics

Volume 37

A.F. Nikiforov
V.G. Novikov
V.B. Uvarov

Quantum-Statistical Models of Hot Dense Matter

Methods for Computation Opacity and Equation of State

Translated from the Russian by Andrei Iacob

Birkhäuser Verlag
Basel · Boston · Berlin

Authors:

Arnold F. Nikiforov
Vladimir G. Novikov
Keldysh Institute of Applied Mathematics
Miusskaya sq., 4
125047 Moscow
Russia
e-mail: arnold@kiam. ru
e-mail: novikov@kiam.ru

Originally published in Russian by Fizmatlit, Physics and Mathematics Publishers Company, Russian Academy of Sciences

2000 Mathematics Subject Classification 80-04, 81-08, 81V45, 82-08, 82D10

A CIP catalogue record for this book is available from the Library of Congress, Washington D.C., USA

Bibliographic information published by Die Deutsche Bibliothek
Die Deutsche Bibliothek lists this publication in the Deutsche Nationalbibliografie; detailed bibliographic data is available in the Internet at <http://dnb.ddb.de>.

ISBN 3-7643-2183-0 Birkhäuser Verlag, Basel – Boston – Berlin

Part of Springer Science+Business Media
Printed on acid-free paper produced of chlorine-free pulp. TCF ∞
Printed in Germany
ISBN-10: 3-7643-2183-0
ISBN-13: 978-3-7643-2183-3

9 8 7 6 5 4 3 2 1 www.birkhauser.ch

Contents

III APPENDIX Methods for solving the Schrödinger and Dirac equations 337

Preface

In the processes studied in contemporary physics one encounters the most diverse conditions: temperatures ranging from absolute zero to those found in the cores of stars, and densities ranging from those of gases to densities tens of times larger than those of a solid body. Accordingly, the solution of many problems of modern physics requires an increasingly large volume of information about the properties of matter under various conditions, including extreme ones. At the same time, there is a demand for an increasing accuracy of these data, due to the fact that the reliability and computational substantiation of many unique technological devices and physical installations depends on them.

The relatively simple models ordinarily described in courses on theoretical physics are not applicable when we wish to describe the properties of matter in a sufficiently wide range of temperatures and densities. On the other hand, experiments aimed at generating data on properties of matter under extreme conditions usually face considerably technical difficulties and in a number of instances are exceedingly expensive. It is precisely for these reasons that it is important to develop and refine in a systematic manner quantum-statistical models and methods for calculating properties of matter, and to compare computational results with data acquired through observations and experiments. At this time, the literature addressing these issues appears to be insufficient. If one is concerned with opacity, which determines the radiative heat conductivity of matter at high temperatures, then one can mention, for example, the books of D. A. Frank-Kamenetskii [67], R. D. Cowan [49], and also the relatively recently published book by D. Salzmann [196]. There are also a number of papers and collections of short conference reports that analyze theoretical models in use and software packages [45, 205, 246, 240, 241]. Let us mention here one of the most perfected software programs, OPAL, and the astrophysical library of opacity coefficients (at the Livermore National Laboratory, USA) that is based on OPAL [92]. A large amount of work on improved models of matter and tables of thermophysical properties was carried out by the T4 group at the Los Alamos National Laboratory, USA. The results of this work are systematized in the SESAME database [246].

The aim of the present book is to give an exposition of a number of quantum-statistical self-consistent field models (Part I) and methods for the computation of properties of matter at high temperatures under conditions of local thermodynamical equilibrium (Part II) — models and methods that recommend themselves well in practice — and also to perform a critical analysis of these approaches, with numerous examples illustrating the effectiveness of the models and numerical methods applied.

In Part I the exposition begins with the very simple and at the same time universal *generalized Thomas-Fermi model* for matter with given density and temperature. This model is then replaced by other, refined ones: the *modified Hartree* and *Hartree-Fock-Slater models*, and also the *relativistic Hartree-Fock-Slater model*. The latter uses the Hartree self-consistent field, an approximation for local exchange that refines the Slater exchange potential, and the relativistic Dirac equation for the radial parts of the wave functions. It is interesting to note that the models mentioned above were first formulated for a free atom at temperature zero, and then generalized to arbitrary temperatures and densities for the so-called *average atom* [188], which corresponds to an ion with average occupation numbers. Thus, for example, a generalization of the Thomas-Fermi model (originally proposed in 1926–1928) was achieved in 1949 by Feynman, Metropolis and Teller in [61].

The quantum-statistical models listed above, among them the Hatree-Fock model for matter with given temperature and density, can be derived by using a unified variational principle, namely, the requirement of minimum of the grand thermodynamic potential, written in the corresponding approximation. This unified approach makes the hierarchic structure of the models transparent and allows one to keep track of the limits of applicability of the various approximations.

The solution of the systems of nonlinear equations arising in the construction of self-consistent field models has required the development of special iteration methods. As an initial approximation for calculating the self-consistent potential a potential found earlier for a less precise model was used. After solving the Schrödinger (or Dirac) equation with the self-consistent potential thus obtained, it became possible to find the energy spectrum of the quantum-mechanical system, the corresponding wave functions, as well as the mean occupation numbers of electron states and the mean degree of ionization of the substance studied. The wide utilization of physical approximations in the iteration process and the special attention paid to the tight spots, which required a large expenditure of computing time, enabled researchers to construct sufficiently efficient and reliable algorithms. Let us point out that reliability of computational methods is extremely important in obtaining tables of thermophysical data, if one takes into account that it may be needed to perform calculations in a wide range of temperatures and densities, for arbitrary substances and mixtures.

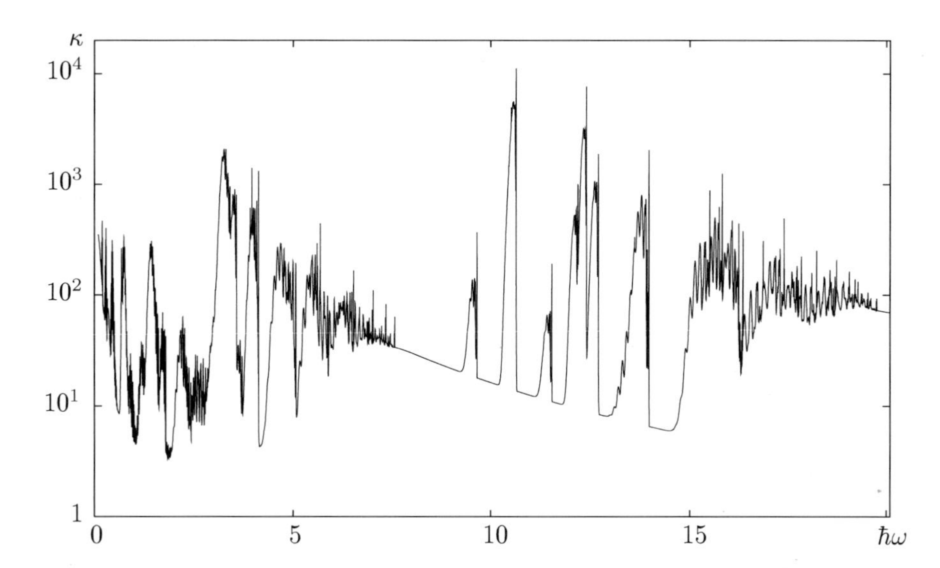

Figure 1: Spectral absorption coefficient κ, in cm^2/g, as a function of the photon energy $\hbar\omega$, in keV, for a gold plasma ($Z = 79$, $T = 1$ keV, $\rho = 0.1$ g/cm^3)

Based on the models considered in Part I, in Part II we propose methods for calculating various characteristics of matter, e.g., spectral photon-absorption coefficients, Rosseland and Planck mean free paths, equations of state, which are necessary in the complex computations that describe hydrodynamic processes with radiative transfer in high-temperature plasmas, in particular, when laser radiation or other sources of energy act on matter. The computational work, which requires accounting for a large number of diverse effects, has a very large volume. To illustrate how complex computations of this kind can be we show here the graph of the spectral photon absorption coefficient in a gold plasma with density $\rho = 0.1$ g/cm^3 at temperature $T = 1$ keV (see Figure 1).

At such high temperatures the transfer of energy is effected mainly by photons. The main processes of interaction of radiation with matter that need to be accounted for here are photon absorption in spectral lines, photoionization, inverse bremsstrahlung, and also Compton scattering.

Despite the fact that the line widths are very small (less than 1 eV), since the number of ion states is huge, especially for matter with high Z, the number of lines can be so large that the plasma heat conductivity in the domain of high temperatures is mainly determined by photon absorption in the spectral lines. For each line one has to take into account the splitting and broadening effects, calculate its profile, determined by the interaction of the ion with electrons and other ions, and compute its intensity and realization probability. It is quite clear that only reliable quantum-statistical models, effective numerical methods and fast computers may help us in understanding the relative roles played by various effects in the investigation of the interaction of radiation with matter and allow us to calculate Rosseland mean opacities in the requisite ranges of temperatures and densities.

The book deals mainly with matter under local thermodynamical equilibrium. Some problems connected with nonequilibrium plasma and methods for deriving its properties are considered in the end of Chapter V.

The calculation of Rosseland mean free paths requires almost always detailed information on ion energy levels and wave functions. At the same time, in the computation of the equation of state one can usually restrict oneself to the average-atom (more precisely — average-ion) model. In Chapter VI formulas for pressure, internal energy and entropy of matter are relatively easily derived in the setting of various models.

A number of problems of general nature, which supplement the traditional course on quantum mechanics, are covered in the Appendix, where we present the methods used in the main text to solve the Schrödinger and Dirac equations for particles moving in a central field. We treat analytic, approximate and numerical methods for the solution of these equations. Although these methods are probably studied well enough, the authors hope that the reader will be attracted not only by the novelty of the methods presented, but also by their effectiveness in applications. Thus, for example, the *phase method* for the numerical integration of the Schrödinger and Dirac equations, which uses considerations connected with the semi-classical limit, enables us to find the energy eigenvalues with high accuracy after only two or three iterations. Moreover, the phase method proved to be little sensitive to the choice of the initial approximation, and hence extremely reliable and sufficiently economical in computating self-consistent potentials in a wide range of temperatures and densities.

In those cases in which the solutions of the Schrödinger and Dirac equations can be found in analytic, closed form, it is recommended to rely not on the study of power series after one extracts the asymptotics, but on a sufficiently general and very simple method of reducing the original equations to *equations of hypergeometric type*. This allows one immediately to obtain the asymptotics and the solution in closed form in terms of *classical orthogonal polynomials.*

The semi-classical approximation is treated in a manner that facilitates its application not only to the Schrödinger equation, but also to the Dirac equation. For the bound-state wave functions it is expedient to use the interpolation form of writing the semi-classical approximation in terms of Bessel functions, which allows one to obtain solutions without angular points in the entire domain of interest. In the numerical integration of the Schrödinger equation for the continuum wave functions, a convenient normalization has been found that uses the semi-classical approximation in the first zero of the wave function.

Over the years the authors of this book took active part in solving problems of modern nuclear physics at the Keldysh Institute of Applied Mathematics. Based on the quantum-statistical models and iteration methods for solving nonlinear systems of equations developed for this purpose, the software package and database THERMOS was created, which allows one to obtain tables of radiative and thermodynamic properties of various substances in a wide range of temperatures and densities. The physical models and computational algorithms used in the THERMOS code, as well as in numerous other applications, constitute the main content of the book we here offer to the reader. The material of the book was used over the years to teach graduate courses for the students of the Physical Faculty of Moscow State University.

The authors are deeply grateful to the scientific workers of the Keldysh Institute of Applied Mathematics and the Russian Federal Nuclear Center, who over many years took part in solving individual problems, in carrying out calculations and experiments, in the discussion and interpretation of the obtained results. It gives us special pleasure to mention here the help and support of Yu. B. Khariton, Ya. B. Zel′dovich, Yu. N. Babaev, V. N. Klimov, A. N. Tikhonov, A. A. Samarskii, Yu. A. Romanov, V. S. Imshennik and A. V. Zabrodin.

The authors wish to express their heartfelt gratitude to their colleagues A. S. Skorobogatova, N. N. Kuchumova, Yu. L. Levitan, N. F. Bitko, V. M. Marchenko, N. I. Leonova, V. V. Dragalov, and A. D. Solomyannaya, with whom they collaborated to solve the problem assigned by the Institute, and to K. V. Ivanova, N. N. Fimin, and V. V. Nagovitsyn, who helped preparing the manuscript for publication.

Invaluable help in the understanding of many subtle problems came from numerous discussions at a number of meetings and conferences on equations of state held near Elbrus (chairman Academician V. E. Fortov), and also at the III-rd and IV-th International Opacity Workshop & Code Comparison Study, which took place in 1994 in Garshing, Germany and in 1997 in Madrid, Spain [241, 242]. Valuable remarks were made by S. J. Rose, B. F. Rozsnyai and C. A. Iglesias. We are grateful to the scientific editor of the book, V. S. Yarunin, who read the manuscript and made a number of very useful recommendations, and also to Yu. F. Smirnov, Yu. A. Danilov and Yu. I. Morozov for discussions on separate chapters of the book.

Tragically, this preface is not signed by one of the authors of the book, Vasili B. Uvarov, who died unexpectedly in 1997. V. B. Uvarov, senior scientist at the Keldsyh Institute of Applied Mathematics, professor at the Moscow State University, laureate of the 1962 Lenin prize, was a multilaterally talented and amazingly modest person. He knew how to find original and, at the same time simple methods for solving many difficult problems of contemporary physics and mathematics; some of these are presented in the book.

Moscow, December 2004 *A. F. Nikiforov, V. G. Novikov*

Part I

Quantum-statistical self-consistent field models

Chapter 1

The generalized Thomas-Fermi model

In highly-ionized hot plasmas electrons and ions interact between themselves mainly through the electrostatic attraction or repulsion of their charges. For this reason, the first and foremost problem in the study of microscopic properties of matter is that of determining the electrostatic potential in the field of which each particle is situated, and also of determining the charge density at any point. Obviously, these two quantities must be consistent, since the electrostatic field determines the charge distribution, which in turn is the source of the field.

The calculation of *steady* (or *stationary*) *states of an electron-ion system* is a very difficult problem. To simplify it one can use the fact that the mass of the electron is many orders of magnitude smaller than the mass of an atomic nucleus, but, on the other hand, the electrons and nuclei are acted upon by forces of the same order of magnitude. Consequently, nuclei move considerably slower than electrons, and to a high degree of accuracy one can assume that with respect to electrons the nuclei are fixed force centers. When nuclei shift position, electrons reorganize themselves rapidly, and so one can consider the equilibrium state of the electron system for a fixed position of the nuclei, assuming that in any macroscopically small volume a charge neutrality is preserved.

We shall consider that the equilibrium state of the system of interacting electrons corresponds to the most probable distribution of electrons over their energies, under the assumption that the total energy of the system and the total number of electrons are preserved. The potential for the most probable electron distribution will satisfy a nonlinear Poisson equation that connects the self-consistent potential with the electron density.

1.1 The Thomas-Fermi model for matter with given temperature and density

1.1.1 The Fermi-Dirac statistics for systems of interacting particles

At high temperatures the *generalized Thomas-Fermi model* is the best and easiest to implement model of dense matter [61]. This model is based on the Fermi-Dirac statistics and the semiclassical approximation for electrons, i.e., it is assumed that the electrons of atoms are continuously distributed in phase space according to the Fermi-Dirac statistics. The Fermi-Dirac distribution is usually derived for the case of an ideal gas in an external field, when one can speak about the energy levels ε_k of an individual particle (see, e.g., [115]). One then assumes that the total energy E of the system is equal to the sum of the energies of the individual particles:

$$E = \sum_k N_k \varepsilon_k.$$

Here the mutual interaction of electrons is neglected and the values of the electron energies ε_k are independent of the occupation numbers N_k. These original assumptions are not applicable to electrons of atoms, because, first, their interaction cannot be neglected and, second, the energy levels of electrons in atoms depend on the occupancy of levels.

Let us derive the Fermi-Dirac distribution taking into account the Coulomb interaction of electrons in the self-consistent field approximation. We shall assume that our system is enclosed in some fixed volume. In the one-dimensional case, where the surface element in phase space is $dx\,dp$ (with x the coordinate and p the momentum), one can easily see, using the Bohr-Sommerfeld quantization rule, that one quantum state requires an area equal to $2\pi\hbar$ [120]. If we take into consideration the spin of the electron, which results in the doubling of the number of states, and use the system of atomic units ($e = 1$, $m = 1$, $\hbar = 1$), then after the natural generalization of the one-dimensional result to the three-dimensional case, we conclude that the number of states in the phase-space volume $d\vec{r}\,d\vec{p}$ is equal to $2\,d\vec{r}\,d\vec{p}/(2\pi)^3$ (here $\vec{p}$ is the electron momentum, $d\vec{r} = dx\,dy\,dz$, $d\vec{p} = dp_x dp_y dp_z$). This quantity corresponds to the statistical weight g_k of the level with energy ε_k.

If $n(\vec{r}, \vec{p})$ denotes the degree of occupation of the phase-volume element $d\vec{r}d\vec{p}$ by electrons, then the total number of electrons in the system is

$$N = \iint n(\vec{r}, \vec{p}) \frac{2d\vec{r}\,d\vec{p}}{(2\pi)^3}. \tag{1.1}$$

On the other hand,

$$N = \int \rho(\vec{r})\,d\vec{r},$$

where $\rho(\vec{r})$ is the density of electrons (more precisely, their concentration). Comparing with (1.1), we get

$$\rho(\vec{r}) = \int n(\vec{r}, \vec{p}) \frac{2d\vec{p}}{(2\pi)^3}. \tag{1.2}$$

Let us find the total energy E of the electron system. The potential energy of the electrons is

$$E_{\mathrm{p}} = -\int \rho(\vec{r}) V_a(\vec{r})\, d\vec{r} + \frac{1}{2} \iint \frac{\rho(\vec{r})\rho(\vec{r}\,')}{|\vec{r} - \vec{r}\,'|}\, d\vec{r}\, d\vec{r}\,',$$

where $V_{\mathrm{a}}(\vec{r})$ is the potential of atomic nuclei. Therefore, in the nonrelativistic approximation the sum of the kinetic and potential energies of the electron system is given by the expression

$$E = \iint \left[\frac{p^2}{2} - V_a(\vec{r})\right] n(\vec{r}, \vec{p}) \frac{2d\vec{r}\, d\vec{p}}{(2\pi)^3} + \frac{1}{2} \iint \frac{\rho(\vec{r})\rho(\vec{r}\,')}{|\vec{r} - \vec{r}\,'|}\, d\vec{r}\, d\vec{r}\,'. \tag{1.3}$$

In accordance with the fundamental principles of statistical thermodynamics, we start from the fact that *a closed system can be found with equal probability in any admissible steady quantum state.* Since we consider that the energy E of the electron system is fixed, the probability of a given electron distribution is proportional to the number of ways in which the given state can be realized for a fixed energy E and total number of electrons N [119].

Let S be the logarithm of the number of admissible states of the system, i.e., the entropy of the system [111]. If one assumes that it makes sense to speak about the state of a single electron in the field $V(\vec{r})$, for which, in accordance with the Pauli principle, $N_k \leq g_k$, then for given occupation numbers N_k we have

$$S = \ln \prod_k \binom{g_k}{N_k} = \sum_k \ln \frac{g_k!}{(g_k - N_k)!\, N_k!}.$$

When this formula is used it is not necessary to assume that the particles do not interact. It suffices that near the equilibrium state a change in the occupation numbers N_k will not result in a change in the number of energy levels and their degeneracy (i.e., their statistical weights).

Using the Stirling formula $n! \approx \sqrt{2\pi n}(n/e)^n$ for the factorial of large numbers, we obtain for $n \gg \ln n$

$$S = \sum_k \left[g_k \ln g_k - (g_k - N_k)\ln(g_k - N_k) - N_k \ln N_k\right] =$$

$$\sum_k \left[g_k \ln \frac{g_k}{g_k - N_k} - N_k \ln \frac{N_k}{g_k - N_k}\right] =$$

$$-\sum_k g_k \left[n_k \ln n_k + (1 - n_k)\ln(1 - n_k)\right],$$

where $n_k = N_k/g_k$. Then the semiclassical approximation for the entropy S yields

$$S = -\iint [n \ln n + (1-n)\ln(1-n)] \frac{2d\vec{r}\,d\vec{p}}{(2\pi)^3}, \tag{1.4}$$

where $n = n(\tilde{r}, \tilde{p})$ and it is assumed that the phase space element $d\vec{r}\,d\vec{p}$ is macroscopically small, but contains a sufficiently large number of particles.

We need to find an occupancy distribution $n(\vec{r}, \vec{p})$ for which the quantity S is maximal for the given total energy E and total number of electrons N. Therefore, the most probable distribution can be found by setting $\delta S = 0$ and varying $n(\vec{r}, \vec{p})$ while keeping E and N fixed. In order to consider all the variations $\delta n(\vec{r}, \vec{p}) = \delta n$ as independent, we will solve the problem with the help of undetermined Lagrange multipliers $\lambda_1 = -1/\theta$ and $\lambda_2 = \mu/\theta$:

$$\delta S - \frac{1}{\theta}\,\delta E + \frac{\mu}{\theta}\,\delta N = 0. \tag{1.5}$$

As it will become clear below, the meaning of the quantities μ and θ is that of chemical potential and temperature, respectively.

Using relations (1.1)–(1.4), let us calculate the variations δS, δE and δN, which figure in (1.5):

$$\delta S = \iint \delta n \ln\left(\frac{1-n}{n}\right) \frac{2d\vec{r}\,d\vec{p}}{(2\pi)^3},$$

$$\begin{aligned}\delta E = \iint \delta n \left(\frac{p^2}{2} - V_a(\vec{r})\right) \frac{2d\vec{r}\,d\vec{p}}{(2\pi)^3} &+ \\ \frac{1}{2}\iint \frac{\delta\rho(\vec{r})\,\rho(\vec{r}\,')}{|\vec{r}-\vec{r}\,'|}\,d\vec{r}\,d\vec{r}\,' + \frac{1}{2}\iint \frac{\delta\rho(\vec{r}\,')\,\rho(\vec{r})}{|\vec{r}-\vec{r}\,'|}\,d\vec{r}\,d\vec{r}\,' &= \\ \iint \delta n \left(\frac{p^2}{2} - V_a(\vec{r})\right) \frac{2d\vec{r}\,d\vec{p}}{(2\pi)^3} + \iint \delta\rho(\vec{r}) \frac{\rho(\vec{r}\,')}{|\vec{r}-\vec{r}\,'|}\,d\vec{r}\,d\vec{r}\,'&.\end{aligned}$$

Since

$$\iint \delta\rho(\vec{r}) \frac{\rho(\vec{r}\,')}{|\vec{r}-\vec{r}\,'|}\,d\vec{r}\,d\vec{r}\,' = \iint \delta n \left[\int \frac{\rho(\vec{r}\,')\,d\vec{r}\,'}{|\vec{r}-\vec{r}\,'|}\right] \frac{2\,d\vec{r}\,d\vec{p}}{(2\pi)^3} = -\iint \delta n V_e(\vec{r}) \frac{2d\vec{r}\,d\vec{p}}{(2\pi)^3},$$

it follows that

$$\delta E = \iint \delta n \left[\frac{p^2}{2} - V(\vec{r})\right] \frac{2\,d\vec{r}\,d\vec{p}}{(2\pi)^3}.$$

Here $V(\vec{r})$ is the potential generated by the electrons and the atomic nuclei:

$$V(\vec{r}) = V_a(\vec{r}) + V_e(\vec{r}), \quad V_e(\vec{r}) = \int \frac{\rho(\vec{r}\,')\,d\vec{r}\,'}{|\vec{r}-\vec{r}\,'|}.$$

The expression that we obtain for the variation δE in a self-consistent field $V(\vec{r})$ is formally identical with that for noninteracting electrons placed in the external field $V(\vec{r})$. Further, we obviously have

$$\delta N = \iint \delta n \, \frac{2d\vec{r}\, d\vec{p}}{(2\pi)^3}.$$

Substitution of the expressions for variations in the extremum condition (1.5) yields

$$\iint \delta n(\vec{r}, \vec{p}) \left[\ln \frac{1-n}{n} - \frac{1}{\theta} \left(\frac{p^2}{2} - V(\vec{r}) - \mu \right) \right] \frac{2d\vec{r}\, d\vec{p}}{(2\pi)^3} = 0.$$

Since the variation $\delta n(\vec{r}, \vec{p})$ is arbitrary, one concludes that

$$\ln \frac{1-n}{n} - \frac{1}{\theta} \left(\frac{p^2}{2} - V(\vec{r}) - \mu \right) = 0,$$

whence

$$n(\vec{r}, \vec{p}) = \frac{1}{1 + \exp\left(\dfrac{p^2/2 - V(\vec{r}) - \mu}{\theta} \right)}. \tag{1.6}$$

We see that when the electrostatic interaction between electrons is taken into account we obtain the Fermi-Dirac distribution (1.6) provided that we use the formula (1.4) for the entropy. Note also that asking that the entropy be maximal for a fixed energy E and given number of electrons N is equivalent to asking that the grand thermodynamic potential $\Omega = E - \theta S - \mu N$ achieves a minimum for the given μ and θ (we have $\delta\Omega = 0$ if (1.5) holds).

1.1.2 Derivation of the Poisson-Fermi-Dirac equation for the atomic potential

Distribution (1.6) allows us to find the electron density from (1.2):

$$\rho(\vec{r}) = \frac{2}{(2\pi)^3} \int_0^\infty \frac{4\pi p^2 \, dp}{1 + \exp\left(\dfrac{p^2/2 - V(\vec{r}) - \mu}{\theta} \right)}.$$

It is convenient to change the integration variable to $y = p^2/(2\theta)$. This yields

$$\rho(\vec{r}) = \frac{(2\theta)^{3/2}}{2\pi^2} I_{1/2} \left(\frac{V(\vec{r}) + \mu}{\theta} \right), \tag{1.7}$$

where

$$I_k(x) = \int_0^\infty \frac{y^k \, dy}{1 + \exp(y - x)}$$

is the Fermi-Dirac integral [122].

Since we now have an expression for $\rho(\vec{r})$, we can write down the equation for the potential $V(\vec{r})$. Indeed, $V(\vec{r})$ satisfies the Poisson equation

$$\Delta V = -4\pi \sum_i Z\delta(\vec{r} - \vec{r}_i) + 4\pi\rho(\vec{r}),$$

where $\vec{r}_i$ is the position vector of the i-th nucleus. Using (1.7), we obtain

$$\Delta V = -4\pi \sum_i Z\delta(\vec{r} - \vec{r}_i) + \frac{2}{\pi}(2\theta)^{3/2} I_{1/2}\left(\frac{V(\vec{r}) + \mu}{\theta}\right). \tag{1.8}$$

Equation (1.8), supplemented by boundary conditions for $V(\vec{r})$, enables us, in principle, to determine the self-consistent potential for any given distribution of nuclei. However, it is clear that it is not possible to solve this problem as stated, and hence that its formulation needs to be simplified. Usually one finds the average potential $\widetilde{V}(\vec{r})$ in some domain near a nucleus, whose position is taken as the origin of coordinates.

To obtain the average potential $\widetilde{V}(\vec{r})$, let us average the potential $V(\vec{r})$ over the different positions of the nuclei. The average potential will be spherically symmetric, provided that there is no distinguished direction in the plasma. The Poisson equation (1.8) for $\widetilde{V}(r)$ reads

$$\Delta\widetilde{V} = -4\pi Z\delta(\vec{r}) + \frac{2}{\pi}(2\theta)^{3/2} I_{1/2}\left(\frac{\widetilde{V}(r) + \mu}{\theta}\right), \tag{1.9}$$

where $0 \le r < r_0$ and the value r_0 is calculated from the charge neutrality condition

$$4\pi \int_0^{r_0} \widetilde{\rho}(r) r^2 \, dr = Z. \tag{1.10}$$

Here $\widetilde{\rho}(r)$ is the average electron density corresponding to the potential $\widetilde{V}(r)$. From condition (1.10) and equation (1.9) we obtain the boundary condition for $\widetilde{V}(r)$:

$$\left.\frac{d\widetilde{V}}{dr}\right|_{r=r_0} = 0. \tag{1.11}$$

Furthermore, since the potential is in fact defined only up to a constant term, we take

$$\widetilde{V}(r_0) = 0. \tag{1.12}$$

As r_0 it is natural to take the radius of the average atom cell, determined by the condition

$$\frac{4}{3}\pi(r_0 a_0)^3 n = 1,$$

where $n = \rho N_A / A$ is the number of nuclei per unit of volume, i.e., their concentration, measured in $1/\text{cm}^3$, ρ is the density of matter in g/cm^3, A is the atomic mass, $N_A = 6.022 \cdot 10^{23}$ is the Avogadro number, and $a_0 = \hbar^2/(me^2) = 0.529 \cdot 10^{-8}$ cm is the atomic unit of length. This yields the value

$$r_0 = \frac{1}{a_0} \left(\frac{3}{4\pi} \frac{A}{\rho N_A} \right)^{1/3} = 1.388 \left(\frac{A}{\rho} \right)^{1/3}. \tag{1.13}$$

Thus, the volume of the average atom cell is assumed to be equal to the volume assigned to one ion in matter with density ρ.*) In what follows, instead of $\widetilde{V}(r)$ and $\widetilde{\rho}(r)$ we will use the notation $V(r)$ and $\rho(r)$. Also, for the electron density $\rho(r)$ we will indicate the dependence on the distance r (in contrast to the density of matter ρ).

1.1.3 Formulation of the boundary value problem

If in equations (1.9)–(1.12) we pass to spherical coordinates, we obtain the equation for the Thomas-Fermi potential $V(r)$:

$$\frac{1}{r} \frac{d^2}{dr^2} (rV) = \frac{2}{\pi} (2\theta)^{3/2} I_{1/2} \left(\frac{V(r) + \mu}{\theta} \right), \quad 0 < r < r_0, \tag{1.14}$$

together with the boundary conditions

$$rV(r)|_{r=0} = Z, \quad V(r_0) = 0, \quad \left. \frac{dV(r)}{dr} \right|_{r=r_0} = 0. \tag{1.15}$$

To solve equation (1.14) it suffices to have two conditions; the third condition serves for determining the value of the chemical potential μ. When equation (1.14) is integrated numerically it is convenient to eliminate μ from its right-hand side and replace the function $V(r)$, which becomes infinite at $r = 0$, by a function that remains bounded as $r \to 0$. To this end we make the substitution

$$\frac{V(r) + \mu}{\theta} = \frac{\Phi(r)}{r},$$

which yields the equation

$$\frac{d^2\Phi}{dr^2} = \frac{4}{\pi} \sqrt{2\theta}\, r I_{1/2} \left(\frac{\Phi}{r} \right), \quad 0 < r < r_0. \tag{1.16}$$

The function $\Phi(r)$ is subject to the boundary conditions

$$\Phi(0) = \frac{Z}{\theta}, \quad \frac{d\Phi}{dr} = \left. \frac{\Phi(r)}{r} \right|_{r=r_0}.$$

*) We use the term *average atom cell* assuming that the neutral spherical cell contains the average ion and free electrons.

It is further convenient to pass from the independent variable r to the dimensionless variable $x = r/r_0$, so that for any density of matter the integration of the equation will be carried out over the same interval $0 < x < 1$. After the change of variables

$$\frac{\Phi(r)}{r} = \frac{\phi(x)}{x}, \quad x = \frac{r}{r_0}$$

we arrive at the equation

$$\frac{d^2\phi}{dx^2} = ax\, I_{1/2}\left(\frac{\phi}{x}\right) \quad \left(a = \frac{4}{\pi}\sqrt{2\theta}\, r_0^2\right) \tag{1.17}$$

with the boundary conditions

$$\phi(0) = \frac{Z}{\theta r_0}, \qquad \phi'(1) = \phi(1). \tag{1.18}$$

The condition $V(r_0) = 0$ allows us, after the boundary value problem (1.17)–(1.18) is solved, to find the chemical potential μ. Indeed, we have

$$\frac{V(r_0) + \mu}{\theta} = \left.\frac{\phi(x)}{x}\right|_{x=1},$$

i.e., $\mu/\theta = \phi(1)$. Together with the chemical potential μ it is convenient to use the corresponding dimensionless quantity

$$\eta = -\frac{\mu}{\theta} = -\phi(1),$$

which is positive for large temperatures (provided that the density of matter is not too large; see Figure 1.6 below).

1.1.4 The Thomas-Fermi potential as a solution of the Poisson equation depending on only two variables

Equation (1.17) with the boundary conditions (1.18) is solved for matter with atomic number Z and atomic weight A under given physical conditions, specified by the temperature T and the density ρ. The quantity T will be measured in keV, and so in atomic units $\theta = 36.75\,T$. It is readily seen that the solution of problem (1.17)–(1.18) is determined by only two quantities, a and $\phi(0)$, which can be expressed in terms of new variables $\sigma = \rho/(AZ)$ and $\tau = T/Z^{4/3}$, which in turn play the role of a reduced density and reduced temperature, respectively. Indeed, by (1.13) and (1.18),

$$a = \frac{4}{\pi}\sqrt{2\theta}\, r_0^2 \sim \left(\frac{AZ}{\rho}\right)^{2/3}\left(\frac{\theta}{Z^{4/3}}\right)^{1/2}, \quad \phi(0) = \frac{Z}{\theta r_0} \sim \left(\frac{\rho}{AZ}\right)^{1/3}\frac{Z^{4/3}}{\theta}.$$

The quantities σ and τ completely determine the function $\phi(x)$ and the dimensionless (reduced) chemical potential $\eta = -\phi(1)$.

Therefore, calculations of the Thomas-Fermi potential carried out for some substance (A_1, Z_1) with temperature T_1 and density ρ_1 can be also used for another substance (A_2, Z_2) with temperature T_2 and density ρ_2, as long as the following relations hold:

$$\frac{\rho_1}{A_1 Z_1} = \frac{\rho_2}{A_2 Z_2}, \qquad \frac{T_1}{Z_1^{4/3}} = \frac{T_2}{Z_2^{4/3}}.$$

Moreover, in this case the chemical potentials μ_1 and μ_2 are connected by the relation $\mu_1/\theta_1 = \mu_2/\theta_2 = -\eta$. This self-similarity property allows one to obtain the necessary data (atomic potential, internal energy, entropy, pressure, and so on) for any substance once calculations were carried out for some substance in a sufficiently wide range of temperatures and densities [98].

1.1.5 Basic properties of the Fermi-Dirac integrals

In order to solve problem (1.17)–(1.18) we must know a number of properties of the Fermi-Dirac integrals

$$I_k(x) = \int_0^\infty \frac{y^k\,dy}{1 + \exp(y - x)} \qquad (k > -1) \tag{1.19}$$

and also know how to calculate them. Since the chemical potential μ may take both positive and negative values, the argument of the integral $I_k(x)$ may vary within wide limits, as seen from (1.14). Hence, it is useful to study the asymptotic behavior of the integrals $I_k(x)$ as $x \to \pm\infty$.

a) Let $x \gg 1$. Since the graph of the function

$$f(y) = \frac{1}{1 + \exp(y - x)}$$

is step-shaped (see Figure 1.1), for a fixed value of $x \gg 1$ we have

$$I_k(x) \approx \int_0^x y^k dy = \frac{x^{k+1}}{k+1}. \tag{1.20}$$

In particular, $I_{1/2}(x) \approx \dfrac{2}{3}\, x^{3/2}, \quad I_{3/2}(x) \approx \dfrac{2}{5} x^{5/2}.$

b) Let $x \ll -1$. Then, neglecting the 1 in the denominator of the integral (1.19), we have

$$I_k(x) \approx \int_0^\infty e^{x-y}\, y^k dy = \Gamma(k+1)\, e^x. \tag{1.21}$$

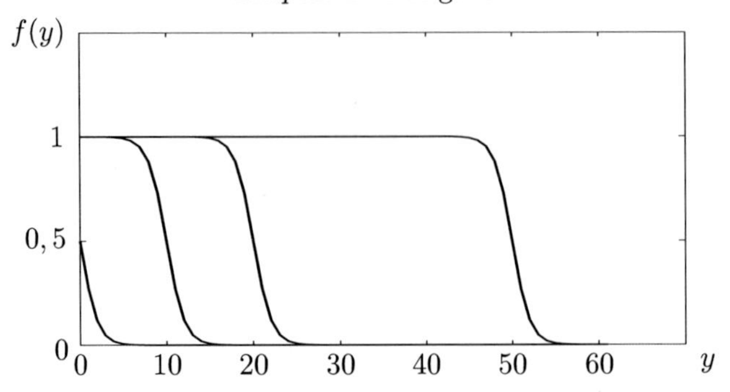

Figure 1.1: The family of curves $f(y) = \dfrac{1}{1+\exp(y-x)}$ for $x = 0,\ 10,\ 20,\ 50$

Consequently, for large in modulus negative values of the argument x

$$I_{1/2}(x) \approx \frac{1}{2}\sqrt{\pi}\, e^{x}, \quad I_{3/2}(x) \approx \frac{3}{4}\sqrt{\pi}\, e^{x}. \tag{1.22}$$

The graph of the integral $I_k(x)$ is sufficiently simple to trace (Figure 1.2).

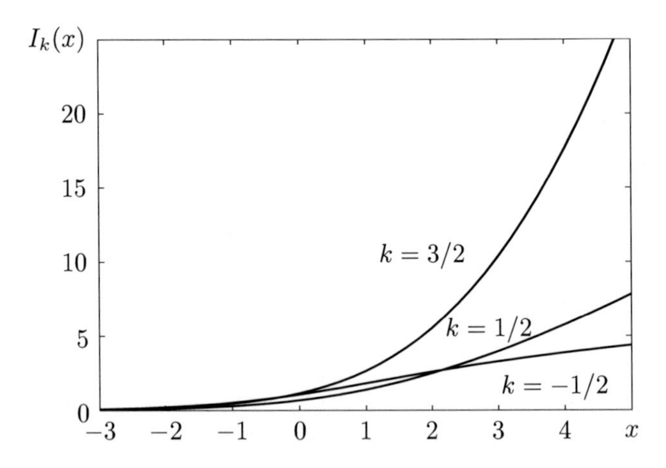

Figure 1.2: The Fermi-Dirac integrals $I_k(x)$, $k = -1/2,\ 1/2,\ 3/2$

The next terms in the asymptotic expansions (1.20) and (1.21), and also sufficiently accurate tables and interpolation formulas for the computation of the integrals $I_k(x)$ can be found in [46].

To conclude this subsection, let us derive the formula for the differentiation of the integrals $I_k(x)$:

$$I'_k(x) = \int_0^\infty y^k \frac{d}{dx}\left(\frac{1}{1+\exp(y-x)}\right) dy = -\int_0^\infty y^k \frac{d}{dy}\left(\frac{1}{1+\exp(y-x)}\right) dy =$$

$$- y^k \frac{1}{1+\exp(y-x)}\bigg|_0^\infty + \int_0^\infty \frac{k y^{k-1}}{1+\exp(y-x)} = k I_{k-1}(x).$$

Thus, for $k > 0$

$$I'_k(x) = k\, I_{k-1}(x). \tag{1.23}$$

1.1.6 The uniform free-electron density model

When the kinetic energy of the electrons is large compared with their potential energy, i.e., $V(r)/\theta \ll 1$ throughout most of the average atom cell, which corresponds to either high temperatures or densities of matter, (1.7) shows that the electron density in the cell is

$$\rho(r) \approx \rho(r_0) = \frac{(2\theta)^{3/2}}{2\pi^2} I_{1/2}(-\eta) = \rho_e. \tag{1.24}$$

If Z_0 is the average ion charge, then obviously

$$\rho_e = \frac{Z_0}{(4/3)\pi r_0^3}. \tag{1.25}$$

Note that formula (1.24) may be valid not only for $V(r)/\theta \ll 1$, but also in the case of a degenerate gas of electrons, when $V(r)/\theta$ is sufficiently large, but $V(r)/\theta \ll -\eta$.

According to (1.24) and (1.25), the average ion charge Z_0 is given by

$$Z_0 = \frac{4\pi}{3} r_0^3 \frac{(2\theta)^{3/2}}{2\pi^2} I_{1/2}(-\eta) = 317.5 \frac{AT^{3/2}}{\rho} \frac{2}{\sqrt{\pi}} I_{1/2}(-\eta) \tag{1.26}$$

Recall that here and in what follows T is the temperature in keV and ρ is the density of matter in g/cm^3. Formula (1.26) shows that the average ionization degree $\alpha = Z_0/Z$ is a function of the variables $\sigma = \rho/(AZ)$ and $\tau = T/Z^{4/3}$, i.e., $\alpha = \alpha(\sigma, \tau)$, which is in agreement with Subsection 1.1.4.

In the limit $\eta \gg 1$ we get the Boltzmann statistics from the Fermi-Dirac statistics. Using the asymptotics of the integral $I_{1/2}(x)$ for large negative x (see (1.22)), we derive from (1.26) the following expression for the average ion charge of a classical ideal plasma:

$$Z_0 = 317.5 \frac{AT^{3/2}}{\rho} e^{-\eta}.$$

If the effective radius of the ion core, $r = r^*$, the magnitude of which can be estimated from the electron density distribution (see Figure 2.3), is much smaller than the dimensions of the average atom cell, then the condition $V(r)/\theta \ll 1$ is satisfied almost everywhere in the cell and we arrive at the *uniform free-electron density model*. The corresponding potential $V(r)$ is determined by the equation $\Delta V = 4\pi\rho_e$ and the boundary conditions

$$rV(r)|_{r=0} = Z_0, \quad V(r_0) = \left.\frac{dV}{dr}\right|_{r=r_0} = 0.$$

This yields

$$V(r) = \frac{Z_0}{r}\left[1 - \frac{3}{2}\frac{r}{r_0} + \frac{1}{2}\left(\frac{r}{r_0}\right)^3\right]. \tag{1.27}$$

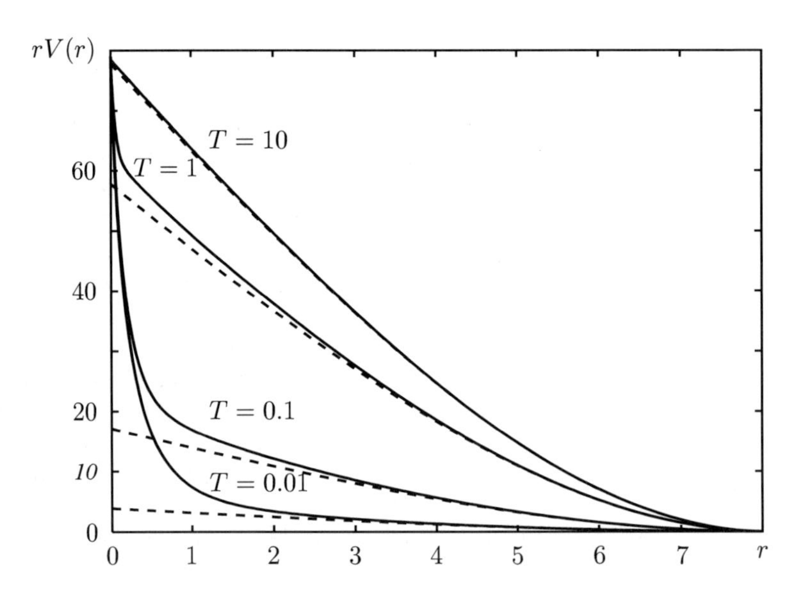

Figure 1.3: The function $rV(r)$ in the Thomas-Fermi model (solid curves) and in the approximation of uniform electron density (dashed curves) for gold with density $\rho = 1$ g/cm^3 and different values of the temperature T in keV

Figure 1.3 shows the curves $rV(r)$ for different values of the temperature for the Thomas-Fermi potential and the potential (1.27); Table 1.1 gives the corresponding values of the chemical potential η, the average ion charge Z_0 (see (1.26)) and the effective radius r^* of the ion core. As expected, the potential (1.27) and the Thomas-Fermi potential are close to one another for $r > r^*$, but differ considerably for $r < r^*$, as clearly displayed by Figure 1.3 and Table 1.1.

Table 1.1: Reduced chemical potential η, average ion charge Z_0, effective radius r^* of the ion core in the Thomas-Fermi model for gold at density $\rho = 1$ g/cm^3 and different values of the temperature in keV (the radius of the atom cell is $r_0 = 8.08$)

T	0.01	0.1	1	10
η	2.78	4.75	6.99	10.1
Z_0	3.80	17.1	57.8	77.9
r^*	2.63	2.58	0.83	0.11

As one can see in Figure 1.3, the uniform free-electron density model may serve as an initial approximation for the Thomas-Fermi model, as, incidentally, the Thomas-Fermi model itself does for quantum-mechanical models of matter considered in Chapter 3.

1.1.7 The Thomas-Fermi model at temperature zero

For small θ, when $(V(r) + \mu)/\theta \gg 1$ throughout most of the cell, the equation of the Thomas-Fermi potential can be obtained by passing to the limit in (1.14) and using (1.20):

$$\frac{1}{r}\frac{d^2}{dr^2}(rV) = \frac{4}{3\pi}(2\theta)^{3/2}\left(\frac{V(r)+\mu}{\theta}\right)^{3/2} \tag{1.28}$$

(as seen from (1.28), the temperature θ gets cancelled in this process). It is convenient to introduce the new function

$$\Phi_0(r) = r\,[V(r) + \mu]\,, \tag{1.29}$$

which satisfies the equation

$$\sqrt{r}\,\frac{d^2\Phi_0}{dr^2} = \frac{8\sqrt{2}}{3\pi}\,\Phi_0^{3/2}, \quad 0 < r < r_0. \tag{1.30}$$

Equation (1.30) was studied in many works [213, 60, 73, 127]. It remains valid also for large values of the chemical potential $\mu\,(\mu \gg \theta)$, which may be the case not only when $\theta \to 0$, but also at relatively high temperatures in the case of strong compression, when the electrons constitute a degenerate Fermi gas.

1.2 Methods for the numerical integration of the Thomas-Fermi equation

1.2.1 The shooting method

As a preliminary step, let us analyze the qualitative behavior of the solution of the Thomas-Fermi equation (1.17)

$$\frac{d^2\phi}{dx^2} = axI_{1/2}\left(\frac{\phi}{x}\right), \quad a = \frac{4}{\pi}\sqrt{2\theta}\, r_0^2, \ \ 0 < x < 1, \tag{1.31}$$

with the boundary conditions

$$\phi(0) = \frac{Z}{\theta r_0}, \quad \phi(1) = \phi'(1). \tag{1.32}$$

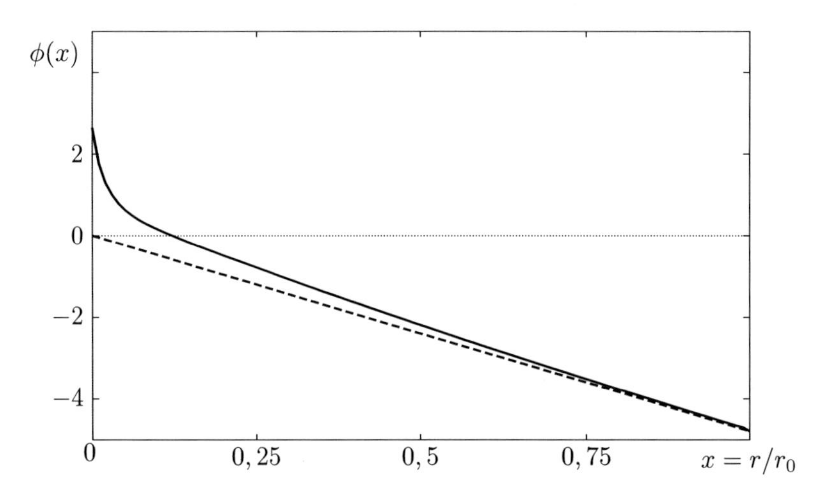

Figure 1.4: Typical profile of the function $\phi(x)$ (solid curve) and the line $y = -\eta x$ (dashed line). The calculations were done for gold at temperature $T = 0.1$ keV and density $\rho = 1$ g/cm^3

Since $\phi'' > 0$, the function $\phi(x)$ is concave on the interval $0 < x < 1$. In the case $\eta > 0$ ($\eta = -\phi(1)$) we have $\phi(1) = \phi'(1) < 0$. Observing that the potential $V(r)$ changes the most for small values of r and that the curve $\phi(x)$ is tangent to the line $y = -\eta x$ at the point $x = 1$, we obtain the typical profile of $\phi(x)$ shown in Figure 1.4.

To integrate equation (1.31) numerically it is convenient to start at the point $x = 1$, move to the left for a given trial value η, and verify that the boundary condition is satisfied at $x = 0$, and then continue, modifying η and "shooting" till one reaches the "target" $\phi(0) = Z/(\theta r_0)$ by some algorithm for the numerical integration of equation (1.31).

Since the function $\phi(x)$ changes the most when x is small, the best way to proceed is to work with a variable step in x, or with a constant step in a new variable, for example, $u = \sqrt{x}$, which makes the profile $\phi(u)$ more shallower. The accuracy of the computations is guaranteed if one uses the Runge-Kutta method with an automatic step choice.

When using the shooting method one can apply Newton's method to find η from the equation $f(\eta) = Z/(\theta r_0)$, where the values of the function $f(\eta) = \phi(x)|_{x=0}$ are calculated by integrating the equation (1.31) with the boundary condition $\phi(1) = \phi'(1) = -\eta$. Since the function $F(\eta) = \ln f(\eta)$ is closer to a linear function than $f(\eta)$, we will start with the equation

$$F(\eta) = \ln \frac{Z}{\theta r_0}, \tag{1.33}$$

which satisfies better the conditions for the applicability of Newton's method. In this manner we obtain the following iteration scheme:

$$F(\eta^{(s)}) + (\eta^{(s+1)} - \eta^{(s)}) \frac{dF}{d\eta}\bigg|_{\eta=\eta^{(s)}} = \ln \frac{Z}{\theta r_0},$$

i.e.,

$$\eta^{(s+1)} = \eta^{(s)} + \frac{\ln \dfrac{Z}{\theta r_0} - F(\eta)}{dF/d\eta}\Bigg|_{\eta=\eta^{(s)}}. \tag{1.34}$$

To find $dF/d\eta$, simultaneously with the Thomas-Fermi equation (1.31) one can solve the equation for the function $\chi(x) = \partial\phi/\partial\eta$:

$$\chi'' = \frac{a}{2} \chi I_{-1/2}\left(\frac{\phi}{x}\right), \quad \chi(1) = \chi'(1) = -1. \tag{1.35}$$

After integrating (1.31) and (1.35), we obtain $dF/d\eta = \chi(0)/\phi(0)$.

In choosing the initial approximation $\eta^{(0)}$ it is convenient to use the uniform electron density model (1.24), (1.25) and the assumption that $Z_0 = Z$, which yields

$$I_{1/2}(-\eta) = \frac{3\pi}{2} \frac{Z}{(2\theta)^{3/2} r_0^3}. \tag{1.36}$$

To calculate the initial value $\eta = \eta^{(0)}$, we replace $I_{1/2}(x)$ in formula (1.36) by the following function:

$$\widetilde{I}_{1/2}(x) = \sqrt{\frac{3}{2}} \left\{ \ln \left[1 + \left(\frac{\pi}{6}\right)^{1/3} \exp\left(\frac{2x}{3}\right) \right] \right\}^{3/2}. \tag{1.37}$$

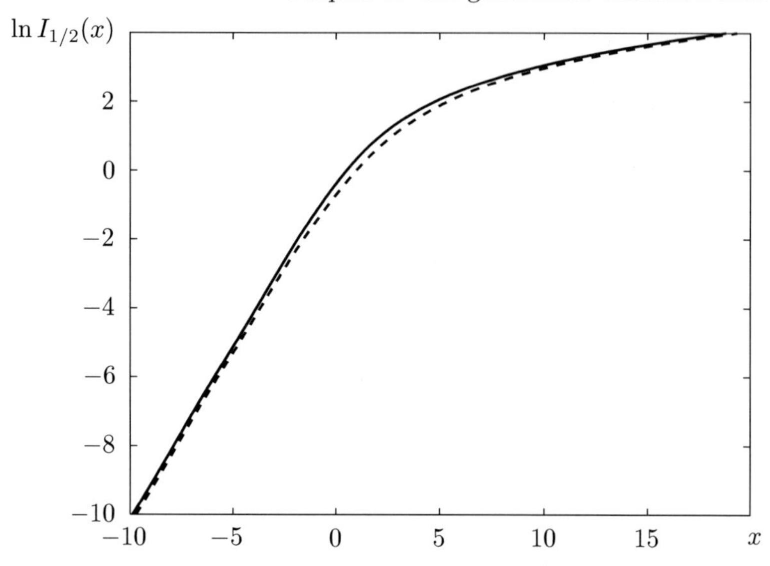

Figure 1.5: Graphs of the functions $\ln I_{1/2}(x)$ (solid curve) and $\ln \widetilde{I}_{1/2}(x)$ (dashed curve)

Indeed, this function has the same asymptotic behaviors as $I_{1/2}(x)$ when $x \to \infty$ and $x \to -\infty$, and differs from the latter at $x = 0$ by less than 20% (Figure 1.5).

In this way we obtain

$$\eta^{(0)} = \frac{1}{2} \ln \frac{\pi}{6} - \frac{3}{2} \ln \left[\exp \left(\sqrt[3]{\frac{2q^2}{3}} \right) - 1 \right], \tag{1.38}$$

where

$$q = \frac{3\pi}{2} \frac{Z}{(2\theta)^{3/2} r_0^3}.$$

This expression can be simplified for small as well as for large values of q:

$$\eta^{(0)} \approx -\left(\frac{3}{2} q \right)^{2/3} \quad \text{for} \quad q \gg 1,$$

$$\eta^{(0)} \approx \ln \left(\frac{\sqrt{\pi}}{2q} \right) \quad \text{for} \quad q \ll 1.$$

Note that the value of η furnished by (1.38) is usually somewhat lower than the true value.

A representation about the dependence of $\phi(0)$ on $\eta = -\phi(1)$ is provided by the Latter graphs [122], which show the curves $y = \phi(0) = f(\eta)$ for different

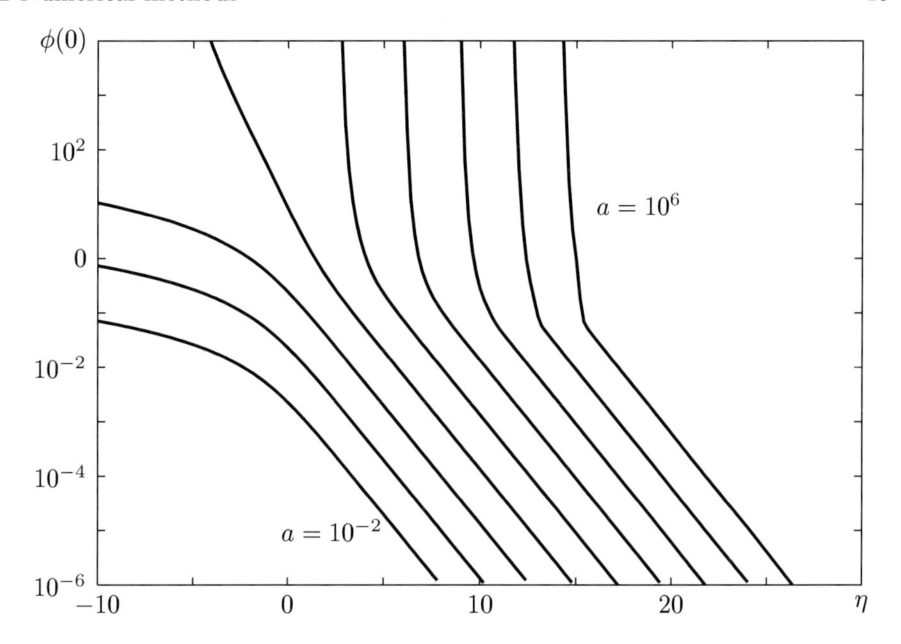

Figure 1.6: Dependence of $\phi(0)$ on $\eta = -\phi(1)$ for fixed values of the parameter a ($a = 10^{-2}, 10^{-1}, \ldots, 10^6$)

values of the coefficient $a = (4\sqrt{2\theta}/\pi)\, r_0^2$ (Figure 1.6). Earlier these graphs were used to obtain the initial approximation $\eta^{(0)}$.

The iterations (1.34) may diverge in some cases; for example, this may happen for very low temperatures. This is usually connected with a bad choice for the initial approximations of $\eta = \eta^{(0)}$. To ensure that the iterations will converge it is useful to combine Newton's method with the bisection method. A method that is faster than the shooting method is the double-sweep method with iterations [153].

1.2.2 Linearization of the equation and a difference scheme

The double-sweep method does not apply directly to problem (1.31)–(1.32) because equation (1.31) is nonlinear in the unknown function $\phi(x)$. For this reason we will first linearize the right-hand side of (1.31). We have

$$I_{1/2}\left(\frac{\phi}{x}\right) \approx I_{1/2}\left(\frac{\widetilde{\phi}}{x}\right) + \frac{\phi - \widetilde{\phi}}{x}\,\frac{d}{d\zeta}\,I_{1/2}(\zeta)\Big|_{\zeta = \widetilde{\phi}/x} = I_{1/2}\left(\frac{\widetilde{\phi}}{x}\right) + \frac{\phi - \widetilde{\phi}}{2x}\,I_{-1/2}\left(\frac{\widetilde{\phi}}{x}\right). \qquad (1.39)$$

Here $\widetilde{\phi}(x)$ is some approximation of $\phi(x)$. Since the potential $V(r)$ varies most rapidly for small r, we can write equation (1.31) in difference form on a nonuniform in r grid $x_i = r_i/r_0$ $(i = 0, 1, 2, \ldots, N;\ 0 \leq x \leq 1)$, using the following approximation for the derivative $\phi''(x)$:

$$\left.\frac{d^2\phi}{dx^2}\right|_{x=x_i} \approx \frac{1}{(x_{i+1}+x_i)/2 - (x_i + x_{i-1})/2}\left(\frac{\phi_{i+1} - \phi_i}{x_{i+1} - x_i} - \frac{\phi_i - \phi_{i-1}}{x_i - x_{i-1}}\right) = \frac{2}{x_{i+1} - x_{i-1}}\left(\frac{\phi_{i+1} - \phi_i}{x_{i+1} - x_i} - \frac{\phi_i - \phi_{i-1}}{x_i - x_{i-1}}\right).$$

Then equation (1.31) becomes

$$\frac{2}{x_{i+1} - x_{i-1}}\left(\frac{\phi_{i+1} - \phi_i}{x_{i+1} - x_i} - \frac{\phi_i - \phi_{i-1}}{x_i - x_{i-1}}\right) = a x_i I_{1/2}\left(\frac{\widetilde{\phi}_i}{x_i}\right) + \frac{a(\phi_i - \widetilde{\phi}_i)}{2} I_{-1/2}\left(\frac{\widetilde{\phi}_i}{x_i}\right). \tag{1.40}$$

The boundary condition (1.32) at $x = 0$ for the difference scheme (1.40) is obvious:

$$\phi_0 = \frac{Z}{\theta r_0}. \tag{1.41}$$

To obtain the boundary condition at $x = 1$ in the difference form we use Taylor's formula:

$$\phi_{N-1} = \phi_N - \phi'_N h_N + \frac{1}{2}\phi''_N h_N^2 + O(h_N^3),\ h_N = x_N - x_{N-1} = 1 - x_{N-1}. \tag{1.42}$$

Here in accordance with (1.32) and (1.39) we must put

$$\phi'_N = \phi_N, \quad \phi''_N = aI_{1/2}(\phi_N) \approx aI_{1/2}(\widetilde{\phi}_N) + \frac{a(\phi_N - \widetilde{\phi}_N)}{2} I_{-1/2}(\widetilde{\phi}_N).$$

Upon replacing in equalities (1.40)–(1.42) $\widetilde{\phi}_i$ by $\phi_i^{(s)}$ and ϕ_i by $\phi_i^{(s+1)}$, where s is the iteration number, we obtain a linear difference scheme for $\phi_i^{(s+1)}$ in dependence on $\phi_i^{(s)}$, which can be solved by a double-sweep method with iterations (see [70], and also [197], p. 34).

1.2.3 Double-sweep method with iterations

Equation (1.40) is of the form

$$a_i y_{i-1} + b_i y_i + c_i y_{i+1} = d_i, \tag{1.43}$$

where $y_i = \phi_i^{(s+1)}$ is the unknown function, $i = 1, \ldots, N-1$,

$$a_i = \frac{1}{x_i - x_{i-1}}, \quad b_i = -a_i - c_i - \frac{a(x_{i+1} - x_{i-1})}{4} I_{-1/2}\left(\frac{\phi_i^{(s)}}{x_i}\right),$$

$$c_i = \frac{1}{x_{i+1} - x_i}, \quad d_i = a(x_{i+1} - x_{i-1}) \left[\frac{x_i}{2} I_{1/2}\left(\frac{\phi_i^{(s)}}{x_i}\right) - \frac{\phi_i^{(s)}}{4} I_{-1/2}\left(\frac{\phi_i^{(s)}}{x_i}\right)\right].$$

We will solve equation (1.43) on the nonuniform grid $x_i = (i/N)^2$ by the double-sweep method, setting

$$y_{i+1} = \alpha_i y_i + \beta_i. \tag{1.44}$$

The coefficients α_i and β_i are calculated for $i = N-1, N-2, \ldots, 1$ by formulas obtained via substitution of expression (1.44) in (1.43):

$$\alpha_{i-1} = -\frac{a_i}{b_i + c_i \alpha_i}, \quad \beta_{i-1} = \frac{d_i - c_i \beta_i}{b_i + c_i \alpha_i}.$$

The values α_{N-1} and β_{N-1} are determined from the boundary condition (1.42):

$$\alpha_{N-1} = \frac{1}{1 - h_N + \frac{h_N^2}{4} a I_{-1/2}\left(\phi_N^{(s)}\right)},$$

$$\beta_{N-1} = \frac{-\frac{h_N^2}{2} a \left[I_{1/2}\left(\phi_N^{(s)}\right) - \frac{\phi_N^{(s)}}{2} I_{-1/2}\left(\phi_N^{(s)}\right)\right]}{1 - h_N + \frac{h_N^2}{4} a I_{-1/2}\left(\phi_N^{(s)}\right)}.$$

Since $y_0 = Z/(\theta r_0)$ is known, formula (1.44) with $i = 0, 1, \ldots, N-1$ yields $y_i = \phi_i^{(s+1)}$.

To start the calculations we need initial approximations for η and $\phi(x)$. If we use the uniform electron density model, then according to expression (1.27) we have

$$\phi^{(0)}(x) = \frac{Z}{\theta r_0}\left(1 - \frac{3}{2}x + \frac{1}{2}x^3\right) - \eta^{(0)} x, \tag{1.45}$$

where $\eta^{(0)}$ is given by (1.38).

The convergence of iterations is illustrated in Figure 1.7, which shows the graphs of the functions $\theta\phi^{(s)}(x)$ $(s = 0, 1, 2, \ldots)$ for gold ($Z = 79$) with density $\rho = 1$ g/cm^3 and for different values of the temperature T in keV. The graphs show that even for $T = 0$, when the initial approximation is rather crude, the function $\phi^{(s)}(x)$ is close to the solution $\phi(x)$ after two iterations.

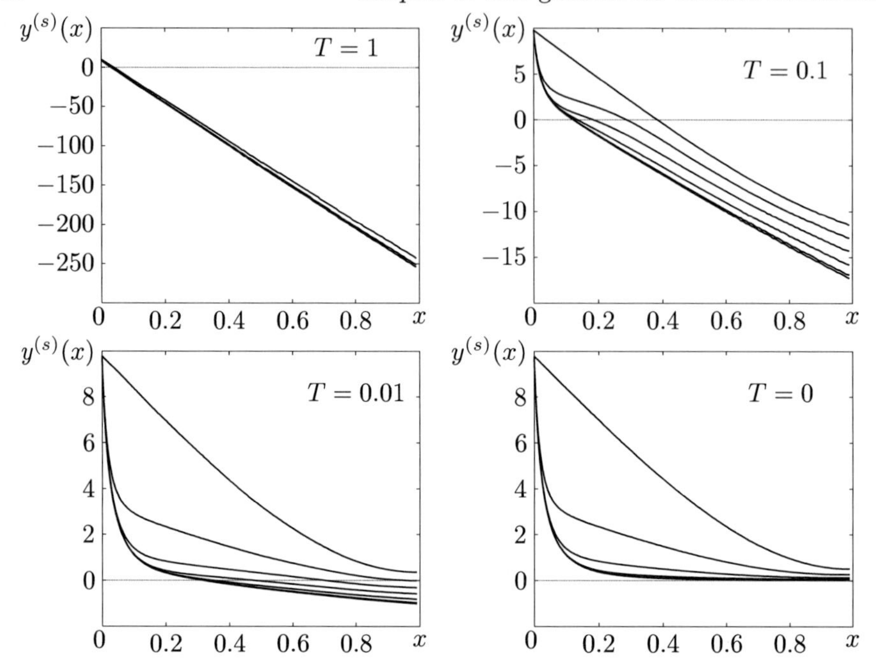

Figure 1.7: Double-sweep method with iterations. Successive approximations $y^{(s)}(x) = \theta\phi^{(s)}(x)$ for gold with density $\rho = 1$ g/cm^3 and for different values of the temperature T in keV. As s grows, the values of $y^{(s)}(x)$ decrease, approaching the function $y(x) = \theta\phi(x)$, where $\phi(x)$ is the solution of (1.31)

1.3 The Thomas-Fermi model for mixtures

1.3.1 Setting up of the problem. Thermodynamic equilibrium condition

In physics and technology one rarely has to deal with pure materials consisting of a single element. For example, stellar matter is a mixture of many elements, ranging from hydrogen to iron. On the other hand, when we study the properties of a pure material we need to keep in mind that even a small impurity may change rather drastically its properties, for instance, its opacity.

Let us consider a mixture of N components with given temperature θ and average density ρ. We denote by m_i the mass fraction of the i-th component ($i = 1, 2, \ldots, N$). As in the case of a single-element substance, we shall assume that the average atom cells are spheres of radii r_{0i}, so that the volume occupied by one cell of the i-th component is $v_i = \frac{4}{3}\,\pi\, r_{0i}^3$.

Let us introduce the intrinsic (partial) density ρ_i of the i-th component as the mass of the component divided by the volume that component occupies. Since the volume of the mixture (which is proportional to $\sum m_i/\rho$) is equal to the sum of the volumes occupied by the individual components of the mixture, we have

$$\frac{1}{\rho}\sum_i m_i = \sum_i \frac{m_i}{\rho_i}.$$

Furthermore, under the conditions of thermodynamic equilibrium for system in contact with diffusion, their chemical potentials should be equal (see [111]).

It follows that the calculation of the Thomas-Fermi potential $V_i(r)$ for different cells of the mixture reduces to solving a system of second order nonlinear differential equations (see (1.17) and (1.18)):

$$\frac{d^2}{dx^2}\phi_i(x) = a_i x\, I_{1/2}\left(\frac{\phi_i}{x}\right), \quad i = 1, 2, \ldots, N, \tag{1.46}$$

with boundary conditions

$$\phi_i(0) = \frac{Z_i}{\theta r_{0i}}, \quad \phi_i(1) = \left.\frac{d\phi_i}{dx}\right|_{x=1} = \frac{\mu_i}{\theta} = -\eta_i.$$

Here

$$x = \frac{r}{r_{0i}}, \quad \frac{\phi_i(x)}{x} = \frac{V_i(r) + \mu_i}{\theta},$$

$$r_{0i} = 1.388\left(\frac{A_i}{\rho_i}\right)^{1/3}, \quad a_i = \frac{4\sqrt{2\theta}}{\pi} r_{0i}^2 \tag{1.47}$$

and A_i and Z_i are the atomic weight and atomic number of the i-th component, respectively. The radii r_{0i} (and the intrinsic densities ρ_i connected with them) must be chosen so that they will obey the relations

$$w\sum_{i=1}^{N} m_i = \sum_{i=1}^{N} m_i w_i \quad \left(w = \frac{1}{\rho}, \quad w_i = \frac{1}{\rho_i}\right), \tag{1.48}$$

$$\eta_i = \eta_j = \eta = -\frac{\mu}{\theta} \quad \text{(for any } i \text{ and } j). \tag{1.49}$$

1.3.2 Linearization of the system of equations

We are dealing here with a nonlinear boundary value problem. To solve it, let us linearize the system (1.46), expanding its right-hand side in the variables ϕ_i, w_i near the approximate solution and retaining only the linear terms. If as approximate values for ϕ_i and w_i we take their values from the preceding iteration, we arrive at the iteration scheme

$$\overset{(s+1)}{\phi_i''} = \overset{(s)}{a_i} x\, I_{1/2}\left(\frac{\overset{(s)}{\phi_i}}{x}\right) + \left.\frac{da_i}{dw_i}\right|_{w_i=\overset{(s)}{w}_i} x\, I_{1/2}\left(\frac{\overset{(s)}{\phi_i}}{x}\right)\left(\overset{(s+1)}{w_i} - \overset{(s)}{w}_i\right) +$$
$$\frac{1}{2}\overset{(s)}{a_i}\, I_{-1/2}\left(\frac{\overset{(s)}{\phi_i}}{x}\right)\left(\overset{(s+1)}{\phi_i} - \overset{(s)}{\phi_i}\right) \tag{1.50}$$

with boundary conditions

$$\overset{(s+1)}{\phi_i}(0) = \frac{Z_i}{\theta \overset{(s)}{r_{0i}}} + \frac{Z_i}{\theta}\left.\frac{d(1/r_{0i})}{dw_i}\right|_{w_i=\overset{(s)}{w_i}}\left(\overset{(s+1)}{w_i} - \overset{(s)}{w_i}\right), \tag{1.51}$$

$$\overset{(s+1)}{\phi_i}(1) = \left.\frac{d}{dx}\overset{(s+1)}{\phi_i}(x)\right|_{x=1} = -\overset{(s+1)}{\eta} \tag{1.52}$$

$(i = 1, 2, \ldots, N)$, where s is the iteration number. Using the relations

$$\frac{da_i}{dw_i} = \frac{2}{3}\frac{a_i}{w_i}, \qquad \frac{d(1/r_{0i})}{dw_i} = -\frac{1}{3r_{0i}\, w_i},$$

we recast (1.50)–(1.52) as

$$\overset{(s+1)}{\phi_i''} = \overset{(s)}{a_i} x\, I_{1/2}\left(\frac{\overset{(s)}{\phi_i}}{x}\right)\left[1 + \frac{2}{3}\frac{\overset{(s+1)}{w_i} - \overset{(s)}{w}_i}{\overset{(s)}{w}_i}\right] + \frac{\overset{(s)}{a_i}}{2} I_{-1/2}\left(\frac{\overset{(s)}{\phi_i}}{x}\right)\left(\overset{(s+1)}{\phi_i} - \overset{(s)}{\phi_i}\right), \tag{1.53}$$

$$\overset{(s+1)}{\phi_i}(0) = \frac{Z_i}{\theta \overset{(s)}{r_{0i}}}\left(1 - \frac{\overset{(s+1)}{w_i} - \overset{(s)}{w}_i}{3\overset{(s)}{w}_i}\right), \tag{1.54}$$

$$\overset{(s+1)}{\phi_i}(1) = \left.\frac{d}{dx}\overset{(s+1)}{\phi_i}(x)\right|_{x=1} = -\overset{(s+1)}{\eta} \tag{1.55}$$

$(i = 1, 2, \ldots, N)$. The system of equations (1.53)–(1.55) contains, in addition to the unknown functions $\overset{(s+1)}{\phi_i}(x)$, the unknowns $\overset{(s+1)}{w_i}$, which must obey condition (1.48). The quantities $\overset{(s)}{a_i}$ and $\overset{(s)}{r_{0i}}$ are expressed in terms of $\overset{(s)}{w_i}$ via (1.47).

1.3.3 Iteration scheme and the double-sweep method

Since the above system (1.53)–(1.55) is linear in the unknowns $\overset{(s+1)}{\phi_i}(x)$, one can explicitly isolate the dependence of $\overset{(s+1)}{\phi_i}(x)$ on $\overset{(s+1)}{w_i}$ and subsequently decouple the iterations for the calculation of $\overset{(s+1)}{\phi_i}(x)$ and $\overset{(s+1)}{w_i}$. Indeed, let us write (1.53) in the form

$$L \overset{(s+1)}{\phi_i} = -\frac{\overset{(s)}{a_i}}{2} I_{-1/2}\left(\frac{\overset{(s)}{\phi_i}}{x}\right)\overset{(s)}{\phi_i} + \left[1 + \frac{2}{3}\,\frac{\overset{(s+1)}{w_i} - \overset{(s)}{w_i}}{\overset{(s)}{w_i}}\right]\overset{(s)}{a_i} x\, I_{1/2}\left(\frac{\overset{(s)}{\phi_i}}{x}\right),$$

where

$$L = \frac{d^2}{dx^2} - \frac{\overset{(s)}{a_i}}{2} I_{-1/2}\left(\frac{\overset{(s)}{\phi_i}}{x}\right).$$

The solution of the linear equation $Ly = C_1 f_1(x) + C_2 f_2(x)$ can be written as $y = C_1 y_1 + C_2 y_2$, where $Ly_1 = f_1$, $Ly_2 = f_2$. Consequently,

$$\overset{(s+1)}{\phi_i}(x) = \overset{(s+1)}{v_i}(x) + \left[1 + \frac{2}{3}\,\frac{\overset{(s+1)}{w_i} - \overset{(s)}{w_i}}{\overset{(s)}{w_i}}\right]\overset{(s+1)}{u_i}(x), \qquad (1.56)$$

where in accordance with (1.54)–(1.56) the functions $\overset{(s+1)}{v_i}(x)$ and $\overset{(s+1)}{u_i}(x)$ must satisfy the following nonhomogeneous equations and boundary conditions:

$$\overset{(s+1)}{v_i''} - \frac{1}{2}\overset{(s)}{a_i} I_{-1/2}\left(\frac{\overset{(s)}{\phi_i}}{x}\right)\overset{(s+1)}{v_i} = -\frac{1}{2}\overset{(s)}{a_i} I_{-1/2}\left(\frac{\overset{(s)}{\phi_i}}{x}\right)\overset{(s)}{\phi}_i,$$

$$\overset{(s+1)}{u_i''} - \frac{1}{2}\overset{(s)}{a_i} I_{-1/2}\left(\frac{\overset{(s)}{\phi_i}}{x}\right)\overset{(s+1)}{u_i} = \overset{(s)}{a_i} x I_{1/2}\left(\frac{\overset{(s)}{\phi_i}}{x}\right),$$

$$(1.57)$$

$$\overset{(s+1)}{v_i}(0) = \frac{3}{2}\,\frac{Z_i}{\theta \overset{(s)}{r_{0i}}}, \qquad \overset{(s+1)}{v_i}(1) = \frac{d}{dx}\overset{(s+1)}{v_i}(x)\bigg|_{x=1},$$

$$\overset{(s+1)}{u_i}(0) = -\frac{1}{2}\,\frac{Z_i}{\theta \overset{(s)}{r_{0i}}}, \qquad \overset{(s+1)}{u_i}(1) = \frac{d}{dx}\overset{(s+1)}{u_i}(x)\bigg|_{x=1}.$$

The boundary conditions for the functions $\overset{(s+1)}{u_i}(x)$ and $\overset{(s+1)}{v_i}(x)$ at $x = 1$ are obvious, while those at $x = 0$ can be derived as follows. By (1.54) and (1.56),

$$\overset{(s+1)}{v_i}(0) + \left[1 + \frac{2}{3}\,\frac{\overset{(s+1)}{w_i} - \overset{(s)}{w_i}}{\overset{(s)}{w_i}}\right]\overset{(s+1)}{u_i}(0) = \frac{Z_i}{\theta \overset{(s)}{r_{0i}}}\left[1 - \frac{1}{3}\,\frac{\overset{(s+1)}{w_i} - \overset{(s)}{w_i}}{\overset{(s)}{w_i}}\right].$$

Putting here $\overset{(s+1)}{v_i}(0) = \alpha Z_i/(\theta \overset{(s)}{r_{0i}})$, $\overset{(s+1)}{u_i}(0) = \beta Z_i/(\theta \overset{(s)}{r_{0i}})$, we deduce that the coefficients α and β satisfy the relations $\alpha + \beta = 1$, $(2/3)\beta = -1/3$, which in turn yield $\alpha = 3/2$, $\beta = -1/2$. Equations (1.57) show that $\overset{(s+1)}{u_i}(x)$ and $\overset{(s+1)}{v_i}(x)$ do not depend on $\overset{(s+1)}{w_i}$.

To solve the boundary value problems (1.57) for the functions $\overset{(s+1)}{u_i}(x)$ and $\overset{(s+1)}{v_i}(x)$ it is convenient to apply the double-sweep method (see Subsection 1.2.3). Once the functions $\overset{(s+1)}{u_i}(x)$ and $\overset{(s+1)}{v_i}(x)$ are calculated, one can find $\overset{(s+1)}{\eta} = -\overset{(s+1)}{\phi_i}(1)$, summing over i with the weight $m_i \overset{(s)}{w_i}/\overset{(s+1)}{u_i}(1)$ the equalities (1.56) at $x = 1$ and making use of condition (1.48), which in the iteration process will be assumed to hold for all values of s:

$$\overset{(s+1)}{\eta} = -\frac{w \sum\limits_{i=1}^{N} m_i + \sum\limits_{i=1}^{N} m_i \overset{(s)}{w}_i \dfrac{\overset{(s+1)}{v_i}(1)}{\overset{(s+1)}{u_i}(1)}}{\sum\limits_{i=1}^{N} m_i \dfrac{\overset{(s)}{w}_i}{\overset{(s+1)}{u_i}(1)}}. \tag{1.58}$$

Once $\overset{(s+1)}{\eta}$ is known, we can find $\overset{(s+1)}{w_i}$, setting $x = 1$ in (1.56):

$$\overset{(s+1)}{w_i} = -\frac{1}{2}\overset{(s)}{w_i}\left[1 + 3\,\frac{\overset{(s+1)}{\eta} + \overset{(s+1)}{v_i}(1)}{\overset{(s+1)}{u_i}(1)}\right]. \tag{1.59}$$

To obtain an initial approximation we will use, as in the case of a single-element matter, the uniform free-electron density model (see formulas (1.24), (1.25)) and the assumption that $Z_{0i} = Z_i$. This yields the relation

$$w_i \approx \frac{3\pi}{4\sqrt{2}\,(1.388)^3}\,\frac{Z_i/A_i}{\theta^{3/2}\,I_{1/2}(-\eta)}, \tag{1.60}$$

from which, using (1.48), we obtain

$$I_{1/2}(-\eta) \approx \frac{3\pi}{4\sqrt{2}\,(1.388)^3}\,\frac{\rho}{\theta^{3/2}}\,\frac{\sum_i m_i Z_i/A_i}{\sum_i m_i}. \tag{1.61}$$

Using (1.60), (1.61), and also (1.37), we finally conclude that

$$\begin{gathered} \overset{(0)}{w_i} = w\frac{Z_i}{A_i}\frac{\sum_i m_i}{\sum_i m_i Z_i/A_i}, \\ \overset{(0)}{\eta} = \frac{1}{2}\ln\frac{\pi}{6} - \frac{3}{2}\ln\left[\exp\left(\sqrt[3]{\frac{2q^2}{3}}\right) - 1\right], \\ q = 2.795 \cdot 10^{-3} \cdot \frac{\rho}{T^{3/2}}\,\frac{\sum_i m_i Z_i/A_i}{\sum_i m_i}, \end{gathered} \tag{1.62}$$

$$\overset{(0)}{\phi}_i(x) = \frac{Z_i}{\theta \overset{(0)}{r}_{0i}} \left(1 - \frac{3}{2}x + \frac{1}{2}x^3\right) - \overset{(0)}{\eta}\, x.$$

The proposed algorithm allows one to carry out calculations for any values of the matter density and temperature, including the temperature zero. By (1.62), for a fixed density ρ and $T \to 0$ the initial approximation $\overset{(0)}{\eta} \to -\infty$, which also holds for the true value of η. When $\eta < -10$ the function $I_{1/2}(\phi/x)$ can be replaced, with good accuracy, by the asymptotic expression $I_{1/2}(\phi/x) \approx 2/3(\phi/x)^{3/2}$, which leads to formulas that are practically identical to those for $T = 0$ (see (1.30) in § 1.1). Using the estimate (1.62) for large q, the condition $\eta < -10$ can be replaced by the more stringent one $T < 0.001\,\rho^{2/3}$ (where T is measured in keV and ρ in g/cm^3). Thus, in order to compute the Thomas-Fermi potential at $T = 0$ by means of the method introduced above it suffices to calculate it for $T = T^* = 0.001\,\rho^{2/3}$ and then perform the corresponding transformations.

The double-sweep method with iterations is of second order of accuracy. If this proves insufficient (which is the case, for example, for very small densities), the one can use a higher-order scheme, applying Newton's method to find the chemical potential [148].

1.3.4 Discussion of computational results

The scheme (1.58)-(1.59) has been used to carry out computations in wide ranges of temperature and density, for various substances and mixtures. To illustrate the convergence of the iterations we give here the results of computations of the quantity $\overset{(s)}{\eta}$ for a mixture of 10 elements with atomic numbers $Z = 10, 20, \ldots, 100$ and $m_i = A_i$ (see Table 1.2). For $m_i \neq A_i$ the iterations converge no slower than for $m_i = A_i$. The number of iterations decreases rapidly when the temperature increases.

Let us remark that when the number of components in the mixture is increased, the number of iterations practically does not increase, remaining roughly the same as in the case of a single-component substance. This feature is an essential advantage of the iteration scheme, because for a single-component substance there are no iterations with respect to the partial density ρ_i.

As one can see in Table 1.2, the numerical method discussed above allows one to calculate the atomic potentials in the Thomas-Fermi model for arbitrary temperatures and densities, for a substance consisting of an arbitrary number of elements. The domain of applicability of the Thomas-Fermi model is determined by how small the quantum, exchange, and oscillation corrections to the potential $V(r)$ are (see Subsection 4.1.2). As a rule, when the temperature and the density are increased, these corrections become smaller, and their relative magnitude is the smaller the larger the atomic numbers Z of the elements forming the mixture.

Table 1.2: Successive iterations $\overset{(s)}{\eta}$ for a mixture of 10 elements for some values of the temperature T in keV and density ρ in g/cm^3, obtained by the double-sweep method

T / ρ	10^{-3}	10^{-2}	10^{-1}	1	10
	0.54991	4.31552	7.79948	11.25669	14.71056
	1.19695	4.67097	8.06290	11.37281	14.71547
	1.80567	4.98772	8.31204	11.38652	14.71548
	2.39004	5.29413	8.56695	11.38660	14.71548
	2.96247	5.62701	8.73447	11.38660	
10^{-2}	3.53084	6.00593	8.76165		
	4.09465	6.40929	8.76153		
	4.61664	6.71097	8.76154		
	4.96164	6.78590			
	5.03472	6.78521			
	5.03344	6.78523			
	5.03346	6.78523			
	5.03346				
	−14.66551	−1,03005	3,12536	6,64434	10,10539
	−7.58149	−0.00109	3.52243	6.85742	10.11872
	−3.64227	0.83856	3.91206	6.91850	10.11881
	−1.61185	1.55669	4.27765	6.92075	10.11881
1	−0.54902	2.13822	4.50923	6.92075	
	−0.14336	2.45274	4.55315		
	−0.12944	2.49462	4.55316		
	−0.13091	2.49378	4.55316		
	−0.13091	2.49378			
	−308.991	−31.2226	−3.2089	1.88920	5.48475
	−173.017	−17.3531	−1.4698	2.32163	5.54920
	−118.438	−11.8071	−0.6193	2.55149	5.55142
10^2	−111.267	−11.0669	−0.4169	2.58634	5.55143
	−111.561	−11.0966	−0.4145	2.58669	5.55143
	−111.563	−11.0967	−0.4146	2.58669	
	−111.563	−11.0967	−0.4146		

Chapter 2

Electron wave functions in a given potential

The statistical Thomas-Fermi method provides us with an approximate way of describing a wide class of spatially nonhomogeneous multi-electron systems and has the advantage of being simple and universal compared with more accurate quantum-mechanical methods. If the Thomas-Fermi potential differs only slightly from the atomic potential in matter, which is true for sufficiently high temperatures or densities, then using this potential one can find, in the one-electron approximation, the electron *energy levels* and *wave functions* corresponding to the atom with average occupation numbers in the Thomas-Fermi approximation, and also find the *average occupation numbers* themselves. This approach yields the simplest approximation of the *average atom*, i.e., the ion with average occupation numbers in an electrically neutral spherical cell. Later we will show how, using the energy levels and wave functions of the average atom, one can carry out calculations for concrete ions in a plasma in the framework of more complex models.

2.1 Description of electron states in a spherical average atom cell

To find the energy levels and wave functions we will solve the Schrödinger equation

$$-\frac{1}{2}\Delta\psi + U(r)\psi = \varepsilon\psi, \tag{2.1}$$

where $U(r)$ is the potential energy of the electron under study, defined by the Thomas-Fermi potential. Separation of variables in spherical coordinates yields

$$\psi(\vec{r}) = \frac{1}{r}R(r)(-1)^m Y_{\ell m}(\vartheta, \varphi),$$

where $Y_{\ell m}(\vartheta, \varphi)$ is a spherical harmonic (see Appendix, A.1.3), and the radial function $R(r) = R_{\varepsilon\ell}(r)$ satisfies the equation

$$-\frac{1}{2}R'' + \left[U(r) + \frac{\ell(\ell+1)}{2r^2}\right] R = \varepsilon R, \tag{2.2}$$

or

$$R'' + 2[\varepsilon - U_\ell(r)]R = 0.$$

For bound states, $R(r) = R_{n\ell}(r)$, $\varepsilon = \varepsilon_{n\ell}$, where $n = n_r + \ell + 1$ is the principal quantum number and n_r is the number of zeroes of the radial function inside the cell (see Subsection 2.1.1 below). As seen from (2.2), the energy eigenvalues $\varepsilon_{n\ell}$ and the corresponding radial functions $R_{n\ell}(r)$ are determined by the function

$$U_\ell(r) = U(r) + \frac{\ell(\ell+1)}{2r^2},$$

which serves as the potential energy in the one-dimensional Schrödinger equation (2.2).

2.1.1 Classification of electron states within the average atom cell

Next let us examine the character of the dependence of the potential energy $U_\ell(r)$ on ℓ, setting $U(r) = -V(r)$, where $V(r)$ is the Thomas-Fermi potential. Figure 2.1 shows the graphs of the functions $U_\ell(r)$ ($\ell = 0, 1, 2, 3$) for gold ($Z = 79$) at temperature $T = 0.1$ keV and density $\rho = 100$ g/cm^3. The temperature and density were chosen so that the figure will display all possible types of behavior of the function $U_\ell(r)$[*)].

For the sake of convenience, in the investigation of the function $U_\ell(r)$ we will replace $\ell(\ell+1)$ by $(\ell+1/2)^2$, in accordance with the semiclassical approximation used in the Thomas-Fermi model (see Appendix, A.4), and introduce the notation

$$y = \left(\ell + \frac{1}{2}\right)^2, \quad U_y(r) = -V(r) + \frac{y}{2r^2}.$$

The extrema of the function $U_y(r)$ for each given y are attained for the values r that are roots of the equation $\frac{d}{dr}U_y(r) = 0$, i.e., the equation $F(r) = y$, where $F(r) = -r^3\frac{d}{dr}V(r)$. As computations show, the graphs of the function $F(r)$ are bell-shaped (see Figure 2.2). The roots of the equation $F(r) = y$ determine the positions of the minimum and the maximum of the $U_y(r)$. The behavior of the curves $F(r)$ is readily understood using the Poisson equation and the boundary conditions for the Thomas-Fermi potential. The boundary condition at $r = 0$ gives

$$F(0) = -r^3\frac{d}{dr}\left(\frac{Z}{r}\right)\bigg|_{r=0} = 0, \quad F'(0) = Z.$$

*) As the density of matter is decreased, the curves $U_\ell(r)$ with different ℓ become practically identical for large values of r.

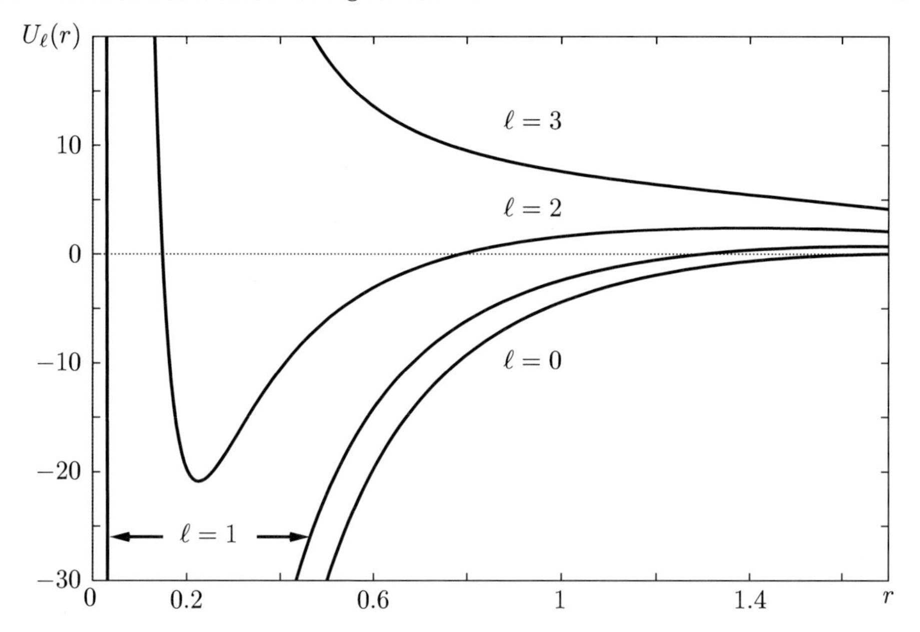

Figure 2.1: The potential energy $U_\ell(r) = \frac{\ell(\ell+1)}{2r^2} - V(r)$ for $\ell = 0, 1, 2, 3$ for gold at temperature $T = 0.1$ keV and density $\rho = 100$ g/cm^3 ($V(r)$ is the Thomas-Fermi potential; the radius of the average atom cell is $r_0 = 1.67$)

The Poisson equation $(r^2V')' = 4\pi r^2\rho(r)$ and the boundary condition $V'(r_0) = 0$ yield

$$F(r_0) = 0, \quad F'(r_0) = -4\,\pi\, r_0^3\, \rho(r_0) \approx -3Z_0 < 0,$$

where Z_0 is the average charge of the ion.

Let us denote the maximum value of $F(r)$ by y^*. Examining Figure 2.2 we see that for $y < y^*$ the equation $F(r) = y$ has two roots,$^{*)}$ the smallest [resp., largest] of which corresponds to the minimum [resp., maximum] of the function $U_y(r)$. As y grows, the maximum and minimum of $U_y(r)$ approach one another, and the magnitude of the maximum increases:

$$\frac{d}{dy}\varepsilon_{\max}(y) = \frac{d}{dy}U_y(\widetilde{r}_y) = \left(\frac{\partial U_y}{\partial y} + \frac{\partial U_y}{\partial \widetilde{r}_y}\frac{\partial \widetilde{r}_y}{\partial y}\right) = \frac{1}{2\widetilde{r}_y^2} > 0$$

(when $r = \widetilde{r}_y$ the function $U_y(r)$ attains its local maximum $\varepsilon_{\max}(y)$). If $y < y^*$, then for $\varepsilon_{\min}(y) < \varepsilon < \varepsilon_{\max}(y)$, where $\varepsilon_{\min}(y)$ is the value of the local minimum of $U_y(r)$, the equation $\varepsilon = U_y(r)$ has at least two roots, $r_1(\varepsilon)$ and $r_2(\varepsilon)$, while for

$^{*)}$ At low temperatures, for some values of ℓ the equation $F(r) = (\ell + 1/2)^2$ may have more than two roots.

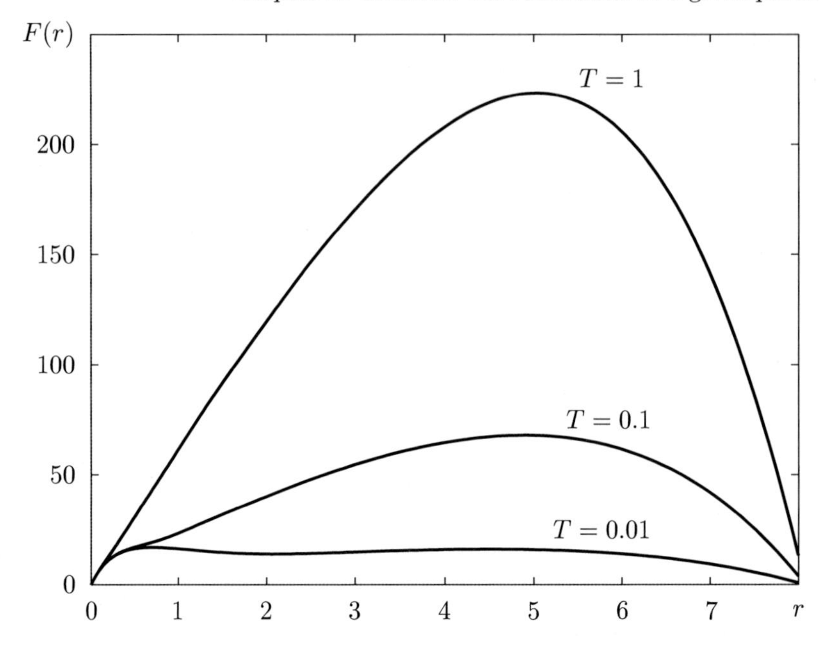

Figure 2.2: The function $F(r) = -r^3\, dV(r)/dr$ for gold at density $\rho = 1$ g/cm^3 and different values of the temperature T in keV

$\varepsilon > \varepsilon_{\max}(y)$ it has exactly one root. In the case $y > y^*$ the equation $\varepsilon = U_y(r)$ has one root for $\varepsilon > U_y(r_0)$, and has no roots in the remaining cases.

The behavior of an electron in the atom cell is determined by the distribution of the roots of the equation

$$\varepsilon - U_y(r) = 0 \tag{2.3}$$

in the interval $0 < r < r_0$; these roots determine the domain of the classical motion of the electron, i.e., the domain where $\varepsilon - U_y(r) > 0$. Three typical cases are possible.

(1) Equation (2.3) has a single root $r = r_1$, i.e., one turning point. The domain of classical motion of the electron with the energy ε, given by the inequality $\varepsilon > U_y(r)$, extends to the boundary of the cell $r = r_0$. Clearly, such states have a *continuous energy spectrum*, i.e., a *continuum* (see, e.g., Figure 2.1 for $\varepsilon > 5$ and $\ell = 0, 1, 2, 3$). The corresponding electron states are not localized and transitions of such electrons from one cell to another are possible.

(2) Equation (2.3) has two roots r_1, r_2 (turning points), with $r_1 < r_2 \ll r_0$ (see Figure 2.1 with $\varepsilon < -5$ for $\ell = 0, 1, 2$). In this case one obtains a bound state, the *energy spectrum is discrete*, and there is no need to specify boundary conditions at $r = r_0$, because the wave function is negligibly small outside the domain of classical motion for $r \gg r_2$.

(3) Equation (2.3) has two or more roots, and the second root is close to the boundary of the cell (see Figure 2.1 with $-1 < \varepsilon < 3$ for $\ell = 2$). Such states can be assigned to the *intermediate* group. As a matter of fact, due to the influence of neighboring atoms the energy spectrum of such electrons consists of a number of bands of allowed energies. In some cases the positioning of the bands can be determined by imposing boundary conditions of a special type: for example, in a solid body one imposes some periodicity conditions. The simplest estimate of the position and width of the bands $\Delta\varepsilon = \varepsilon_1 - \varepsilon_2$ can be obtained by solving the Schrödinger equation for the given n and ℓ, with the conditions $R_{\varepsilon_1\ell}(r_0) = 0$ and $d\left(r^{-1}R_{\varepsilon_2\ell}(r)\right)/dr\big|_{r=r_0} = 0$ (see [188], and also § 4.1).

Thus, we see that for the most important cases, (1) and (2), the energy levels and wave functions can be found by solving the Schrödinger equation (2.2) with the methods considered in the Appendix. The states in the intermediate group require further investigation (see §§ 4.1 and 4.2 below). However, in many situations, especially at high temperatures, only very few electrons are found in the states of the intermediate group and in such situations there is no need to account for the band structure of the energy spectrum.

Remark. Generally speaking, when one solves the Schrödinger equation (2.2), one should eliminate from the Thomas-Fermi potential $V(r)$ the proper electrostatic energy of the electron in question ("self-interaction"). Hence, in equation (2.2) instead of $U(r) = -V(r)$ one can take

$$U(r) = -\frac{Z-1}{Z}V(r) - \frac{1}{r}$$

(this includes the Fermi-Amaldi correction [45]). Such corrections obviously modify the behavior of the potential for $r \sim r_0$ and change the energy spectrum of the electrons in the intermediate group.

2.1.2 Model of an atom with average occupation numbers

Once the energy levels $\varepsilon = \varepsilon_{n\ell}$ of electrons in a central field characterized by the Thomas-Fermi potential are calculated, one can find the average occupation numbers of electron states, using the Pauli principle and the Fermi-Dirac statistics. Here the number $N_{n\ell m}$ of *bound* electrons (i.e., electrons corresponding to the discrete spectrum) in states with quantum numbers n, ℓ, m and energy $\varepsilon_{n\ell}$ within one atom cell is given by the formula

$$N_{n\ell m} = \frac{2}{1 + \exp\left(\dfrac{\varepsilon_{n\ell} - \mu}{\theta}\right)}. \tag{2.4}$$

Obviously, the average number of electrons with quantum numbers n and ℓ is

$$N_{n\ell} = \sum_{m=-\ell}^{\ell} N_{n\ell m} = \frac{2(2\ell+1)}{1+\exp\left(\frac{\varepsilon_{n\ell}-\mu}{\theta}\right)}, \tag{2.5}$$

where $2(2\ell+1)$ is the degree of degeneracy (i.e., the multiplicity) of the energy level $\varepsilon_{n\ell}$.

To calculate the number of electrons of the *continuum*, i.e., of the *free electrons*, we will use the semiclassical approximation. The number of possible states of an electron with quantum numbers ℓ, m, with fixed projection of the spin and with energy in the interval $(\varepsilon, \varepsilon + d\varepsilon)$, belonging to the phase-space volume element $dp_{\varepsilon\ell}\, dr$ is

$$\frac{1}{\pi}\, dp_{\varepsilon\ell}\, dr,$$

where

$$p_{\varepsilon\ell} = p_{\varepsilon\ell}(r) = \sqrt{2\left[\varepsilon + V(r) - \frac{(\ell+1/2)^2}{2r^2}\right]}.$$

Summing over the two possible projections of the spin, we find that in the semiclassical approximation the average number of electrons with quantum numbers ℓ, m and energy in the interval $(\varepsilon, \varepsilon + d\varepsilon)$ and which are located in the spherical layer $(r, r+dr)$ is given by

$$\frac{1}{1+\exp\left(\frac{\varepsilon-\mu}{\theta}\right)}\, \frac{2}{\pi}\, dp_{\varepsilon\ell}\, dr = \frac{2}{\pi p_{\varepsilon\ell}}\, \frac{d\varepsilon\, dr}{1+\exp\left(\frac{\varepsilon-\mu}{\theta}\right)} = N(\varepsilon, r)\, d\varepsilon\, dr, \tag{2.6}$$

where

$$N(\varepsilon, r) = \frac{2}{\pi p_{\varepsilon\ell}(r)\left[1+\exp\left(\frac{\varepsilon-\mu}{\theta}\right)\right]}.$$

The semiclassical approximation (2.6), used for all electrons, including the bound states, leads to an expression for the electron density $\rho(r)$ in the Thomas-Fermi model. In this case the number of electrons in the spherical layer $(r, r+dr)$ is given by

$$4\pi r^2 \rho(r)\, dr = \left[\sum_{\ell,m} \int N(\varepsilon, r)\, d\varepsilon\right] dr,$$

whence

$$\rho(r) = \frac{1}{4\pi r^2}\left[\sum_{\ell} \sum_{m=-\ell}^{\ell} \int_{\tilde{\varepsilon}_0(r)}^{\infty} \frac{2\, d\varepsilon}{\pi p_{\varepsilon\ell}(r)\left[1+\exp\left(\frac{\varepsilon-\mu}{\theta}\right)\right]}\right]. \tag{2.7}$$

In (2.7) the integration is carried out over the domain of classical motion of the electron, given by $p^2_{\varepsilon\ell}(r) = 2\varepsilon + 2V(r) - (\ell+1/2)^2/r^2 \geq 0$, i.e., for

$$\varepsilon > \widetilde{\varepsilon}_0(r) = \frac{(\ell+1/2)^2}{2r^2} - V(r). \tag{2.8}$$

Summing over m, replacing summation with respect to ℓ by integration with respect to $y = (\ell+1/2)^2$ and switching the order of integration, we obtain

$$\rho(r) = \frac{1}{2\pi^2 r^2} \int\limits_{-V(r)}^{\infty} \frac{d\varepsilon}{1+\exp\left(\dfrac{\varepsilon-\mu}{\theta}\right)} \int\limits_0^{2r^2[\varepsilon+V(r)]} \frac{dy}{\sqrt{2[\varepsilon+V(r)]-y/r^2}} =$$

$$\frac{\sqrt{2}}{\pi^2} \int\limits_{-V(r)}^{\infty} \frac{\sqrt{\varepsilon+V(r)}}{1+\exp\left(\dfrac{\varepsilon-\mu}{\theta}\right)}\, d\varepsilon = \frac{(2\theta)^{3/2}}{2\pi^2} \int\limits_0^{\infty} \frac{\sqrt{x}\, dx}{1+\exp\left(x - \dfrac{V(r)+\mu}{\theta}\right)},$$

or

$$\rho(r) = \frac{(2\theta)^{3/2}}{2\pi^2} I_{1/2}\left(\frac{V(r)+\mu}{\theta}\right), \tag{2.9}$$

which coincides with formula (1.7).

2.1.3 Derivation of the expression for the electron density by means of the semiclassical approximation for wave functions

Formula (2.9) for the electron density $\rho(r)$ can also be derived directly from the quantum-mechanical expression

$$\rho(\vec{r}) = \sum_{\substack{n,\ell,m \\ (\varepsilon_{n\ell}<\varepsilon_0)}} \frac{2}{1+\exp\left(\dfrac{\varepsilon_{n\ell}-\mu}{\theta}\right)} |\psi_{n\ell m}(\vec{r})|^2 +$$

$$\sum_{\ell,m} \int\limits_{\varepsilon>\varepsilon_0} \frac{2}{1+\exp\left(\dfrac{\varepsilon-\mu}{\theta}\right)} |\psi_{\varepsilon\ell m}(\vec{r})|^2\, d\varepsilon, \tag{2.10}$$

where $\varepsilon_0 = \varepsilon_0(\ell)$ is the boundary of the continuum for electrons with given ℓ, and $\psi_{n\ell m}(\vec{r})$ and $\psi_{\varepsilon\ell m}(\vec{r})$ are the wave functions for the bound and free electron states, respectively:

$$\psi_{n\ell m}(\vec{r}) = \frac{1}{r} R_{n\ell}(r)(-1)^m Y_{\ell m}(\vartheta,\varphi), \quad \psi_{\varepsilon\ell m}(\vec{r}) = \frac{1}{r} R_{\varepsilon\ell}(r)(-1)^m Y_{\ell m}(\vartheta,\varphi). \tag{2.11}$$

In the domain of classical motion we use for the radial parts of the wave functions the semiclassical approximation (which, strictly speaking, is valid far from the turning points):

$$R_{n\ell}(r) = \frac{C_1}{\sqrt{p_{\varepsilon\ell}(r)}} \sin\left(\xi + \frac{\pi}{4}\right), \quad \varepsilon = \varepsilon_{n\ell}, \tag{2.12}$$

$$R_{\varepsilon\ell}(r) = \frac{C_2}{\sqrt{p_{\varepsilon\ell}(r)}} \sin\left(\xi + \frac{\pi}{4}\right). \tag{2.13}$$

Here $\xi = \xi(r) = \int\limits_{r_1}^{r} p_{\varepsilon\ell}(r')\,dr'$ (with r_1 denoting the first turning point), and the constants C_1 and C_2 are determined from the normalization conditions. Outside the domain of classical motion and near the turning points we set the wave functions equal to zero.

For the normalization constant C_1 the semiclassical approximation yields the relation

$$\int\limits_0^{r_0} R_{n\ell}^2(r)\,dr \approx C_1^2 \int\limits_{r_1}^{r_2} \frac{1}{p_{\varepsilon\ell}(r)} \frac{1+\sin 2\xi(r)}{2}\,dr \approx \frac{C_1^2}{2} \int\limits_{r_1}^{r_2} \frac{dr}{p_{\varepsilon\ell}(r)} = 1. \tag{2.14}$$

For the continuum electrons we require that the following condition be satisfied when $r_0 \to \infty$:

$$\int\limits_0^{r_0} R_{\varepsilon\ell}(r) R_{\varepsilon'\ell}(r)\,dr = \delta(\varepsilon - \varepsilon').$$

This yields $C_2 = \sqrt{2/\pi}$ (see § 2.3).

Substituting the wave functions (2.11) in (2.10) and using the addition theorem for spherical harmonics:

$$\sum_m |Y_{\ell m}(\vartheta, \varphi)|^2 = \frac{2\ell+1}{4\pi},$$

we obtain

$$\rho(\vec{r}) = \rho(r) = \frac{1}{4\pi r^2} \left\{ \sum_{\substack{n,\ell \\ (\varepsilon_{n\ell} < \varepsilon_0(\ell))}} \frac{2(2\ell+1)}{1+\exp\left(\frac{\varepsilon_{n\ell}-\mu}{\theta}\right)} R_{n\ell}^2(r) + \right.$$

$$\left. \sum_\ell \int\limits_{\varepsilon_0(\ell)}^{\infty} \frac{2(2\ell+1)}{1+\exp\left(\frac{\varepsilon-\mu}{\theta}\right)} R_{\varepsilon\ell}^2(r)\,d\varepsilon \right\}. \tag{2.15}$$

To carry out the summation over n we use the Bohr-Sommerfeld rule

$$\int_{r_1}^{r_2} p_{\varepsilon\ell}(r)\,dr = \pi\left(n_r + \frac{1}{2}\right), \qquad n_r = n - \ell - 1, \tag{2.16}$$

which yields

$$\pi\frac{dn_r}{d\varepsilon} = \int_{r_1}^{r_2} \frac{dr}{p_{\varepsilon\ell}(r)}, \tag{2.17}$$

i.e.,

$$dn_r = \frac{d\varepsilon}{\pi}\int_{r_1}^{r_2} \frac{dr}{p_{\varepsilon\ell}(r)}.$$

Next, let us insert the semiclassical wave functions (2.12)–(2.13) in formula (2.15) and pass from summation over n to summation over n_r, and then replace the latter by integration with respect to ε:

$$\rho(r) = \frac{1}{4\pi r^2}\sum_\ell 2(2\ell+1)\times$$
$$\left\{\int_{\tilde{\varepsilon}_0(r)}^{\varepsilon_0(\ell)} \frac{1}{1+\exp\left(\dfrac{\varepsilon-\mu}{\theta}\right)}\left(\frac{C_1^2}{\pi p_{\varepsilon\ell}(r)}\int_{r_1}^{r_2}\frac{dr'}{p_{\varepsilon\ell}(r')}\right)\sin^2\left(\xi+\frac{\pi}{4}\right)d\varepsilon + \right.$$
$$\left.\int_{\varepsilon_0(\ell)}^{\infty} \frac{1}{1+\exp\left(\dfrac{\varepsilon-\mu}{\theta}\right)}\frac{C_2^2}{p_{\varepsilon\ell}(r)}\sin^2\left(\xi+\frac{\pi}{4}\right)d\varepsilon\right\}.$$

Here $\tilde{\varepsilon}_0(r)$ and $\varepsilon_0(\ell)$ determine the domain of classical motion of the electron (see (2.8)) and the boundary of the continuum, respectively. Combining the integrals we obtain

$$\rho(r) = \frac{1}{4\pi r^2}\sum_\ell 2(2\ell+1)\left\{\int_{\tilde{\varepsilon}_0(r)}^{\infty} \frac{1}{1+\exp\left(\dfrac{\varepsilon-\mu}{\theta}\right)}\frac{2}{\pi\, p_{\varepsilon\ell}(r)}\sin^2\left(\xi+\frac{\pi}{4}\right)d\varepsilon\right\}. \tag{2.18}$$

If in (2.18) we replace $\sin^2(\xi+\pi/4) = (1+\sin 2\xi)/2$ by $1/2$ and then pass from summation over ℓ to integration with respect to $y = (\ell+1/2)^2$ and carry out the latter first, then we obtain an expression identical to (2.9).

As the derivation of formula (2.18) shows, in the Thomas-Fermi model oscillations of wave functions, and hence oscillations of the electron density are neglected. The Thomas-Fermi method describes only the average behavior of physical quantities and does not reproduce at all their oscillations, which are connected with the

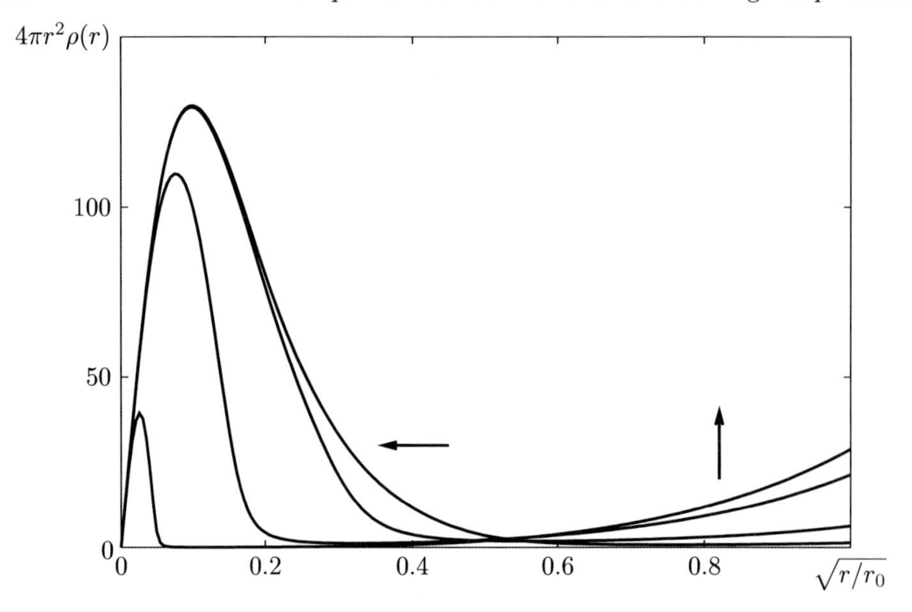

Figure 2.3: Dependence of the radial electron density $4\pi r^2\rho(r)$ on $x = \sqrt{r/r_0}$ according to the Thomas-Fermi model for gold at density $\rho = 1$ g/cm^3 and different temperatures $T = 1$, 10, 100, 1000 eV (the arrows indicate the direction in which the curves change when the temperature is increased)

shell structure. Thus, the electron density of the average atom according to the Thomas-Fermi model and its potential obtained by solving the Poisson equation turn out to be consistent to one another, but the electron wave functions obtained in this potential and the potential calculated by means of these functions will not be so.

In addition, let us remark that the Thomas-Fermi model reproduces incorrectly the behavior of the electron density in the vicinity of the nucleus, i.e., for $r \sim 0$. From the behavior of the function $I_{1/2}[(V(r) + \mu)/\theta]$ for small r it follows that $\rho(r) \sim 1/r^{3/2} \to \infty$ as $r \to 0$ (see formula (2.9)), whereas the electron density must tend to some finite value. Indeed, for $r = 0$ only the electrons with $\ell = 0$ contribute to the electron density:

$$\rho(0) = \sum_{\alpha} N_\alpha |\psi_\alpha(0)|^2 = \sum_{n} C_{n0}^2 \, |Y_{00}(\vartheta, \varphi)|^2,$$

where C_{n0} are certain constants, determined by the occupation numbers of the levels with $\ell = 0$. Therefore, for the electron density calculated by means of wave functions one has $\rho(0) < \infty$, in contrast to the Thomas-Fermi model. Note that this peculiarity of the Thomas-Fermi model is practically inconsequential if one is not interested in $\rho(r)$ for $r \sim 0$.

As formula (1.7) shows, we have $\rho'(r) = 0$ on the boundary of the average atom cell, i.e., for $r = r_0$, which is in agreement with the uniform free-electron density model. Note that the behavior of $\rho(r)$ does not reveal the shell structure of the atom. More accurate quantum-mechanical calculations (according to the Hartree-Fock model, for example) also do not allow one to extract the shell structure from the graph of $\rho(r)$. This structure can be seen from the graph of the radial density $4\pi r^2 \rho(r)$ (see Figure 3.6).

The dependence of the distribution of the radial electron density $4\pi r^2 \rho(r)$ on $x = \sqrt{r/r_0}$, obtained using formula (1.7), is shown in Figure 2.3 for gold with fixed density of matter $\rho = 1$ g/cm^3 and different values of the temperature. As seen in this, there exists a point $r = r^*$ in which the function $r^2 \rho(r)$ attains its minimum. The value r^* gives information on the effective radius of the ion: in the domain $r > r^*$ one finds mainly the free electrons, while for $r < r^*$ one finds the bound electrons, which form the ion core. For the case considered here, at temperature $T = 1$ eV, $r^* = r_0(0.5)^2 \approx 2$ a.u., while for $T = 1$ keV, $r^* = r_0(0.05)^2 = 0.02$ a.u. ($r_0 = 8.08$ a.u.).

2.1.4 Average degree of ionization

Using the argument that has led to formula (2.9), let us calculate the number of free electrons Z_0 in one average atom cell. Here we assign to the continuum the electron states with one turning point (see Figure 2.1). Then

$$Z_0 = Z_0^{(1)} - Z_0^{(2)},$$

where

$$Z_0^{(1)} = \int_0^{r_0} dr \int_0^{\infty} d\varepsilon \int_0^{2r^2[\varepsilon + V(r)]} \frac{2\,dy}{\pi p_{\varepsilon\ell}(r)\left[1 + \exp\left(\dfrac{\varepsilon - \mu}{\theta}\right)\right]}, \tag{2.19}$$

$$Z_0^{(2)} = \int_0^{r_0} dr \int d\varepsilon \int \frac{2\,dy}{\pi p_{\varepsilon\ell}(r)\left[1 + \exp\left(\dfrac{\varepsilon - \mu}{\theta}\right)\right]}. \tag{2.20}$$

The domain of integration in (2.20) is defined by the inequalities

$$0 \le y \le 2r^2(\varepsilon + V(r)) \quad (0 \le y \le y^*), \qquad U_y(r) \le \varepsilon \le \varepsilon_y \quad (\varepsilon \ge 0),$$

i.e.,

$$0 \le y \le \min\{y^*,\, 2r^2[\varepsilon + V(r)]\}, \quad 0 \le \varepsilon \le \varepsilon_{y^*}.$$

Incidentally, without substantially lowering the physical accuracy of the calculations, we can obtain a simpler expression for $Z_0^{(2)}$ if for given ℓ we set the boundary of the continuum equal to $\varepsilon = y/(2r_0^2)$. Then the domain of integration in (2.20) is specified by the inequalities

$$0 \le \varepsilon \le \frac{y}{2r_0^2}, \quad 0 \le y \le 2r^2[\varepsilon + V(r)],$$

i.e.,

$$2r_0^2\varepsilon \le y \le 2r^2[\varepsilon + V(r)], \quad 0 \le \varepsilon \le \frac{V(r)}{(r_0/r)^2 - 1}.$$

This yields for the values of $Z_0^{(1)}$ and $Z_0^{(2)}$ the expressions

$$Z_0^{(1)} = \frac{4\sqrt{2}}{\pi} \int_0^{r_0} r^2\,dr \int_0^{\infty} d\varepsilon \frac{\sqrt{\varepsilon + V(r)}}{1 + \exp\left(\dfrac{\varepsilon - \mu}{\theta}\right)} =$$

$$\frac{8\sqrt{2}}{\pi} \int_0^{r_0} r^2[V(r)]^{3/2} dr \int_1^{\infty} \frac{t^2\,dt}{1 + \exp\left[(t^2 - 1)\dfrac{V(r)}{\theta} + \eta\right]}, \tag{2.21}$$

$$Z_0^{(2)} = \frac{4\sqrt{2}}{\pi} \int_0^{r_0} r^2\,dr \int_0^{V(r)/[(r_0/r)^2 - 1]} \frac{\sqrt{V(r) - [(r_0/r)^2 - 1]\,\varepsilon}}{1 + \exp\left(\dfrac{\varepsilon - \mu}{\theta}\right)} d\varepsilon =$$

$$\frac{8\sqrt{2}}{\pi} \int_0^{r_0} \frac{r^2[V(r)]^{3/2}}{(r_0/r)^2 - 1} dr \int_0^1 \frac{t^2\,dt}{1 + \exp\left[\dfrac{1 - t^2}{(r_0/r)^2 - 1}\dfrac{V(r)}{\theta} + \eta\right]} \tag{2.22}$$

(in (2.21) $t = \sqrt{1 + \varepsilon/V(r)}$, while in (2.22) $t = \sqrt{1 - [(r_0/r)^2 - 1]\,\varepsilon/V(r)}$).

Calculations show that the quantity $Z_0^{(2)}$ serves as a correction, that is, usually $Z_0^{(2)} \ll Z_0^{(1)}$. Let us remark that $Z_0^{(1)}$ gives the number of free electrons with energy $\varepsilon > 0$ per one atom cell. The inner integrals in (2.21) and (2.22) can be calculated by Simpson's formula, and for the outer ones one can use the trapezoid formula, since the values of the potential $V(r)$ are usually known on a nonuniform mesh.

The average degree of ionization $\alpha = Z_0/Z$, like other quantities in the Thomas-Fermi model, depends only on the reduced density $\sigma = \rho/(AZ)$ and reduced temperature $\tau = T/Z^{4/3}$. The curves $\alpha =$ const in the plane (τ, σ)-plane are shown in Figure 2.4.

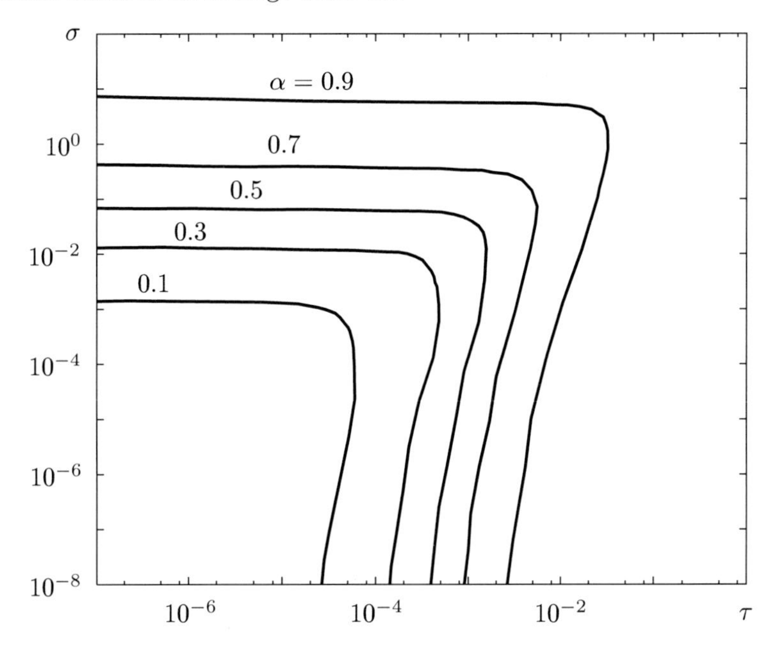

Figure 2.4: Curves of constant degree of ionization $\alpha = Z_0/Z = $const in the (τ, σ)-plane, where $\tau = T/Z^{4/3}$ and $\sigma = \rho/(AZ)$

2.1.5 Corrections to the Thomas-Fermi model

Due to its simplicity and universality, the Thomas-Fermi (TF) model found wide applications in the description of properties of hot plasmas. Soon after the TF model was formulated for temperature $T = 0$ [213, 60] and generalized for $T \neq 0$ [61], the question of finding its domain of applicability and the related issue of constructing refined models arose. In this direction, already in the original model for $T = 0$ it was proposed to include exchange and correlation effects [54, 225]. To improve the generalized TF model for matter with given density and temperature, researchers have resorted to the expansion of the Hamiltonian of the electron system in powers of the Planck constant $\hbar$. This has led to the TF model with quantum and exchange corrections, known as the TFC model [103, 104, 98, 99].

The correction $\delta V(r)$ to the TF potential in the TFC model can be expressed as follows (see § 6.1):

$$\delta V(r) = \frac{\sqrt{2}}{6\pi\sqrt{\theta}} \left[r_0 \chi(x)/r + \frac{1}{2} I_{-1/2}\left(\frac{\phi(x)}{x}\right) - \frac{1}{2} I_{-1/2}[\phi(1)] - \chi(1) \right], \quad (2.23)$$

where $x = r/r_0$, $\phi(x)$ is the solution of the TF model (1.17) and $\chi(x)$ is the solution of the nonhomogeneous linear equation

$$\chi''(x) = \frac{\sqrt{2\theta}\, r_0^2}{\pi} \left\{ 2I_{-1/2}\left(\frac{\phi(x)}{x}\right)\chi(x) + 7x\left[I_{-1/2}\left(\frac{\phi(x)}{x}\right)\right]^2 - xI_{1/2}\left(\frac{\phi(x)}{x}\right) I_{-3/2}\left(\frac{\phi(x)}{x}\right)\right\} \tag{2.24}$$

with boundary conditions $\chi(0) = 0$, $\chi(1) = \chi'(1)$.

As it turned out, it is not enough to include quantum, exchange, and correlation corrections, since in addition to the regular corrections, i.e., those obtained by expanding the Hamiltonian in a series of powers of the Planck constant $\hbar$, the ignored corrections connected with the shell structure (the so-called oscillation corrections) may be very important. Indeed, it is precisely the shell effects that explain the oscillations of the value of the normal density of various substances in dependence of their atomic number Z. Recall that when we calculated the electron density in the TF model by means of expression (2.18) we replaced $\sin^2(\xi + \pi/4)$ by 1/2. If we do not make this substitution, then we can approximately account for the shell structure and obtain analytic formulas for calculating the electron density which reproduce qualitatively its oscillations [106].

The range of applicability of the models listed above is limited by the smallness of the corrections. Moreover, the atomic potential, wave functions and corresponding electron density furnished by models with corrections remain inconsistent. A more consistent approach, which does not require that the corrections be small, is an approach that takes into consideration the shell structure of the atom, i.e., is based on the calculation of electron wave functions and the application of self-consistent field models.

2.2 Bound-state wave functions

At temperature $T \sim 1$ keV, knowledge of the Thomas-Fermi potential is sufficient for solving many problems in the physics of hot plasma. At low temperatures, too, the Thomas-Fermi potential does not differ too much from the Hartree-Fock-Slater potential, especially for high-Z elements. As a consequence, using the Thomas-Fermi potential one can easily find various characteristics of matter, study the qualitative behavior of these characteristics in their dependence on density and temperature, and so on. Furthermore, the Thomas-Fermi potential practically always serves as a good initial approximation in calculations of the self-consistent field by means of more complex quantum-mechanical models.

To obtain the one-electron wave functions one needs to solve the Schrödinger equation with some self-consistent central potential $V(r)$. Various methods — analytical, approximate and numerical—for solving the Schrödinger equation are considered in the Appendix at the end of the book. In the next section we discuss the application of these methods in practice and analyze their effectiveness.

2.2.1 Numerical methods for solving the Schrödinger equation

When iterations are carried out to obtain self-consistent atomic potentials one has to solve hundreds, and sometimes even thousands of times the Schrödinger equation with the given potential $V(r)$. As a consequence, the effectiveness of a method for calculating self-consistent potentials is to a large degree determined by the effectiveness of the method used to solve the Schrödinger equation

$$-\frac{1}{2}R''_{n\ell} + \left[-V(r) + \frac{\ell(\ell+1)}{2r^2}\right] R_{n\ell} = \varepsilon_{n\ell} R_{n\ell}, \qquad 0 < r < r_0. \tag{2.25}$$

The boundary conditions for the radial function $R_{n\ell}(r)$ at $r = 0$ are specified by the requirement that the function $R_{n\ell}(r)/r$ be bounded at zero, which gives $R_{n\ell}(0) = 0$. Depending on the physical situation, the boundary conditions at $r = r_0$ can be quite different: $R_{n\ell}(r_0) = 0$, $(R_{n\ell}(r)/r)'\big|_{r=r_0} = 0$, or periodicity conditions (a version of such conditions is given in § 4.1).

As a rule, to solve the equation (2.25) in §§ 2 and 3 of the this chapter we will take for $V(r)$ the Thomas-Fermi potential. We will address first the problem of finding the discrete energy spectrum of equation (2.25) under the condition that the wave function and its derivative vanish as $r_0 \to \infty$. Some of the first calculations for a free atom were performed by D. Hartree [82]. To solve the Schrödinger equation Numerov's method and some iteration scheme for finding the energy eigenvalues $\varepsilon_{n\ell}$ have been used (see also [62]).

A more effective method is the *phase method* [216, 155], which is based on the semiclassical approximation and a generalization of the Bohr-Sommerfeld condition (see Appendix, § A.5). The phase method allows one to find the energy eigenvalues with high accuracy in only two or three iterations; moreover, it is little sensitive to the choice of the initial approximation, and hence exceptionally reliable and sufficiently economical in the computation of self-consistent potentials for a wide range of densities and temperatures.

As a comparison, the *method of reverse iterations* [141] has been tried as well; it turned out to be faster, but less reliable, because it requires a sufficiently accurate initial value of the energy, which is not always possible to achieve in massive computations.

2.2.2 Hydrogen-like and semiclassical wave functions

In addition to direct numerical methods, different approximate methods for solving equation (2.25) which rely on the analytic representation of wave functions are available. The most convenient of these from the point of view of their further utilization are the *hydrogen-like* (or *hydrogenic*) *wave functions*

$$R^{\mathrm{H}}_{n\ell}(r) = \frac{\sqrt{Z_{n\ell}}}{n}\sqrt{\frac{(n-\ell-1)!}{(n+\ell)!}}\, e^{-x/2}\, x^{\ell+1}\, L^{2\ell+1}_{n-\ell-1}(x), \tag{2.26}$$

where $x = 2Z_{n\ell}r/n$, $L^{\alpha}_{m}(x)$ are the Laguerre polynomials, and the value of the effective charge $Z_{n\ell}$ is chosen in a special way, for example, by employing the *method of the trial potential* (see Appendix, A.3.3)

Using the hydrogen-like wave functions one can readily calculate quantities such as oscillator strengths, the probabilities of various radiative and collisional processes, and other characteristics of atoms. The energy eigenvalue corresponding to the wave function (2.26) is given by

$$\varepsilon^{\mathrm{H}}_{n\ell} = -\frac{Z^2_{n\ell}}{2n^2} + A_{n\ell}, \tag{2.27}$$

where $A_{n\ell}$ is a constant that describes external screening.

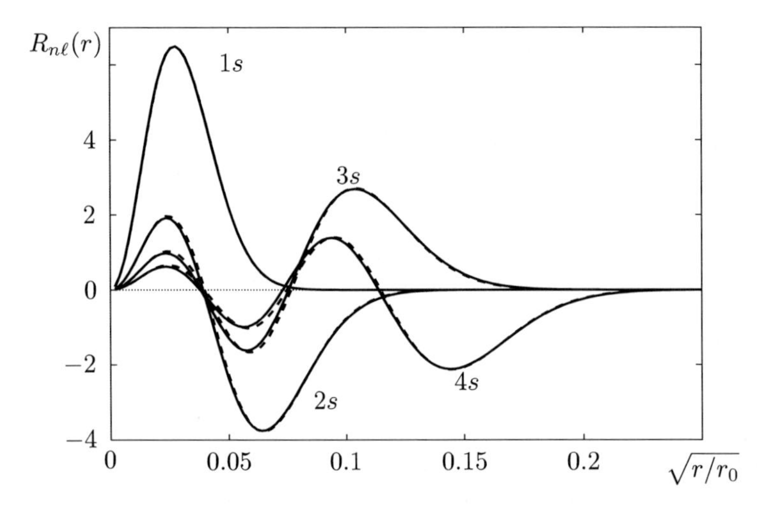

Figure 2.5: Radial wave functions $R_{n\ell}(r)$ as functions of $x = \sqrt{r/r_0}$, calculated in the hydrogen-like approximation (dashed curves) and by solving numerically the Schrödinger equation (solid curves). The graphs of $R_{n\ell}(r)$ are shown for the principal quantum numbers $n = 1, 2, 3, 4$ and $\ell = 0$ (the number of zeroes of the corresponding function is $n-1$). The calculations were done using the TF potential for gold at $T = 1$ keV, $\rho = 0.1$ g/cm^3

Hydrogen-like wave functions and numerical solutions of the Schrödinger equation with the Thomas-Fermi potential, obtained for gold ($T = 1$ keV, $\rho = 0.1$ g/cm^3), are shown in Figure 2.5. To save space the radial functions are shown only

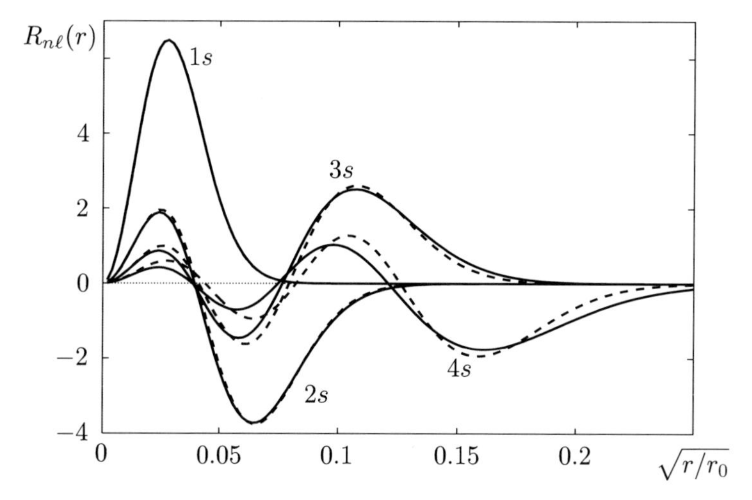

Figure 2.6: Hydrogen-like and numerical wave functions for $n = 1, 2, 3, 4$, $\ell = 0$ in the TF potential for gold at $T = 0.01$ k.eV, $\rho = 0.1$ g/cm^3

for $\ell = 0$, since for $\ell \neq 0$ the agreement between the approximate functions and the exact ones is far better. At low temperatures the results may be less good (see Figure 2.6), because the hydrogen-like approximation works only when the atomic potential is close to the Coulomb potential in some region that is essential for the wave function. The higher the temperature of the substance, the more adequate the hydrogen-like wave functions prove to be. However, in some cases — for instance, for the first shell — the hydrogen-like approximation is practically always applicable, except, possibly, in the case of low-Z elements when the first shell is fully occupied.

Incidentally, it may not be enough that the graphs of the approximate wave functions are close to the graphs of the exact ones. More reliable indicators of the quality of the approximate functions are the accuracy with which the orthogonality relations are satisfied, what errors arise in the computation of matrix elements, and so on. For example, the orthogonality conditions for the one-electron wave functions for different values of the principal quantum number n and equal values of ℓ lead to the relation $\int R_{n\ell}(r) R_{n'\ell}(r)\, dr = 0 \ (n \neq n')$, which the hydrogen-like functions for $Z_{n\ell} \neq Z_{n'\ell}$ satisfy only approximately. At the same time, for different values of ℓ the wave functions are orthogonal thanks to the properties of the spherical harmonics $Y_{\ell m}(\vartheta, \varphi)$.

More accurate than the hydrogen-like wave functions are the wave functions provided by the *improved semiclassical approximation*, which relies on Bessel functions (see Appendix, A.4); we will use superscript "s" to indicate semiclassical quantities, though we could have also used the notation "WKB". The semiclassical wave functions can be used in a wider range of temperatures and densities. One should keep in mind, however, that the semiclassical functions are less convenient than the hydrogen-like ones when it comes to deriving analytic formulas for calculating cross-sections of radiative processes.

In the semiclassical approximation the value of the energy $\varepsilon = \varepsilon^{s}_{n\ell}$ is calculated by means of the Bohr-Sommerfeld formula

$$\int_{r_1}^{r_2} p(r)dr = \pi\left(n - \ell - \frac{1}{2}\right), \tag{2.28}$$

where

$$p(r) = \sqrt{2\left[\varepsilon + V(r) - \frac{(\ell + 1/2)^2}{2r^2}\right]}, \tag{2.29}$$

and r_1, r_2 are turning points ($p(r_1) = p(r_2) = 0$, $r_2 > r_1$).

The radial wave functions have the form

$$R(r) = \begin{cases} \dfrac{C}{\pi}\sqrt{\dfrac{\xi}{|p|}}K_{1/3}(\xi), & \text{if } p^2(r) \leq 0, \\ \dfrac{C}{\sqrt{3}}\sqrt{\dfrac{\xi}{p}}\left[J_{-1/3}(\xi) + J_{1/3}(\xi)\right], & \text{if } p^2(r) \geq 0, \end{cases} \tag{2.30}$$

where

$$p = p(r), \quad \xi = \xi(r) = \left|\int_{\widetilde{r}}^{r} p(r')dr'\right|,$$

$\widetilde{r}$ is the turning point, $\varepsilon = \varepsilon^{s}_{n\ell}$ and C is a normalization constant.

Let us examine some specific aspects of the application of formulas (2.28)–(2.30), depending on the choice of the turning point $\widetilde{r}$. Set

$$R^{(1)}_{n\ell}(r) = \begin{cases} \dfrac{C_1}{\pi}\sqrt{\dfrac{\xi}{|p|}}K_{1/3}(\xi), & \text{if } r \leq r_1, \\ \dfrac{C_1}{\sqrt{3}}\sqrt{\dfrac{\xi}{p}}\left[J_{-1/3}(\xi) + J_{1/3}(\xi)\right], & \text{if } r_1 \leq r < r_2, \end{cases}$$

where

$$\xi = \left|\int_{r_1}^{r} p(r')\,dr'\right|;$$

$$R^{(2)}_{n\ell}(r) = \begin{cases} \dfrac{C_2}{\sqrt{3}}\sqrt{\dfrac{\xi}{p}}\left[J_{-1/3}(\xi) + J_{1/3}(\xi)\right], & \text{if } r_1 < r \leq r_2, \\ \dfrac{C_2}{\pi}\sqrt{\dfrac{\xi}{|p|}}K_{1/3}(\xi), & \text{if } r \geq r_2, \end{cases}$$

where

$$\xi = \left| \int_{r_2}^{r} p(r')\, dr' \right| .$$

The normalization constant C_1 is usually taken to be positive, while the sign of C_2 is determined by the number of zeroes of the wave function. The function $R_{n\ell}^{(1)}(r)$ [resp., $R_{n\ell}^{(2)}(r)$] has a singularity at $r = r_2$ [resp., $r = r_1$]. To obtain a solution that is smooth for all values of r, we must match the functions $R_{n\ell}^{(1)}(r)$ and $R_{n\ell}^{(2)}(r)$ by choosing the value of the energy $\varepsilon = \varepsilon_{n\ell}$.

For a given value of the energy—in particular, for the value obtained from the Bohr-Sommerfeld condition — we use the folowing interpolation on the interval (r_1, r_2):

$$R_{n\ell}^{\mathrm{S}}(r) = [1 - a(r)] R_{n\ell}^{(1)}(r) + a(r) R_{n\ell}^{(2)}(r), \quad a(r) = \frac{\int_{r_1}^{r} p(r)\, dr}{\int_{r_1}^{r_2} p(r)\, dr}. \tag{2.31}$$

Here the interpolation coefficient $a(r)$ is chosen in such a way that the wave function and its first derivative will be continuous in the turning points r_1 and r_2. Note that in some energy interval the radial function furnished by formula (2.31) will have no points of inflexion regardless of whether the Bohr-Sommerfeld condition is satisfied or not. Therefore, if a more accurate value of the energy than $\varepsilon_{n\ell}^{\mathrm{s}}$ is known, it can be used to obtain the wave function by means of formula (2.31).

In practical computations it is convenient to use tables of the functions $K_{1/3}(\xi)$ and $J_{-1/3}(\xi) + J_{1/3}(\xi)$. Moreover, in order to avoid dividing the small quantities $\xi(r)$ and $p(r)$ one by another in the vicinity of the turning points, the wave function can be represented for $\xi \sim 0$ in the form

$$R_{n\ell}^{\mathrm{S}}(r) = A_1 + A_2 \xi^{2/3} \quad (\xi \le 0.1), \tag{2.32}$$

where the coefficients A_1 and A_2 are calculated using (2.30), $A_2 = \pm 0.955275\, A_1$ and the different signs correspond to the right and respectively the left side relative to the first turning point and reverse signs for the second turning point.

At temperature $T = 1$ keV and density $\rho = 0.1$ g/cm^3 the graphs of the semi-classical and exact wave functions practically coincide for all considered values of n and ℓ. Good agreement is observed also at lower temperatures (see Figure 2.7 below), in contrast to the hydrogen-like wave functions, which describe the spectrum only for the inner shells with small quantum numbers.

Table 2.1: Relative differences $\delta = (\varepsilon_{n\ell} - \widetilde{\varepsilon}_{n\ell})/\varepsilon_{n\ell} \cdot 100\%$ between the approximate energy levels, and the exact values $\varepsilon_{n\ell}$ in the TF potential for gold at $T = 1$ keV and $\rho = 0.1$ g/cm^3. Also shown are the deviations Δ of the approximate wave functions from the numerical solutions of the Schrödinger equation ($\Delta = \Delta_{n\ell}$ is calculated via (2.33) and multiplied by 100 %)

		$\lvert\varepsilon_{n\ell}\rvert$	δ %		Δ %	
n	ℓ	Schröd.	H	WKB	H	WKB
1	0	77532	0.01	1.73	0.29	1.95
2	0	16992	−0.03	−0.12	2.68	1.35
2	1	16573	0.05	0.03	1.29	2.67
3	0	7039	−0.07	0.06	4.56	2.08
3	1	6883	0.03	−0.12	3.54	2.02
3	2	6645	0.04	−0.09	1.33	2.35
4	0	3794	−0.09	−0.10	5.20	0.83
4	1	3728	0.04	0.04	3.97	0.73
4	2	3636	0.04	−0.07	1.74	0.75
4	3	3576	0.00	0.03	0.44	1.64
5	0	2336	−0.11	−0.05	5.50	0.88
5	1	2302	0.03	0.06	4.21	0.43
5	2	2257	0.05	−0.09	1.98	1.19
5	3	2228	0.01	−0.04	0.52	0.44
5	4	2220	0.00	0.03	0.10	1.84
6	0	1558	−0.14	0.01	5.65	1.22
6	1	1538	0.02	0.13	4.35	0.86
6	2	1513	0.03	−0.02	2.09	1.10
6	3	1496	0.01	−0.06	0.60	0.36
6	4	1491	0.00	−0.02	0.15	0.53
6	5	1490	0.00	−0.01	0.05	1.89

Tables 2.1 and 2.2 list the relative errors δ for energy levels in the TF potential obtained by the various methods discussed above. Also listed are the mean square deviations $\Delta = \Delta_{n\ell}$ of the approximate wave functions from the numerical solutions, calculated according to the formula

$$\Delta_{n\ell} = \sqrt{\int_0^{r_0} \left[\widetilde{R}_{n\ell}(r) - R_{n\ell}(r)\right]^2 dr}, \qquad (2.33)$$

where $\widetilde{R}_{n\ell}(r)$ is the corresponding semiclassical or hydrogen-like wave function.

Table 2.2: Difference between the approximate energies and wave functions and the exact ones for gold at $T = 0.01$ keV and $\rho = 0.1$ g/cm^3 (same notations as in Table 2.1)

		$\lvert\varepsilon_{n\ell}\rvert$	δ %		Δ %	
n	ℓ	Schröd.	H	WKB	H	WKB
1	0	72242	0.00	1.83	0.31	1.95
2	0	12015	−0.10	0.19	3.71	1.69
2	1	11523	0.09	0.38	1.95	2.93
3	0	2854	−0.59	0.02	11.99	0.57
3	1	2627	0.15	−0.03	11.03	0.72
3	2	2208	1.31	0.00	6.16	1.88
4	0	672.5	−1.48	0.33	27.44	0.56
4	1	576.3	1.46	0.37	27.46	0.56
4	2	401.3	6.47	0.46	27.21	0.59
4	3	169.7	–	1.29	–	1.20
5	0	143.5	–	0.01	–	0.91
5	1	111.4	–	0.12	–	1.06
5	2	59.84	–	−0.84	–	1.35
6	0	38.31	–	−0.34	–	0.79

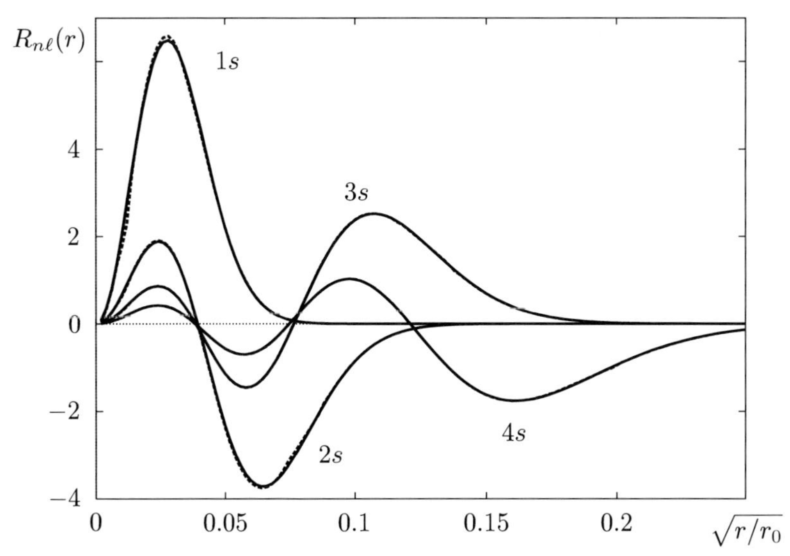

Figure 2.7: Semiclassical and numerical radial functions for $n = 1, 2, 3, 4$, $\ell = 0$ in the TF potential for gold with $T = 0.01$ keV and $\rho = 0.1$ g/cm^3 (compare with Figure 2.6)

The quantity $\Delta^2_{n\ell}$ corresponds to the square of the relative error, averaged with the weight $R^2_{n\ell}(r)$:

$$\Delta^2_{n\ell} = \int_0^{r_0} \left[\frac{\widetilde{R}_{n\ell}(r) - R_{n\ell}(r)}{R_{n\ell}(r)} \right]^2 R^2_{n\ell}(r)\, dr.$$

An analysis of the results of numerous computations, only a small part of which are shown in tables 2.1, 2.2 and figures 2.5–2.7, leads to the following conclusions. At high temperatures ($T \sim 1$ keV) the hydrogen-like wave functions practically coincide with the exact ones (see Table 2.1). For the inner shells the method of the trial potential yields good results for any temperatures and densities (as well as for the free atom). For outer shells with number of electrons $N_{n\ell} \geq 1$, at temperatures $T \sim 0.01$ keV and below, the radial functions are far from the hydrogen-like ones.

Incidentally, for high-Z elements all the methods for solving the Schrödinger equation considered here have practically the same physical accuracy, as a comparison with the corresponding numerical solutions of the Dirac equation shows: the difference between the relativistic and nonrelativistic radial functions is considerably larger than the error in the solutions of the Schrödinger equation by approximate methods.

2.2.3 Relativistic wave functions

For high-Z elements it is necessary to account for relativistic effects and so, instead of the Schrödinger equation (2.25), one needs to solve the system of equations (see Appendix, A.2):

$$\begin{cases} \dfrac{d}{dr}F(r) + \dfrac{\kappa}{r}F(r) = \alpha \left[\varepsilon_{n\ell j} + V(r) + \dfrac{2}{\alpha^2} \right] G(r), \\[2ex] \dfrac{d}{dr}G(r) - \dfrac{\kappa}{r}G(r) = -\alpha[\varepsilon_{n\ell j} + V(r)]F(r), \end{cases} \tag{2.34}$$

where $F(r) = F_{n\ell j}(r) = rf(r)$ and $G(r) = G_{n\ell j}(r) = rg(r)$ are the so-called *Dirac spin orbitals*; $f(r)$ and $g(r)$ are the *large* and respectively the *small radial components* of the wave function, ℓ is the orbital quantum number, and $j = \ell \pm 1/2$ is the total momentum quantum number ($j = 1/2$ for $\ell = 0$); $\kappa = -2(j - \ell)(j + 1/2) = \pm(j + 1/2)$, or $\kappa = -\ell - 1$ for $j = \ell + 1/2$ and $\kappa = \ell$ for $j = \ell - 1/2$.

A *numerical method* for solving the system of equations (2.34) can be constructed by using methods for solving the Schrödinger equation, in particular, the phase method (see Appendix, A.5). To this end it suffices to make the substitution [80, 236]:

$$F(r) = \sqrt{\eta(r)}\, P(r), \quad G(r) = \sqrt{\chi(r)}\, Q(r), \tag{2.35}$$

where

$$\eta(r) = \alpha \left[\varepsilon + V(r) + \frac{2}{\alpha^2} \right], \quad \chi(r) = \alpha[\varepsilon + V(r)]. \tag{2.36}$$

This yields an equation of Schrödinger type for each of the functions $P(r)$ and $Q(r)$. Thus, for $P(r)$ we have

$$P'' + k^2(r)P = 0, \tag{2.37}$$

with $k^2(r)$ given by

$$k^2(r) = \eta\chi - \frac{3}{4}\left(\frac{\eta'}{\eta}\right)^2 + \frac{1}{2}\frac{\eta''}{\eta} - \frac{\kappa}{r}\frac{\eta'}{\eta} - \frac{\kappa(\kappa+1)}{r^2}. \tag{2.38}$$

The equation for the small component $Q(r)$ is similar with (2.37)–(2.38), and in fact can be obtained from the latter by making the change of variables $P(r) \to Q(r)$, $\eta(r) \to \chi(r)$, $\chi(r) \to \eta(r)$, $\kappa \to -\kappa$.

To obtain the *relativistic hydrogen-like radial functions* $\widetilde{F}(r) = F^{\mathrm{H}}_{n\ell j}(r)$ and $\widetilde{G}(r) = G^{\mathrm{H}}_{n\ell j}(r)$ we will proceed by analogy with the nonrelativistic case and use the solution of system (2.34) with the potential $\widetilde{V}(r) = Z_{n\ell j}/r - A_{n\ell j}$. The effective charge $Z_{n\ell j}$ and the external screening constant $A_{n\ell j}$ are calculated by the method of the trial potential from the minimum condition for the functional

$$J(Z_{n\ell j}) = \int_0^{r_0} [rV(r) - r\widetilde{V}(r)]^2 \left[\widetilde{F}^2(r) + \widetilde{G}^2(r)\right] dr, \tag{2.39}$$

(see [142]), under the supplementary condition

$$\int_0^{r_0} [V(r) - \widetilde{V}(r)] \left[\widetilde{F}^2(r) + \widetilde{G}^2(r)\right] dr = 0. \tag{2.40}$$

The relativistic hydrogen-like radial functions $\widetilde{F}(r)$ and $\widetilde{G}(r)$ are given by the expressions (see Appendix, A.2):

$$\begin{aligned} \widetilde{F}(r) = F^{\mathrm{H}}_{n\ell j}(r) &= C_{n\ell j}\, x^\nu e^{-x/2} \left[f_1\, x\, L^{2\nu+1}_{n-j-3/2}(x) + f_2\, L^{2\nu-1}_{n-j-1/2}(x) \right], \\ \widetilde{G}(r) = G^{\mathrm{H}}_{n\ell j}(r) &= C_{n\ell j}\, x^\nu e^{-x/2} \left[g_1\, x\, L^{2\nu+1}_{n-j-3/2}(x) + g_2\, L^{2\nu-1}_{n-j-1/2}(x) \right], \end{aligned} \tag{2.41}$$

$$x = 2ar/\alpha, \quad \nu = \sqrt{(j+1/2)^2 - \zeta^2}, \quad \zeta = \alpha\, Z_{n\ell j},$$

$$a = \frac{\zeta}{\sqrt{(n-j-1/2+\nu)^2 + \zeta^2}},$$

$$C_{n\ell j}=\frac{a}{2\nu(\kappa-\nu)\alpha^{1/2}\zeta}\sqrt{\frac{(\kappa-\nu)(n-j-1/2)!\,[a\kappa(n-j-1/2+\nu)-\zeta\nu]}{\Gamma\,(n-j-1/2+2\nu)}},$$

$$f_1=\frac{a\zeta^2}{a\kappa(n-j-1/2+\nu)-\zeta\nu},\quad f_2=\kappa-\nu,\quad g_1=\frac{\kappa-\nu}{\zeta}f_1,\quad g_2=\zeta.$$

For $j=n-1/2$ in (2.41) we must put $L_{-1}^{2\nu+1}(x)=0$.

The energy levels $\varepsilon_{n\ell j}^{\mathrm{H}}$ are given by the formulas

$$\varepsilon_{n\ell j}^{\mathrm{H}}=-\frac{Z_{n\ell j}^2/\widetilde{n}^2}{1+\zeta^2/\widetilde{n}^2+\sqrt{1+\zeta^2/\widetilde{n}^2}}+A_{n\ell j}, \tag{2.42}$$

where

$$\widetilde{n}=n-\frac{\zeta^2}{(j+1/2)+\sqrt{(j+1/2)^2-\zeta^2}},$$

$$A_{n\ell j}=\int_0^{r_0}\left[\frac{Z_{n\ell j}}{r}-V(r)\right]\left[F_{n\ell j}^2(r)+G_{n\ell j}^2(r)\right]dr.$$

As formulas (2.41) show, the dependence of the relativistic hydrogen-like functions on the effective charge $Z_{n\ell j}$ of the nucleus is rather complex compared to that of the nonrelativistic functions. Consequently, the search for the minimum of the functional (2.39) by an iterative method similar to method (A.76) in section A.3 of the Appendix leads to complicated formulas. As it turned out, in practice it is more convenient to compute $Z_{n\ell j}$ directly from the condition of minimum for the integral in the right-hand side of (2.39), using the method of tangent parabolas.

The typical graph of the functional $J(Z_{n\ell j})$ at high temperatures is shown in Figure 2.8 for gold ($T=1$ keV, $\rho=0.1$ g/cm^3). For the given n,ℓ,j the functional has a unique minimum, which is always the case at high temperatures. At lower temperatures, for large values of the principal quantum number n there may arise stationary segments with subsequent vanishing of the minimum. This means that for such values of n there are no hydrogen-like states. Results of the computation of the effective charges $Z_{n\ell j}$ and $Z_{n\ell}$ are given in Table 2.3.

The relativistic and nonrelativistic wave functions are compared in Figure 2.9. Then Figure 2.10 compares the hydrogen-like wave functions with numerical solutions of the Dirac equation. The difference between the hydrogen-like approximation and the numerical solution of the Dirac equation may depend on n in a non-monotone fashion. For example, for $n=3$ the difference between the relativistic and nonrelativistic functions is more notable than for $n=1$ (see Figure 2.9).

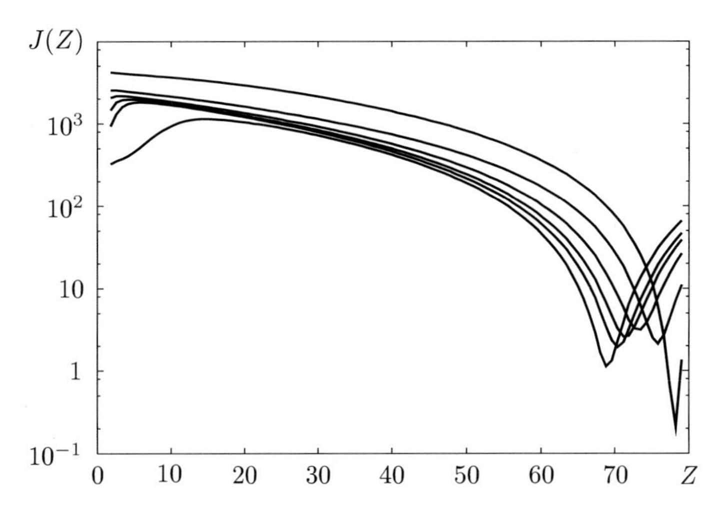

Figure 2.8: Dependence of the functional J on $Z = Z_{n\ell j}$ in the TF potential at $T = 1$ keV, $\rho = 0.1$ g/cm^3. Graphs are shown for the principal quantum numbers $n = 1, 2, 3, 4, 5, 10$, $\ell = 0$ and $j = 1/2$. The top (resp. bottom) curve corresponds to $n = 1$ (resp. $n = 10$)

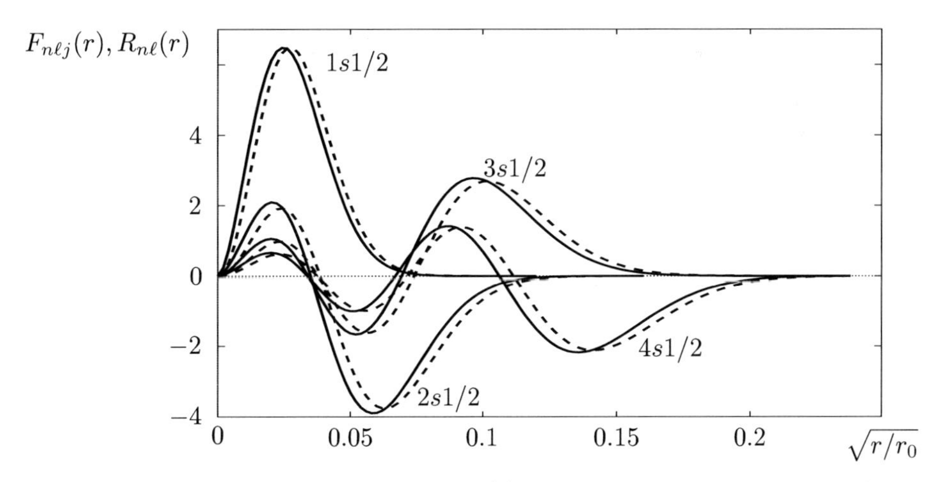

Figure 2.9: The large components $F_{n\ell j}(r)$ of the relativistic radial wave functions (solid curves) and solutions $R_{n\ell}(r)$ of the Schrödinger equation (dashed curves). The computations were carried out in the TF potential for gold at for $T = 1\,\text{keV}$, $\rho = 0.1$ g/cm^3

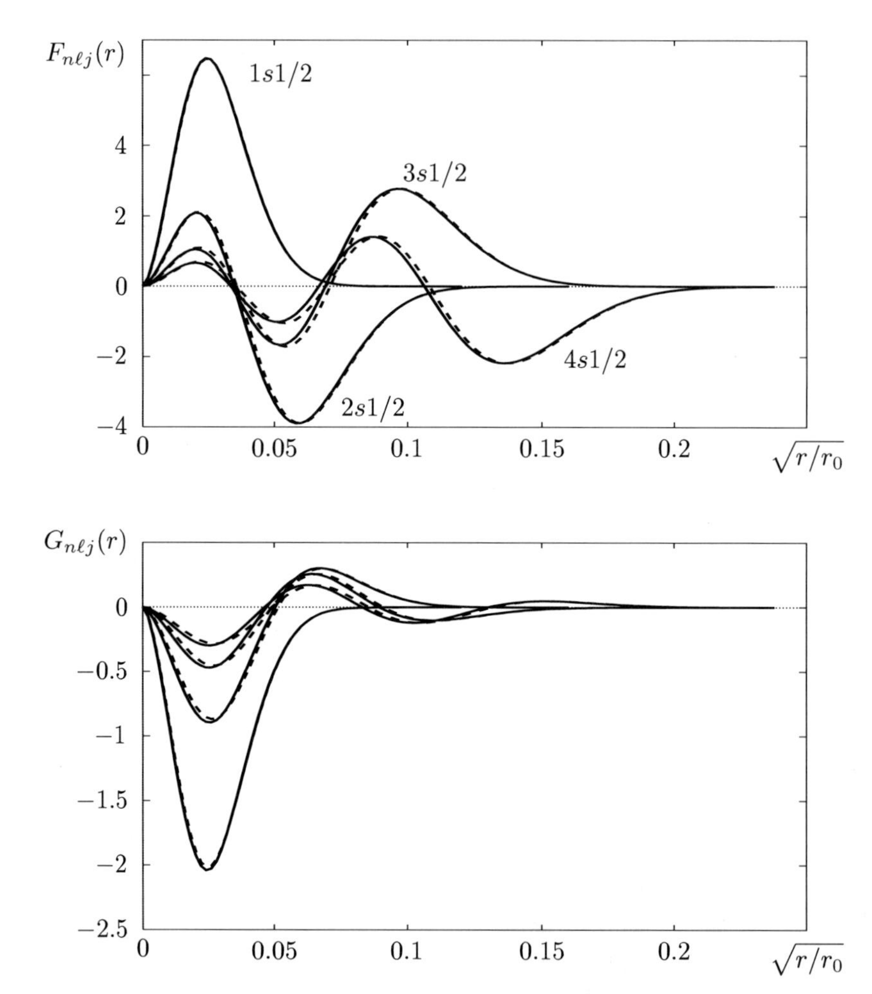

Figure 2.10: The large ($F_{n\ell j}(r)$) and the small ($G_{n\ell j}(r)$) components of the hydrogen-like radial wave functions (dashed curves) and numerical solutions of the Dirac equation (solid curves). The graphs are shown for the principal quantum numbers $n = 1, 2, 3, 4$, $\ell = 0$, $j = 1/2$ (the main maximum decreases as n grows). The calculations were carried out in the TF potential for gold at $T = 1$ keV, $\rho = 0.1$ g/cm^3

Table 2.3: Effective charges in the TF potential for gold at $T = 1$ keV, $\rho = 0.1$ g/cm^3, obtained by the method of the trial potential

n	ℓ	j	$Z_{n\ell j}$	$Z_{n\ell}$
1	0	1/2	78.09	77.87
2	0	1/2	75.91	74.95
2	1	1/2	74.66	73.45
2	1	3/2	73.65	
3	0	1/2	73.34	72.24
3	1	1/2	72.10	70.92
3	1	3/2	71.17	
3	2	3/2	68.75	68.49
3	2	5/2	68.56	
4	0	1/2	71.54	70.58
4	1	1/2	70.43	69.44
4	1	3/2	69.68	
4	2	3/2	68.05	67.83
4	2	5/2	67.91	
4	3	5/2	66.65	66.61
4	3	7/2	66.63	
5	0	1/2	70.42	69.62
5	1	1/2	69.46	68.65

n	ℓ	j	$Z_{n\ell j}$	$Z_{n\ell}$
5	1	3/2	68.85	
5	2	3/2	67.57	67.40
5	2	5/2	67.46	
5	3	5/2	66.57	66.53
5	3	7/2	66.54	
5	4	7/2	66.26	66.25
5	4	9/2	66.25	
6	0	1/2	69.68	68.99
6	1	1/2	68.85	68.15
6	1	3/2	68.33	
6	2	3/2	67.28	67.13
6	2	5/2	67.18	
6	3	5/2	66.48	66.45
6	3	7/2	66.46	
6	4	7/2	66.23	66.23
6	4	9/2	66.23	
6	5	9/2	66.17	66.17
6	5	11/2	66.17	

Computations have shown also that for $\ell = 0$ the role of relativistic effects is notable even for large n; this is connected with the behavior of the relativistic functions for $r = 0$. As ℓ grows, the difference between wave functions decreases rapidly.

To calculate the relativistic wave functions in the *semiclassical approximation* it suffices to use formulas (2.35)–(2.36) and the WKB method for the equation (2.37). Energy eigenvalues obtained by various methods are listed in tables 2.4 and 2.5. As these tables show, the errors for the energy levels obtained by solving the Dirac equation numerically, and the errors of the hydrogen-like approximation do not exceed 1%, while the relativistic values differ from the nonrelativistic ones by more than 10%.

The errors of the relativistic wave functions in the hydrogen-like and semiclassical approximation are similar to those for the Schrödinger equation, shown in tables 2.1 and 2.2. A comparison of just the solutions of the Schrödinger and Dirac equations reveals that the relativistic corrections are considerably larger than the errors of the approximate methods.

Table 2.4: Energy levels in eV in the TF potential for gold with $T = 1$ keV, $\rho = 0.1$ g/cm^3, calculated in various approximations

				$\lvert\varepsilon_{n\ell j}\rvert$		$\lvert\varepsilon_{n\ell}\rvert$
n	ℓ	j	Dirac	H	WKB	Schröd.
1	0	1/2	85819	85824	87191	77532
2	0	1/2	19314	19317	19406	16992
2	1	1/2	18824	18812	18655	16573
2	1	3/2	16946	16939	16901	
3	0	1/2	7755	7756	7784	7039
3	1	1/2	7569	7561	7521	6883
3	1	3/2	7071	7068	7077	
3	2	3/2	6814	6811	6810	6645
3	2	5/2	6698	6696	6699	
4	0	1/2	4100	4100	4113	3794
4	1	1/2	4022	4017	4001	3728
4	1	3/2	3821	3820	3817	
4	2	3/2	3722	3720	3719	3636
4	2	5/2	3674	3672	3669	
4	3	5/2	3609	3609	3605	3576
4	3	7/2	3590	3590	3584	
5	0	1/2	2493	2493	2497	2336
5	1	1/2	2454	2451	2443	2302
5	1	3/2	2354	2353	2352	
5	2	3/2	2305	2304	2303	2257
5	2	5/2	2281	2280	2280	
5	3	5/2	2250	2250	2248	2228
5	3	7/2	2240	2240	2237	
5	4	7/2	2230	2231	2229	2220
5	4	9/2	2225	2225	2222	
6	0	1/2	1649	1649	1651	1558
6	1	1/2	1627	1625	1620	1538
6	1	3/2	1570	1569	1569	
6	2	3/2	1542	1542	1541	1513
6	2	5/2	1528	1528	1528	
6	3	5/2	1511	1511	1509	1496
6	3	7/2	1505	1505	1505	
6	4	7/2	1499	1499	1498	1491
6	4	9/2	1496	1496	1495	
6	5	9/2	1495	1495	1494	1490
6	5	11/2	1493	1493	1492	

Table 2.5: Energy levels in eV in the TF potential for gold with $T = 0.01$ keV, $\rho = 0.1$ g/cm^3, calculated in various approximations

				$\lvert\varepsilon_{n\ell j}\rvert$		$\lvert\varepsilon_{n\ell}\rvert$
n	ℓ	j	Dirac	H	WKB	Schröd.
1	0	1/2	80522	80534	81885	72242
2	0	1/2	14275	14283	14327	12015
2	1	1/2	13715	13701	13518	11523
2	1	3/2	11883	11871	11887	
3	0	1/2	3432	3436	3455	2854
3	1	1/2	3174	3160	3121	2627
3	1	3/2	2771	2766	2760	
3	2	3/2	2334	2306	2317	2208
3	2	5/2	2247	2222	2241	
4	0	1/2	824.4	821.6	826.4	672.5
4	1	1/2	713.5	694.9	700.2	576.3
4	1	3/2	615.6	605.6	612.1	
4	2	3/2	432.9	407.3	427.8	401.3
4	2	5/2	413.4	390.8	410.0	
4	3	5/2	176.1	151.7	173.9	169.7
4	3	7/2	172.1	148.2	170.0	
5	0	1/2	178.6	172.3	178.3	143.5
5	1	1/2	139.1	125.2	136.6	111.4
5	1	3/2	119.5	110.0	119.1	
5	2	3/2	64.22	59.38	64.10	59.84
5	2	5/2	61.64	57.94	61.60	
5	3	5/2	15.73	4.06	16.76	15.51
5	3	7/2	15.61	4.06	16.61	
6	0	1/2	44.88	50.08	45.36	38.31
6	1	1/2	34.54	49.54	34.36	29.62
6	1	3/2	31.10	45.27	31.19	
6	2	3/2	17.39	2.18	17.34	16.60
6	2	5/2	16.94	2.18	16.88	

2.3 Continuum wave functions

2.3.1 The Schrödinger equation

In order to calculate the continuum wave function of an electron with energy ε in the *nonrelativistic approximation* we need to solve the Schrödinger equation

$$-\frac{1}{2}R''_{\varepsilon\ell} + \left[-V(r) + \frac{\ell(\ell+1)}{2r^2}\right] R_{\varepsilon\ell} = \varepsilon R_{\varepsilon\ell}, \quad 0 < r < r_0, \tag{2.43}$$

with the boundary condition $R_{\varepsilon\ell}(r)\big|_{r=0} = 0$ and some normalization condition. In practical computations performed within the limits of the average atom cell, the normalization of the function $R_{\varepsilon\ell}(r)$ is usually specified by the condition that the quantity $R^2_{\varepsilon\ell}(r)\,dr d\varepsilon$ gives the number of electron states in the energy interval $(\varepsilon, \varepsilon + d\varepsilon)$ that belong to the spherical layer $(r, r + dr)$, so that

$$\int_0^{r_0} R^2_{\varepsilon\ell}(r)\,dr = w(\varepsilon), \tag{2.44}$$

where $w(\varepsilon)$ is the density of electron states with given ℓ and a definite direction of the spin, counted for the unit energy interval.

In the semiclassical approximation (see formula (1.59))

$$R^2_{\varepsilon\ell}(r)\,dr\,d\varepsilon = \frac{2}{\pi\,p_{\varepsilon\ell}(r)}\sin^2\left(\xi + \frac{\pi}{4}\right)dr\,d\varepsilon \approx \frac{dr\,d\varepsilon}{\pi p_{\varepsilon\ell}(r)} = \frac{dr\,dp}{\pi},$$

$$p = p_{\varepsilon\ell}(r) = \sqrt{2\varepsilon + 2V(r) - \frac{(\ell + 1/2)^2}{r^2}},$$

which coincides with the number of possible states of the electron (without accounting for spin), assigned to the corresponding phase-space element $drdp$; we have

$$w(\varepsilon) \approx \frac{1}{\pi}\int_{\tilde r}^{r_0} \frac{dr}{p_{\varepsilon\ell}(r)} \qquad (\tilde r = \tilde r_{\varepsilon\ell} \text{ is the turning point}). \tag{2.45}$$

The above expression for $w(\varepsilon)$ can be used to normalize the solutions of equation (2.43) in accordance with condition (2.44). As seen from (2.44), the dimensional representation of the continuum radial function differs from that of the bound radial function by a factor of $1/\sqrt{\varepsilon}$, and so when one employs the continuum function it is usually necessary to carry out the corresponding integration over the energy ε.

In the improved semiclassical approximation

$$R_{\varepsilon\ell}(r) = \begin{cases} \dfrac{1}{\pi}\sqrt{\dfrac{\xi}{|p|}}K_{1/3}(\xi), & \text{if } r \le \tilde{r}, \\[2ex] \dfrac{1}{\sqrt{3}}\sqrt{\dfrac{\xi}{p}}\left[J_{-1/3}(\xi) + J_{1/3}(\xi)\right], & \text{if } r \ge \tilde{r}, \end{cases} \tag{2.46}$$

where

$$\xi = \xi(r) = \left| \int_{\tilde{r}}^{r} p(r')\, dr' \right|$$

(see Appendix, § A.4). Using the asymptotics of the Bessel functions together with interpolation in the vicinity of the turning point $\tilde{r}$, one can recast formula (2.46) in the computationally more convenient form

$$R_{\varepsilon\ell}(r) = \begin{cases} \dfrac{1}{\sqrt{2\pi|p|}}\, e^{-\xi}, & \text{if } \xi \ge 0.5 \text{ and } r < \tilde{r}, \\[2ex] \dfrac{\sqrt{2}}{3}\left[b(r)(r-\tilde{r}) + \dfrac{0.826994}{b(r)}\right], & \text{if } \xi < 0.5, \\[2ex] \sqrt{\dfrac{2}{\pi p}}\sin\left(\xi + \dfrac{\pi}{4}\right), & \text{if } \xi \ge 0.5 \text{ and } r > \tilde{r}, \end{cases} \tag{2.47}$$

where

$$b(r) = 1.11985\left\{\frac{2}{9r^2}\left[\varepsilon(r+\tilde{r}) + \tilde{r}V(\tilde{r})\right]\right\}^{1/6}.$$

When $r_0 \gg \tilde{r}$ the semiclassical functions (2.46) and (2.47) satisfy the normalization conditions (2.44)–(2.45). Indeed,

$$\int_0^{r_0} R_{\varepsilon\ell}^2(r)\, dr \approx \int_{\tilde{r}}^{r_0} \frac{2}{\pi p}\sin^2\left(\xi + \frac{\pi}{4}\right) dr \approx \frac{1}{\pi}\int_{\tilde{r}}^{r_0} \frac{dr}{p_{\varepsilon\ell}(r)}.$$

Moreover, letting $r_0 \to \infty$ we obtain

$$\int_0^{\infty} R_{\varepsilon\ell}(r) R_{\varepsilon'\ell}(r)\, dr = \delta(\varepsilon - \varepsilon'). \tag{2.48}$$

This corresponds to the normalization usually adopted for the continuum wave functions, which for $r \to \infty$ have the asymptotics

$$R_{\varepsilon\ell}(r) \simeq \sqrt{\frac{2}{\pi k}}\sin\left(kr + \varphi_0\right), \tag{2.49}$$

where $k = \sqrt{2\varepsilon}$ and φ_0 is a phase shift that is determined by the form of the potential $V(r)$ and depends on ℓ and ε.

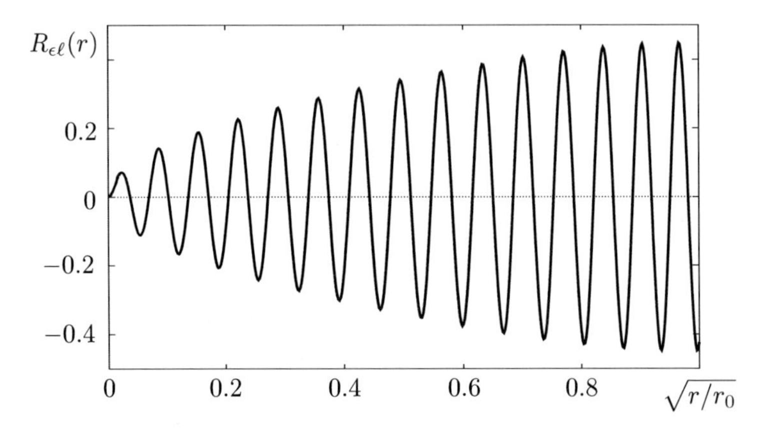

Figure 2.11: Graphs of the numerical solution of the Schrödinger equation and of the semiclassical approximation (2.47) for the continuum radial wave functions as functions of the variable $x = \sqrt{r/r_0}$ (in the case considered here the curves were found to be identical). The computations were carried out in the TF potential for gold at $T = 1$ keV, $\rho = 0.1$ g/cm^3, $\varepsilon = 5$ a.u. $= 136$ eV, $\ell = 0$

In accordance to (2.48)–(2.49), when equation (2.43) is integrated numerically one often uses the δ-function normalization with respect to energy, verifying that the asymptotics (2.49) holds for sufficiently large r (see, e.g., [45]). However, it is far more efficient to normalize numerical solutions of (2.43) already for small values of r, using, for example, the derivative $R'_{\varepsilon\ell}(r)$ in the first zero $r = r^*$ of the numerical solution $R_{\varepsilon\ell}(r)$ and the semiclassical approximation (2.46) for $\xi = \xi^*$, where ξ^* is defined by the condition $J_{-1/3}(\xi^*) + J_{1/3}(\xi^*) = 0$. The formulas giving the normalization constant have a simple form (see formula (A.112) in the Appendix):

$$\begin{cases} \left.\dfrac{dR_{\varepsilon\ell}(r)}{dr}\right|_{r=r^*} = -0.801952\sqrt{p_{\varepsilon\ell}(r^*)}, \\ R_{\varepsilon\ell}(r^*) = 0. \end{cases} \tag{2.50}$$

A comparison of the continuum wave functions obtained by various method for $\ell = 0$ and $\varepsilon = 5$ a.u. $= 136$ eV is made in Figure 2.11. One can see that the function given by formulas (2.47) practically coincides with the exact one. For that reason in practice one can confine oneself to the approximation (2.47). Notice that with the increase of ℓ and ε the application of the semiclassical approximation becomes more and more justified.

2.3.2 The Dirac equations

The *relativistic radial functions* are obtained by solving the Dirac equations (see (2.34))

$$\begin{cases} \dfrac{d}{dr}F_\varepsilon(r) + \dfrac{\kappa}{r}F_\varepsilon(r) = \alpha\left[\varepsilon + V(r) + \dfrac{2}{\alpha^2}\right]G_\varepsilon(r), \\ \\ \dfrac{d}{dr}G_\varepsilon(r) - \dfrac{\kappa}{r}G_\varepsilon(r) = -\alpha[\varepsilon + V(r)]F_\varepsilon(r), \end{cases} \tag{2.51}$$

with the boundary conditions $F_\varepsilon(0) = 0$, $G_\varepsilon(0) = 0$, where $F_\varepsilon(r) = F_{\varepsilon\ell j}(r)$, $G_\varepsilon(r) = G_{\varepsilon\ell j}(r)$, $\kappa = -2(j-\ell)(j+1/2)$. To solve the system (2.51) numerically for a given value of ε one resorts to the same methods as those used for the bound states (for example, the Runge-Kutta method or the Numerov method).

In correspondence with the nonrelativistic case, we impose the normalization condition

$$\int_0^{r_0} [F_\varepsilon^2(r) + G_\varepsilon^2(r)]\, dr = \tilde{w}(\varepsilon). \tag{2.52}$$

The expression for the relativistic density of states $\tilde{w}(\varepsilon)$ can be obtained by resorting to the semiclassical approximation for the function $P(r) = F_\varepsilon(r)/\sqrt{\eta}$ (where $\eta = \eta(r) = \alpha[\varepsilon + V(r) + 2/\alpha^2]$) with the help of equation (2.37), which allows one to write the expressions for $F_\varepsilon(r)$ and $G_\varepsilon(r)$ in a form similar to (2.46). In doing so one has to keep in mind that when $r_0 \to \infty$ the normalization condition (2.48) for the relativistic case becomes

$$\int_0^{\infty} [F_\varepsilon(r)F_{\varepsilon'}(r) + G_\varepsilon(r)G_{\varepsilon'}(r)]\, dr = \delta(\varepsilon - \varepsilon'). \tag{2.53}$$

By formula (2.37), the semiclassical approximation for the large component $F_\varepsilon(r)$ may be recast in a form analogous to (2.47):

$$F_\varepsilon(r) = \sqrt{\eta}\, P(r) \simeq C\sqrt{\frac{\eta}{\bar{k}}}\sin\left(\xi + \frac{\pi}{4}\right), \quad r > \tilde{r}. \tag{2.54}$$

Here $\xi = \xi(r) = \int_{\tilde{r}}^{r} \bar{k}(r')dr'$, $\bar{k}^2(r) = k^2(r) - 1/(4r^2)$, $\tilde{r}$ is a turning point (i.e., $\bar{k}(\tilde{r}) = 0$),

$$k^2(r) = \eta\chi - \frac{3}{4}\left(\frac{\eta'}{\eta}\right)^2 + \frac{1}{2}\frac{\eta''}{\eta} - \frac{\kappa}{r}\frac{\eta'}{\eta} - \frac{\kappa(\kappa+1)}{r^2}, \quad \chi = \chi(r) = \alpha[\varepsilon + V(r)].$$

The semiclassical approximation for the small component $G_\varepsilon(r)$ can be obtained by analogy with (2.54), using the equation for $Q(r) = G_\varepsilon(r)/\sqrt{\chi}$ (see the remark after (2.38) in § 2.2). However, it is better to start directly with equation

(2.51), recalling that in the semiclassical approximation $P'(r) \simeq C\sqrt{\bar{k}}\cos(\xi+\pi/4)$. We have

$$G_\varepsilon(r) = \frac{1}{\eta}\left[\frac{d}{dr}F_\varepsilon + \frac{\kappa}{r}F_\varepsilon\right] \simeq \frac{C}{\sqrt{\eta\bar{k}}}\left[\bar{k}\cos\left(\xi+\frac{\pi}{4}\right) + \left(\frac{\eta'}{2\eta}+\frac{\kappa}{r}\right)\sin\left(\xi+\frac{\pi}{4}\right)\right],$$

that is,

$$G_\varepsilon(r) = C\sqrt{\frac{\bar{k}^2 + [\eta'/(2\eta)+\kappa/r]^2}{\eta\bar{k}}}\cos\left(\xi+\frac{\pi}{4}-\delta\right), \tag{2.55}$$

where

$$\tan\delta = \frac{\eta'/(2\eta)+\kappa/r}{\bar{k}}.$$

To calculate the normalization constant C we use (2.53) and the asymptotics of the radial function for $r \to \infty$. First let us derive a useful relation for the normalization integral. To this end we multiply the first equation of the system (2.51) by $G_{\varepsilon'}$ and then subtract from the result the same equation, but with the places of ε and ε' switched. Similarly, we multiply the second equation in (2.51) by $F_{\varepsilon'}$, then switch the places of ε and ε' and subtract the resulting equations from one another. This yields

$$\begin{cases} G_{\varepsilon'}\dfrac{d}{dr}F_\varepsilon - G_\varepsilon\dfrac{d}{dr}F_{\varepsilon'} + \dfrac{\kappa}{r}(F_\varepsilon G_{\varepsilon'} - F_{\varepsilon'}G_\varepsilon) = \alpha(\varepsilon-\varepsilon')G_\varepsilon G_{\varepsilon'}, \\[2ex] F_{\varepsilon'}\dfrac{d}{dr}G_\varepsilon - F_\varepsilon\dfrac{d}{dr}G_{\varepsilon'} - \dfrac{\kappa}{r}(G_\varepsilon F_{\varepsilon'} - G_{\varepsilon'}F_\varepsilon) = -\alpha(\varepsilon-\varepsilon')F_\varepsilon F_{\varepsilon'}. \end{cases}$$

Next, let us subtract the second of the above equations from the first and integrate the result from 0 to some value r:

$$\int_0^r [F_\varepsilon(r')F_{\varepsilon'}(r') + G_\varepsilon(r')G_{\varepsilon'}(r')]\,dr' =$$

$$\frac{1}{\alpha(\varepsilon-\varepsilon')}\left[F_\varepsilon(r)G_{\varepsilon'}(r) - F_{\varepsilon'}(r)G_\varepsilon(r)\right]. \tag{2.56}$$

Now let us require that, when $r \to \infty$, relation (2.56) goes over into the normalization condition (2.53). The asymptotic behavior of $F_\varepsilon(r)$ and $G_\varepsilon(r)$ as $r \to \infty$ is readily derived from (2.54), (2.55) recalling that $V(r) \to 0$ when $r \to \infty$:

$$\begin{cases} F_\varepsilon(r) \simeq C\sqrt{\dfrac{\widetilde{k}}{\alpha\varepsilon}}\sin(\widetilde{k}r + \widetilde{\varphi}_0), \\[2ex] G_\varepsilon(r) \simeq C\sqrt{\dfrac{\alpha\varepsilon}{\widetilde{k}}}\cos(\widetilde{k}r + \widetilde{\varphi}_0), \end{cases} \tag{2.57}$$

where $\widetilde{k} = \widetilde{k}(\varepsilon) = \sqrt{\varepsilon(2+\alpha^2\varepsilon)}$, $\widetilde{\varphi}_0$ is a phase shift.

Substituting expression (2.57) in the right-hand side of (2.56) and letting $r \to \infty$ we obtain

$$\int_0^\infty [F_\varepsilon(r)F_{\varepsilon'}(r) + G_\varepsilon(r)G_{\varepsilon'}(r)]\,dr = \frac{C^2}{\alpha(\varepsilon-\varepsilon')}\times$$

$$\lim_{r\to\infty}\left\{\sqrt{\frac{\widetilde{k}\varepsilon'}{\varepsilon\widetilde{k}'}}\sin(\widetilde{k}r+\widetilde{\varphi}_0)\cos(\widetilde{k}'r+\widetilde{\varphi}_0) - \sqrt{\frac{\widetilde{k}'\varepsilon}{\varepsilon'\widetilde{k}}}\sin(\widetilde{k}'r+\widetilde{\varphi}_0)\cos(\widetilde{k}r+\widetilde{\varphi}_0)\right\} =$$

$$\frac{C^2}{\alpha}\lim_{r\to\infty}\left\{\frac{\sin(\widetilde{k}-\widetilde{k}')r}{\varepsilon-\varepsilon'} + \left[\frac{\sqrt{\varepsilon'\widetilde{k}}-\sqrt{\varepsilon\widetilde{k}'}}{\sqrt{\varepsilon\widetilde{k}'}(\varepsilon-\varepsilon')}\sin(\widetilde{k}r+\widetilde{\varphi}_0)\cos(\widetilde{k}'r+\widetilde{\varphi}_0) + \frac{\sqrt{\varepsilon'\widetilde{k}}-\sqrt{\varepsilon\widetilde{k}'}}{\sqrt{\varepsilon\widetilde{k}'}(\varepsilon-\varepsilon')}\sin(\widetilde{k}'r+\widetilde{\varphi}_0)\cos(\widetilde{k}r+\widetilde{\varphi}_0)\right]\right\}. \quad (2.58)$$

Since

$$\lim_{r\to\infty}\frac{\sin(\widetilde{k}-\widetilde{k}')r}{\pi(\widetilde{k}-\widetilde{k}')} = \delta(\widetilde{k}-\widetilde{k}'), \qquad (\widetilde{k}-\widetilde{k}')\delta(\widetilde{k}-\widetilde{k}') = (\varepsilon-\varepsilon')\delta(\varepsilon-\varepsilon'),$$

we have

$$\lim_{r\to\infty}\frac{\sin(\widetilde{k}-\widetilde{k}')r}{\varepsilon-\varepsilon'} = \pi\delta(\varepsilon-\varepsilon').$$

Now let us show that the expression inside the square brackets in formula (2.58) can be neglected when $r \to \infty$. To do this it suffices to examine the first term, because the second term coincides with the first after we replace k by k' and ε by ε'. The first term inside the square brackets gives

$$\lim_{r\to\infty}\frac{\sqrt{\varepsilon'\widetilde{k}}-\sqrt{\varepsilon\widetilde{k}'}}{\sqrt{\varepsilon\widetilde{k}'}(\varepsilon-\varepsilon')}\sin(\widetilde{k}r+\widetilde{\varphi}_0)\cos(\widetilde{k}'r+\widetilde{\varphi}_0) =$$

$$\frac{1}{2}\left[\frac{\sqrt{\varepsilon'\widetilde{k}}-\sqrt{\varepsilon\widetilde{k}'}}{\sqrt{\varepsilon\widetilde{k}'}}\lim_{r\to\infty}\frac{\sin(\widetilde{k}-\widetilde{k}')r}{\varepsilon-\varepsilon'} + \frac{\sqrt{\varepsilon'\widetilde{k}}-\sqrt{\varepsilon\widetilde{k}'}}{\sqrt{\varepsilon\widetilde{k}'}(\varepsilon-\varepsilon')}\lim_{r\to\infty}\sin\left[(\widetilde{k}+\widetilde{k}')r+2\widetilde{\varphi}_0\right]\right].$$

In the expression obtained the first term is equal to zero, because the factor in front of the limit vanishes for $\varepsilon = \varepsilon'$, and by a basic property of the δ-function $f(\varepsilon')\delta(\varepsilon-\varepsilon') = f(\varepsilon)\delta(\varepsilon-\varepsilon')$. In the second term $\lim_{r\to\infty}\sin[(\widetilde{k}+\widetilde{k}')r+2\widetilde{\varphi}_0]$ is also equal to zero, being the limit of a rapidly oscillating function. Summing up, we have

$$\int_0^\infty [F_\varepsilon(r)F_{\varepsilon'}(r) + G_\varepsilon(r)G_{\varepsilon'}(r)]\,dr = \frac{C^2}{\alpha}\pi\delta(\varepsilon-\varepsilon'),$$

which in conjunction with (2.53) yields $C = \sqrt{\alpha/\pi}$.

Substituting the semiclassical functions $F_\varepsilon(r)$ and $G_\varepsilon(r)$ given by (2.54) and (2.55) in (2.52) and replacing $\sin^2(\xi + \pi/4)$ and $\cos^2(\xi + \pi/4 - \delta)$ by 1/2, we have

$$\widetilde{w}(\varepsilon) \approx \int_0^{r_0} \frac{\alpha\eta}{2\pi\bar{k}} \left(1 + \frac{\bar{k}^2 + [\eta'/(2\eta) + \kappa/r]^2}{\eta^2}\right) dr. \tag{2.59}$$

The relativistic and nonrelativistic continuum radial functions within the limits of the average atom cell, obtained by numerical integration of the corresponding equations (2.43) and (2.51) for a given energy ε, are shown in Figure 2.12.

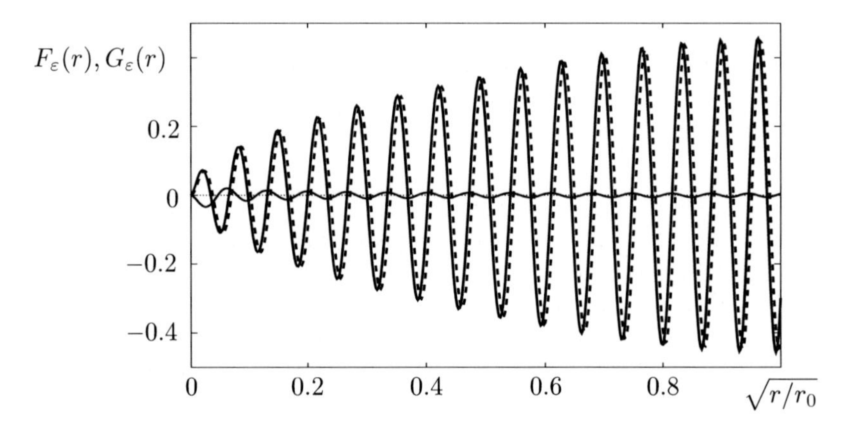

Figure 2.12: The large and small components of the relativistic continuum radial wave functions: numerical solutions of the Dirac equation (solid curves), and the numerical solution of the Schrödinger equation (dashed curve). The computations were carried out in the TF potential for gold at $T = 1$ keV, $\rho = 0.1$ g/cm^3, $\varepsilon = 136$ eV, $\ell = 0$, $j = 1/2$

The methods for the calculation of one-electron wave functions of the bound and continuum states considered in Chapter II are used in computations of self-consistent potentials for various models of matter, for the calculation of the photon absorption cross-sections and cross-sections of other processes in plasma.

Chapter 3

Quantum-statistical self-consistent field models

In a quantum-mechanical system of many interacting particles the motion of any particle in the system is related in a rather complex way with the motion of all the other particles. As a consequence, each particle is not in a determined state and cannot be described by means of its one-particle wave function. The state of the system as a whole is described by a wave function that depends on the coordinates and spin variables of all the particles that form the system. The self-consistent field method starts from the assumption that *for an approximate description of the system one can introduce a wave function for each particle of the system.* Then the interaction with the remaining particles is approximately accounted for by introducing a field, averaged over the motion of those particles by means of their one-particle wave functions. The one-particle wave functions must be consistent in the sense that, on one hand, they are solutions of the Schrödinger equation for a single particle moving in the averaged field produced by the other particles, and, on the other hand, these one-particle wave functions themselves define the averaged potential in which the particles move; whence the term "self-consistent field". The simplest method of introducing a self-consistent field, in which one defines not wave functions, but the density of the spatial distribution of electrons, is the Thomas-Fermi method proposed by L. Thomas (1926) and independently by E. Fermi (1928). The subsequent development of the self-consistent field approach has led to the elaboration of the Hartree and Hartree-Fock models.

If we consider the electron system at nonzero temperature the concept of the self-consistent field must incorporate elements of statistics. One has to pay attention to the distribution of ions over different states, taking into account also their interaction with the free electrons and other ions. In the first approximation this leads to the concept of the *average atom*, when an ion with the average occupation numbers and free electrons are considered in an *electrically neutral spherical cell*, i.e., an *average atom cell*. The average atom may be considered in the Thomas-Fermi model, as well as in more accurate approximations of the Hartree or Hartree-Fock type or their modifications, including relativistic ones. These and other quantum-statistical self-consistent field models for an average atom can be derived from a unified variational principle, the requirement of a minimum for the grand thermodynamic potential.

3.1 Quantum-mechanical refinement of the generalized Thomas-Fermi model for bound electrons

3.1.1 The Hartree self-consistent field for an average atom

The Thomas-Fermi model gives an electron density $\rho(r)$ that is consistent with the potential $V(r)$ in the framework of the semiclassical approximation. However, the wave functions calculated for the Thomas-Fermi potential yield the electron density $\rho(\vec{r}) = \sum_\alpha N_\alpha |\psi_\alpha(\vec{r})|^2$, which does not coincide with the original electron density for the TF model (here α denotes the set of quantum numbers that specify the state of the electron). An analysis of the calculations carried out in Subsection 2.1.3 has shown that in order to obtain the electron density in the TF model we essentially use only the component of $\rho(\vec{r}) = \sum_\alpha N_\alpha |\psi_\alpha(\vec{r})|^2$, which does not include the oscillations of wave functions.

At high temperatures the contribution of the oscillations is the most important for the bound electrons with small values of the principal quantum number n. For electrons with large quantum numbers and free electrons the contribution of the oscillations is essentially smoothed out and so for them one can use the semiclassical approximation (2.7). Accordingly, let us represent the electron density as a sum of two terms [151, 188]:

$$\rho(\vec{r}) = \rho_1(\vec{r}) + \rho_2(\vec{r}), \tag{3.1}$$

where

$$\rho_1(\vec{r}) = \sum_{\varepsilon_\alpha < \varepsilon_0} N(\varepsilon_\alpha)|\psi_\alpha(\vec{r})|^2, \qquad \rho_2(\vec{r}) = \sum_{\varepsilon_\alpha > \varepsilon_0} N(\varepsilon_\alpha)|\psi_\alpha(\vec{r})|^2.$$

By appropriately choosing the energy ε_0, which plays the role of an effective boundary of the continuum, one can ensure that in the expression for $\rho_2(\vec{r})$ the summation is carried out only over free-electron states and over those bound states for which $N(\varepsilon_\alpha) \ll 1$. Moreover, in the expression for $\rho_2(\vec{r})$ summation can be replaced by integration, as we did when we derived the Thomas-Fermi model in Subsection 2.1.3.

Since it is natural to consider that the potential for the average atom is spherically symmetric, calculation of $\rho_2(\vec{r})$ in the semiclassical approximation yields

$$\rho_2(\vec{r}) = \rho_2(r) = \frac{1}{\pi^2} \int \frac{p^2\, dp}{1 + \exp\left(\dfrac{p^2/2 - V(r) - \mu}{\theta}\right)}, \tag{3.2}$$

where the integral is taken over the domain

$$\frac{p^2}{2} - V(r) > \varepsilon_0 \quad (p \geq 0).$$

The density of the bound electrons $\rho_1(\vec{r})$ is calculated directly in terms of the wave functions

$$\psi_\alpha(\vec{r}) = \psi_{n\ell m}(\vec{r}) = \frac{1}{r} R_{n\ell}(r)(-1)^m Y_{\ell m}(\vartheta, \varphi),$$

where the radial function $R_{n\ell}(r)$ satisfies the Schrödinger equation

$$-\frac{1}{2}R''_{n\ell} + \left[-V(r) + \frac{\ell(\ell+1)}{2r^2}\right] R_{n\ell}(r) = \varepsilon_{n\ell} R_{n\ell}(r). \tag{3.3}$$

Here we use the potential $V(r)$ produced by the nucleus and all the electrons in the average atom cell, which is justified for sufficiently high Z, when one can neglect the "self-action" (see the Remark in Subsection 2.1.1). Generally speaking, when one solves the Schrödinger equation (3.3) it is necessary to eliminate from the potential $V(r)$ the contribution of the electron under consideration. However, it is more convenient to take into account the effect of the self-interaction later, when we refine the Hartee model (see § 3.2 and § 3.3 below).

We will classify as bound states those states for which the classical domain of motion ($r_1 < r < r_2$) lies inside the average atom cell and $r_2 \ll r_0$ (here r_0 is the radius of the cell and r_1 and r_2 are turning points). For such states the wave function $R_{n\ell}(r)$ should be negligibly small on the boundary of the cell, and hence it can be assumed to be zero for $r = r_0$. This function has $n_r = n - \ell - 1$ zeroes in the interval (r_1, r_2). Then for $\rho_1(\vec{r})$ we obtain the expression

$$\rho_1(\vec{r}) = \sum_{\substack{n,\,\ell,\,m \\ (\varepsilon_{n\ell}<\varepsilon_0)}} \frac{2}{1 + \exp\left(\dfrac{\varepsilon_{n\ell} - \mu}{\theta}\right)} \left(\frac{R_{n\ell}(r)}{r}\right)^2 |Y_{\ell m}(\vartheta, \varphi)|^2.$$

By the addition theorem for spherical harmonics,

$$\sum_{m=-\ell}^{\ell} Y^*_{\ell m}(\vartheta, \varphi) Y_{\ell m}(\vartheta, \varphi) = \frac{2\ell + 1}{4\pi}.$$

Therefore, the density of the bound electrons in the *average atom* is given by

$$\rho_1(\vec{r}) = \rho_1(r) = \frac{1}{4\pi r^2} \sum_{\varepsilon_{n\ell}<\varepsilon_0} N_{n\ell} R^2_{n\ell}(r), \quad N_{n\ell} = \frac{2(2\ell+1)}{1 + \exp\left(\dfrac{\varepsilon_{n\ell} - \mu}{\theta}\right)} \tag{3.4}$$

(compare this with the derivation of formula (2.15)).

As in the Thomas-Fermi model, we shall assume that the potential $V(r)$ satisfies the Poisson equation

$$\frac{1}{r}\frac{d^2}{dr^2}(rV) = 4\pi[\rho_1(r) + \rho_2(r)] \tag{3.5}$$

together with the boundary conditions $rV(r)|_{r=0} = Z$, $V(r_0) = 0$. The chemical potential μ is determined from the condition of charge neutrality for the average atom cell,

$$4\pi \int_0^{r_0} \rho(r) r^2\, dr = Z. \tag{3.6}$$

3.1.2 Computational algorithm

Equations (3.1)–(3.6) constitute a rather complicated nonlinear system for the functions $R_{n\ell}(r)$, the energy levels $\varepsilon_{n\ell}$, the occupation numbers $N_{n\ell}$, the potential $V(r)$ and the chemical potential μ. This system can be solved by an iteration method. In contrast to the Thomas-Fermi model, the model (3.1)–(3.6) reproduces the shell structure of the atom. As expected, at high temperatures the electron density $\rho(r)$ and the potential $V(r)$ turned out to be only weakly sensitive to the variation of the effective boundary of the continuum, ε_0. In practice the quantity ε_0 must be selected so that the occupation numbers $N_{n\ell}$ are small for levels with energy $\varepsilon_{n\ell} > \varepsilon_0$ (see subsections 3.3.3 and 3.3.4 below).

Clearly, the average-atom model for matter with given temperature and density considered above is a modification of the Hartree model. The model (3.1)–(3.6) allows one to carry out, in a unified manner, calculations for any temperature, and in fact the higher the temperature, the faster the iterations converge and the shorter the time needed for the computations. For heavy elements, when relativistic corrections are essential, one needs to pass from the Schrödinger equation to the Dirac equation (see Subsection 3.1.4 below).

The calculations can be done according the following scheme. As the initial approximation for the potential it is natural to take the Thomas-Fermi potential for the given temperature and density and the corresponding chemical potential $\eta = -\mu/\theta$. Then one solves the Schrödinger equation for bound states (by the phase method, for example; see Appendix, § A.5), which yields the energy levels $\varepsilon_{n\ell}$ and the radial functions $R_{n\ell}(r)$.

To get a more accurate chemical potential η let us use the charge neutrality condition (3.6), written in the form

$$f(\eta) = \sum_{\substack{n,\ell \\ (\varepsilon_{n\ell}<\varepsilon_0)}} \frac{2(2\ell+1)}{1+\exp\left(\dfrac{\varepsilon_{n\ell}}{\theta}+\eta\right)} + 4\pi\int_0^{r_0} r^2\rho_2(r)\, dr = Z, \tag{3.7}$$

where

$$\rho_2(r) = \frac{(2\theta)^{3/2}}{2\pi^2} \int\limits_{y_0}^{\infty} \frac{y^{1/2}\,dy}{1+\exp\left(y - \dfrac{V(r)}{\theta} + \eta\right)} =$$

$$\frac{(2\theta)^{3/2}}{2\pi^2} I_{1/2}\left(\frac{V(r)}{\theta} - \eta\right) - \frac{(2\theta)^{3/2}}{2\pi^2} \int\limits_{0}^{y_0} \frac{y^{1/2}\,dy}{1+\exp\left(y - \dfrac{V(r)}{\theta} + \eta\right)}, \tag{3.8}$$

$$y_0 = y_0(r) = \max\left\{0; \frac{V(r)+\varepsilon_0}{\theta}\right\}.$$

Finally, using the value of η found in this way and relations (3.4), (3.8), we obtain the occupation numbers $N_{n\ell}$, as well as $\rho_1(r)$ and $\rho_2(r)$.

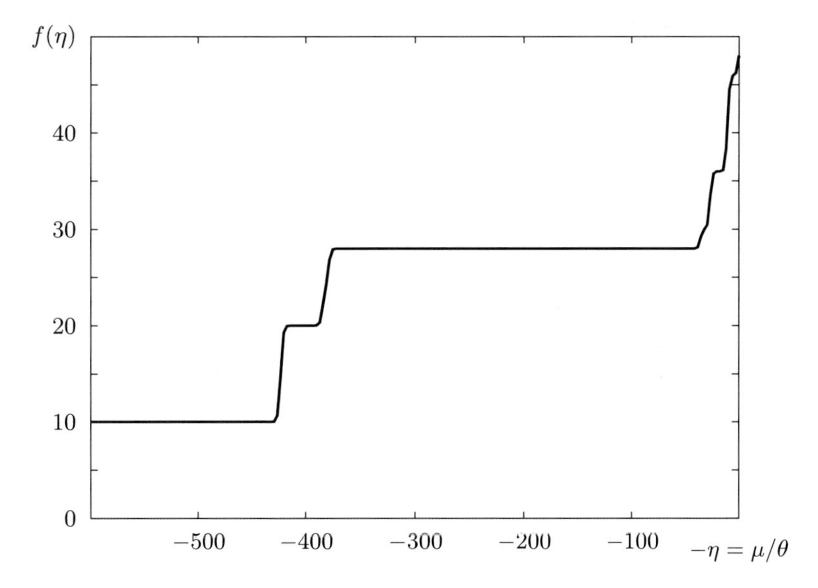

Figure 3.1: The function $f(\eta)$ (see equation (3.7))

Equation (3.7) is solved by Newton's method. Let us note that at low temperatures the Newton iterations may diverge, since in these circumstances the function $f(\eta)$ has step-like segments (see, e.g., Figure 3.1, where $f(\eta)$ is calculated for iron ($Z = 26$) with $T = 1$ eV and $\rho = 0.1$ g/cm^3). In this case one combines Newton method with the bisection method, which is applied whenever the next Newton iteration gives a value of η that lies outside the previously determined interval where the solution is to be found. Practice has shown that the bisection method applied in this manner is more effective than the secant method.

Once the refined density $\rho(r) = \rho_1(r) + \rho_2(r)$ is known, one can compute the new potential by means of the Poisson equation (3.5). In the case of the average atom the potential $V(\vec{r})$ is spherically symmetric and the ordinary differential equation (3.5) is easily solved. However, it is better to calculate the expression for $V(r)$ using the known solution of the Poisson equation in the three-dimensional case:

$$V(\vec{r}) = \frac{Z}{r} - \int \frac{\rho(r')r'^2\,dr'\,d\Omega'}{|\vec{r} - \vec{r}\,'|}.$$

To this end we use the expansion

$$\frac{1}{|\vec{r} - \vec{r}\,'|} = \sum_{s=0}^{\infty} \frac{4\pi}{2s+1} \sum_{m=-s}^{s} \frac{r_<^s}{r_>^{s+1}} Y_{sm}^*(\vartheta, \varphi) Y_{sm}(\vartheta', \varphi'),$$

where $r_< = \min(r, r')$, $r_> = \max(r, r')$. Integration with respect to the angles ϑ' and φ' yields

$$\int Y_{sm}(\vartheta', \varphi')\,d\Omega' = \delta_{m0}\,\delta_{s0} \int Y_{00}(\vartheta', \varphi')\,d\Omega'.$$

As a result, the sum over s and m for $V(r)$ reduces to a single term, corresponding to $s = 0$. Therefore,

$$V(r) = \frac{Z}{r} - 4\pi \int_0^{r_0} \frac{r'^2 \rho(r')\,dr'}{r_>} = \frac{Z}{r} - 4\pi \left[\frac{1}{r} \int_0^r r'^2 \rho(r')\,dr' + \int_r^{r_0} r' \rho(r')\,dr' \right]. \qquad (3.9)$$

After the new potential is obtained by means of formula (3.9), the iterations are repeated the necessary number of times.

We have described a simple iteration process $V^{(s+1)}(r) = F[V^{(s)}(r)]$, where $F[V(r)]$ designates the procedure by which the wave functions, the electron density and the new potential are obtained, starting with $V(r)$. At high temperatures, when the main contribution to the self-consistent potential comes from the central field of the nucleus (for example, for $Z = 79$ this is the case when $T > 0.1$ keV), the simple iterations converge rapidly. However, as a rule such iterations diverge at low temperatures. To improve the convergence we shall use a linear combination of the obtained potential and the potential furnished by the preceding iteration:

$$V_\alpha(r) = (1 - \alpha)V^{(s)}(r) + \alpha F[V^{(s)}(r)]. \qquad (3.10)$$

In his calculations D. Hartree used the values $\alpha = 0.3 \div 0.5$ (see § 3 of Chapter 5 in [83]). In our calculations the parameter α was chosen based on the following considerations. Let us call mismatch $\Delta = \Delta(\alpha)$ the quantity

$$\Delta(\alpha) = \{V_\alpha(r) - F[V_\alpha(r)]\}\Big|_{r=\tilde{r}}, \qquad (3.11)$$

where $r = \tilde{r}$ is the point in which the difference $V_\alpha(r) - F[V_\alpha(r)]$ is maximal in absolute value.

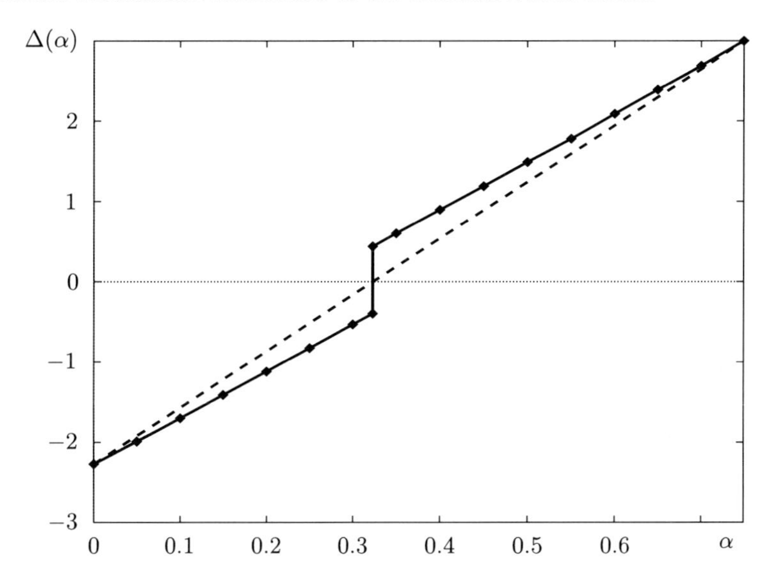

Figure 3.2: Graph of the dependence of the mismatch $\Delta(\alpha)$ of the potential on the parameter α for gold at temperature $T = 0.01$ keV and density $\rho = 1$ g/cm^3 (rhombuses). The minimum of the mismatch is attained for $\alpha' = 0.320$. By drawing a straight line (dashed) through the points corresponding to the values Δ for $\alpha = 0$ and $\alpha = 0.75$, we obtain $\alpha = \alpha^* = 0.323$

Computations have shown that the dependence $\Delta(\alpha)$ has a piecewise-linear character (Figure 3.2). For some value $\alpha = \alpha'$ one observes a jump of $\Delta(\alpha)$ — the quantity $\Delta(\alpha)$ changes sign. The smallest mismatch is attained precisely for this value of α.

The calculation of an accurate value of α' require a large volume of additional computations, and so in practice instead of α' it is convenient to use a value $\alpha^\star$ close to α', obtained by drawing a line through the values of $\Delta(\alpha)$ for $\alpha = 0$ and some other $\alpha = \alpha$:

$$\frac{\alpha^\star}{\Delta(0)} = \frac{\bar{\alpha}}{\Delta(0) - \Delta(\bar{\alpha})}. \tag{3.12}$$

Since the quantity $\Delta(0)$ is known from the preceding iteration, one needs only one more computation for $\alpha = \bar{\alpha}$. The value of $\bar{\alpha}$ is chosen by experimenting (in our calculations we took $\bar{\alpha} = 0.75$). If the value $\alpha^\star$ obtained from formula (3.12), is too small or too large (lies outside the interval $[0.1, 1]$), then $\alpha^\star$ is chosen equal to 0.1 or 1, respectively.

Thus, we obtain the iteration process

$$V^{(s+1)}(r) = V_{\alpha^*}(r) = (1 - \alpha^*)V^{(s)}(r) + \alpha^* F[V^{(s)}(r)] \quad (0.1 \le \alpha^* \le 1).$$

The convergence of the iterations is determined by the condition

$$|\Delta^{(s)}| = |\Delta(\alpha^*)| < \varepsilon$$

(where, e.g., $\varepsilon = 10^{-6}$). As a rule, no more than 10 iterations are needed; moreover, with the growth of the temperature the number of iterations drops rapidly and the value of α^* approaches 1.

3.1.3 Analysis of computational results for iron

Let us examine the graphs obtained by calculations for iron with temperature $T = 0.1$ keV and density $\rho = 0.1$ g/cm^3. Here the radius of the average atom cell is $r_0 = 11.4$ and the chemical potential is $\eta = 6.058$ ($\eta_{\text{TF}} = 6.039$).

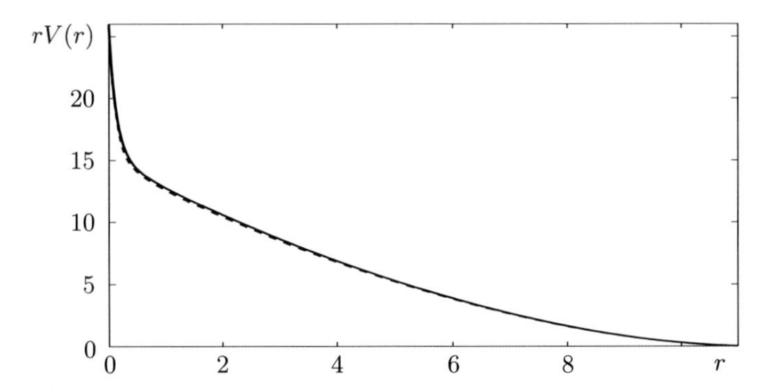

Figure 3.3: The potential $rV(r)$ in the Hartree (solid curve) and Thomas-Fermi models (dashed curve) for iron: $Z = 26$, $T = 0.1$ keV, $\rho = 0.1$ g/cm^3

Figure 3.3 compares the functions $rV(r)$ calculated for the Hartree potential and the Thomas-Fermi potential. Examining the graphs one concludes that the Thomas-Fermi potential can be applied successfully for high temperatures, and, in addition, is a good initial approximation for quantum-mechanical computations.

To analyze how the choice of the effective boundary of the continuum ε_0 influences the energy levels and the occupation numbers, computations were carried out with a different numbers of bound states: for $\varepsilon_0 = -5.5$ eV, taking into account the shells with principal quantum numbers $n \le 7$, and also for $\varepsilon_0 = -25$ eV, $n \le 6$ and $\varepsilon_0 = -50$ eV, $n \le 5$ (see Table 3.1).

Table 3.1 shows a weak dependence of the potential $V(r)$ on the parameter ε_0, which is confirmed by the coincidence of the occupation numbers $N_{n\ell}$ and the closeness of the energy levels $\varepsilon_{n\ell}$. The graphs of the function $rV(r)$ for the three indicated cases are practically indistinguishable. This is connected with the fact that for the states with energies $\varepsilon_{n\ell} > -100$ eV ($n > 4$) the average occupation numbers $N_{n\ell}$ are very small due to strong ionization. Let us note that for higher densities the dependence on ε_0 may turn out to be essential.

Table 3.1: Energy levels $\varepsilon_{n\ell}$ and occupation numbers $N_{n\ell}$ for iron at $T = 0.1$ keV, $\rho = 0.1$ g/cm^3. The calculations were carried out for three values of the effective boundary of the continuum: $\varepsilon_0 = -5.5$ eV ($\eta = 6.06538$), $\varepsilon_0 = -25$ eV ($\eta = 6.06537$) and $\varepsilon_0 = -50$ eV ($\eta = 6.06253$)

		$\varepsilon_0 = -5.5$		$\varepsilon_0 = -25$		$\varepsilon_0 = -50$	
n	ℓ	$\varepsilon_{n\ell}$	$N_{n\ell}$	$\varepsilon_{n\ell}$	$N_{n\ell}$	$\varepsilon_{n\ell}$	$N_{n\ell}$
1	0	−7120.4	2.000	−7120.3	2.000	−7120.2	2.000
2	0	−1126.8	1.989	−1126.6	1.989	−1126.6	1.989
2	1	−1018.4	5.903	−1018.2	5.903	−1018.2	5.903
3	0	−376.92	0.182	−376.74	0.182	−376.73	0.183
3	1	−344.33	0.405	−344.15	0.405	−344.14	0.406
3	2	−296.34	0.429	−296.16	0.429	−296.14	0.430
4	0	−167.25	0.024	−167.07	0.024	−167.11	0.024
4	1	−154.39	0.064	−154.21	0.064	−154.25	0.065
4	2	−136.48	0.090	−136.30	0.090	−136.34	0.090
4	3	−129.01	0.117	−128.83	0.117	−128.86	0.117
5	0	−80.914	0.010	−80.758	0.010	−80.839	0.010
5	1	−74.684	0.029	−74.530	0.029	−74.612	0.029
5	2	−66.114	0.045	−65.961	0.045	−66.043	0.045
5	3	−62.266	0.060	−62.107	0.060	−62.185	0.060
5	4	−60.520	0.076	−60.348	0.076	−60.420	0.076
6	0	−38.114	0.007	−38.007	0.007		
6	1	−34.717	0.020	−34.613	0.020		
6	2	−30.036	0.031	−29.936	0.031		
6	3	−27.756	0.043	−27.651	0.043		
6	4	−26.527	0.054	−26.409	0.054		
6	5	−25.474	0.066	−25.336	0.066		
7	0	−14.918	0.005				
7	1	−12.967	0.016				
7	2	−10.261	0.026				
7	3	−8.786	0.035				
7	4	−7.804	0.045				
7	5	−6.834	0.055				
7	6	−5.754	0.064				

Figures 3.4 and 3.5 display the density of the bound electrons $\rho_1(r)$ and the density of the free electrons $\rho_2(r)$. We see that the electrons remaining on the shells are located sufficiently close to the nucleus (see Figure 3.4). At the same time the continuum electrons are uniformly distributed in space from the boundary of the cell $r_0 = 11.4$ up to the value $r \sim 2$ with a considerably smaller density (see Figure 3.5), which allows one to use the uniform free-electron density model for $\rho_2(r)$ at high temperatures. The distribution of $\rho_2(r)$ (Figure 3.5), calculated by means of formula (3.8), has an integrable singularity near zero, in agreement with the Thomas-Fermi model (see § 2.1). Let us emphasize however that this singularity has practically no influence on the computational results.

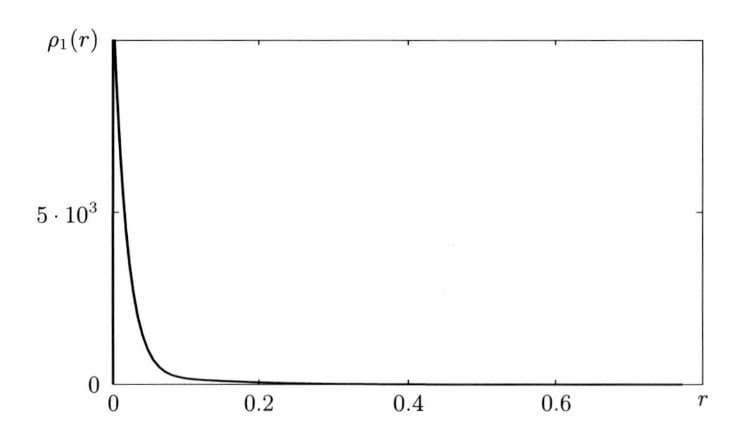

Figure 3.4: Density of the bound electrons $\rho_1(r)$ in the Hartree model for iron ($Z = 26$, $T = 0.1$ keV, $\rho = 0.1$ g/cm^3; $r_0 = 11.4$, $Z_0 \approx 15$)

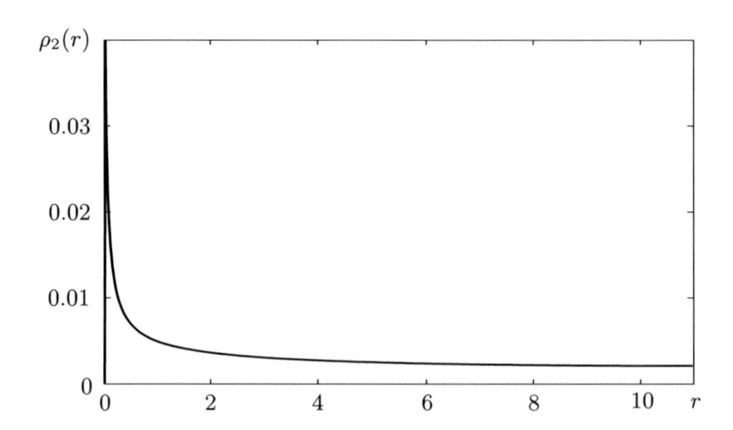

Figure 3.5: Free-electron density $\rho_2(r)$ in the Hartree model for iron ($Z = 26$, $T = 0.1$ keV, $\rho = 0.1$ g/cm^3; $r_0 = 11.4$, $Z_0 \approx 15$)

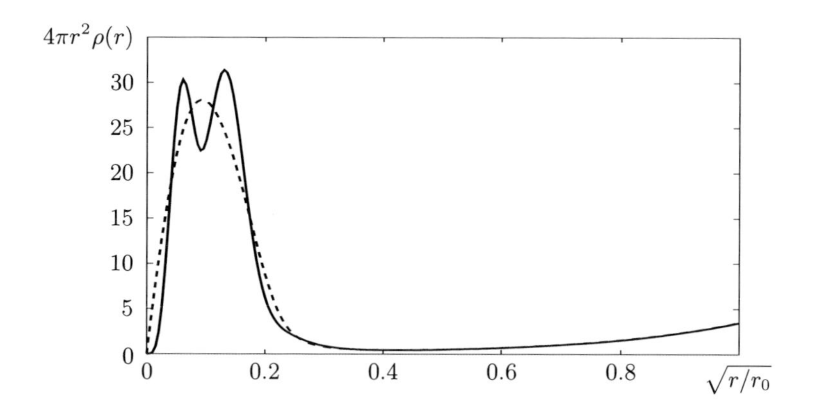

Figure 3.6: Dependence of the radial electron density $4\pi r^2\rho(r)$ on $x = \sqrt{r/r_0}$ in the Hartree model (solid curve) and the Thomas-Fermi model (dashed curve), calculated for $Z = 26$, $T = 0.1$ keV, $\rho = 0.1$ g/cm^3 (compare this with the difference in the corresponding potentials shown in Figure 3.3)

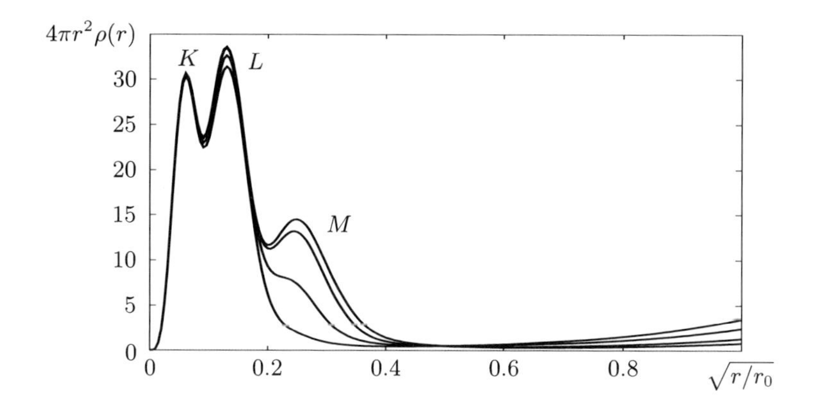

Figure 3.7: Dependence of the radial electron density $4\pi r^2\rho(r)$ on $x = \sqrt{r/r_0}$ in the Hartree model for iron with density $\rho = 0.1$ g/cm^3 at different temperatures: $T = 10$, 20, 50 and 100 eV. One sees how when the temperature is increased the M-shell is ionized and simultaneously the density of the free electrons increases

Figure 3.6 above shows the radial electron density $4\pi r^2\rho(r)$ as a function of the variable $x = \sqrt{r/r_0}$. One sees that, compared with the Thomas-Fermi model, the Hartree model has one different qualitative feature: it renders correctly the oscillations of the radial electron density. As one can see from the graph, at temperature $T = 0.1$ keV the shell with principal quantum number $n = 3$ is practically unoccupied (cf. the average occupation numbers $N_{n\ell}$ in Table 3.1). The two sharp maxima correspond to fully occupied K- and L-shells; the M-shell is not revealed. When the temperature is decreased the degree of occupation of the M-shell increases, which is clearly seen in the graphs of the electron density shown in Figure 3.7 above.

3.1.4 The relativistic Hartree model

For large values of Z in the Hartree model instead of the Schrödinger equation we must use the Dirac equations. Then the 4-component wave function of an electron with total momentum j and projection of the momentum on the z-axis m can be written in the form (see Appendix, § A.2):

$$\psi_{n\ell jm}(\vec{r}) = (-1)^{m+1/2}\, i^{\ell} \begin{pmatrix} -C^{jm}_{j\mp\frac{1}{2},m-\frac{1}{2};\frac{1}{2},\frac{1}{2}} \dfrac{F_{n\ell j}(r)}{r} Y_{\ell,m-\frac{1}{2}}(\vartheta,\varphi) \\ C^{jm}_{j\mp\frac{1}{2},m+\frac{1}{2};\frac{1}{2},-\frac{1}{2}} \dfrac{F_{n\ell j}(r)}{r} Y_{\ell,m+\frac{1}{2}}(\vartheta,\varphi) \\ \mp iC^{jm}_{j\pm\frac{1}{2},m-\frac{1}{2};\frac{1}{2},\frac{1}{2}} \dfrac{G_{n\ell j}(r)}{r} Y_{\ell\pm1,m-\frac{1}{2}}(\vartheta,\varphi) \\ \pm iC^{jm}_{j\pm\frac{1}{2},m+\frac{1}{2};\frac{1}{2},-\frac{1}{2}} \dfrac{G_{n\ell j}(r)}{r} Y_{\ell\pm1,m+\frac{1}{2}}(\vartheta,\varphi) \end{pmatrix}.$$

Here the upper [resp., lower] sign is taken for $\ell = j - 1/2$ [resp., $\ell = j + 1/2$]. The values of the Clebsch-Gordan coefficients $C^{jm}_{\ell,m-m_s;\frac{1}{2},m_s}$ are given in the Appendix, Table A.1.

The radial functions $F_{n\ell j}(r) = F(r)$ and $G_{n\ell j}(r) = G(r)$ are solutions of the Dirac equations with the potential $V(r)$:

$$\begin{cases} \dfrac{d}{dr}F(r) + \dfrac{\kappa}{r}F(r) = \alpha\left[\varepsilon_{n\ell j} + V(r) + \dfrac{2}{\alpha^2}\right]G(r), \\ \dfrac{d}{dr}G(r) - \dfrac{\kappa}{r}G(r) = -\alpha[\varepsilon_{n\ell j} + V(r)]F(r), \end{cases} \tag{3.13}$$

$$\kappa = -2(j-\ell)(j+1/2).$$

The boundary conditions for the bound states read

$$F(0) = 0, \quad G(0) = 0, \quad F(r_0) = 0, \quad G(r_0) = 0.$$

The density of the bound electrons for $\varepsilon_{n\ell j} < \varepsilon_0$ is given by the expression

$$\rho_1(\vec{r}) = \sum_{n,\ell,j,m} \frac{1}{1+\exp\left(\dfrac{\varepsilon_{n\ell j}-\mu}{\theta}\right)} |\psi_{n\ell jm}(\vec{r})|^2 = \sum_{n,\ell,j} \frac{1}{1+\exp\left(\dfrac{\varepsilon_{n\ell j}-\mu}{\theta}\right)} \sum_m |\psi_{n\ell jm}(\vec{r})|^2.$$

Let us consider the contribution in $\rho_1(\vec{r})$ for fixed quantum numbers n, j and $\ell = j - 1/2$:

$$\sum_m |\psi_{n\ell jm}(\vec{r})|^2 = \frac{F^2_{n\ell j}(r)}{r^2}\left[\sum_{m=-j}^{j} \frac{j+m}{2j} \left|Y_{\ell,m-\frac{1}{2}}(\vartheta,\varphi)\right|^2 + \sum_{m=-j}^{j} \frac{j-m}{2j} \left|Y_{\ell,m+\frac{1}{2}}(\vartheta,\varphi)\right|^2\right] + \frac{G^2_{n\ell j}(r)}{r^2}\left[\sum_{m=-j}^{j} \frac{j-m+1}{2j+2} \left|Y_{\ell+1,m-\frac{1}{2}}(\vartheta,\varphi)\right|^2 + \sum_{m=-j}^{j} \frac{j+m+1}{2j+2} \left|Y_{\ell+1,m+\frac{1}{2}}(\vartheta,\varphi)\right|^2\right].$$

Changing the summation index m to $m+1$ in the first sum in each of the square brackets, we obtain

$$\sum_m |\psi_{n\ell jm}(\vec{r})|^2 = \frac{F^2_{n\ell j}(r)}{r^2}\left[\sum_{m=-j-1}^{j-1} \frac{j+m+1}{2j} \left|Y_{\ell,m+\frac{1}{2}}(\vartheta,\varphi)\right|^2 + \sum_{m=-j}^{j} \frac{j-m}{2j} \left|Y_{\ell,m+\frac{1}{2}}(\vartheta,\varphi)\right|^2\right] + \frac{G^2_{n\ell j}(r)}{r^2}\left[\sum_{m=-j-1}^{j-1} \frac{j-m}{2j+2} \left|Y_{\ell+1,m+\frac{1}{2}}(\vartheta,\varphi)\right|^2 + \sum_{m=-j}^{j} \frac{j+m+1}{2j+2} \left|Y_{\ell+1,m+\frac{1}{2}}(\vartheta,\varphi)\right|^2\right].$$

Now let us adjust the limits of the sum over m in accordance with the parameters of the spherical harmonics. To this end we omit [resp., add] the terms equal to zero in the first [resp., second] set of brackets:

$$\sum_m |\psi_{n\ell jm}(\vec{r})|^2 = \frac{F^2_{n\ell j}(r)}{r^2}\left[\sum_{m=-j}^{j-1} \frac{j+m+1}{2j}\left|Y_{\ell,m+\frac{1}{2}}(\vartheta,\varphi)\right|^2 + \sum_{m=-j}^{j-1} \frac{j-m}{2j}\left|Y_{\ell,m+\frac{1}{2}}(\vartheta,\varphi)\right|^2\right] +$$
$$\frac{G^2_{n\ell j}(r)}{r^2}\left[\sum_{m=-j-1}^{j} \frac{j-m}{2j+2}\left|Y_{\ell+1,m+\frac{1}{2}}(\vartheta,\varphi)\right|^2 + \sum_{m=-j-1}^{j} \frac{j+m+1}{2j+2}\left|Y_{\ell+1,m+\frac{1}{2}}(\vartheta,\varphi)\right|^2\right].$$

Combining the terms in each of the square brackets we obtain

$$\sum_m |\psi_{n\ell jm}(\vec{r})|^2 = \frac{F^2_{n\ell j}(r)}{r^2}\frac{2j+1}{2j}\sum_{m_\ell=-\ell}^{\ell} |Y_{\ell,m_\ell}(\vartheta,\varphi)|^2 + \frac{G^2_{n\ell j}(r)}{r^2}\frac{2j+1}{2j+2}\sum_{m_\ell=-\ell-1}^{\ell+1} |Y_{\ell+1,m_\ell}(\vartheta,\varphi)|^2.$$

As in the nonrelativistic case, we use the addition theorem for spherical harmonics, which yields

$$\sum_m |\psi_{n\ell jm}(\vec{r})|^2 = \frac{2j+1}{4\pi r^2}[F^2_{n\ell j}(r) + G^2_{n\ell j}(r)]. \tag{3.14}$$

This expression was obtained for $\ell = j-1/2$. In much the same manner one verifies that equality (3.14) holds also for $\ell = j+1/2$. Finally, we have

$$\rho_1(r) = \frac{1}{4\pi r^2}\sum_{n,\ell,j} N_{n\ell j}[F^2_{n\ell j}(r) + G^2_{n\ell j}(r)], \tag{3.15}$$

where

$$N_{n\ell j} = \frac{2j+1}{1+\exp\left(\dfrac{\varepsilon_{n\ell j}-\mu}{\theta}\right)}$$

and the sum is taken over the states with energy $\varepsilon_{n\ell j} < \varepsilon_0$.

For $\rho_2(r)$ in the relativistic Hartree model one can retain the preceding formula (3.8), because the relativistic corrections for the continuum electrons are usually small. The calculation of the potential $V(r)$ is carried out as in the nonrelativistic model, using formula (3.9), where the contribution of the bound electrons

Table 3.2: Energy levels in eV and occupation numbers for gold with $T = 0.1$ keV and $\rho = 0.1$ g/cm^3, computed in the relativistic and nonrelativistic variants of the Hartree model ($n \leq n_{\max} = 9$). In the relativistic approach $\eta = 6.830$, $Z_0 = 22.99$; in the nonrelativistic approach $\eta = 6.822$, $Z_0 = 23.11$

			Relativistic computation			Nonrelativistic computation	
n	ℓ	j	$\|\varepsilon_{n\ell j}\|$	$N_{n\ell j}$	$\sum_j N_{n\ell j}$	$\|\varepsilon_{n\ell}\|$	$N_{n\ell}$
1	0	1/2	80523	2.000	2.000	73022	2.000
2	0	1/2	14661	2.000	2.000	12639	2.000
2	1	1/2	14054	2.000	6.000	12151	6.000
2	1	3/2	12261	4.000			
3	0	1/2	3953.5	2.000	2.000	3435.8	2.000
3	1	1/2	3687.7	2.000	6.000	3211.1	6.000
3	1	3/2	3293.8	4.000			
3	2	3/2	2858.4	4.000	10.000	2794.9	10.000
3	2	5/2	2774.5	6.000			
4	0	1/2	1279.0	1.995	1.995	1175.4	1.986
4	1	1/2	1163.7	1.984	5.923	1079.3	5.888
4	1	3/2	1100.9	3.939			
4	2	3/2	918.77	3.653	9.062	904.68	9.018
4	2	5/2	904.82	5.409			
4	3	5/2	677.93	2.920	6.733	675.74	6.756
4	3	7/2	673.85	3.813			
5	0	1/2	564.95	0.469	0.469	543.42	0.398
5	1	1/2	519.05	0.325	0.903	503.02	0.854
5	1	3/2	505.31	0.578			
5	2	3/2	431.70	0.300	0.738	431.57	0.752
5	2	5/2	428.83	0.438			
5	3	5/2	341.74	0.191	0.445	344.04	0.459
5	3	7/2	341.03	0.254			
5	4	7/2	265.62	0.121	0.272	268.24	0.282
5	4	9/2	265.46	0.151			
	...		...	...	...	...	...

to $\rho(r)$ is calculated by means of formula (3.15) and using the Dirac equation (3.13).

Figure 3.8 shows the graphs of the total electron density for gold at temperature $T = 0.1$ keV and density $\rho = 0.1$ g/cm^3, calculated in the relativistic and nonrelativistic Hartree models. As the graphs reveal, the influence of the relativistic effects for high-Z elements can be substantial for the first three shells (see also Table 3.2 for the energy levels $\varepsilon_{n\ell}$, $\varepsilon_{n\ell j}$ and occupation numbers $N_{n\ell}$, $N_{n\ell j}$).

The relativistic Hartree model, derived by simple arguments, is not sufficiently rigorous, yet it allows one to account for the main relativistic effects. What this model leaves out are the exchange and correlation effects, as well as the Breit and a few other relativistic corrections. Let us point out that, in contrast to the average atom, in the consideration of individual ion states it is necessary to take into account also the non-centrally-symmetric part of the potential and refine the

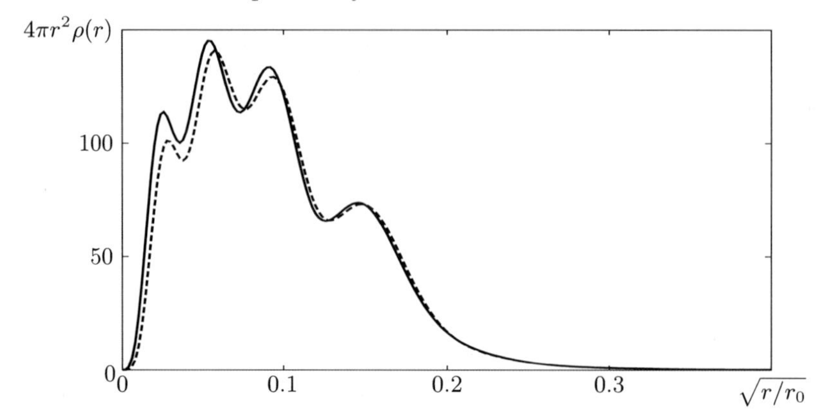

Figure 3.8: Dependence of the radial electron density $4\pi r^2\rho(r)$ on $x = \sqrt{r/r_0}$ in the Hartree model for gold ($Z = 79$, $T = 0.1$ keV, $\rho = 0.1$ g/cm^3), calculated by using the relativistic (solid curve) and nonrelativistic (dashed curve) radial functions

effects of the spin-orbit interaction [74]. A more rigorous approach to the derivation of quantum-statistical self-consistent field models can be realized by means of variational principles.

3.2 The Hartree-Fock self-consistent field model for matter with given temperature and density

3.2.1 Variational principle based on the minimum condition for the grand thermodynamic potential

As above, we shall assume that the substance is in local thermodynamic equilibrium, and that the temperature is high enough so that the substance is a plasma consisting of electrons and ions of various multiplicities. In Chapter I, to derive the equations of the Thomas-Fermi model we used the simplest variational principle for a closed system — the requirement that the entropy be maximal. A more general quantum-statistical description of the state of matter requires the introduction of the mathematical apparatus of the density matrix and the application of various statistical approximations. Thus, for example, one often resorts to the approximation of the grand canonical ensemble [17]. Based on the approximation selected one formulates and solves a variational problem for the wave functions and the self-consistent potential.

The most general variational principle that is usually employed under equilibrium conditions for the grand canonical ensemble is the condition of minimum for the grand thermodynamic potential $\Omega = E - \mu N - \theta S$. The application of this principle for a plasma with significant interaction between particles is practically impossible, and for this reason one resorts to various approximate approaches. One of the most productive turned out to be the density-functional method [113, 133, 176, 168, 171]. In this method one constructs a certain approximate *electron density functional*, which is interpreted, e.g., as the grand thermodynamic potential of electrons. Then one solves the variational problem on the minimum of this functional, which allows one to derive various models describing the state of matter.

We will use a somewhat more general variational principle, without requiring, in particular, that the functional to be minimized depends only on the electron density. The justification for the application of the proposed method is that it allows us to easily derive known equations, among them the equation for the generalized Thomas-Fermi model, the Hartree-Fock equations for the average atom and the Hartree-Fock-Slater equations.

The application of variational methods directly to a system of interacting ions and electrons (see [171]) runs into a number of difficulties that have not been circumvented to this time. For this reason one often adopts a simpler approach, based on the application of the adiabatic approximation. In the adiabatic approximation the ions are regarded as classical particles, which move considerably slower than the electrons, so that one can assume that for each new position of the ions the electrons manage to reach thermodynamic equilibrium.

The average value of a physical variable F in some state of the electron system, described by the density matrix $\widehat{W}$, can be represented in the form [17]

$$\langle \widehat{F} \rangle = \mathrm{Tr}(\widehat{W}\widehat{F}), \tag{3.16}$$

where $\widehat{W}$ is the corresponding statistical operator (density matrix operator) of the system. We shall assume that at equilibrium the functional Ω attains its minimum, where Ω has the meaning of the grand thermodynamic potential of electrons and is defined, in accordance with (3.16), by the formula

$$\Omega = \langle \widehat{\Omega} \rangle = \mathrm{Tr}\left[\widehat{W}(\widehat{H} - \mu\widehat{N} + \theta \ln \widehat{W})\right]. \tag{3.17}$$

Here $\widehat{H}$ is the Hamiltonian of the electron system, $\widehat{N}$ is the particle number operator, θ is the temperature, μ is the chemical potential, and $\widehat{W}$ is the statistical operator corresponding to the density matrix.

In the approximation of the grand canonical ensemble the equilibrium density matrix $\widehat{W}$ obeys the Gibbs distribution

$$\widehat{W} = \frac{\exp\left(-\dfrac{\widehat{H} - \mu\widehat{N}}{\theta}\right)}{\mathrm{Tr}\exp\left(-\dfrac{\widehat{H} - \mu\widehat{N}}{\theta}\right)}, \tag{3.18}$$

which is defined by the Hamiltonian $\widehat{H}$.

Let us consider the expression for the Hamiltonian of the system of electrons in a given field of ions. If one takes into account that the speed of propagation of the interaction is finite, then already the classical energy of this interaction depends on the whole prehistory of the motion of particles, rather than being determined solely by the position of the particles at the given instance. If the relative velocities of particles in the system are small compared with the speed of light, then the distribution of particles in space changes little over the time necessary for the transmission of the interaction between particles. In this case one can define the classical Hamiltonian, up to terms of order $(v/c)^2$, as a function of only the coordinates and momenta of the particles in the system. Then the Hamilton operator can be written in the form

$$\widehat{H} = \sum_i \frac{\widehat{\vec{p}_i}^{\,2}}{2m_i} + U(\vec{r}_1, \vec{r}_2, \ldots, \vec{r}_N) + \widehat{H}_1,$$

where U is the potential energy of inter-particle interaction, which depends on the mutual disposition of the particles and $\widehat{H}_1$ is an operator that includes the spin-orbit interaction as well as the part of the potential energy that depends on the momenta of the particles and accounts for the retardation effects. The operator $\widehat{H}_1$ is of order $(v/c)^2$ and in the nonrelativistic theory can be accounted for by perturbation theory.

In the nonrelativistic approximation, if one neglects all spin interactions of electrons and all nuclear effects (such as the dimensions and masses of nuclei being finite rather than infinitesimal), the Hamiltonian of the electron system reads

$$\widehat{H} = \sum_i \left[-\frac{1}{2}\Delta_i + U_a(\vec{r}_i) \right] + \frac{1}{2} \sum_{i \neq j} \frac{1}{|\vec{r}_i - \vec{r}_j|}, \tag{3.19}$$

where $U_a(\vec{r}_i)$ is the potential energy of the ith electron with coordinates $\vec{r}_i$ in a given potential $V_a(\vec{r})$ of atomic nuclei, i.e., $U_a(\vec{r}_i) = -V_a(\vec{r}_i)$.

In the chosen representation for the given density matrix $\widehat{W}$ the average value of the operator $\widehat{\Omega} = \widehat{H} - \mu\widehat{N} + \theta \ln \widehat{W}$ is calculated by means of formula (3.16):

$$\langle \widehat{\Omega} \rangle = \mathrm{Tr}(\widehat{W}\widehat{\Omega}) = \sum_n w_n \int \Psi_n^*(Q)\widehat{\Omega}\Psi_n(Q)\, dQ, \tag{3.20}$$

where $w_n = w(E_n, N_n)$ is the statistical probability of the system of electrons with given energy $E = E_n$ and particle number $N = N_n$, $\Psi_n(Q)$ is the wave function of such a state, Q is the set of coordinates q_i of the electrons, where q_i incorporates the space variable $\vec{r}_i$ and the spin variable σ_i of the electron labelled i. Integration over the whole configuration space $Q \equiv \{q_i\}$ means integration with respect to all space variables $\vec{r}_i$ and summation over all spin variables σ_i.

As is clearly seen from (3.20), the average value of any quantity in such an approach is calculated in two steps: first one calculates the average value of the quantity in the state with wave function $\Psi_n(Q)$ for given energy E_n and particle number N_n, and then the resulting values are averaged over the different states, with a weight equal to the probabilities w_n of these states. Note that the result does not change if instead of $\Psi_n(Q)$ one takes any complete system of functions.

For the electron density $\rho(q)$ we obtain

$$\rho(q) = \mathrm{Tr}\left(\widehat{W} \sum_i \delta(q_i - q)\right) = \sum_n w_n \times$$
$$\sum_{i=1}^{N_n} \int \left|\Psi_n(q_1, q_2, \ldots, q_{i-1}, q, q_{i+1}, \ldots, q_{N_n})\right|^2 dq_1\, dq_2 \ldots dq_{i-1}\, dq_{i+1} \ldots dq_{N_n}. \tag{3.21}$$

Here and in what follows the Dirac delta function $\delta(q_i - q) = \delta_{\sigma_i \sigma} \delta(\vec{r}_i - \vec{r})$ is used to simplify notation and to carry out the integration with respect to all variables q_i (i.e., integration with respect to $\vec{r}_1, \vec{r}_2, \ldots, \vec{r}_N$ and summation with respect to $\sigma_1, \sigma_2, \ldots, \sigma_N$).

3.2.2 The self-consistent field equation in the Hartree-Fock approximation

Here we shall use the one-particle approximation, i.e., we shall assume that, for any given energy and number of particles the wave function $\Psi(Q)$ of the system is represented as an anti-symmetric sum of products of one-particle wave functions, taken over the occupied states ν_i (the Slater determinant)

$$\Psi(Q) = \frac{1}{\sqrt{N!}} \begin{vmatrix} \psi_{\nu_1}(q_1) & \psi_{\nu_1}(q_2) & \ldots & \psi_{\nu_1}(q_N) \\ \psi_{\nu_2}(q_1) & \psi_{\nu_2}(q_2) & \ldots & \psi_{\nu_2}(q_N) \\ \ldots & \ldots & \ldots & \ldots \\ \psi_{\nu_N}(q_1) & \psi_{\nu_N}(q_2) & \ldots & \psi_{\nu_N}(q_N) \end{vmatrix}, \tag{3.22}$$

or as a linear combination of such determinats.

The system of functions $\psi_\nu(q)$ is assumed to be orthonormal. Using the orthogonality of the $\psi_\nu(q)$'s, one can express the average value of the original Hamiltonian $\widehat{H}$ over coordinates for a one-determinant function $\Psi(Q)$ as follows (see Subsection 3.2.3 below, and also [83, 64]):

$$\langle\Psi|\widehat{H}|\Psi\rangle = \sum_{\nu} \int \psi_\nu^*(q) \left[-\frac{1}{2}\Delta - V_a(\vec{r})\right] \psi_\nu(q)\, dq + \frac{1}{2}\sum_{\nu,\lambda} \iint \frac{|\psi_\nu(q)|^2 |\psi_\lambda(q')|^2\, dq\, dq'}{|\vec{r}-\vec{r}\,'|} - \frac{1}{2}\sum_{\nu,\lambda} \iint \frac{\psi_\nu^*(q)\psi_\nu(q')\psi_\lambda^*(q')\psi_\lambda(q)\, dq\, dq'}{|\vec{r}-\vec{r}\,'|}, \tag{3.23}$$

where the summation is carried over all states ν and λ that appear in the expression (3.22) for the function $\Psi(Q)$.

Next, let us assume that for a temperature different from zero the one-particle states ν are occupied with some probabilities n_ν ($0 \le n_\nu \le 1$), which near the equilibrium state can be approximately considered to be independent [105]. Then the total probability of the given state is equal to the product of the probabilities of the corresponding one-particle states. For example, the probability w_n of the state n, in which the first N_n states are occupied and the remaining ones are not, is given by the formula

$$w_n = n_{\nu_1} \cdot n_{\nu_2} \cdot \ldots \cdot n_{\nu_{N_n}} \cdot (1 - n_{\nu_{N_n+1}}) \cdot (1 - n_{\nu_{N_n+2}}) \cdot \ldots . \tag{3.24}$$

Let us remark that, by virtue of their definition, the n_ν are the average occupation numbers of the one-particle states ν.

In the approximation described above one can calculate all average values of operators that we need. Thus, using induction, one can show that in formula (3.17)

$$\mathrm{Tr}(\widehat{W} \ln \widehat{W}) = \sum_n w_n \ln w_n = \sum_\nu \left[n_\nu \ln n_\nu + (1 - n_\nu) \ln (1 - n_\nu)\right]. \tag{3.25}$$

Indeed, for a single one-particle state ($\nu = 1$) we have two possibilities: $w_1 = n_1$ for the occupied state and $w_2 = 1 - n_1$ for the unoccupied one, and so formula (3.25) is obviously valid.

Suppose (3.25) is valid for k states ($\nu = 1, 2, \ldots k$), i.e.,

$$\sum_n w_n \ln w_n = \sum_{\nu=1}^{k} [n_\nu \ln n_\nu + (1 - n_\nu) \ln(1 - n_\nu)].$$

Now add a $(k+1)$st state. The number of possible states of the system doubles, and the probabilities of the new states are expressed in terms of w_n as follows:

$$w_n^{(1)} = n_{k+1} w_n, \quad w_n^{(2)} = (1 - n_{k+1}) w_n.$$

Consequently, we have

$$\sum_n w_n^{(1)} \ln w_n^{(1)} + \sum_n w_n^{(2)} \ln w_n^{(2)} =$$

$$\sum_n n_{k+1} w_n (\ln n_{k+1} + \ln w_n) + \sum_n (1 - n_{k+1}) w_n [\ln(1 - n_{k+1}) + \ln w_n] =$$

$$n_{k+1} \ln n_{k+1} + (1 - n_{k+1}) \ln(1 - n_{k+1}) + \sum_n w_n \ln w_n =$$

$$\sum_{\nu=1}^{k+1} [n_\nu \ln n_\nu + (1 - n_\nu) \ln(1 - n_\nu)]$$

(here we used the fact that $\sum w_n = 1$). By the induction step, formula (3.25) is valid for arbitrary k.

Note that formula (3.25) corresponds to formula (1.4) for the entropy in the Thomas-Fermi approximation. From the orthogonality of the functions $\psi_\nu(q)$ and formulas (3.21) and (3.24) it follows that

$$\rho(q) = \sum_\nu n_\nu |\psi_\nu(q)|^2. \tag{3.26}$$

Now using (3.17), (3.23) and (3.25) we obtain an expression for the thermodynamic potential Ω in the one-particle approximation, in the sense indicated above (we denote it by Ω^{HF}):

$$\Omega^{\mathrm{HF}} = \sum_\nu n_\nu \int \psi_\nu^*(q) \left[-\frac{1}{2}\Delta - V_a(\vec{r}) \right] \psi_\nu(q)\, dq +$$

$$\frac{1}{2} \sum_{\nu,\lambda} n_\nu n_\lambda \iint \frac{|\psi_\nu(q)|^2 |\psi_\lambda(q')|^2 \, dq\, dq'}{|\vec{r} - \vec{r}\,'|} -$$

$$\frac{1}{2} \sum_{\nu,\lambda} n_\nu n_\lambda \iint \frac{\psi_\nu^*(q)\psi_\nu(q')\psi_\lambda^*(q')\psi_\lambda(q)\, dq\, dq'}{|r' - r''|} +$$

$$\theta \sum_\nu [n_\nu \ln n_\nu + (1 - n_\nu) \ln(1 - n_\nu)] - \mu \sum_\nu n_\nu. \tag{3.27}$$

Let us find the minimum condition for Ω^{HF} for arbitrary n_ν ($0 \le n_\nu \le 1$) and an arbitrary orthonormal system of functions $\psi_\nu(q)$. The normalization conditions $\int |\psi_\nu(q)|^2 \, dq = 1$ and the orthogonality conditions $\int \psi_\nu^*(q)\psi_\lambda(q)\, dq = 0$ for $\lambda \neq \nu$ are accounted for via Lagrange multipliers $\Lambda_{\nu\lambda}$. Then the minimum condition for Ω^{HF} reads

$$\delta\Omega^{\mathrm{HF}} + \delta \left[\sum_{\nu,\lambda} \Lambda_{\nu\lambda} \int \psi_\nu^*(q)\psi_\lambda(q)\, dq \right] = 0.$$

Substituting here the expression (3.27) for Ω^{HF} and calculating the variation, we have

$$\delta\Omega^{\mathrm{HF}} + \delta\left[\sum_{\nu,\lambda}\Lambda_{\nu\lambda}\int\psi_\nu^*(q)\psi_\lambda(q)\,dq\right] = \sum_\nu \delta n_\nu \int \psi_\nu^*(q)\widehat{H}_0\psi_\nu(q)\,dq +$$
$$\sum_\nu n_\nu \int \psi_\nu^*(q)\widehat{H}_0\delta\psi_\nu(q)\,dq + \sum_\nu n_\nu \int \delta\psi_\nu^*(q)\widehat{H}_0\psi_\nu(q)\,dq + \theta\sum_\nu \delta n_\nu \ln\frac{n_\nu}{1-n_\nu} -$$
$$\mu\sum\delta n_\nu + \sum_{\nu,\lambda}\Lambda_{\nu\lambda}\left[\int\delta\psi_\nu^*(q)\psi_\lambda(q)\,dq + \int\psi_\nu^*(q)\delta\psi_\lambda(q)\,dq\right] = 0,$$

where

$$\widehat{H}_0\psi_\nu(q) = \left[-\frac{1}{2}\Delta - V_a(\vec{r}) + \sum_\lambda n_\lambda \int\frac{|\psi_\lambda(q')|^2\,dq'}{|\vec{r}-\vec{r}\,'|}\right]\psi_\nu(q) -$$
$$\sum_\lambda n_\lambda \int\frac{\psi_\lambda^*(q')\psi_\nu(q')\,dq'}{|\vec{r}-\vec{r}\,'|}\psi_\lambda(q) \quad (3.28)$$

(in deriving (3.28), in the calculation of the double sums and double integrals we replaced λ by ν, q by q', and so on).

Since the variations $\delta\psi_\nu$, $\delta\psi_\nu^*$, and also δn_ν, are independent, the minimum condition for Ω^{HF} reads

$$\widehat{H}_0\psi_\nu(q) + \frac{1}{n_\nu}\sum_{\lambda\neq\nu}\Lambda_{\nu\lambda}\psi_\lambda(q) = -\frac{\Lambda_{\nu\nu}}{n_\nu}\psi_\nu(q) \equiv \varepsilon_\nu\psi_\nu(q), \quad (3.29)$$

$$n_\nu = n(\varepsilon_\nu) = \frac{1}{1+\exp\left(\dfrac{\varepsilon_\nu-\mu}{\theta}\right)}. \quad (3.30)$$

Equations (3.28)–(3.30) are the well-known Hartee-Fock (HF) equations for matter with given temperature and density [105, 129].

Therefore, to obtain the Hartree-Fock equations (3.28)–(3.30) it suffices to impose the minimum condition for the thermodynamic potential $\Omega = \Omega^{\mathrm{HF}}$ in the one-particle approximation. Note that Ω^{HF} is expressed in terms of the one-particle wave functions and occupation numbers, and, generally speaking, is not a functional of the electron density $\rho(q)$.

3.2.3 The Hartree-Fock equations for a free ion

At temperature zero*) the average occupation numbers n_ν can take the value 0 or 1 depending on the difference between the energy of the level ε_ν and the Fermi

*) The content of Subsection 3.2.3 is important for the understanding of § 5.3, but it may be skipped at a first reading.

energy μ. If the energy levels are degenerate (i.e., several energies ε_ν coincide for different values of ν), then it is necessary to be more specific about the values n_ν, so that the number of electrons in one atom will not exceed Z (except for negative ions). Before we specify the set of occupation numbers n_ν and the one-particle wave functions $\psi_\nu(q)$, let us show how to obtain the average value of the Hamiltonian (3.19) with single-determinant wave functions (3.22) and derive a formula analogous to (3.23).

To calculate the contribution to the average value of the Hamiltonian depending on the coordinates of one electron, for example, for

$$V_1 = -\frac{1}{2}\Delta_1 - V_a(\vec{r}_1),$$

let us represent $\Psi(Q)$ as the expansion with respect to the first column of the determinant (3.22):

$$\Psi(Q) = \frac{1}{\sqrt{N!}} \sum_\alpha \psi_\alpha(q_1) \widetilde{\Psi}_\alpha(q_2, q_3, \dots q_N),$$

where $\widetilde{\Psi}_\alpha$ is the cofactor of order $N-1$ obtained by deleting the first column and the α-th row in (3.22). Since

$$\int \widetilde{\Psi}^*_\alpha(q_2, \dots, q_N)\, \widetilde{\Psi}_\beta(q_2, \dots, q_N)\, dq_2 \dots dq_N = (N-1)!\, \delta_{\alpha\beta}\ , \tag{3.31}$$

thanks to the orthogonality of the one-particle wave functions, the matrix element $\langle \Psi \,|\, V_1 \,|\, \Psi \rangle$ is given by

$$\langle \Psi \,|\, V_1 \,|\, \Psi \rangle = \frac{1}{N} \sum_\alpha \int \psi^*_\alpha(q_1)\, V_1\, \psi_\alpha(q_1)\, dq_1. \tag{3.32}$$

Further, since the expression (3.32) does not depend on the number of the electron chosen, we obtain

$$\langle \Psi \Big| \sum_{i=1}^N V_i \Big| \Psi \rangle = \sum_\alpha \int \psi^*_\alpha(q) \left[-\frac{1}{2}\Delta - V_a(\vec{r}) \right] \psi_\alpha(q)\, dq. \tag{3.33}$$

To calculate the contribution of the two-electron interactions, let explain how to calculate a matrix element for the operator

$$V_{12} = \frac{1}{|\vec{r}_1 - \vec{r}_2\,|}.$$

Using Laplace's theorem [94], we represent the N-particle wave function $\Psi(Q)$ as the expansion by minors of order 2 constructed from the first two columns in (3.22) with one-particle wave functions depending of q_1 and q_2:

$$\Psi(Q) = \frac{1}{\sqrt{N!}} \sum_{\alpha<\beta} \begin{vmatrix} \psi_\alpha(q_1) & \psi_\alpha(q_2) \\ \psi_\beta(q_1) & \psi_\beta(q_2) \end{vmatrix} \widetilde{\Psi}_{\alpha\beta}(q_3, q_4, \dots, q_N);$$

here α and β run over all occupied states ($\alpha \neq \beta$), and the function $\widetilde{\Psi}_{\alpha\beta}$ is the cofactor of order $N-2$ obtained by deleting the first two columns and the rows α and β in (3.22).

The functions $\widetilde{\Psi}_{\alpha\beta}$ obey an orthogonality relation similar to (3.31):

$$\int \widetilde{\Psi}^*_{\alpha\beta}(q_3, q_4, \ldots, q_N)\widetilde{\Psi}_{\gamma\delta}(q_3, q_4, \ldots, q_N)\, dq_3\, dq_4 \ldots dq_N = (N-2)!\, \delta_{\alpha\gamma}\, \delta_{\beta\delta}. \tag{3.34}$$

Using (3.34), we obtain the following expression for the average value of the operator V_{12}:

$$\begin{aligned} \langle \Psi | V_{12} | \Psi \rangle = & \\ \frac{1}{N(N-1)} \sum_{\alpha<\beta} \iint & \begin{vmatrix} \psi_\alpha(q_1) & \psi_\alpha(q_2) \\ \psi_\beta(q_1) & \psi_\beta(q_2) \end{vmatrix}^* \frac{1}{|\vec{r}_1 - \vec{r}_2|} \begin{vmatrix} \psi_\alpha(q_1) & \psi_\alpha(q_2) \\ \psi_\beta(q_1) & \psi_\beta(q_2) \end{vmatrix} dq_1\, dq_2 = \\ & \frac{2}{N(N-1)} \sum_{\alpha<\beta} \left[\iint \frac{|\psi_\alpha(q_1)|^2 |\psi_\beta(q_2)|^2}{|\vec{r}_1 - \vec{r}_2|} dq_1\, dq_2 - \right. \\ & \left. \iint \frac{\psi^*_\alpha(q_1)\psi_\beta(q_1)\psi^*_\beta(q_2)\psi_\alpha(q_2)}{|\vec{r}_1 - \vec{r}_2|} dq_1\, dq_2 \right]. \end{aligned} \tag{3.35}$$

Observing that (3.35) does not depend of the choice of the pair, summing over all pairs yields

$$\begin{aligned} \left\langle \Psi \left| \sum_{i<j} \frac{1}{|\vec{r}_i - \vec{r}_j|} \right| \Psi \right\rangle = & \sum_{\alpha<\beta} \iint \frac{|\psi_\alpha(q)|^2 |\psi_\beta(q')|^2}{|\vec{r} - \vec{r}\,'|} dq\, dq' - \\ & \sum_{\alpha<\beta} \iint \frac{\psi^*_\alpha(q)\psi_\beta(q')\psi^*_\beta(q)\psi_\alpha(q')}{|\vec{r} - \vec{r}\,'|} dq\, dq'. \end{aligned} \tag{3.36}$$

The expressions (3.33) and (3.36) for the one-particle and two-particle operators, respectively, prove the validity of formula (3.23).

Next let us consider some variants of the Hartree-Fock model. For an ion with closed shells, for which $N_{n\ell} = 2(2\ell+1)$, the occupation numbers are $n_\nu = 1$ for all participating levels ν. In the central-field approximation,

$$\psi_\nu(q) = \frac{R_{n\ell}(r)}{r} (-1)^m Y_{\ell m}(\vartheta, \varphi)\, \chi_{\frac{1}{2} m_s}(\sigma), \tag{3.37}$$

where $\chi_{\frac{1}{2} m_s}(\sigma)$ is a two-component spinor and $q \equiv \{r, \vartheta, \varphi, \sigma\}$. To each ν we associate a set of quantum numbers n, ℓ, m and m_s, which completely determines the state of an electron (relativistic effects are neglected). The energy levels ε_ν are independent of the projection of the orbital momentum m and the projection of the spin m_s, i.e., the multiplicity g_ν of each degenerate level ν equals $2(2\ell+1)$.

From (3.28), (3.29) one can obtain, using again (3.37), the Hartree-Fock equations for a free ion in the single-determinant approximation of a central field, setting $V_a(\vec{r}) = Z/r$. To that end it suffices to integrate with respect to q' in (3.28), i.e., integrate with respect to r', ϑ', φ' and sum with respect to the spin variable σ', taking into account the orthogonality property of the one-particle functions ψ_ν.

First let us consider the term in the square brackets in (3.28), which corresponds to the Coulomb interaction (see also Appendix, § A.1, Example 3):

$$Y(r) = \sum_\lambda n_\lambda \int \frac{|\psi_\lambda(q')|^2\, dq'}{|\vec{r}-\vec{r}'|} =$$

$$\sum_{n',\ell',m',m'_s} n_{n'\ell'm'm'_s} \sum_{\sigma'} \int \sum_{s,\widetilde{m}} \frac{4\pi}{2s+1} \frac{r_<^s}{r_>^{s+1}} R^2_{n'\ell'}(r')\, dr' \times$$

$$\int \left[Y^*_{\ell'm'}(\vartheta',\varphi') Y_{\ell'm'}(\vartheta',\varphi') Y^*_{s\widetilde{m}}(\vartheta,\varphi) Y_{s\widetilde{m}}(\vartheta',\varphi') \chi^\dagger_{\frac{1}{2}m'_s}(\sigma') \chi_{\frac{1}{2}m'_s}(\sigma') \right] d\Omega';$$

here superscript "†" denotes the conjugate transpose. Since for the closed shells one has $n_{n'\ell'm'm'_s} = 1$, and since

$$\sum_{\sigma'} \chi^\dagger_{\frac{1}{2}m_s}(\sigma')\, \chi_{\frac{1}{2}m'_s}(\sigma') = \delta_{m_s m'_s},$$

$$\int Y_{\ell_1 m_1}(\vartheta,\varphi)\, Y_{\ell_2 m_2}(\vartheta,\varphi)\, Y^*_{\ell_3 m_3}(\vartheta,\varphi)\, d\Omega =$$

$$\sqrt{\frac{(2\ell_1+1)(2\ell_2+1)}{4\pi(2\ell_3+1)}}\, C^{\ell_3 0}_{\ell_1 0 \ell_2 0}\, C^{\ell_3 m_3}_{\ell_1 m_1 \ell_2 m_2},$$

where $C^{\ell_3 m_3}_{\ell_1 m_1 \ell_2 m_2}$ are Clebsch-Gordan coefficients, we obtain

$$Y(r) = \sum_{n',\ell',m',m'_s} \sum_{s,\widetilde{m}} \sqrt{\frac{4\pi}{2s+1}}\, X^s_{n'\ell',n'\ell'}(r)\, C^{\ell'0}_{\ell'0s0}\, C^{\ell'm'}_{\ell'm's\widetilde{m}} Y^*_{s\widetilde{m}}(\vartheta,\varphi).$$

Here we used the notation

$$X^s_{n\ell,n'\ell'}(r) = \int_0^\infty \frac{r_<^s}{r_>^{s+1}} R_{n\ell}(r') R_{n'\ell'}(r')\, dr'.$$

The coefficients $C^{\ell'm'}_{\ell'm's\widetilde{m}}$ are different from zero only for $\widetilde{m} = 0$. Using the equalities

$$Y_{00}(\vartheta,\varphi) = 1/\sqrt{4\pi}, \quad C^{\ell'0}_{\ell'0\,00} = 1$$

and the orthogonality property

$$\sum_{m'} C^{\ell' m'}_{\ell' m' s0} = (2\ell' + 1)\delta_{s0}$$

we get

$$Y(r) = \sum_{n',\ell'} N_{n'\ell'}\, X^0_{n'\ell',n'\ell'}(r), \quad \text{where} \quad N_{n'\ell'} = 2(2\ell' + 1).$$

For the exchange term in (3.28),

$$X_{n\ell}(r) = \sum_\lambda n_\lambda \int \frac{\psi^*_\lambda(q')\psi_\nu(q')\,dq'}{|\vec{r} - \vec{r}\,'|}\,\psi_\lambda(q),$$

we have

$$X_{n\ell}(r) = (-1)^m \sum_{n',\ell',m',m'_s} n_{n'\ell' m' m'_s} \times$$

$$\sum_{\sigma'} \int dr' \int d\Omega' R_{n'\ell'}(r') R_{n\ell}(r') Y^*_{\ell' m'}(\vartheta', \varphi') Y_{\ell m}(\vartheta', \varphi') \chi^\dagger_{\frac{1}{2} m'_s}(\sigma') \chi_{\frac{1}{2} m'_s}(\sigma') \times$$

$$\sum_{s,\tilde{m}} \frac{4\pi}{2s+1} \frac{r^s_<}{r^{s+1}_>} Y^*_{s\tilde{m}}(\vartheta, \varphi)\, Y_{s\tilde{m}}(\vartheta', \varphi')\, \frac{R_{n'\ell'}(r)}{r}\, Y_{\ell' m'}(\vartheta, \varphi) \chi_{\frac{1}{2} m'_s}(\sigma) =$$

$$(-1)^m \sum_{n',\ell',m',m'_s,s,\tilde{m}} \sqrt{\frac{4\pi}{2s+1}}\, X^s_{n\ell,n'\ell'}(r) \frac{R_{n'\ell'}(r)}{r} \sqrt{\frac{2\ell+1}{2\ell'+1}}\, C^{\ell' 0}_{\ell 0 s 0} C^{\ell' m'}_{\ell m s \tilde{m}} \times$$

$$Y_{\ell' m'}(\vartheta, \varphi) Y^*_{s\tilde{m}}(\vartheta, \varphi) \delta_{m_s m'_s} \chi_{\frac{1}{2} m_s}(\sigma).$$

The product $Y_{\ell' m'}(\vartheta, \varphi)\, Y^*_{s\tilde{m}}(\vartheta, \varphi)$ admits an expansion into harmonics [220]:

$$Y_{\ell' m'}(\vartheta, \varphi)\, Y^*_{s\tilde{m}}(\vartheta, \varphi) =$$

$$(-1)^{\tilde{m}} \sum_{\bar{\ell},\bar{m}} \sqrt{\frac{(2\ell' + 1)(2s + 1)}{4\pi(2\bar{\ell} + 1)}}\, C^{\bar{\ell} 0}_{\ell' 0 s 0}\, C^{\bar{\ell}\bar{m}}_{\ell' m' s\, -\tilde{m}}\, Y_{\bar{\ell}\bar{m}}(\vartheta, \varphi).$$

Next, using the orthogonality property,

$$\sum_{m',\tilde{m}} (-1)^{\tilde{m}} C^{\bar{\ell}\bar{m}}_{\ell' m' s - \tilde{m}} C^{\ell' m'}_{\ell m s \tilde{m}} = (-1)^{\ell - \ell'} \sqrt{\frac{2\ell' + 1}{2\ell + 1}}\, \delta_{\ell\bar{\ell}}\, \delta_{m\bar{m}},$$

interchanging the upper and lower indices $\ell' 0$ and $\ell 0$ in one of the coefficients $\left(C^{\ell' 0}_{\ell 0 s 0} = (-1)^s \sqrt{\dfrac{2\ell + 1}{2\ell' + 1}}\, C^{\ell 0}_{\ell' 0 s 0} \right)$ and observing that the number $\ell - \ell' + s$ must be even, we obtain

$$X_{n\ell}(r) = (-1)^m \sum_{n',\ell'} \frac{N_{n'\ell'}}{4\ell + 2} \sum_{s=|\ell-\ell'|}^{\ell+\ell'} \left(C^{\ell 0}_{\ell' 0 s 0}\right)^2 X^s_{n\ell,n'\ell'}(r)\, \frac{R_{n'\ell'}(r)}{r}.$$

After all these manipulations, the Hartree-Fock equations (3.29) for the radial functions $R_{n\ell}(r)$ in the case of closed shells take on the form

$$\left[-\frac{1}{2}\frac{d^2}{dr^2}+\frac{\ell(\ell+1)}{2r^2}-\frac{Z}{r}+\sum_{n',\ell'} N_{n'\ell'}\, X^0_{n'\ell',n'\ell'}(r)\right] R_{n\ell}(r)-$$
$$\sum_{n',\ell'}\frac{N_{n'\ell'}}{4\ell+2}\sum_{s=|\ell-\ell'|}^{\ell+\ell'}\left(C^{\ell 0}_{\ell' 0\; s0}\right)^2 X^s_{n\ell,n'\ell'}(r)\, R_{n'\ell'}(r)=$$
$$\varepsilon_{n\ell}\, R_{n\ell}(r)+\sum_{n'\neq n}\varepsilon_{n'\ell,n\ell}\, R_{n'\ell}(r). \quad (3.38)$$

We should mention that when the one-particle functions are chosen in the form (3.37) it suffices to require the orthogonality of the radial parts with different values of n, since with respect to the other quantum numbers the orthogonality condition is automatically satisfied. This means that the non-diagonal Lagrange multipliers are different from zero only for $\ell' = \ell$.

For ions with non-closed shells, instead of one determinant (3.22) one usually employs a representation of the wave function as a linear combination of determinants, corresponding to a specific coupling scheme for momenta [62, 83, 156]. Note that the wave functions and energy levels obtained by solving these equations require further refinements, in particular, to account for the spin-orbit interaction and the non-centrally-symmetric part of the Coulomb interaction. In many cases it suffices to solve simpler equations to obtain $R_{n\ell}(r)$, using the obtained wave functions to construct the basis functions and applying the Ritz method to calculate the ion energy levels of many-electron ions [212].

Let us consider one such variant of the HF method, in which the coupling of momenta can be approximately replaced by a special choice of non-integer occupation numbers n_ν ($\nu \equiv nlmm_{\rm s}$) such that $\sum_{m,m_{\rm s}} n_\nu = N_{n\ell}$. In the simplest approximation of the average configuration that is consistent with the notion of average atom we take

$$n_\nu = \frac{N_{n\ell}}{2(2\ell+1)}.$$

The Hartree-Fock equations (3.38) for the closed shells remain valid in this approximation, and the total energy of the ion is given by the formula [62]:

$$E=\sum_{n,\ell} N_{n\ell}\, I_{n\ell}+\frac{1}{2}\sum_{n,\ell} N_{n\ell}(N_{n\ell}-1)\, H_{n\ell,n\ell}+\sum_{n,\ell}\sum_{n'\ell'>n\ell} N_{n\ell}N_{n'\ell'}\, H_{n\ell,n'\ell'}, \quad (3.39)$$

where we have introduced the kinetic energy and the energy of the interaction of the electron with the nucleus,

$$I_{n\ell}=\int R_{n\ell}(r)\left(-\frac{1}{2}\frac{d^2}{dr^2}+\frac{\ell(\ell+1)}{2r^2}-\frac{Z}{r}\right) R_{n\ell}(r)\, dr, \quad (3.40)$$

as well as the matrix elements of the interaction between electrons,

$$H_{n\ell,n\ell} = F^0_{n\ell,n\ell} - \frac{1}{4\ell+1} \sum_{s=2}^{2\ell} \left(C^{\ell 0}_{\ell 0\, s0}\right)^2 F^s_{n\ell,n\ell}\,, \tag{3.41}$$

$$H_{n\ell,n'\ell'} = F^0_{n\ell,n'\ell'} - \frac{1}{4\ell+2} \sum_{s=|\ell-\ell'|}^{\ell+\ell'} \left(C^{\ell 0}_{\ell' 0\, s0}\right)^2 G^s_{n\ell,n'\ell'}. \tag{3.42}$$

The Slater integrals $F^s_{n\ell,n'\ell'}$ and $G^s_{n\ell,n'\ell'}$ are calculated by the formulas:

$$F^s_{n\ell,n'\ell'} = \iint \frac{r^s_<}{r^{s+1}_>} R^2_{n\ell}(r_1)\, R^2_{n'\ell'}(r_2)\, dr_1\, dr_2, \tag{3.43}$$

$$G^s_{n\ell,n'\ell'} = \iint \frac{r^s_<}{r^{s+1}_>} R_{n\ell}(r_1)\, R_{n'\ell'}(r_1)\, R_{n\ell}(r_2)\, R_{n'\ell'}(r_2)\, dr_1\, dr_2. \tag{3.44}$$

Concerning methods for calculating the Slater integrals, see, e.g., [83].

3.3 The modified Hartree-Fock-Slater model

3.3.1 Semiclassical approximation for the exchange interaction

The numerical integration of the Hartree-Fock equations (3.28)–(3.30) runs into certain difficulties, connected with the integral term—the last term in the right-hand side of (3.28), associated with the exchange effects. If in (3.28) we neglect the exchange, then we obtain from (3.29) a Schrödinger equation for the functions $\psi_\nu(q)$, which describe the electron states in the self-consistent Coulomb field of the electrons and nuclei (Hartree model, § 3.1). Being solutions of an equation with a given potential and the corresponding boundary conditions, the wave functions in this model are mutually orthogonal. However, it turns out that the self-consistent Hartree field incorporates also the Coulomb contribution of the electron under investigation, known as self-action. The self-action can be easily eliminated if one subtracts from the self-consistent Hartree potential the contribution of the electron in question in the state with wave function $\psi_\nu(q)$, but then the resulting wave functions are no longer mutually orthogonal.

The semiclassical approximation can be used to approximately account for the exchange effects, thereby refining the equations of the Hartree model. Since this yields a unique effective potential for all electrons, the corresponding system of wave functions will be orthogonal. Exchange effects were first accounted for in a local approximation for the free atom at temperature $T = 0$ by Slater [203]. His results were subsequently revised in [113] (see also [78]), and then extended to arbitrary temperatures in [152] and [79].

To obtain an approximate expression for the exchange term via a variational principle, we use as trial functions $\psi_\nu(q)$ the normalized solutions of the Schrödinger equation

$$\widehat{H}_{\text{ef}}\psi_\nu(q) = \varepsilon_\nu \psi_\nu(q), \quad \widehat{H}_{\text{ef}} = -\frac{1}{2}\Delta - V(\vec{r}), \tag{3.45}$$

describing the motion of electrons in a field with some effective potential $V(\vec{r})$, which accounts for the exchange in a local manner. Since in the present case the operator $\widehat{H}_{\text{ef}}$ is assumed to be Hermitian, the system of functions $\psi_\nu(q)$ obtained by solving problem (3.45) will be orthogonal.

The variational principle allows us to choose the best form of the potential $V(\vec{r})$ for which the solutions of equation (3.45) will be as close as possible to the solutions of the Hartree-Fock equations (3.29)–(3.30). In doing so it makes sense to retain the expression (3.30) for the occupation numbers, because that formula does not depend explicitly on the specific approximation of the exchange effects.

Let us write the exchange energy (see the last term in (3.23))

$$E_{\text{ex}} = -\frac{1}{2}\sum_{\nu,\lambda} n_\nu n_\lambda \iint \frac{\psi_\nu^*(q)\,\psi_\nu(q')\,\psi_\lambda^*(q')\,\psi_\lambda(q)}{|\vec{r}-\vec{r}\,'|}\,dq\,dq'$$

in the form

$$E_{\text{ex}} = -\frac{1}{2}\iint \frac{|\rho(q,q')|^2}{|\vec{r}-\vec{r}\,'|}\,dq\,dq', \tag{3.46}$$

where

$$\rho(q,q') = \sum_\nu n_\nu \psi_\nu(q)\psi_\nu^*(q') = n(\widehat{H}_{\text{ef}})\sum_\nu \psi_\nu(q)\psi_\nu^*(q').$$

Here we have introduced the so-called occupation number operator $n(\widehat{H}_{\text{ef}})$, whose eigenvalues are n_ν and whose eigenfunctions coincide with the eigenfunctions of the operator $\widehat{H}_{\text{ef}}$, which acts only on functions of the variable $\vec{r}$ (see the expression (3.45) for $\widehat{H}_{\text{ef}}$). Note that according to (3.26) we have $\rho(q) = \rho(q,q')|_{q'=q}$.

The system of functions $\psi_\nu(q)$ is complete, and so

$$\sum_\nu \psi_\nu(q)\psi_\nu^*(q') = \delta(q-q') = \delta_{\sigma\sigma'}\,\delta(\vec{r}-\vec{r}\,').$$

Hence, assuming that the operator $\widehat{H}_{\text{ef}}$ does not act on spin variables, we obtain

$$\rho(q,q') = \delta_{\sigma\sigma'} n(\widehat{H}_{\text{ef}})\,\delta(\vec{r}-\vec{r}\,').$$

Now let us use the Fourier integral representation of the δ-function:

$$\delta(\vec{r}-\vec{r}\,') = \frac{1}{(2\pi)^3}\int e^{i\vec{k}(\vec{r}-\vec{r}\,')}\,d\vec{k}.$$

This decomposition of the δ-function in plane waves $e^{i\vec{k}\vec{r}}$ enables us to simplify considerably the calculations. It is important to emphasize that here we are using only the completeness property of the system of plane waves, and we do not assume that plane waves are taken as approximations for the electron wave functions.

The occupation number operator $n(\widehat{H}_{\mathrm{ef}})$ is a function of the sum of two non-commuting operators, $(1/2)\,\Delta$ and $V(\vec{r})$. The first [resp., second] operator acts on $e^{i\vec{k}\vec{r}}$ as multiplication by $-k^2/2$ [resp., by $V(\vec{r})$]. In the semiclassical approximation one can neglect that the operators $(1/2)\,\Delta$ and $V(\vec{r})$ do not commute [68, 105]. Then in accordance with formula (3.30), we obtain

$$n(\widehat{H}_{\mathrm{ef}})\, e^{i\vec{k}(\vec{r}-\vec{r}\,')} = f(k^2,\vec{r}\,)\, e^{i\vec{k}(\vec{r}-\vec{r}\,')},$$

where

$$f(k^2,\vec{r}\,) = \frac{1}{1+\exp\left(\dfrac{k^2}{2\theta} - \dfrac{V(\vec{r}\,)+\mu}{\theta}\right)}.$$

Substituting the expressions obtained in (3.46), we have

$$E_{\mathrm{ex}} \approx -\frac{1}{2}\sum_{\sigma,\sigma'}\delta^2_{\sigma\sigma'}\frac{1}{(2\pi)^6}\iiiint \frac{f(k^2,\vec{r}\,)f(k'^2,\vec{r}\,)e^{i(\vec{k}-\vec{k}\,')(\vec{r}-\vec{r}\,')}d\vec{k}\,d\vec{k}'\,d\vec{r}\,d\vec{r}\,'}{|\vec{r}-\vec{r}\,'|}\;.$$

The integration with respect to $\vec{r}\,'$ can be carried out in analytic form because the volume occupied by the system of electrons is assumed to be sufficiently large. Directing the axis z' along the vector $\vec{k}-\vec{k}\,'$, we obtain

$$\int \frac{e^{i(\vec{k}-\vec{k}\,')(\vec{r}-\vec{r}\,')}}{|\vec{r}-\vec{r}\,'|}\,d\vec{r}\,' = \lim_{\alpha\to 0}\int \frac{e^{-\alpha R+i(\vec{k}-\vec{k}\,')\vec{R}}}{R}\,d\vec{R} =$$

$$\lim_{\alpha\to 0} 2\pi\int_0^\infty e^{-\alpha R}R\,dR\int_0^\pi e^{i|\vec{k}-\vec{k}\,'|R\cos\vartheta}\sin\vartheta\,d\vartheta =$$

$$\frac{4\pi}{|\vec{k}-\vec{k}\,'|}\lim_{\alpha\to 0}\int_0^\infty e^{-\alpha R}\sin(|\vec{k}-\vec{k}\,'|R)\,dR = \frac{4\pi}{|\vec{k}-\vec{k}\,'|^2}.$$

Since $\sum_{\sigma,\sigma'}\delta^2_{\sigma\sigma'} = 2$, using spherical coordinates for the integration with respect to $\vec{k}$ and $\vec{k}'$ we get

$$E_{\mathrm{ex}} = -\frac{2}{(2\pi)^3}\int d\vec{r}\int_0^\infty\!\!\int_0^\infty f(k^2,\vec{r})f(k'^2,\vec{r})k^2\,dk\,k'^2dk'\int_0^\pi\!\!\int_0^\pi \frac{\sin\vartheta\;d\vartheta\sin\vartheta' d\vartheta'}{k^2-2kk'\cos\vartheta+k'^2}$$

$$= -\frac{1}{2\pi^3}\int d\vec{r}\int_0^\infty\!\!\int_0^\infty f(k^2,\vec{r})f(k'^2,\vec{r})\ln\left|\frac{k+k'}{k-k'}\right| k\,dk\,k'\,dk'.$$

Next, let us introduce the new integration variables $\alpha = k^2/(2\theta)$, $\beta = k'^2/(2\theta)$ and denote

$$\xi = \xi(\vec{r}) = \frac{V(\vec{r}) + \mu}{\theta},$$

$$F(\xi) = \int_0^\infty d\alpha \int_0^\infty d\beta \left[\frac{1}{1+\exp(\alpha-\xi)} \cdot \frac{1}{1+\exp(\beta-\xi)} \ln \frac{\sqrt{\alpha}+\sqrt{\beta}}{|\sqrt{\alpha}-\sqrt{\beta}|} \right].$$

Then

$$E_{\text{ex}} = -\frac{\theta^2}{2\pi^3} \int F(\xi)\, d\vec{r}, \tag{3.47}$$

where $F(\xi)$ can be expressed in terms of the Fermi-Dirac integral $I_k(x)$. Indeed, for the derivative $F'(\xi)$ we have

$$\frac{d}{d\xi} F(\xi) = \int_0^\infty\int_0^\infty d\alpha\, d\beta \ln \frac{\sqrt{\alpha}+\sqrt{\beta}}{|\sqrt{\alpha}-\sqrt{\beta}|} \times$$
$$\left\{ -\frac{d}{d\alpha}\left(\frac{1}{1+\exp(\alpha-\xi)} \right) \cdot \frac{1}{1+\exp(\beta-\xi)} - \frac{d}{d\beta}\left(\frac{1}{1+\exp(\beta-\xi)} \right) \cdot \frac{1}{1+\exp(\alpha-\xi)} \right\},$$

whence, upon integrating by parts,

$$F'(\xi) = -\int_0^\infty \frac{d\alpha}{1+\exp(\alpha-\xi)} \int_0^\infty \frac{d\beta}{1+\exp(\beta-\xi)} \left[\frac{1}{\alpha-\beta}\sqrt{\frac{\beta}{\alpha}} + \frac{1}{\beta-\alpha}\sqrt{\frac{\alpha}{\beta}} \right] =$$
$$\int_0^\infty \frac{\alpha^{-1/2}\, d\alpha}{1+\exp(\alpha-\xi)} \int_0^\infty \frac{\beta^{-1/2}\, d\beta}{1+\exp(\beta-\xi)} = I^2_{-1/2}(\xi).$$

We conclude that

$$E_{\text{ex}} = -\frac{\theta^2}{2\pi^3} \int d\vec{r} \int_{-\infty}^{(V(\vec{r})+\mu)/\theta} \left[I_{-1/2}(t) \right]^2 dt. \tag{3.48}$$

In the semiclassical approximation we can express E_{ex} through the electron density $\rho(q)$. To this end, let us clarify the connection between $\rho(q)$ and $V(\vec{r})$, using the semiclassical approximation and a Fourier integral in much the same way as we proceeded above to derive (3.47). We have

$$\begin{aligned}\rho(q) =&\rho(q,q')|_{q'=q} = \delta_{\sigma\sigma'}\, n(\widehat{H}_{\rm ef})\,\delta(\vec{r}-\vec{r}\,')|_{q'=q} = \\ &\frac{1}{(2\pi)^3}\int e^{-i\vec{k}\vec{r}} n(\widehat{H}_{\rm ef})\, e^{i\vec{k}\vec{r}}\, d\vec{k} \approx \frac{1}{(2\pi)^3}\int f(k^2,\vec{r})\, d\vec{k} = \\ \frac{1}{(2\pi)^3}&\int_0^\infty \frac{4\pi k^2\, dk}{1+\exp\left(\dfrac{k^2}{2\theta}-\dfrac{V(\vec{r})+\mu}{\theta}\right)} = \frac{(2\theta)^{3/2}}{4\pi^2} I_{1/2}\left(\frac{V(\vec{r})+\mu}{\theta}\right), \end{aligned} \tag{3.49}$$

which obviously coincides with formula (1.7), since $\rho(q) = \rho(\vec{r})/2$. Therefore, we can write

$$E_{\rm ex} = -\int \varphi[\rho(\vec{r}),\theta]\, d\vec{r}, \tag{3.50}$$

where

$$\varphi[\rho(\vec{r}),\theta] \equiv \varphi(\rho,\theta) = \frac{\theta^2}{2\pi^3}\int_{-\infty}^{\xi} \left[I_{-1/2}(t)\right]^2 dt. \tag{3.51}$$

Here, in accordance with (3.48), (3.49), in the semiclassical approximation instead of $\xi = (V(\vec{r})+\mu)/\theta$ we take the root of the equation $\rho(\vec{r}) = (2\theta)^{3/2} I_{1/2}(\xi)/(2\pi^2)$. Obviously, ξ is a function of $\rho(\vec{r})/\theta^{3/2}$, where the electron density $\rho(\vec{r})$ is calculated by using the wave functions $\psi_\nu(q)$.

3.3.2 The equations of the Hartree-Fock-Slater model

Thus, for the thermodynamic potential Ω, which in the approximation considered above we denote by $\Omega^{\rm HFS}$, we have

$$\begin{aligned}\Omega^{\rm HFS} =& \sum_\nu n_\nu \int \psi_\nu^*(q)\left[-\frac{1}{2}\Delta - V_a(\vec{r})\right]\psi_\nu(q)\, dq + \\ &\frac{1}{2}\sum_{\nu\lambda} n_\nu n_\lambda \iint \frac{|\psi_\nu(q)|^2|\psi_\lambda(q')|^2\, dq\, dq'}{|\vec{r}-\vec{r}\,'|} - \int \varphi[\rho(\vec{r}),\theta]\, d\vec{r} + \\ &\theta\sum_\nu [n_\nu \ln n_\nu + (1-n_\nu)\ln(1-n_\nu)] - \mu\sum_\nu n_\nu,\end{aligned}$$

where

$$\rho(\vec{r}) = \sum_\sigma \rho(q), \quad \rho(q) = \sum_\nu n_\nu |\psi_\nu(q)|^2.$$

Proceeding as in the case of the Hartree-Fock equations, let us require that the minimum condition

$$\delta\Omega^{\rm HFS} + \delta\left[\sum_\nu \Lambda_\nu \int \psi_\nu^*(q)\psi_\nu(q)\, dq\right] = 0$$

be satisfied, where the Lagrange multipliers Λ_ν are introduced to account for the normalization conditions for $\psi_\nu(q)$. They correspond to the diagonal Lagrange multipliers $\Lambda_{\nu\nu}$ in equations (3.29). Note that in the present case it is not necessary to require that the wave functions $\psi_\nu(q)$ be orthogonal, because the solutions of equation (3.45) are orthogonal to one another. Taking the variation, we obtain the equations of the Hartree-Fock-Slater (HFS) model:

$$\left[-\frac{1}{2}\Delta - V_a(\vec{r}) + \sum_\lambda n_\lambda \int \frac{|\psi_\lambda(q')|^2\, dq'}{|\vec{r}-\vec{r}\,'|} - \frac{\partial\varphi}{\partial\rho}\right]\psi_\nu(q) = \varepsilon_\nu\psi_\nu, \tag{3.52}$$

$$n_\nu = \frac{1}{1+\exp\left(\dfrac{\varepsilon_\nu-\mu}{\theta}\right)}, \tag{3.53}$$

where the eigenvalues are $\varepsilon_\nu = -\Lambda_\nu/n_\nu$.

This shows that the effective potential in (3.45) is equal to

$$V(\vec{r}) = V_c(\vec{r}) + V_{ex}(\vec{r}).$$

The Coulomb part of the potential, $V_c(\vec{r})$, is calculated as in the Hartree model:

$$V_c(\vec{r}) = V_a(\vec{r}) - \sum_\lambda n_\lambda \int \frac{|\psi_\lambda(q')|^2\, dq'}{|\vec{r}-\vec{r}\,'|}. \tag{3.54}$$

By (3.51) and (3.52), the exchange correction $V_{ex}(\vec{r})$ is given by

$$V_{ex}(\vec{r}) = \frac{\partial\varphi}{\partial\rho} = \frac{\partial\varphi}{\partial\xi}\cdot\frac{\partial\xi}{\partial\rho} = \frac{\theta^{1/2}}{\pi\sqrt{2}} I_{-1/2}(\xi), \tag{3.55}$$

where ξ is the root of the equation (see (3.49))

$$I_{1/2}(\xi) = \frac{2\pi^2}{(2\theta)^{3/2}}\rho(\vec{r}). \tag{3.56}$$

The calculation of exchange corrections can be simplified by using interpolation formulas. In particular, the exchange energy E_{ex} is given by the integral $\int_{-\infty}^{\xi(\rho,\theta)} \left[I_{-1/2}(t)\right]^2 dt$, which is convenient to represent, using (3.56), as a function of one variable

$$\zeta = \frac{\rho}{\theta^{3/2}} = \frac{\sqrt{2}}{\pi^2} I_{1/2}(\xi). \tag{3.57}$$

Let us construct an interpolation formula for the exchange integral, which gives the true asymptotics for $\zeta \to 0$ and $\zeta \to \infty$:

$$\int_{-\infty}^{\xi} \left[I_{-1/2}(t)\right]^2 dt \simeq \frac{\pi^4\zeta^2}{\left[1 + a\zeta + (8\pi^4/81)\zeta^2\right]^{1/3}}. \tag{3.58}$$

Here we have introduced a free parameter a, which can be found from the condition that the error be minimal. As a result, using (3.51) in conjunction with (3.58) we obtain for the function $\varphi(\rho, \theta)$ the expression

$$\varphi(\rho,\theta) \approx \frac{\pi\rho^2}{2\theta}\left[1 + 4.9\,\frac{\rho}{\theta^{3/2}} + \frac{8\pi^4}{81}\cdot\frac{\rho^2}{\theta^3}\right]^{-1/3}, \tag{3.59}$$

where $\rho = \rho(\vec{r})$ is the electron density. The error of formula (3.59) does not exceed 1%.

In much the same way as we derived the formula (3.58), we can obtain the approximate formula

$$I_{-1/2}(\xi) \simeq \frac{\sqrt{2}\,\pi^2\zeta}{[1 + 5.7\zeta + (\pi^4/3)\zeta^2]^{1/3}},$$

where ζ is given by formula (3.57). This yields a convenient interpolation formula for $\partial\varphi/\partial\rho$, with an error of about 1.5% [152]. Finally, we have

$$V_{\text{ex}}(\vec{r}) = \frac{\pi\rho(\vec{r})}{\theta}\left[1 + 5.7\,\frac{\rho(\vec{r})}{\theta^{3/2}} + \frac{\pi^4}{3}\cdot\frac{\rho^2(\vec{r})}{\theta^3}\right]^{-1/3}, \tag{3.60}$$

where $\rho(\vec{r})$ is the electron density. Since $V_{\text{ex}}(\vec{r}) > 0$, exchange effects lead to a decrease of the Coulomb repulsion in the potential.

The behavior of the exchange correction (3.60) as a function of the electron density $\rho = \rho(\vec{r})$ for various values of the temperature is shown in Figure 3.9. For comparison, the figure also shows the results obtained by using the interpolation formula from [188],

$$V_{\text{ex}}^{\text{R}}(\vec{r}) = [1 - \lambda^2(r)]\,\frac{3}{2}\left(\frac{3\rho(r)}{\pi}\right)^{1/3} + \lambda^2(r)\,\frac{\pi\rho(r)}{\theta}, \tag{3.61}$$

where $\lambda(r) = \max\{1, 2\theta(3\pi^2\rho(r))^{-2/3}\}$. The formula (3.61) for $\theta = 0$ gives the expression V_{ex}^{S}, obtained by Slater [203]:

$$V_{\text{ex}}^{\text{S}} = \frac{3}{2}\left(\frac{3\rho}{\pi}\right)^{1/3}.$$

From (3.60) with $\theta = 0$ we have $V_{\text{ex}} = (3\rho/\pi)^{1/3}$, which agrees with results of W. Kohn et al. [87, 113] and differs from Slater's expression by the factor 3/2. This is a consequence of the application of the variational principle for obtaining V_{ex}, in contrast to [203], where the exchange term was approximated directly in the Hartree-Fock equations. An analysis of various approximations for the exchange correction at temperature zero is contained in the book by Slater [204], as well as in that by Cowan [49], Chapter 8. As one can see in Figure 3.9, when the

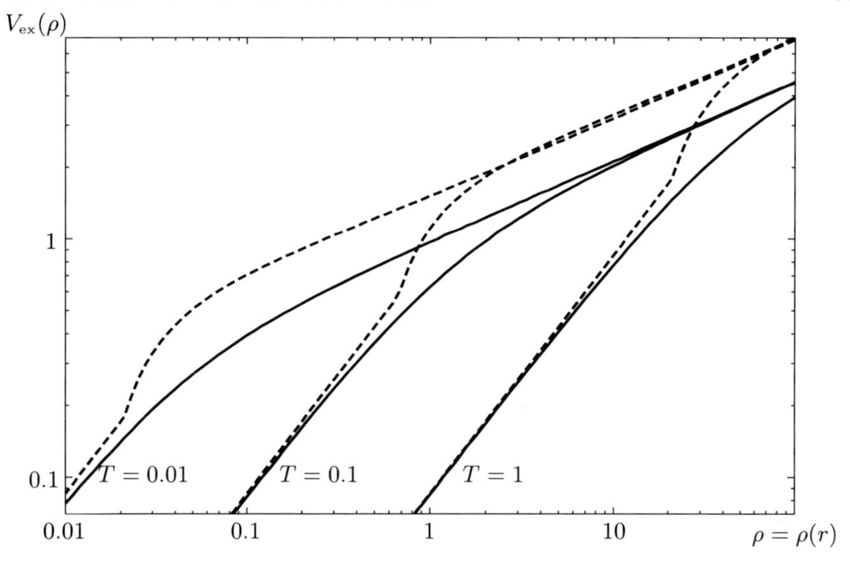

Figure 3.9: Dependence of the exchange correction $V_{ex}(\rho)$ to the potential on the electron density $\rho = \rho(r)$ for three different temperatures: $T = 0.01$, 0.1, 1 keV. The solid [resp., dashed] curve represents expression (3.60) [resp., (3.61)]

temperature is different from zero, formula (3.60) gives a smoother dependence of V_{ex} on $\rho(r)$ than the interpolation (3.61); at high temperatures, in both cases $V_{ex} \simeq \pi\rho/\theta$.

As we will show below, the variational principle yields wave functions that are closer to the solutions of the Hartree-Fock equation than the ones given by other approximation methods. Let us remark that for the one-particle energy levels the result may be different, but the utilization of the wave functions obtained in the Hartree-Fock-Slater model and of the Hartree-Fock approximation for the energy levels of ions allows one to obtain reliable results (see Subsection 4.4.4).

3.3.3 The equations of the Hartree-Fock-Slater model in the case when the semiclassical approximation is used for continuum electrons

When one solves the system of equations (3.52)–(3.56) it is required, generally speaking, to calculate the wave functions of all possible electron states, including those for the continuous spectrum. The number of such functions may be very large. Since the semiclassical approximation gives good results for the electrons with large quantum numbers, it is natural to use it for the electron states whose energy is larger than some value ε_0, as we did in the Hartree model. The quantity ε_0 can be regarded as the effective boundary of the continuum.

For large temperatures, due to ionization the average occupation numbers for highly excited states are very small and if, in addition, the radius of the ion core is much smaller than the dimensions of the average atom cell, i.e., $r^* \ll r_0$, then the issue of how to choose ε_0 in practical calculations is readily settled. Namely, after one calculates the average occupation numbers $N_{n\ell}$, the parameter ε_0 is chosen from the condition that $N_{n\ell}$ be small for the upper shells (see Subsection 3.1.3).

When the variational principle considered in § 3.2 is used, it suffices to stipulate that for the choice of ε_0 the number of states that are not associated to the continuum does not change when the variation is performed, i.e., that the bound states do not cross the fixed edge ε_0 for small variations of the occupation numbers and wave functions, or, in other words, that no additional discrete spectrum states arise for $\varepsilon < \varepsilon_0$.

Let us examine this question in more detail, following [157]. The expression for the one-particle energy ε_ν in the Hartree-Fock-Slater model is

$$\varepsilon_\nu = \int \psi_\nu^*(q) \left[-\frac{1}{2}\Delta - V(\vec{r}) \right] \psi_\nu(q)\, dq,$$

where according to (3.54)–(3.55)

$$V(\vec{r}) = V_a(\vec{r}) - \int \frac{\rho(q')\, dq'}{|\vec{r} - \vec{r}'|} + \frac{\partial \varphi}{\partial \rho}. \tag{3.62}$$

In the semiclassical approximation the energy of an electron is given by the formula $\varepsilon_\nu \cong \varepsilon(q, \vec{p}) \equiv \varepsilon = p^2/2 - V(\vec{r})$, where $\vec{p}$ is the momentum of the electron.

Let us use the indicated approximation to calculate the thermodynamic potential Ω^{HFS}, considering that the semiclassical states with energy $\varepsilon > \varepsilon_0$ are occupied with a distribution function $n(q, \vec{p})$, $0 \le n(q, \vec{p}) \le 1$, which for the moment is not known. In connection with this, in the ensuing formulas summation with respect to the discrete states will be carried out for the states with energy $\varepsilon_\nu < \varepsilon_0$, while the integration with respect to q and $\vec{p}$ will be carried out for states with energy $\varepsilon = \varepsilon(q, \vec{p}) > \varepsilon_0$. Note that in what follows, when we perform the variation, although ε_0 is fixed, some terms in the sum over ν may be added or may disappear, because the condition $\varepsilon_\nu < \varepsilon_0$ is replaced by $\varepsilon_\nu + \delta\varepsilon_\nu < \varepsilon_0$. Here the variation $\delta\varepsilon_\nu$ is determined by the variations of $\psi_\nu(q)$, n_ν and $n(q, \vec{p})$. To formally carry out the summation and integration over all states, we will simplify notation by using the step function

$$\beta(x) = \begin{cases} 1, & \text{if } x > 0, \\ 0, & \text{if } x \le 0. \end{cases}$$

We denote the corresponding thermodynamic potential by $\Omega^{\mathrm{HFS}}(\varepsilon_0)$. Applying to the thermodynamic potential (3.27) the semiclassical approximation for $\varepsilon > \varepsilon_0$, we obtain

$$\Omega^{\mathrm{HFS}}(\varepsilon_0)=\sum_\nu \beta(\varepsilon_0-\varepsilon_\nu)n_\nu \int \psi_\nu^*(q)\left(-\frac{1}{2}\Delta\right)\psi_\nu(q)\,dq+$$
$$\iint[1-\beta(\varepsilon_0-\varepsilon)]n(q,\vec{p})\,\frac{p^2}{2}\frac{d\vec{p}}{(2\pi)^3}\,dq-\int \rho(q)V_a(\vec{r})\,dq+$$
$$\frac{1}{2}\iint \frac{\rho(q)\rho(q')\,dq\,dq'}{|\vec{r}-\vec{r}\,'|}-\int \varphi[\rho(\vec{r}),\theta]\,d\vec{r}+$$
$$\theta\sum_\nu \beta(\varepsilon_0-\varepsilon_\nu)[n_\nu\ln n_\nu+(1-n_\nu)\ln(1-n_\nu)]+$$
$$\theta\iint[1-\beta(\varepsilon_0-\varepsilon)]\,[n\ln n+(1-n)\ln(1-n)]\frac{d\vec{p}}{(2\pi)^3}\,dq-\mu\int \rho(q)\,dq.$$

Here $\psi_\nu(q)$, n_ν and $n=n(q,\vec{p})$ are unknown functions that need to be determined, and

$$\rho(q)=\sum_\nu \beta(\varepsilon_0-\varepsilon_\nu)n_\nu|\psi_\nu(q)|^2+\int[1-\beta(\varepsilon_0-\varepsilon)]\,n(q,\vec{p})\frac{d\vec{p}}{(2\pi)^3}. \tag{3.63}$$

Taking the variation of $\Omega^{\mathrm{HFS}}(\varepsilon_0)$ subject to the constraint $\int|\psi_\nu(q)|^2\,dq=1$ and using formula (3.63) to calculate $\delta\rho(q)$, we obtain

$$\delta\Omega^{\mathrm{HFS}}(\varepsilon_0)+\delta\sum_\nu \Lambda_\nu\int \psi_\nu^*(q)\psi_\nu(q)\,dq=$$
$$\sum_\nu n_\nu\beta(\varepsilon_0-\varepsilon_\nu)\int \delta\psi_\nu^*(q)\left[-\frac{1}{2}\Delta-V(\vec{r})+\frac{\Lambda_\nu}{n_\nu}-\mu\right]\psi_\nu(q)\,dq+$$
$$\sum_\nu n_\nu\beta(\varepsilon_0-\varepsilon_\nu)\int \psi_\nu^*(q)\left[-\frac{1}{2}\Delta-V(\vec{r})+\frac{\Lambda_\nu}{n_\nu}-\mu\right]\delta\psi_\nu(q)\,dq+$$
$$\sum_\nu \delta n_\nu\beta(\varepsilon_0-\varepsilon_\nu)\left[\int \psi_\nu^*(q)\left(-\frac{1}{2}\Delta-V(\vec{r})\right)\psi_\nu(q)\,dq+\theta\ln\frac{n_\nu}{1-n_\nu}-\mu\right]+$$
$$\iint \delta n[1-\beta(\varepsilon_0-\varepsilon)]\left[\frac{p^2}{2}-V(\vec{r})+\theta\ln\frac{n}{1-n}-\mu\right]\frac{d\vec{p}}{(2\pi)^3}\,dq+$$
$$\sum_\nu \delta\beta(\varepsilon_0-\varepsilon_\nu)\,[\theta n_\nu\ln n_\nu+\theta(1-n_\nu)\ln(1-n_\nu)-\mu n_\nu]+$$
$$\sum_\nu \delta\beta(\varepsilon_0-\varepsilon_\nu)n_\nu\int \psi_\nu^*(q)\left(-\frac{1}{2}\Delta-V(\vec{r})\right)\psi_\nu(q)\,dq-$$
$$\iint \delta\beta(\varepsilon_0-\varepsilon)\,[\theta n\ln n+\theta(1-n)\ln(1-n)-\mu n]\,\frac{d\vec{p}}{(2\pi)^3}\,dq-$$
$$\iint \delta\beta(\varepsilon_0-\varepsilon)\left(\frac{p^2}{2}-V(\vec{r})\right)n(q,\vec{p})\,\frac{d\vec{p}}{(2\pi)^3}\,dq=0. \tag{3.64}$$

The quantity $\delta\beta(\varepsilon_0-\varepsilon_\nu)$ may be different from zero in the interval $|\varepsilon_\nu-\varepsilon_0| < |\delta\varepsilon_\nu|$. For example, $\delta\beta = 1$ corresponds to an additional term in the sum with respect to ν which arises because the energy ε_ν changes under variation (if this term vanishes, $\delta\beta = -1$). Correspondingly, the quantity $\delta\beta(\varepsilon_0 - \varepsilon)$ is different from zero in the interval $\varepsilon_0 - \delta\varepsilon < \varepsilon < \varepsilon_0$.

In order to satisfy the minimum condition for $\Omega^{\mathrm{HFS}}(\varepsilon_0)$, we require that the factors in front of the variations $\delta\psi^*$, δn_ν and δn vanish and that the expression containing $\delta\beta$ vanishes as well. Setting the factor in front of $\delta\psi_\nu^*(q)$ (or $\delta\psi_\nu(q)$) equal to zero we obtain the equations for electrons with energy $\varepsilon_\nu < \varepsilon_0$, which coincide with (3.52):

$$\left[-\frac{1}{2}\Delta - V(\vec{r})\right] \psi_\nu(q) = \varepsilon_\nu \psi_\nu(q). \tag{3.65}$$

Here $\varepsilon_\nu = -\Lambda_\nu/n_\nu + \mu$, the potential $V(\vec{r})$ is given by formula (3.62), and for electrons with energy $\varepsilon > \varepsilon_0$ the semiclassical approximation is used.

Next, the requirement that the factors in front of δn_ν and $\delta n(q,\vec{p})$ be equal to zero leads to the relations

$$n_\nu = \frac{1}{1+\exp\left(\dfrac{\varepsilon_\nu - \mu}{\theta}\right)}, \tag{3.66}$$

$$n(q,\vec{p}) = \frac{1}{1+\exp\left(\dfrac{\varepsilon - \mu}{\theta}\right)}, \qquad \varepsilon = \frac{p^2}{2} - V(\vec{r}). \tag{3.67}$$

Finally, setting the remaining terms in (3.64) equal to zero, let us use the equalities (3.65)–(3.67), as well as the relations

$$n_\nu \ln n_\nu + (1-n_\nu)\ln(1-n_\nu) = (1-n_\nu)\frac{\varepsilon_\nu - \mu}{\theta} + \ln n_\nu,$$

$$n \ln n + (1-n)\ln(1-n) = (1-n)\frac{\varepsilon - \mu}{\theta} + \ln n,$$

which follows from (3.66) and (3.67). Then for the remaining terms in (3.64) we have

$$\sum_\nu \delta\beta(\varepsilon_0 - \varepsilon_\nu)(\varepsilon_\nu - \mu + \theta \ln n_\nu) - \iint \delta\beta(\varepsilon_0 - \varepsilon)\left[\varepsilon - \mu + \theta \ln n(q,\vec{p})\right] \frac{d\vec{p}}{(2\pi)^3}\, dq = 0. \tag{3.68}$$

Since the variations $\delta\beta(\varepsilon_0 - \varepsilon_\nu)$ and $\delta\beta(\varepsilon_0 - \varepsilon)$ can be different from zero only for $\varepsilon_\nu = \varepsilon_0$ and respectively $\varepsilon = \varepsilon_0$, in (3.68) we can take out of the brackets the common factor

$$A(\varepsilon_0) = \varepsilon_0 - \mu + \theta \ln \frac{1}{1 + \exp\left(\dfrac{\varepsilon_0 - \mu}{\theta}\right)}.$$

This yields

$$A(\varepsilon_0)\delta \left[\sum_\nu \beta(\varepsilon_0 - \varepsilon_\nu) - \iint \beta(\varepsilon_0 - \varepsilon) \frac{d\vec{p}}{(2\pi)^3}\, dq \right] = 0. \tag{3.69}$$

3.3.4 The thermodynamic consistency condition

The first term in the square brackets in (3.69) is the quantum mechanical expression for the number of one-electron states with energies $\varepsilon_\nu < \varepsilon_0$:

$$f_1(\varepsilon_0) = \sum_\nu \beta(\varepsilon_0 - \varepsilon_\nu).$$

Here, in accordance with the definition of the function β, the sum is taken over the states ν for which $\varepsilon_\nu < \varepsilon_0$. In particular, in the nonrelativistic approximation without accounting for bands we have

$$f_1(\varepsilon_0) = \sum_{n,\ell} 2(2\ell + 1) \quad (\varepsilon_{n\ell} < \varepsilon_0).$$

The second term gives the corresponding expression for the number of states in the semiclassical approximation:

$$f_2(\varepsilon_0) = \iint \beta(\varepsilon_0 - \varepsilon) \frac{d\vec{p}}{(2\pi)^3}\, dq = \iint\limits_{p^2/2 - V(r) < \varepsilon_0} \frac{2 d\vec{p}\, d\vec{r}}{(2\pi)^3}.$$

To understand the physical meaning of condition (3.69), recall that the quantity ε_0 was introduced to simplify the calculations. However, as follows from (3.69), in order to satisfy the minimum condition for the thermodynamic potential $\Omega^{\mathrm{HFS}}(\varepsilon_0)$, the number of states accounted for in various approximations for the description of electrons must not change when ε_0 is changed. In other words, when the number of states treated quantum-mechanically increases, the number of states accounted for in the semiclassical approximation must decrease by the same amount, and conversely.

Thus, in accordance with (3.69), in addition to the equations (3.65)–(3.67) of the HFS model with the semiclassical approximation we require for the electrons with energy $\varepsilon < \varepsilon_0$ that $f_1(\varepsilon_0) = f_2(\varepsilon_0)$, which gives an equation for the determination of ε_0:

$$\sum_\nu \beta(\varepsilon_0 - \varepsilon_\nu) = \iint\limits_{p^2/2 - V(r) < \varepsilon_0} \frac{2 d\vec{p}\, d\vec{r}}{(2\pi)^3}. \tag{3.70}$$

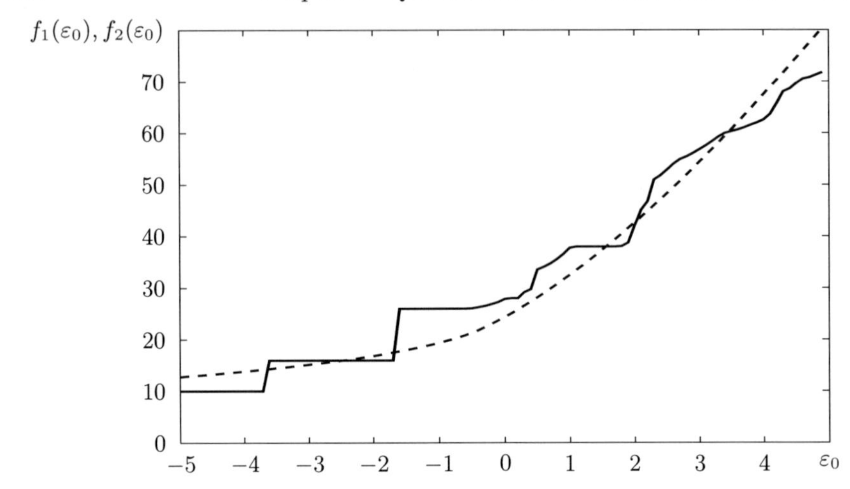

Figure 3.10: The number of states with energy $\varepsilon < \varepsilon_0$: quantum-mechanical calculation $f_1(\varepsilon_0)$ (solid curve); semiclassical calculation $f_2(\varepsilon_0)$ (dashed curve)

The typical behavior of the functions $f_1(\varepsilon_0)$ and $f_2(\varepsilon_0)$ is shown in Figure 3.10 (where $Z = 26$, $T = 15$ eV, $\rho = 15.64$ g/cm^3). The solutions of equation (3.70) correspond to the intersection points of the graphs of f_1 and f_2. There may be many such solutions, but there always are two basic solutions. The first one, $\varepsilon_0 = \infty$, corresponds to the purely quantum mechanical calculation, when for all electron states one solves the Schrödinger equation (3.65) and calculates the wave functions. The second one, $\varepsilon_0 = -\infty$, corresponds to the Thomas-Fermi approximation (with the exchange accounted for via formula (3.55)), when for all electron states one uses the semiclassical approximation. If exchange is neglected, then from (3.62), (3.63) and (3.67) one obtains the equations of the Thomas-Fermi model.

The practical value of condition (3.70) becomes evident when computations are performed in a wide range of temperatures T and densities ρ [145]. In this case if one does not require that the thermodynamic consistency condition be satisfied, then for some values of T and ρ discrete energy levels may pass through $\varepsilon = \varepsilon_0$, which results in a sharp change in thermodynamic functions. This is particularly strongly manifested for the pressure, which is the derivative of the free energy. Such an irregular behavior of thermodynamic functions has no physical interpretation and is caused only by the application of various approximations in the model for $\varepsilon > \varepsilon_0$ and $\varepsilon < \varepsilon_0$. When condition (3.70) is satisfied such incorrect behavior is eliminated and the behavior of the thermodynamic functions becomes regular. Figure 3.11 shows examples of isotherms of the electron pressure of iron, calculated with and without employing equation (3.70).

For relatively small densities of matter the quantity ε_0 can be determined not from equation (3.70), but from the condition $A(\varepsilon_0) \simeq 0$ (see (3.69)). From

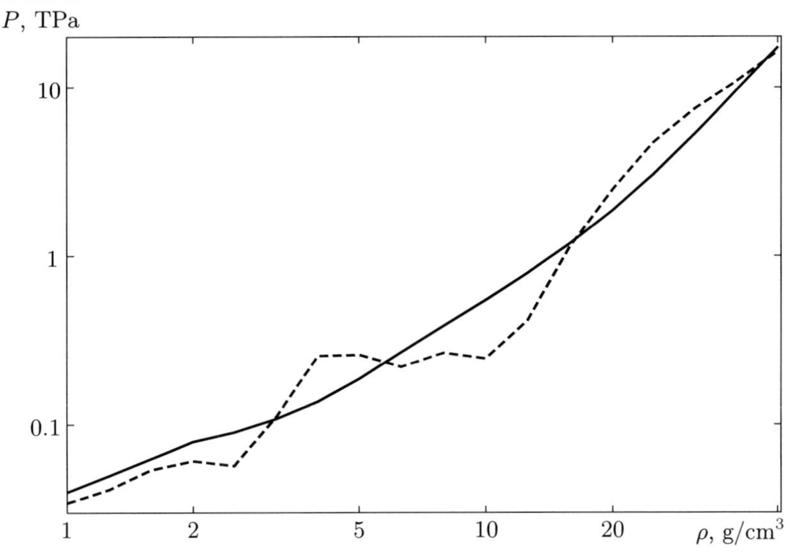

Figure 3.11: The electron pressure for iron in the HFS model (cf. § 6.4) with temperature $T = 10$ eV and different densities; for the continuous curve ε_0 was calculated with the help of equation (3.70), while the dashed curve was computed for $\varepsilon_0 = 0$

this last condition it follows that $(\varepsilon_0 - \mu)/\theta \gg 1$, and using formula (3.66) we conclude that the average occupation numbers of the levels with energy $\varepsilon_\nu \geq \varepsilon_0$ are such that $n_\nu \ll 1$. In this case, as we have seen in Subsection 3.1.3 for iron with $T = 100$ eV and $\rho = 0.1$ g/cm^3, the choice of different values of the parameter ε_0 has practically no influence on computational results.

Thus, based on the variational principle used in this chapter, we succeeded in refining and generalizing the Hartree model introduced in § 3.1. The resulting Hartee-Fock-Slater self-consistent field model for the average atom has a considerably wider domain of applicability than the Thomas-Fermi model. Let us add that starting with the Hartee-Fock-Slater model for the average atom and using the one-electron wave functions to construct a basis for the wave functions of many-electron ions one can refine the model further when considering the individual states of ions in plasma.

Chapter 4

The Hartree-Fock-Slater model for the average atom

The approaches formulated in Chapter III allow one to obtain different models of matter, and then refine them according to the necessities. At high temperatures good results are given by the simplest approach of the *average atom*, in which the self-consistent potential is calculated for an ion with average occupation numbers of the energy levels together with the free electrons in the charge-neutral spherical cell. If the dimensions of the ion cores are small compared with the distances between ions, then the average-atom model allows one to obtain, in a relatively simply manner, the degree of ionization and the equation of state of a substance. In the description of photon absorption processes, which require the consideration of concrete (individual) states of ions, this can be done, in the first approximation, by resorting to perturbation theory [45].

Further refinements of models of matter are connected, first, with refining the description of individual states of ions and, second, with accounting for their interaction at high densities by employing boundary conditions of periodic type.

4.1 The Hartree-Fock-Slater system of equations in a spherical cell

4.1.1 The Hartree-Fock-Slater field

The system of equations (3.65)–(3.67), (3.70) allows us, in principle, to obtain the wave functions $\psi_\alpha(q)$ for a given position of the nuclei and given value of the chemical potential μ, which is determined from the condition of charge neutrality of the system of electrons and ions in matter. However, to actually carry the calculations for the original system of equations is clearly impossible. For that reason, as in the Thomas-Fermi and Hartree models, instead of the potential $V(\vec{r})$ for a given position of the nuclei one usually looks for the average potential near the nucleus in question, averaging $V(\vec{r})$ over the different positions of the other nuclei. If there is no preferred direction, it is natural to consider that the average potential is spherically symmetric. Assuming that in the average-atom approximation the one-electron states with opposite directions of the projection of the spin are occupied in identical manner, in formulas of Chapter III we can carry out the summation with respect to σ, which leads to replacing q by $\vec{r}$, dq by $2d\vec{r}$ and $\rho(q)$ by $\rho(\vec{r})/2$.

The Schrödinger equation in a central potential $V(r)$ admits for a given value $\varepsilon_\alpha = \varepsilon$ the particular solutions

$$\Psi_{\varepsilon\ell m}(\vec{r}) = \frac{1}{r} R_{\varepsilon\ell}(r)(-1)^m Y_{\ell m}(\vartheta, \varphi), \tag{4.1}$$

where the radial function $R_{\varepsilon\ell}(r)$ satisfies the equation

$$-\frac{1}{2} R''_{\varepsilon\ell}(r) + \left[-V(r) + \frac{\ell(\ell+1)}{2r^2}\right] R_{\varepsilon\ell}(r) = \varepsilon R_{\varepsilon\ell}(r). \tag{4.2}$$

For $r = 0$ the radial function must satisfy the boundary condition $R_{\varepsilon\ell}(0) = 0$. To formulate the conditions on the boundary of the average atom cell one can proceed in different ways. Let us consider first the case of high temperatures and relatively small densities, when the ion core takes only a small part of the atom cell and the states of the electrons in the so-called intermediate group are practically not occupied. In this case the boundary conditions for $r = r_0$ are imposed for the bound and free states exactly as in §§ 2.2, 2.3. Such an approach allows one to carry out computations of Rosseland mean opacities, spectral absorption coefficients and equations of state in a wide range of temperatures and densities within the Hartree-Fock-Slater model, using the computational algorithm of the Hartree model presented in Subsection 3.1.2. To do this, it suffices to add the exchange correction $V_{\text{ex}}(r)$ in the formula (3.9) for the potential $V(r)$.

Let us write the Hartree-Fock-Slater system of equations for the average atom in the case of high temperatures and low densities. For the electron density $\rho(r)$, the contribution of bound states with $\varepsilon < \varepsilon_0$ will be calculated using wave functions; for the remaining states we will use the semiclassical approximation:

$$\rho(r) = \rho_1(r) + \rho_2(r), \tag{4.3}$$

where

$$\rho_1(r) = \frac{1}{4\pi r^2} \sum_{\varepsilon_{n\ell} < \varepsilon_0} N_{n\ell} R^2_{n\ell}(r), \quad N_{n\ell} = \frac{2(2\ell+1)}{1 + \exp\left(\dfrac{\varepsilon_{n\ell} - \mu}{\theta}\right)}, \tag{4.4}$$

$$\rho_2(r) = \frac{(2\theta)^{3/2}}{2\pi^2} \int_{y_0}^{\infty} \frac{y^{1/2}\,dy}{1+\exp\left(y - \dfrac{V(r)+\mu}{\theta}\right)}, \tag{4.5}$$

$$y_0 = y_0(r) = \max\left\{0;\ \frac{V(r)+\varepsilon_0}{\theta}\right\}.$$

The chemical potential μ is determined from the condition of charge neutrality of the average atom cell:

$$4\pi \int_0^{r_0} \rho(r) r^2\,dr = Z. \tag{4.6}$$

The potential $V(r)$ includes, in addition to the Coulomb part $V_c(r)$, the exchange correction $V_{ex}(r)$:

$$V(r) = V_c(r) + V_{ex}(r), \tag{4.7}$$

where

$$V_c(r) = \frac{Z}{r} - 4\pi\left[\frac{1}{r}\int_0^r r'^2\rho(r')\,dr' + \int_r^{r_0} r'\rho(r')\,dr'\right], \tag{4.8}$$

$$V_{ex}(r) = \frac{\pi\rho(r)}{\theta}\left[1 + 5.7\,\frac{\rho(r)}{\theta^{3/2}} + \frac{\pi^4}{3}\frac{\rho^2(r)}{\theta^3}\right]^{-1/3}. \tag{4.9}$$

The effective boundary of the continuum ε_0 is determined from the thermodynamic consistency condition (3.70):

$$\frac{8\sqrt{2}}{3\pi}\int_0^{r_0} [\max\{0, \varepsilon_0 + V(r)\}]^{3/2} r^2\,dr = \sum_{n,\ell} 2(2\ell+1) \quad (\varepsilon_{n\ell} < \varepsilon_0). \tag{4.10}$$

Compared with the Hartree model considered earlier, the Hartree-Fock-Slater model incorporates exchange effects and has the advantage that the additional condition (4.10) allows one to choose an optimal—from the point of view of the volume of computations and the resulting accuracy—value of ε_0, because when ε_0 is decreased the volume of computations decreases as well, though the accuracy may worsen. Condition (4.10) also enables one to calculate more correctly the thermodynamic functions of matter, which are derivatives of the thermodynamic potential Ω.

In the low-density approximation considered here, the thermodynamic consistency condition (4.10) will be satisfied whenever ε_0 satisfies the inequality

$$\varepsilon_{n\ell} < \varepsilon_0 < \varepsilon_{n+1,\ell'} \tag{4.11}$$

for some value of $n = n_{\max}$ and arbitrary $\ell < n$, $\ell' < n + 1$. This follows from the fact that the left-hand side of equation (4.10) is a smooth monotone function of ε_0, whereas the right-hand side depends on ε_0 in a step-like manner (cf. Figure 3.10). Hence, for ε_0 in the interval (4.11) equation (4.10) always has a root. Note that in addition to (4.11), it is necessary that $N_{n\ell} \ll 1$. In the computational process the selected value $n = n_{\max}$ does not change.

To simplify formulas we have considered the nonrelativistic variant of the HFS model. The relativistic approximation is readily obtained if for the density of the bound electrons one uses the equations (3.13) and (3.15).

Table 4.1: The energy levels $\varepsilon_{n\ell j}$ in eV and the average occupation numbers $N_{n\ell j}$ according to the relativistic Hartree and Hartree-Fock-Slater models for an iron plasma with temperature $T = 100$ eV and density $\rho = 0.1$ g/cm^3

			$\|\varepsilon_{n\ell j}\|$		$N_{n\ell j}$	
n	ℓ	j	Hartree	HFS	Hartree	HFS
1	0	1/2	7194	7534	2.000	2.000
2	0	1/2	1142	1247	1.991	1.997
2	1	1/2	1031	1144	1.972	1.991
2	1	3/2	1020	1132	3.937	3.979
3	0	1/2	379.6	395.9	0.188	0.215
3	1	1/2	346.3	363.1	0.139	0.160
3	1	3/2	344.7	361.0	0.273	0.313
3	2	3/2	296.3	313.6	0.173	0.201
3	2	5/2	296.0	313.1	0.259	0.300
4	0	1/2	168.1	171.7	0.025	0.025
4	1	1/2	155.0	158.8	0.022	0.022
4	1	3/2	154.5	158.2	0.043	0.044
4	2	3/2	136.5	140.5	0.036	0.037
4	2	5/2	136.4	140.4	0.054	0.056
4	3	5/2	128.9	127.0	0.050	0.049
4	3	7/2	128.8	126.9	0.067	0.065
5	0	1/2	81.31	82.37	0.010	0.010
5	1	1/2	75.01	76.24	0.010	0.010
5	1	3/2	74.79	76.00	0.020	0.020
5	2	3/2	66.18	67.57	0.018	0.018
5	2	5/2	66.12	67.51	0.027	0.027
5	3	5/2	62.26	61.17	0.026	0.025
5	3	7/2	62.24	61.14	0.035	0.034
5	4	7/2	60.49	58.29	0.034	0.033
5	4	9/2	60.47	58.28	0.043	0.041
...	...	...	...	...	...	...

Table 4.1 lists energy levels and average occupation numbers calculated in the relativistic variants of the Hartree and Hartree-Fock-Slater models for an iron plasma with temperature $T = 100$ eV and density $\rho = 0.1$ g/cm^3. The table shows that in the present case accounting for the exchange effects leads to a change of about 15 % in the average occupation numbers for the first nonfilled shell with $n = 3$. The average charge of the ion in the Hartree model is $Z_0 = 14.06$, while in

the Hartree-Fock-Slater model it is $Z_0 = 13.85$.

4.1.2 Periodic boundary conditions in the average spherical cell approximation

When the density of matter is increased, a certain number of electrons may occupy states of the intermediate group (see § 2.1), and so in the general case one has to take into account the band structure of the energy spectrum of electrons. The bands were approximately accounted for in the average-atom model by B. F. Rozsnyai [188] via two boundary conditions, which define the widths of the bands:

$$\begin{cases} R_{n\ell}^{(1)}(r_0) = 0 & \text{for the lower edge of the band,} \\ \left.\dfrac{d}{dr}\left(\dfrac{R_{n\ell}^{(2)}(r)}{r}\right)\right|_{r=r_0} = 0 & \text{for the upper edge of the band.} \end{cases} \tag{4.12}$$

In addition to conditions (4.12), we need to know the distribution of electrons over energies within the band, i.e., the *density of states*, which we will denote by $w(\varepsilon)$ (see § 2.3). In [188], in addition to (4.12), it was assumed that

$$w(\varepsilon) = \frac{dN(\varepsilon)}{d\varepsilon} \sim \sqrt{\varepsilon - \varepsilon_1}, \tag{4.13}$$

where ε_1 is the lower edge of the band.

The approximation (4.12), (4.13) is valid only for sufficiently narrow bands, i.e., when their width is considerably smaller than the distance between neighboring levels. In the case of wide bands this approximation may lead to incorrect results for the position of the bands, as well as for their occupancy (see [152]).

More precise results are provided by an approach based on the Wigner-Zeitz cell method, well known in solid state physics [239, 110]. The utilization of this approach is justified for sufficiently high densities. Let us remark that for small densities the results are practically independent of the boundary conditions obtained in this manner, and hence these conditions can be used for all values of the temperature and density. If the calculations yield a band width that is much smaller than the distance between energy levels, then such states can be regarded as discrete.

Since the physical conditions do not change from cell to cell, in the formulation of the boundary conditions corresponding to the average potential $V(r)$ it is natural to start from the assumption that the structure is periodic in any distinguished direction and take as boundary conditions for the electron wave functions $\Psi_\alpha(\vec{r})$ the conditions resulting from translational symmetry. For $r = r_0$ we have

$$|\Psi_\alpha(\vec{r})|^2 = |\Psi_\alpha(-\vec{r})|^2.$$

It follows that

$$\Psi_\alpha(\vec{r}) = e^{i\vec{k}\vec{r}}\,\Phi_{\alpha\vec{k}}(\vec{r}),$$

where $\vec{k}$ is an arbitrary real constant vector (quasi-momentum) and $\Phi_{\alpha\vec{k}}(\vec{r})$ is a periodic function. From the periodicity condition for the function $\Phi_{\alpha\vec{k}}(\vec{r})$ and the continuity of its derivative we obtain the boundary conditions for the function $\Psi_\alpha(\vec{r})$ (see also [170, 223]):

$$\begin{cases} e^{-i\vec{k}\vec{r}}\,\Psi_\alpha(\vec{r}) = e^{i\vec{k}\vec{r}}\,\Psi_\alpha(-\vec{r}), & \text{if } |\vec{r}| = r_0, \\ e^{-i\vec{k}\vec{r}}\,\dfrac{\partial}{\partial r}\Psi_\alpha(\vec{r}) = -e^{i\vec{k}\vec{r}}\,\dfrac{\partial}{\partial r}\Psi_\alpha(-\vec{r}), & \text{if } |\vec{r}| = r_0. \end{cases} \tag{4.14}$$

The effective reciprocal cell in $\vec{k}$-space is also spherical in this approximation, and its volume is equal to $(2\pi)^3/\left(\frac{4}{3}\pi r_0^3\right)$. It follows that the quantity $|\vec{k}|\cdot r_0$, henceforth denoted by k, can take values in the interval

$$0 \le k \le k_0 = \left(\frac{9\pi}{2}\right)^{1/3} \approx 2.418.$$

We will seek the wave function $\Psi_\alpha(\vec{r})$ for a given value of k as a superposition of functions (4.1), setting $\varepsilon = \varepsilon(k)$ and taking into account that the projection m of the orbital momentum ℓ on the z-axis is conserved (here it is convenient to direct the polar axis z along the quasi-momentum vector $\vec{k}$):

$$\Psi_\alpha(\vec{r}) = \sum_{\ell' \ge |m|} i^{\ell'} A_{n\ell m,\ell'}(k)\widetilde{\Psi}_{\varepsilon\ell' m}(\vec{r}). \tag{4.15}$$

Here the symbol $\alpha = n\ell m$ denotes the set of quantum numbers that determine the state of the electron for the given value of k (the numbers n and ℓ determine the energy level to which the band shrinks when the density of matter is decreased, see below); the factor $i^{\ell'}$ is introduced to make the coefficients $A_{n\ell m,\ell'}(k)$ real.

As computations have shown, the coefficients $A_{n\ell m,\ell'}(k)$ in the expansion (4.15) decrease when ℓ' increases (see, e.g., Table 4.2). In practice one can neglect the electron states with orbital momentum ℓ', for which there is no classical domain of motion within the limits of the average atom cell. Accordingly, in what follows we shall take $\ell' \le \ell_{\max}$.

The wave functions $\Psi_\alpha(\vec{r})$ are normalized as usual

$$\int\limits_{(|\vec{r}|<r_0)} |\Psi_\alpha(\vec{r})|^2\, d\vec{r} = 1.$$

This condition can be satisfied by requiring that

Table 4.2: Dependence of the coefficients $A_{n\ell m,\ell'}(k)$ on ℓ' for various values of k. The results are shown for iron in the TF potential with $T = 0$ and $\rho = 7.85$ g/cm^3 ($m = 0,\ n = 4,\ \ell = 2$)

ℓ' \ k	0.000	0.500	1.000	1.500	2.000	2.418
0	0.000	−0.081	−0.270	0.490	0.720	0.900
1	0.000	0.390	0.620	−0.660	−0.530	−0.084
2	1.000	−0.890	−0.660	0.400	0.170	0.170
3	0.000	−0.190	−0.330	0.380	0.360	0.330
4	0.000	0.017	0.065	−0.130	−0.190	−0.210
5	0.000	−0.001	−0.008	0.030	0.067	0.092
6	0.000	0.000	0.000	−0.005	−0.017	−0.028
7	0.000	0.000	0.000	0.000	0.004	0.009
8	0.000	0.000	0.000	0.000	0.000	0.001

$$\int_0^{r_0} R^2_{\varepsilon\ell'}(r)\,dr = 1, \qquad \sum_{\ell'=|m|}^{\ell_{\max}} A^2_{n\ell m,\ell'} = 1.$$

The coefficients $A_{n\ell m,\ell'}(k)$ can be found from the boundary conditions (4.14). In the manipulations below we will use the fact that

$$\widetilde{\Psi}_{\varepsilon\ell' m}(-\vec{r}) = \frac{(-1)^m}{r} R_{\varepsilon\ell'}(r) Y_{\ell' m}(\pi - \vartheta, \pi + \varphi) = \frac{(-1)^{\ell'+m}}{r} R_{\varepsilon\ell'}(r) Y_{\ell' m}(\vartheta, \varphi).$$

Substituting the expansion (4.15) in the conditions (4.14), multiplying the result by $i^{-\ell''} Y^*_{\ell'' m}(\vartheta, \varphi)$ and integrating over the angular variables, we obtain

$$\sum_{\ell''=|m|}^{\ell_{\max}} A_{n\ell m,\ell''}(k) a_{m\ell'\ell''}(k) g_{\ell'\ell''}(\varepsilon) = 0 \quad (\ell' = |m|, |m|+1, \ldots, \ell_{\max}), \tag{4.16}$$

where

$$g_{\ell'\ell''}(\varepsilon) = \begin{cases} R_{\varepsilon\ell''}(r_0), & \text{if } \ell' \text{ is odd,} \\ \dfrac{d}{dr}\left(\dfrac{R_{\varepsilon\ell''}(r)}{r}\right)\Bigg|_{r=r_0}, & \text{if } \ell' \text{ is even.} \end{cases} \tag{4.17}$$

The coefficients

$$a_{m\ell'\ell''}(k) = i^{\ell''-\ell'} \int Y_{\ell'' m}(\vartheta, \varphi) Y^*_{\ell' m}(\vartheta, \varphi) e^{-ik\cos\vartheta}\, d\Omega \tag{4.18}$$

can be expressed in terms of the Clebsch-Gordan coefficients by decomposing $e^{-ik\cos\vartheta}$ into spherical harmonics [220] (see also Subsection 4.2.2).

The system (4.16) has nontrivial solutions if its determinant is equal to zero. This condition yields the possible values of the energy $\varepsilon_{n\ell m}(k)$. Since for k=0 we have

$$a_{m\ell'\ell''}(0) = \delta_{\ell'\ell''},$$

equation (4.16) with k=0 will be satisfied if $R_{\varepsilon\ell'}(r_0) = 0$ (when ℓ' is odd) or $d(r^{-1}R_{\varepsilon\ell'}(r))/dr)\big|_{r=r_0} = 0$ (when ℓ' is even) for some $\ell' = \ell$ (and then we have $A_{n\ell m,\ell''}(0) = \delta_{\ell\ell''}$). The values of ε that satisfy these conditions will be denoted by $\varepsilon_{n\ell}$, where n is the principal quantum number, corresponding to the fact that function $R_{\varepsilon\ell}(r)$ has $n-\ell-1$ zeroes in the interval $(0, r_0)$. In connection with this the quantum numbers n and ℓ determine the energy $\varepsilon_{n\ell}$, which corresponds to the edge of the band at $k = 0$. It is precisely these quantum numbers that define the discrete level to which the band shrinks when the density is decreased.

4.1.3 The electron density and the atomic potential in the Hartree-Fock-Slater model with bands

For given values of n, ℓ and m, the number $N(\varepsilon, k)$ of electrons found in states ranging from k to $k+dk$ is equal—up to a normalization constant—to $N(\varepsilon)4\pi k^2\, dk$, where

$$N(\varepsilon) = \frac{2}{1 + \exp\left(\dfrac{\varepsilon - \mu}{\theta}\right)}, \qquad \varepsilon = \varepsilon_{n\ell m}(k).$$

Recalling the normalization in the $\vec{k}$-phase space, we finally obtain

$$N(\varepsilon, k) = N(\varepsilon)\frac{4\pi k^2\, dk}{(4/3)\pi k_0^3}.$$

Since $\varepsilon_{n\ell m} = \varepsilon_{n\ell,-m}$ and $A_{n\ell m,\ell'} = A_{n\ell,\,-m,\ell'}$, it is sufficient to carry out the calculations only for $m \geq 0$.

Thus, in order to approximately account for the band structure of the energy spectrum of the intermediate group of electrons it is necessary to modify the HFS model introduced in Subsection 4.1.1. The calculation of the energy spectrum and wave functions by means of formulas (4.15), (4.16) is done for values of the electron energy $\varepsilon < \varepsilon_0$. In the calculation of the electron density $\rho_1(r)$ the summation over bound states is carried out according to formula (4.4). Summation over the states of the intermediate group must be replaced by summation over the quantum numbers n, ℓ, m, ℓ' and integration over the quasi-momentum $\vec{k}$. The formula for $\rho_1(r)$ takes on the form

$$\rho_1(r) = \frac{1}{4\pi r^2}\left[\sum_{n,\ell} N_{n\ell}R_{n\ell}^2(r) + \sum_{n,\ell,m,\ell'} \int_0^{k_0} N(\varepsilon)A_{n\ell m,\ell'}^2(k)R_{\varepsilon\ell'}^2(r)\frac{3k^2\, dk}{k_0^3}\right]. \tag{4.19}$$

Here in the first term in the square brackets summation over n, ℓ is carried out over the bound states with energy $\varepsilon_{n\ell} < \varepsilon_0$. The summation over n, ℓ, m, ℓ' and the integration over k in the second term is carried out for the intermediate-group states with energies $\varepsilon_{n\ell m}(k) < \varepsilon_0$.

The formulas for $\rho_2(r)$ and $V(r)$ remain the same as above (see formulas (4.5)–(4.9)), whereas equation (4.10) for the determination of ε_0 becomes somewhat more complex:

$$\frac{8\sqrt{2}}{3\pi}\int_0^{r_0}[\max\{0,\varepsilon_0+V(r)\}]^{3/2}r^2\,dr = \sum_{n,\ell}2(2\ell+1)+\sum_{n,\ell,m}\int_0^{k_0}\frac{6k^2\,dk}{k_0^3}. \tag{4.20}$$

Here, as in (4.19), summation over the quantum numbers $n\ell$, $n\ell m$ and integration over the quasi-momentum k are carried out over states with energy $\varepsilon < \varepsilon_0$.

4.1.4 The relativistic Hartree-Fock-Slater model

To account for relativistic effects, instead of the Hamiltonian (3.18) of the electron system we follow I. P. Grant's work [74] and consider the Hamiltonian

$$\widehat{H} = \sum_i \widehat{H}_i^{\mathrm{D}} + \frac{1}{2}\sum_{i\neq j}\left[\frac{1}{|\vec{r}_i-\vec{r}_j|}+\widehat{B}_{ij}\right]+\widehat{H}_{\mathrm{em}}+\widehat{H}_{\mathrm{int}}, \tag{4.21}$$

where

$$\widehat{H}_i^{\mathrm{D}} = c\widehat{\vec{\alpha}}\widehat{\vec{p}}+c^2\widehat{\beta}-V_a(\vec{r}_i),$$

$$\widehat{B}_{ij} = -\frac{1}{2|\vec{r}_i-\vec{r}_j|}\left[\widehat{\vec{\alpha}}_i\cdot\widehat{\vec{\alpha}}_j+\frac{\left(\widehat{\vec{\alpha}}_i\cdot(\vec{r}_i-\vec{r}_j)\right)\left(\widehat{\vec{\alpha}}_j\cdot(\vec{r}_i-\vec{r}_j)\right)}{|\vec{r}_i-\vec{r}_j|^2}\right].$$

The corrections $\widehat{H}_{\mathrm{em}}$ and $\widehat{H}_{\mathrm{int}}$ are connected with the effects of vacuum polarization and interaction with the radiation field. Also, $\widehat{\vec{p}} = -i\vec{\nabla}$ and the operators $\widehat{\vec{\alpha}}$ and $\widehat{\beta}$ can be expressed in terms of the Pauli matrices $\vec{\sigma}$ (see Appendix, § A.2).

The utilization of the one-particle approximation in relativistic quantum theory runs into serious methodological difficulties, since, rigorously speaking, the concept of a single particle in a relativistic system with interaction cannot be defined. Nevertheless, by introducing additional approximations one can use a one-particle approximation analogous to that considered in § 3.2. In this approach the multi-electron wave functions are represented as an expansions in one-electron wave functions with determined values of the total momentum j and its projection m on the z-axis.

To obtain Hartree-Fock-Slater equations incorporating relativistic effects it is necessary to proceed in much the same way as in Chapter III, replacing the operator $-(1/2)\Delta$ by $-ic\widehat{\vec{\alpha}}\vec{\nabla}+c^2\widehat{\beta}$, in accordance with (4.21). In the rather crude approximation of the average atom it does not make sense to introduce corrections connected with magnetic interactions (the Breit corrections $\widehat{B}_{ij}$), or the polarization and radiative corrections. If necessary, these corrections can be incorporated in the framework of perturbation theory when computations of detailed characteristics of an atom are carried out, for example, in the computation of the positions of spectral lines.

Proceeding as indicated, the equations of the relativistic Hartree-Fock-Slater model are derived from the equations (3.63)–(3.67) upon replacing the Schrödinger equation by the Dirac equation

$$\left[-ic\widehat{\vec{\alpha}}\vec{\nabla}+c^2\widehat{\beta}-V(\vec{r})\right]\psi_\nu(q)=\varepsilon_\nu\psi_\nu(q), \tag{4.22}$$

where the symbol ν stands for the set of quantum numbers n, ℓ, j, m and $\psi_\nu(q)$ is a 4-component spinor. For n_ν one has the same formula (3.66).

The distribution function $n(q,\vec{p})$ of the continuum electrons also becomes relativistic; however, in it, as in the exchange term, we will neglect relativistic effects, since the influence of the latter on these quantities are small.

According to (A.41)-(A.42), in a spherical cell we have

$$\psi_{n\ell jm}(q)=\frac{1}{r}\begin{cases} F_{n\ell j}(r)\sum_{m_\ell,m_s} C^{jm}_{\ell m_\ell\,\frac{1}{2}m_s} i^\ell(-1)^{m_\ell}Y_{\ell m_\ell}(\vartheta,\varphi)\chi_{sm_s}(\sigma), \\[2ex] (-1)^k G_{n\ell j}(r)\sum_{m_{\ell'},m_s} C^{jm}_{\ell' m_{\ell'}\,\frac{1}{2}m_s} i^{\ell'}(-1)^{m_{\ell'}}Y_{\ell' m_{\ell'}}(\vartheta,\varphi)\chi_{sm_s}(\sigma), \end{cases}$$

where $F_{n\ell j}(r)$ and $G_{n\ell j}(r)$ are the large and respectively the small radial parts of the wave function, j is the momentum quantum number of the electron and m is the projection of j on the z-axis ($\ell=j\pm 1/2$, $\ell'=2j-\ell$, $k=(\ell-\ell'+1)/2$).

As a rule, in computations one can solve the Dirac equation for the inner shells, where relativistic effects are the most significant, while in the remaining cases it is convenient to use the Pauli approximation [49]. This allows one to use the same method for imposing periodic boundary conditions as in Subsection 4.1.2. Indeed, in the Pauli approximation for the radial parts of the wave functions, like in the nonrelativistic case, one solves the Schrödinger equation by incorporating relativistic corrections in the potential $V(r)$. To deal with the inner electrons, for which one uses the Dirac equation, one can consider that the relativistic one-electron energy levels are discrete and neglect the band structure of the spectrum for these levels.

Following the approach described above, for the inner shells the Schrödinger equation (4.2) is replaced by the Dirac equations

$$\begin{cases} \dfrac{dF_{n\ell j}(r)}{dr}=-\dfrac{\kappa}{r}F_{n\ell j}(r)+\alpha\left[\dfrac{2}{\alpha^2}+\varepsilon_{n\ell j}+V(r)\right]G_{n\ell j}(r), \\[3ex] \dfrac{dG_{n\ell j}(r)}{dr}=\dfrac{\kappa}{r}G_{n\ell j}(r)-\alpha\left[\varepsilon_{n\ell j}+V(r)\right]F_{n\ell j}(r), \end{cases} \tag{4.23}$$

where $\kappa = -2(j - \ell)(j + 1/2)$. For higher levels, among them the states of the intermediate group, one applies the Pauli-type approximation (see [49], Chapter 7):

$$-\frac{1}{2}R''_{\varepsilon\ell}(r) + \left[-V(r) - V_{\mathrm{P}}(r) + \frac{\ell(\ell+1)}{2r^2}\right] R_{\varepsilon\ell}(r) = \varepsilon R_{\varepsilon\ell}(r), \tag{4.24}$$

$$V_{\mathrm{P}}(r) = \frac{\alpha^2}{2}\left\{\left[\varepsilon + V(r)\right]^2 - \frac{\delta_{\ell 0}}{2}\,\frac{V'(r)\left[R'_{\varepsilon\ell}(r)/R_{\varepsilon\ell}(r) - 1/r\right]}{1 + \alpha^2\left[\varepsilon + V(r)\right]/2}\right\}. \tag{4.25}$$

4.2 An iteration method for solving the Hartree-Fock-Slater system of equations

4.2.1 Algorithm basics

In practice it may be required to carry out computations of the equation of state and the photon absorption coefficients in a wide range of temperatures and densities (about 500–1000 (T, ρ)-cases for each substance). Therefore, it is very important to construct an efficient algorithm for solving the system of equations of the Hartree-Fock-Slater model or of its modifications considered in § 4.1.

In the solution of the Hartree-Fock-Slater system of equations the iteration process that computes the successive values $V^{(s)}(r)$ is similar to the iteration process used to solve the Hartree system of equations, described in § 3.1. Let us mention some specific aspects of the computations.

The computations with no accounting for the band structure present no difficulties, and the efficiency of the computational algorithms is not inferior to that of the algorithms for the Hartree model. For very small densities one has to pay some attention to the choice of the grid; however, the computations can be simplified considerably if one resorts to the uniform electron density approximation for the electrons of continuum (see Subsection 4.2.4 below). The most labor-consuming is the computation of the band structure of the energy spectrum and wave functions for the electrons of the intermediate group of states. The corresponding algorithms will be described in detail in Subsection 4.2.2.

An important factor for the convergence of the iterations for the potential in the Hartree model, as well as for the Hartree-Fock-Slater system of equations is the condition of preservation of charge neutrality of the average atom cell throughout the iteration process. An analogous role in the convergence of iterations is played by the supplementary condition (4.10) for determining ε_0 (or condition (4.20 for the variant of the model that accounts for the band structure). The role played by these conditions in the convergence of the iterations becomes clear in the case of small densities. In particular, it turns out that the number of discrete levels intervening in formula (4.4) must not change in the iteration process.

4.2.2 Computation of the band structure of the energy spectrum

In the approximation adopted above, the band structure of the energy spectrum of electrons in matter, i.e., the dependence of allowed energy values ε on the quasi-momentum k, is determined, for given n, ℓ and m, from the condition that the system of equations $(m \geq 0)$

$$\sum_{\ell''=m}^{\ell_{\max}} A_{n\ell m,\ell''} a_{m\ell'\ell''}(k) g_{\ell'\ell''}(\varepsilon) = 0 \quad (\ell' = m, m+1, \ldots, \ell_{\max}) \tag{4.26}$$

admits a nontrivial solution $A_{n\ell m,\ell'}$. Such a condition is that the determinant of this system be equal to zero

$$D(k, \varepsilon(k)) = \det \|a_{m\ell'\ell''}(k) g_{\ell'\ell''}(\varepsilon)\| = 0. \tag{4.27}$$

Before we look at a method for calculating $\varepsilon(k)$, let us describe the process of computing the quantities $a_{m\ell'\ell''}(k)$ and $g_{\ell'\ell''}(\varepsilon)$. To compute the coefficients $a_{m\ell'\ell''}(k)$ by means of formula (4.18), we use the expansion of a plane wave into Legendre polynomials, which yields

$$e^{-ik\cos\vartheta} = \sqrt{\frac{\pi}{2k}} \sum_{L=0}^{\infty} (-i)^L (2L+1) P_L(\cos\vartheta)\, J_{L+1/2}(k), \tag{4.28}$$

where $P_L(x)$ are Legendre polynomials and $J_{L+1/2}(k)$ are Bessel functions.

Substituting expression (4.28) in formula (4.18) we obtain

$$a_{m\ell'\ell''}(k) = \sqrt{\frac{\pi}{2k}} \sum_L (-1)^{\frac{L+\ell''-\ell'}{2}} (2L+1) J_{L+1/2}(k) D_{m\ell'\ell''L}, \tag{4.29}$$

where

$$D_{m\ell'\ell''L} = \int P_L(\cos\vartheta) Y^*_{\ell'm}(\vartheta,\varphi) Y_{\ell''m}(\vartheta,\varphi)\, d\Omega = \sqrt{\frac{(2L+1)(2\ell''+1)}{(2\ell'+1)}}\, C^{\ell'0}_{L0\,\ell''0} C^{\ell'm}_{L0\,\ell''m}.$$

Formulas for $C^{\ell'm}_{L0\,\ell''m}$ can be found, for example, in [220]. In the computation of $D_{m\ell'\ell''L}$ one replaces $n!$ by the preliminarily computed numbers $n!/10^n$, which accelerates the computations and allows one to avoid operating with very large numbers.

For the summation over L in (4.29) it is better to go from higher to lower values of L, starting with $L = \ell' + \ell''$. This choice of summation order reduces the computational errors, since the terms of the series (4.29) decrease when L

increases. In this connection, for the calculation of the Bessel functions $J_{L+1/2}(k)$ it is convenient to apply the recursion relation

$$J_{n-1/2}(k) = \frac{2n+1}{k} J_{n+1/2}(k) - J_{n+3/2}(k) \quad (n = \ell' + \ell'' - 1, \ell' + \ell'' - 2, \ldots). \tag{4.30}$$

The calculation using (4.30) is stable against round-off errors. To be able to apply relation (4.30) one needs to know the values of $J_{L+1/2}(k)$ for $L = \ell' + \ell''$ and $L = \ell' + \ell'' - 1$. The latter can be computed by using series expansions for the Bessel functions. Note that, for large L and $k \ll 1$, the quantity $J_{L+1/2}(k)$ may be negligibly small, because for small k

$$J_{L+1/2}(k) \cong \left(\frac{k}{2}\right)^{L+1/2} \frac{(L+1)!}{\sqrt{\pi}\,(2L+2)!}.$$

In this case formula (4.30) is applied for $L \leq L_{\max} < \ell' + \ell''$, and the values of $J_{L+1/2}(k)$ for $L > L_{\max}$ are taken to be zero (the value of $L = L_{\max}$ is determined, for example, by the condition $J_{L+1/2}(k) < 10^{-20}$).

When one solves the Hartree-Fock-Slater system of equations it is necessary to know how to compute the wave functions $R_{\varepsilon\ell}(r)$ and their derivatives $R'_{\varepsilon\ell}(r)$ in a wide range of values of ε and ℓ. To compute $R_{\varepsilon\ell}(r)$ we use quadratic interpolation with respect to energy, based on the wave functions computed earlier on some mesh with respect to ε. To choose the mesh we use the results of the computation of $\varepsilon(k)$ in the preceding iteration. Thus, if the previously computed energy values $\varepsilon^{(s)}(k)$ of the s-th iteration with respect to the potential obey the inequality

$$\varepsilon_a^{(s)} \leq \varepsilon^{(s)}(k) \leq \varepsilon_b^{(s)},$$

then to carry out the $(s+1)$-st iteration we choose the mesh values ε_i so as to guarantee sufficient accuracy in the interval $(\varepsilon_a^{(s)} - \Delta\varepsilon, \varepsilon_b^{(s)} + \Delta\varepsilon)$, where one denotes $\Delta\varepsilon = 0.1|\varepsilon_b^{(s)} - \varepsilon_a^{(s)}|$.

For the first iteration, the values $\varepsilon_a^{(0)}$ and $\varepsilon_b^{(0)}$ are determined using the boundary conditions in the approximation (4.12) and the phase method for solving the Schrödinger equation (see Appendix, § A.5). One of these two values ($\varepsilon_a^{(0)}$ or $\varepsilon_b^{(0)}$, depending of the orbital number ℓ) coincides with $\varepsilon_{n\ell m}^{(0)}(0)$.

As ℓ grows, the dependence of the wave functions $R_{\varepsilon\ell}(r)$ on the energy ε becomes simpler, and consequently for large ℓ it is reasonable to use simpler formulas. As it turns out, in this case linear interpolation with respect to energy provides sufficient accuracy.

To compute the allowed energy values $\varepsilon_{n\ell m}(k)$, i.e., to solve equation (4.27), the secant method proved to be the most efficient. The determinant $D(k,\varepsilon)$ was calculated by the Gauss elimination method with choice of the principal element [222]. The high efficiency of the method was achieved thanks to the simplicity of the computation of the determinant $D(k,\varepsilon)$ as a result of using the previously computed coefficients $a_{m\ell'\ell''}(k)$ and the availability of simple interpolation formulas for $R_{\varepsilon\ell}(r)$.

4.2.3 Computational results

Table 4.3 lists values of residuals in the computation of the Hartree-Fock-Slater potential for aluminum with density $\rho = 0.271$ g/cm^3 and various values of the temperature. The residual was calculated by means of the formula $\Delta^s = \max_r |V^{(s)}(r) - F[V^{(s)}(r)]|$ (compare with (3.11)). For the temperature $T = 1$ eV and the same density ($\rho = 0.271$ g/cm^3), Figure 4.1 shows the graphs of the potential $V^{(s)}(r)$, multiplied by the radius r, for the iteration number $s = 0, 1, 2, \ldots$.

Table 4.3: Residual in the iterations for the Hartree-Fock-Slater potential for aluminum with density $\rho = 0.271$ g/cm^3 and various temperatures: $T = 0$ and T =1; 10; 100; 1000 eV

s	$T = 0$	$T = 1$ eV	$T = 10$ eV	$T = 100$ eV	$T = 1000$ eV
0	$7.996 \cdot 10^{-1}$	$8.391 \cdot 10^{-1}$	$1.108 \cdot 10^{0}$	$3.229 \cdot 10^{-1}$	$1.099 \cdot 10^{-1}$
1	$2.337 \cdot 10^{-1}$	$2.229 \cdot 10^{-1}$	$3.438 \cdot 10^{-1}$	$3.005 \cdot 10^{-2}$	$1.916 \cdot 10^{-2}$
2	$5.506 \cdot 10^{-2}$	$4.537 \cdot 10^{-2}$	$8.002 \cdot 10^{-2}$	$6.754 \cdot 10^{-3}$	$1.022 \cdot 10^{-4}$
3	$2.403 \cdot 10^{-2}$	$2.409 \cdot 10^{-2}$	$2.850 \cdot 10^{-2}$	$2.429 \cdot 10^{-3}$	$1.372 \cdot 10^{-5}$
4	$1.030 \cdot 10^{-2}$	$9.990 \cdot 10^{-3}$	$1.449 \cdot 10^{-2}$	$1.728 \cdot 10^{-4}$	$6.851 \cdot 10^{-7}$
5	$5.548 \cdot 10^{-3}$	$5.032 \cdot 10^{-3}$	$7.691 \cdot 10^{-3}$	$4.129 \cdot 10^{-5}$	
6	$2.281 \cdot 10^{-3}$	$2.220 \cdot 10^{-3}$	$3.967 \cdot 10^{-3}$	$1.175 \cdot 10^{-6}$	
7	$9.943 \cdot 10^{-4}$	$1.087 \cdot 10^{-3}$	$2.015 \cdot 10^{-3}$		
8	$5.050 \cdot 10^{-4}$	$4.865 \cdot 10^{-4}$	$1.016 \cdot 10^{-3}$		
9	$2.425 \cdot 10^{-4}$	$2.516 \cdot 10^{-4}$	$5.093 \cdot 10^{-4}$		
10	$1.176 \cdot 10^{-4}$	$1.116 \cdot 10^{-4}$	$2.547 \cdot 10^{-4}$		
11	$5.127 \cdot 10^{-5}$	$5.300 \cdot 10^{-5}$	$1.272 \cdot 10^{-4}$		
12	$2.572 \cdot 10^{-5}$	$2.429 \cdot 10^{-5}$	$6.350 \cdot 10^{-5}$		
13	$1.210 \cdot 10^{-5}$	$1.268 \cdot 10^{-5}$	$3.171 \cdot 10^{-5}$		
14	$6.434 \cdot 10^{-6}$	$6.124 \cdot 10^{-6}$	$4.524 \cdot 10^{-6}$		

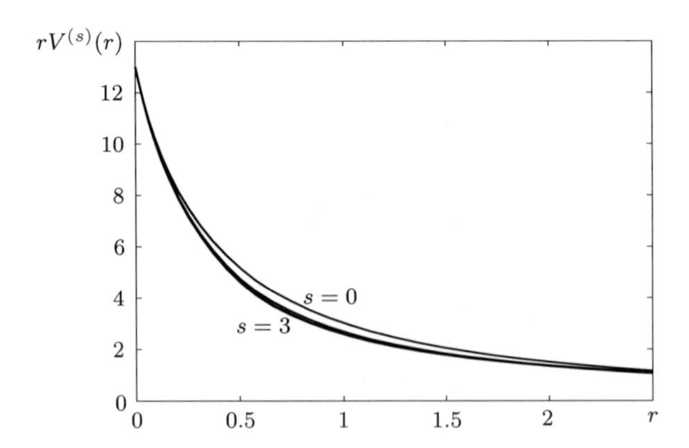

Figure 4.1: Successive iterations of the potential $rV^{(s)}(r)$ for aluminum with temperature $T = 1$ eV and density $\rho = 0.271$ g/cm^3

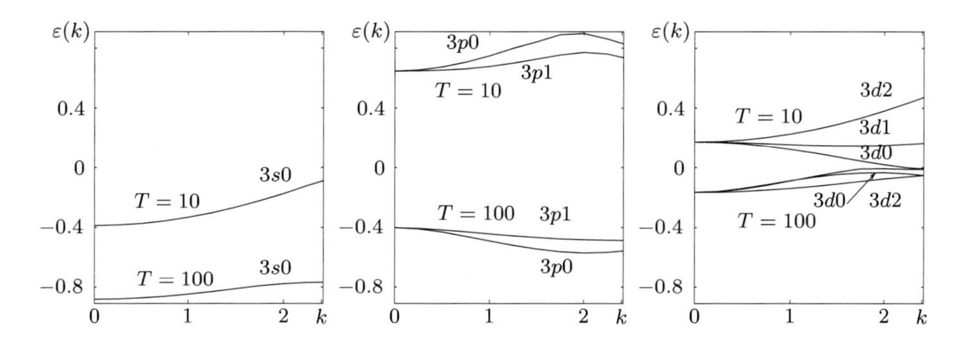

Figure 4.2: Dependence of the allowed energies $\varepsilon(k) = \varepsilon_{n\ell m}(k)$, in atomic units, on the quasi-momentum k for aluminum with density $\rho = 2.71$ g/cm^3 and temperature $T = 10$ eV and $T = 100$ eV

Practically, already the third iteration yields values of the potential close to the solution. A high accuracy ($\Delta^s < 10^{-5}$) is required only for low temperatures in the calculation of internal energy of matter, which in this case is determined by a difference of large quantities.

Figures 4.2 and 4.3 display results of the computation of allowed energies $\varepsilon_{n\ell m}(k)$ for aluminum for various temperatures and densities. As one can see from these figures, the computation of the band structure is necessary only for high densities and low temperatures.

This is particularly transparent in Figure 4.4, which shows the upper and lower edges of the allowed energy bands for aluminum, i.e., $\min_{m,k} \varepsilon_{n\ell m}(k)$ and $\max_{m,k} \varepsilon_{n\ell m}(k)$, in their dependence on the density ρ in g/cm^3. The computations

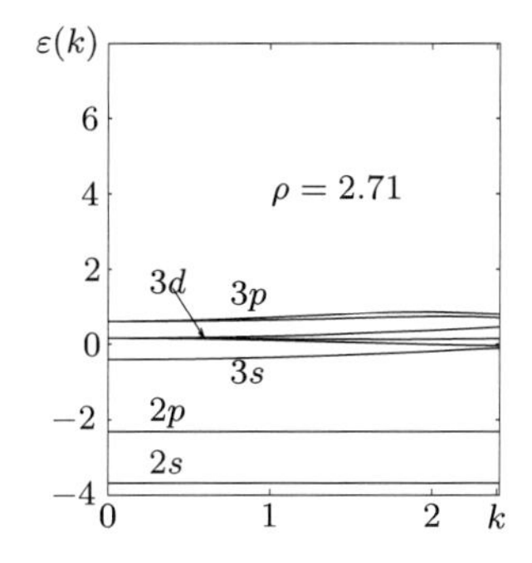

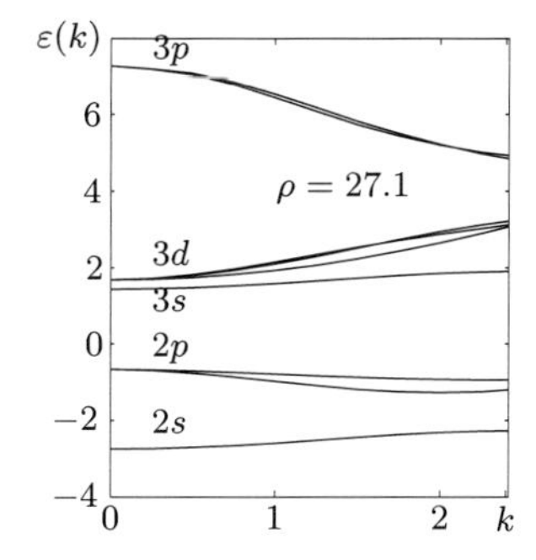

Figure 4.3: Allowed energy values $\varepsilon(k) = \varepsilon_{n\ell m}(k)$ for aluminum with temperature $T = 1$ eV and densities $\rho = 2.71$ g/cm^3 and $\rho = 27.1$ g/cm^3

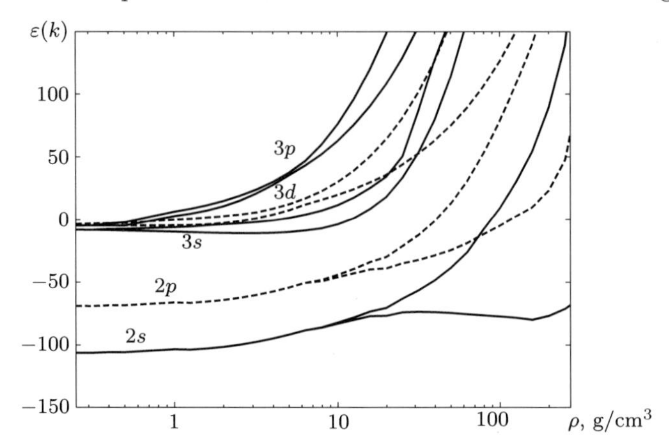

Figure 4.4: Dependence of the upper and lower edges of the allowed energy bands for aluminum ($\min_{m,k} \varepsilon_{n\ell m}(k)$ and $\max_{m,k} \varepsilon_{n\ell m}(k)$, in atomic units) on the density ρ in g/cm^3 at temperature $T = 0$

were carried out for temperature $T = 0$. When the temperature is increased the bands shrink rapidly, and the corresponding levels become discrete.

4.2.4 The uniform-density approximation for free electrons in the case of a rarefied plasma

For low densities of matter practically the entire average atom cell is filled by a homogeneous gas of electrons and in order to calculate self-consistent atomic potential one needs to use special grids, sufficiently fine near the nucleus for $r < \widetilde{r}$, where $\widetilde{r}$ is the dimension of the ion core, and less fine in the remaining domain $\widetilde{r} < r < r_0$ (where r_0 is the radius of the average atom cell). Utilization of special grids complicates the computational scheme and require a persistent error control, which makes massive computations difficult. Moreover, the constraint on the density is not removed. Therefore, it is reasonable to use approximate analytic solutions which are valid in the present case.

Considering that in a plasma with small density the free electrons are uniformly distributed far from nuclei, we set the density of electrons $\rho(r)$ for $r > \tilde{r}$ equal to a constant value $\rho(r) = \tilde{\rho} = \rho(\tilde{r})$.

The magnitude of $\tilde{r}$ is determined by the degree of ionization of the plasma and the occupation of the ion shells. One can set approximately (see (A.36))

$$\tilde{r} = \frac{2n_{\max}^2}{Z_0},$$

where $n_{\max}$ is the principal quantum number of the last partially filled shell and Z_0 is average ion charge.

Near the ion core, i.e., for $r < \tilde{r}$, the distribution of electrons and the self-consistent potential will be calculated by means of formulas (4.3)–(4.9). In the domain $r > \tilde{r}$, using the constancy of the electron density, we have

$$V(r) = V_{\rm c}(r) + V_{\rm ex}(r), \quad \tilde{r} < r < r_0, \tag{4.31}$$

where

$$V_{\rm c}(r) = \frac{4\pi}{3}\frac{\tilde{\rho} r_0^3}{r}\left[1 - \frac{3}{2}\frac{r}{r_0} + \frac{1}{2}\left(\frac{r}{r_0}\right)^3\right]. \tag{4.32}$$

The exchange correction $V_{\rm ex}$ is calculated using formula (4.9) with $\rho(r) = \tilde{\rho}$.

Thus, we must find the solution of the Hartree-Fock-Slater equation only for $r < \tilde{r}$, taking into account the fact that in this domain

$$\int_0^{\tilde{r}} 4\pi r^2 \rho(r)\, dr = Z - \tilde{Z},$$

where $\tilde{Z} = 4\pi(r_0^3 - \tilde{r}^3)\tilde{\rho}/3$.

Expressions (4.31) and (4.32) will be used below to calculate matrix elements at low densities whenever we need to know the electron density and the atomic potential in the entire atom cell.

4.3 Solution of the Hartree-Fock-Slater system of equations for a mixture of elements

4.3.1 Problem setting

For a mixture of N components with mass fractions m_i $(i = 1, 2, \dots N)$ at a specified temperature θ and average density ρ, the radius of the average atom cell of the i-th component is given by the formula (see Subsection 1.3.1):

$$r_{0i} = 1.388\left(\frac{A_i}{\rho_i}\right)^{1/3},$$

where ρ_i is the intrinsic density and A_i is the atomic mass. The concentration of the i-th component (relative fraction of ions) is

$$c_i = \frac{m_i/A_i}{\sum m_i/A_i}.$$

By formula (1.48),

$$\sum_{i=1}^{N}\frac{m_i}{\rho_i} = \frac{1}{\rho}\sum_{i=1}^{N} m_i, \quad \text{or} \quad \sum_{i=1}^{N}\frac{m_i}{A_i} r_{0i}^3 = (1.388)^3\,\frac{1}{\rho}\sum_{i=1}^{N} m_i. \tag{4.33}$$

Furthermore, we must require the equality of the chemical potentials or electron pressures for each of the components in the mixture. Note that for sufficiently high temperatures and not too large densities it suffices to require the equality of chemical potentials, and then other equilibrium characteristics, namely, the electron pressures and the electron densities on the boundaries of the atom cells will also coincide.

To simplify the exposition, we will consider the equations for a mixture and the computational algorithm for the case of relatively low densities, for which it is not necessary to impose the periodicity conditions (4.14). At the same time, the conditions for the applicability of the cell approximation must be satisfied, i.e., the degree of ionization must be sufficiently high.

In each cell the calculation of the potential $V_i(r)$ for the i-th component of the mixture is done using the electron density

$$\rho_i(r) = \rho_{1i}(r) + \rho_{2i}(r), \tag{4.34}$$

where

$$\rho_{1i}(r) = \sum_{n,\ell,j} \frac{1}{4\pi r^2} N_{n\ell j,i} \left[F^2_{n\ell j,i}(r) + G^2_{n\ell j,i}(r)\right] \quad (\varepsilon_{n\ell j,i} < \varepsilon_{0i}),$$

$$\rho_{2i}(r) = \frac{(2\theta)^{3/2}}{2\pi^2} \int\limits_{y_{0i}}^{\infty} \frac{y^{1/2}\,dy}{1 + \exp\left(y - \dfrac{V_i(r) + \mu_i}{\theta}\right)},$$

$$N_{n\ell j,i} = \frac{2j+1}{1 + \exp\left(\dfrac{\varepsilon_{n\ell j,i} - \mu_i}{\theta}\right)}, \qquad y_{0i} = \max\left\{0, \frac{V_i(r) + \varepsilon_{0i}}{\theta}\right\}.$$

The wave functions $F_{n\ell j,i}(r)$ and $G_{n\ell j,i}(r)$ and the energy values $\varepsilon_{n\ell j,i}$ are obtained by solving the Dirac equations (4.23) with the potentials $V_i(r)$ defined by formulas (4.7)–(4.9), in which one uses the electron density $\rho_i(r)$ and the nucleus charge Z_i for the i-th component of the mixture.

Moreover, for each average atom cell the charge neutrality condition is satisfied:

$$Z_i - Z_{0i}(\mu_i) = \sum_{n,\ell,j} N_{n\ell j,i}(\mu_i) \quad (\varepsilon_{n\ell j,i} < \varepsilon_{0i}), \tag{4.35}$$

where

$$Z_{0i}(\mu_i) = 4\pi \int\limits_0^{r_{0i}} \rho_{2i}(r) r^2\,dr.$$

The thermodynamic equilibrium conditions read

$$\mu_i = \mu_j = \mu \qquad (i \neq j).$$

In the low-density approximation considered here the thermodynamic consistency condition (4.10) will be satisfied provided that the quantities ε_{0i} obey for some $n = n_{\max,i}$ the inequalities

$$\varepsilon_{n,i} < \varepsilon_{0i} < \varepsilon_{n+1,i},$$

where

$$\varepsilon_{n,i} = \frac{\sum\limits_{\ell,j} N_{n\ell j,i}\, \varepsilon_{n\ell j,i}}{\sum\limits_{\ell,j} N_{n\ell j,i}}$$

and the numbers n are such that $N_{n,i} = \sum\limits_{\ell,j} N_{n\ell j,i} \ll 1$ for all components of the mixture ($i = 1, 2, \ldots, N$). The chosen values $n_{\max,i}$ remain the same throughout the computational process (see Subsection 4.1.1).

4.3.2 Iteration scheme

As in the case of a substance consisting of atoms of a single element, for a mixture of elements (see (4.33)–(4.35)) the unknowns of the Hartree-Fock-Slater system of equations include, in addition to $V_i(r)$, $\rho_i(r)$, $\varepsilon_{n\ell j,i}$, $\mu_i = \mu$, the radii r_{0i} of the cells, which complicates the problem considerably.

The HFS system of equations for a mixture can be solved by calculating the self-consistent potentials for each element of the mixture, and then ensuring that the chemical potentials coincide. This may be achieved by using interpolation of preliminarily calculated tables for single elements. Such a scheme, however, is not economic enough, since the volume of computations grows rapidly with the number of components in the mixture.

A more economical scheme can be obtained if, at each iteration carried out to compute the self-consistent atomic potentials, one simultaneously adjusts the values of the radii r_{0i} so that when the self-consistent potentials for the elements of the mixture are obtained, their chemical potentials will be simultaneously adjusted as well [160].

In constructing the iteration scheme we will start from the requirement that the charge neutrality condition (4.35) be satisfied during the iteration process. Starting with given values $r_{0i}^{(s)}$ and $V_i^{(s)}(r)$, we can compute the corresponding energy values $\varepsilon_{n\ell j,i}^{(s)}$. Before we calculate the new electron densities and the new atomic potentials $V_i^{(s+1)}(r)$, we will find new values of $r_{0i}^{(s+1)}$ that are closer to the sought solution, the way we did in the Thomas-Fermi model (see § 1.3).

Let $x_i = r_{0i}^3$; then for any values of x_i that we may choose the charge neutrality condition (4.35) and condition (4.33) must be satisfied, i.e.,

$$B_i(\mu_i, x_i) + Z_{0i}(\mu_i, x_i) - Z_i = 0, \quad i = 1, 2, \ldots, N, \tag{4.36}$$

$$\sum_{i=1}^{N} \frac{m_i}{A_i} x_i - (1.388)^3 \frac{1}{\rho} \sum_{i=1}^{N} m_i = 0, \tag{4.37}$$

where

$$Z_{0i}(\mu_i, x_i) = \frac{2(2\theta)^{3/2}}{\pi} \int_0^{\sqrt[3]{x_i}} r^2\, dr \int_{y_{0i}^{(s)}}^{\infty} \frac{y^{1/2}\, dy}{1 + \exp\left(y - \dfrac{V_i^{(s)}(r) + \mu_i}{\theta}\right)}, \tag{4.38}$$

$$y_{0i}^{(s)} = \max\left\{0; \frac{V_i^{(s)}(r) + \varepsilon_{0i}^{(s)}}{\theta}\right\},$$

$$B_i(\mu_i, x_i) = \sum_{n\ell j} N_{n\ell j,i}(\varepsilon_{n\ell j,i}^{(s)}, \mu_i) \qquad (\varepsilon_{n\ell j,i}^{(s)} < \varepsilon_{0i}^{(s)}), \tag{4.39}$$

$$N_{n\ell j,i}(\varepsilon, \mu) = \frac{2j+1}{1 + \exp\left(\dfrac{\varepsilon - \mu}{\theta}\right)}.$$

Since the potentials $V_i^{(s)}(r)$ are calculated for $r \le r_{0i}^{(s)}$, in order that the expression (4.38) be defined for all x_i, we put $V_i^{(s)}(r) = V_i^{(s)}(r_{0i}^{(s)})$ for $r > r_{0i}^{(s)}$.

We shall solve the system (4.36)–(4.37) in the linear approximation, expanding B_i and Z_{0i} near $x_i = \bar{x}_i = (r_{0i}^{(s)})^3$ and $\mu_i = \bar{\mu}_i$. The quantities $\bar{x}_i$, $\bar{\mu}_i$ satisfy the equations (4.36), (4.37) for the given $V_i^{(s)}(r)$, $\varepsilon_{n\ell j,i}^{(s)}$, $\varepsilon_{0i}^{(s)}$, but fail to satisfy the equality of chemical potentials for all elements in the mixture. In the linear approximation we have

$$B_i(\mu_i, x_i) \cong B_i(\bar{\mu}_i, \bar{x}_i) + \frac{\partial B_i}{\partial \mu_i}(\mu_i - \bar{\mu}_i) + \frac{\partial B_i}{\partial x_i}(x_i - \bar{x}_i),$$

$$Z_{0i}(\mu_i, x_i) \cong Z_{0i}(\bar{\mu}_i, \bar{x}_i) + \frac{\partial Z_{0i}}{\partial \mu_i}(\mu_i - \bar{\mu}_i) + \frac{\partial Z_{0i}}{\partial x_i}(x_i - \bar{x}_i), \tag{4.40}$$

where the derivatives are calculated for $\mu_i = \bar{\mu}_i$ and $x_i = \bar{x}_i$:

$$\frac{\partial Z_{0i}}{\partial \mu_i} = \frac{2(2\theta)^{3/2}}{\pi\theta} \int_0^{\sqrt[3]{\bar{x}_i}} r^2\, dr \times$$

$$\left\{ \frac{1}{2} \int_{y_{0i}^{(s)}}^{\infty} \frac{y^{-1/2} dy}{1 + \exp\left(y - \dfrac{V_i^{(s)}(r) + \bar{\mu}_i}{\theta}\right)} + \frac{(y_{0i}^{(s)})^{1/2}}{1 + \exp\left(y_{0i}^{(s)} - \dfrac{V_i^{(s)}(r) + \bar{\mu}_i}{\theta}\right)} \right\},$$

$$\frac{\partial Z_{0i}}{\partial x_i} = \frac{2(2\theta)^{3/2}}{3\pi} \int\limits_{y_{0i}^{(s)}}^{\infty} \frac{y^{1/2}dy}{1+\exp\left(y - \dfrac{V_i^{(s)}(r)+\bar{\mu}_i}{\theta}\right)},$$

$$\frac{\partial B_i}{\partial \mu_i} = \frac{1}{\theta}\sum_{n\ell j} N_{n\ell j,i}\left[1 - \frac{N_{n\ell j,i}}{2j+1}\right] \quad (\varepsilon_{n\ell j,i}^{(s)} < \varepsilon_{0i}^{(s)}),$$

$$\frac{\partial B_i}{\partial x_i} = -\frac{\partial B_i}{\partial \mu_i}\sum_{n\ell j}\frac{\partial \varepsilon_{n\ell j,i}^{(s)}}{\partial x_i} \quad (\varepsilon_{n\ell j,i}^{(s)} < \varepsilon_{0i}^{(s)}).$$

To compute $\partial\varepsilon_{n\ell j,i}^{(s)}/\partial x_i$ we shall make a further approximation, assuming that when x_i changes, the energy value $\varepsilon_{n\ell j,i}^{(s)}$ changes only as a result of the external screening due to the free electrons. In this approximation

$$\varepsilon_{n\ell j,i}^{(s)} \cong \bar{\varepsilon}_{n\ell j,i}^{(s)} + \frac{3}{2}\frac{Z_{0i}}{r_{0i}}, \tag{4.41}$$

where $\bar{\varepsilon}_{n\ell j,i}^{(s)}$ does not depend on r_{0i}. In view of (4.41),

$$\frac{\partial B_i}{\partial x_i} = -\frac{1}{2}\frac{\partial B_i}{\partial \mu_i}\frac{1}{\sqrt[3]{x_i}}\left(\frac{Z_{0i}}{x_i} - 3\frac{\partial Z_{0i}}{\partial x_i}\right)\Bigg|_{\mu=\bar{\mu}_i, x=\bar{x}_i}.$$

Substituting the expansions (4.40) in (4.36), we obtain

$$\left(\frac{\partial B_i}{\partial x_i} + \frac{\partial Z_{0i}}{\partial x_i}\right)(x_i - \bar{x}_i) = Z_i - B_i(\bar{\mu}_i, \bar{x}_i) - Z_{0i}(\bar{\mu}_i, \bar{x}_i) -$$
$$\left(\frac{\partial B_i}{\partial \mu_i} + \frac{\partial Z_{0i}}{\partial \mu_i}\right)(\mu_i - \bar{\mu}_i) = -\left(\frac{\partial B_i}{\partial \mu_i} + \frac{\partial Z_{0i}}{\partial \mu_i}\right)(\mu_i - \bar{\mu}_i), \tag{4.42}$$

because $B_i(\bar{\mu}_i, \bar{x}_i) + Z_{0i}(\bar{\mu}_i, \bar{x}_i) = Z_i$, in accordance with the choice of $\bar{\mu}_i, \bar{x}_i$. We are seeking solutions of equation (4.42) that satisfy the self-consistency condition for our problem, i.e., such that $\mu_i = \mu$. Then for x_i we get the expression

$$x_i = \bar{x}_i\left\{1 - \frac{(\partial B_i/\partial\mu_i + \partial Z_{0i}/\partial\mu_i)(\mu - \bar{\mu}_i)}{(\partial B_i/\partial x_i + \partial Z_{0i}/\partial x_i)\bar{x}_i}\right\}. \tag{4.43}$$

Now let us multiply (4.43) by m_i/A_i and sum over i, observing that

$$\sum_{i=1}^{N}\frac{m_i}{A_i}x_i = \sum_{i=1}^{N}\frac{m_i}{A_i}\bar{x}_i.$$

We have

$$\mu = \sum_{i=1}^{N}\frac{m_i}{A_i}\left(\frac{\partial B_i/\partial\mu_i + \partial Z_{0i}/\partial\mu_i}{\partial B_i/\partial x_i + \partial Z_{0i}/\partial x_i}\right)\bar{\mu}_i \Big/ \sum_{i=1}^{N}\frac{m_i}{A_i}\left(\frac{\partial B_i/\partial\mu_i + \partial Z_{0i}/\partial\mu_i}{\partial B_i/\partial x_i + \partial Z_{0i}/\partial x_i}\right). \tag{4.44}$$

Computations have shown that using formulas (4.43), (4.44) to get the new values of μ and x_i at temperatures $T \geq 10$ eV leads to a sufficiently rapidly converging iteration process. However, at lower temperatures the iterations may diverge, since the initial approximation, obtained within the Thomas-Fermi model, turns out to be rather crude.

To ensure the convergence of the iterations, we will use a method similar to that discussed in Subsection 3.1.2 for a single-component substance. Looking only in the direction of change of x_i, we set

$$x_i^{(s+1)} = \bar{x}_i + \xi^{(s)}(x_i - \bar{x}_i), \tag{4.45}$$

where x_i is calculated via formula (4.43) and the iteration parameter $\xi^{(s)}$ ($0 < \xi^{(s)} < 1$) is determined from the conditions of convergence of the iterations. Specifically, we increase or decrease $\xi^{(s)}$ during the iteration process depending on the quantity

$$\Delta^{(s)} = |\mu^{(s+1)} - \mu^{(s)}| + \sum_i \left(|x_i - \bar{x}_i| + |\Delta V_i^{(s)}| \right),$$

where $\mu^{(s)}$, $\mu^{(s+1)}$ are the values of μ furnished by formula (4.44) at the $(s+1)$-st and s-th iterations and $\Delta V_i^{(s)}$ is the residual of the potential of the i-th component at the s-th iteration (see formula (3.11)).

For the first iteration we put $\xi^{(0)} = 0.5$. After that the change of $\xi^{(s)}$ depends on the change of $\Delta^{(s)}$. If $\Delta^{(s)} < \Delta_{\min}$ (in computations we took $\Delta_{\min} = 0.1N$) and $\Delta^{(s)} < \Delta^{(s-1)}$, we put $\xi^{(s+1)} = 1.2\,\xi^{(s)}$. If $\Delta^{(s)} > \Delta^{(s-1)}$, we put $\xi^{(s+1)} = 0.8\,\xi^{(s)}$. In addition, the value of $\xi^{(s)}$ must be confined to some interval (generally, we took $0.1 < \xi^{(s)} < 0.75$).

The iteration runs as follows. Computing $x_i^{(s+1)}$ by means of formula (4.45), we find the corresponding values of the chemical potentials $\mu_i^{(s+1)}$, which satisfy the charge neutrality conditions (4.36). With the resulting values $x_i^{(s+1)}$, $\mu_i^{(s+1)}$ we calculate the electron density (4.34) for each component of the mixture and the corresponding potentials $V_i^{(s+1)}(r)$. This iteration step is repeated until the condition $\Delta^{(s)} < \epsilon$ is satisfied, where ϵ is the prescribed accuracy (e.g., $\epsilon = 10^{-6}$).

The iteration scheme described above is readily modified when, instead of the equality of chemical potentials, one imposes other equilibrium conditions, for example, the equality of electron pressures [159]. Since the electron pressure P_{ei} of the i-th component for the given temperature is a single-valued function of μ_i and x_i, to obtain the new iteration formulas it suffices to replace in (4.43) and (4.44) the quantities μ and μ_i by P_e and P_{ei}, respectively, and in the calculation of the derivatives $\partial B_i/\partial P_{ei}$, $\partial Z_{0i}/\partial P_{ei}$ use the representation $\dfrac{\partial B_i}{\partial P_{ei}} = \dfrac{\partial B_i}{\partial \mu_i} \Big/ \dfrac{\partial P_{ei}}{\partial \mu_i}$ and the analogous representation for the derivative of Z_{0i}.

4.3.3 Examples of computations

To illustrate the convergence of the iteration scheme computations were carried out for a mixture of 5 elements with atomic numbers $Z_i = 20 \cdot i - 10$ and mass fractions $m_i = A_i$, for the average density of matter $\rho = 1$ g/cm^3 and temperatures $T = 1,\ 10,\ 100,\ 1000$ eV. The successive values of $\mu^{(s)}$ are given in Table 4.4, which also lists the value of the chemical potential $\tilde{\mu}$, obtained without accounting for relativistic corrections. Table 4.4 demonstrates that the chemical potential μ, and hence the average occupation numbers of the electron states, are considerably better than those produced by the TF model for mixtures at low temperatures ($\mu_{\mathrm{TF}} = \mu^{(0)}$).

Table 4.4: Successive iterations of the chemical potential μ for a mixture of 5 elements with $Z_i = 20 \cdot i - 10$, $m_i = A_i$ at various temperatures and with density $\rho =1$ g/cm^3; s is the iteration number. Also shown are the values of the chemical potential $\mu = \tilde{\mu}$ calculated without accounting for relativistic effects

s	$T = 1$	$T = 10$	$T = 100$	$T = 1000$
0	0.009527	−0.89123	−16.5737	−253.798
1	−0.452196	−1.06334	−16.7666	−254.312
2	−0.435761	−1.08857	−16.7659	−254.306
3	−0.380846	−1.07340	−16.7660	−254.308
4	−0.276685	−1.07994	−16.7662	−254.308
5	−0.229350	−1.07795	−16.7663	
6	−0.224280	−1.07864	−16.7663	
7	−0.211847	−1.07813		
8	−0.203926	−1.07837		
9	−0.190126	−1.07823		
10	−0.182146	−1.07835		
11	−0.181542	−1.07829		
12	−0.181299	−1.07829		
13	−0.181475			
14	−0.181249			
15	−0.181299			
16	−0.181093			
17	−0.180894			
18	−0.180739			
19	−0.180614			
20	−0.180649			
21	−0.180632			
22	−0.180633			
$\tilde{\mu}$	−0.177880	−1.09428	−16.7732	−254.121

Table 4.5 gives more detailed results of computations of the self-consistent potential for a mixture of bromate and agar ($C_{190}H_{190}O_{20}Br$), which is used in laser targets [242]. The table shows how the radii of the atom cells r_{0i}, the average ion charges Z_{0i} and the chemical potentials μ_i change in the iteration process. The computations were carried out for the temperature T=10 eV and the density $\rho = 0.1$ g/cm^3.

The approximate description of a mixture of elements described above assumes that the temperature is sufficiently high, so that each component of the mixture is in ionized state. Possible molecular formations of ions (see, e.g., [196]) were neglected.

Table 4.5: Successive iterations for $C_{190}H_{190}O_{20}Br$ at temperature T=10 eV and density ρ =0.1 g/cm^3 (s is the iteration number).

s		H	C	O	Br
0	r_{0i}	4.679940	6.486292	6.717963	7.676817
	Z_{0i}	0.728529	2.097298	1.425806	1.377438
	μ_i	−1.174687	−1.121720	−1.304069	−1.465673
1	r_{0i}	4.515875	6.584268	6.155903	7.407173
	Z_{0i}	0.728529	2.053058	1.817736	2.946070
	μ_i	−1.174687	−1.129704	−1.211395	−1.178415
2	r_{0i}	4.487279	6.575564	6.388097	7.367198
	Z_{0i}	0.711451	2.067104	1.893517	2.904261
	μ_i	−1.146903	−1.142695	−1.105721	−1.146844
3	r_{0i}	4.483484	6.572734	6.433727	7.367306
	Z_{0i}	0.708375	2.065190	1.941147	2.904244
	μ_i	−1.142000	−1.141674	−1.134700	−1.141254
4	r_{0i}	4.483516	6.573425	6.426683	7.368448
	Z_{0i}	0.707964	2.065036	1.939742	2.904784
	μ_i	−1.141348	−1.141255	−1.142349	−1.141198
5	r_{0i}	4.483258	6.573039	6.431701	7.368129
	Z_{0i}	0.707968	2.065064	1.942909	2.904869
	μ_i	−1.141353	−1.141359	−1.140592	−1.141347
6	r_{0i}	4.483762	6.573742	6.422386	7.368866
	Z_{0i}	0.707940	2.065023	1.936132	2.904849
	μ_i	−1.141309	−1.141305	−1.142718	−1.141305
7	r_{0i}	4.483489	6.573333	6.427720	7.368478
	Z_{0i}	0.707994	2.065095	1.940392	2.904945
	μ_i	−1.141396	−1.141403	−1.140385	−1.141396
8	r_{0i}	4.483417	6.573249	6.428882	7.368382
	Z_{0i}	0.707965	2.065052	1.940996	2.904901
	μ_i	−1.141349	−1.141346	−1.141126	−1.141348
9	r_{0i}	4.483433	6.573284	6.428465	7.368403
	Z_{0i}	0.707957	2.065048	1.940477	2.904885
	μ_i	−1.141336	−1.141334	−1.141414	−1.141336
10	r_{0i}	4.483399	6.573229	6.429167	7.368345
	Z_{0i}	0.707959	2.065042	1.941081	2.904876
	μ_i	−1.141339	−1.141340	−1.141229	−1.141341
11	r_{0i}	4.483399	6.573229	6.429167	7.368345
	Z_{0i}	0.707955	2.065041	1.940948	2.904896
	μ_i	−1.141333	−1.141332	−1.141331	−1.141330

4.4 Accounting for the individual states of ions

The average-atom model considered in §§ 4.1–4.3 provides an effective description of the atom cell in plasma for an ion with average occupation numbers and free electrons. A real plasma involves a large number of diverse ion states. As we will see in Subsection 5.5.2, a detailed configuration accounting is rather crucial in opacity calculations. To account for individual states of ions in computations one usually relies on perturbation theory, and one assumes that the perturbation of the potential is connected with the change of the electron density as a result of changes in the occupation numbers of the electron shells (fluctuations of the occupation numbers) [150, 45].

The main inaccuracy in perturbation methods is connected with the fact that the wave functions obtained for the average ion will not be consistent with the atomic potential of the ion under consideration. This inaccuracy can be eliminated by accounting for the ion's occupation numbers corresponding to a given configuration.

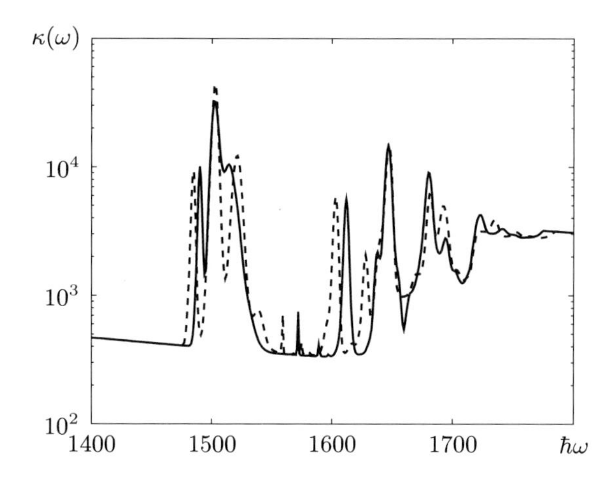

Figure 4.5: Dependence of the absorption coefficient (measured in cm^2/g) in aluminum at temperature $T = 22$ eV and density $\rho = 0.01$ g/cm^3 on the energy of photons (in eV). The solid [resp., dashed] curve was computed using the wave functions of the average atom [resp., of the ions]

To give an example, Figure 4.5 compares the absorption spectra of the aluminum, calculated using the wave functions of the average atom, as well as the wave functions of ions. One can see that an improved accuracy in locating the positions of the lines may turn out to be quite essential in the description of details of the spectrum.

4.4.1 Density functional of the electron system with the individual states of ions accounted for

In a high-temperature plasma, especially for high-Z elements, a very large number of qualitatively different ion states occur, and accounting for all possible states requires laborious computations. However, such accounting can be achieved if we use a detailed description only for the most important part of the ions states and apply the average-atom approximation for the remaining states. The choice of the states that will be subject to a detailed description is determined by the formulation of the problem. The least labor-consuming is the method in which for a group of states whose ionization degrees are close to one another we choose only one representative.

To simplify the exposition, we will consider here the case in which such states are determined by the occupation pattern of the first non-closed shell (or several such shells). Accordingly, we divide all electron states into two groups. In the first group we include the states $\overline{\nu}$ which determine the basic ion configurations that we singled out. The symbol $\overline{\nu}$ stands for the set of quantum numbers that define such electron states, the occupation numbers of which are fixed (given). The remaining states ν, belonging to the continuous and the discrete spectrum, are assigned to the second group and will be described in the average-occupation-numbers approximation. In this approximation it is assumed that the electron populations of the upper levels are in equilibrium with the continuum. It is further assumed that the densities of these populations are small and that they have only a weak influence on the inner structure of the ion levels. The effective configuration corresponding to the given occupation numbers of the electron states $\overline{\nu}$ and some average occupation numbers of the states ν will be denoted by Q.

Suppose that, as in § 3.2, for a temperature $T \neq 0$ the one-particle states ν are occupied with some probabilities n_ν (the values n_ν coincide with the average occupation numbers of the state ν and satisfy the condition $0 \leq n_\nu \leq 1$), and each set of numbers $n_{\overline{\nu}}^Q$ ($n_{\overline{\nu}}^Q = 1$ or $n_{\overline{\nu}}^Q = 0$), determining an effective configuration Q, occurs with probability W_Q ($\sum_Q W_Q = 1$). Considering that the occupation probabilities of the shells near the equilibrium state are independent (see the derivation of formula (3.19)), we conclude that the probability of the given state is calculated as the product of W_Q and the values n_ν or $1 - n_\nu$ depending on whether the state ν is occupied or not. For example, for given occupation numbers of states $\overline{\nu}$ for the configuration Q (this configuration is realized with probability W_Q), with occupied states $\nu_1, \nu_2, \ldots, \nu_n$ and unoccupied remaining states $\nu_{\overline{N}_n+1}, \nu_{\overline{N}_n+2}, \ldots$, we have

$$W_n = W_Q n_{\nu_1} n_{\nu_2} \cdot \ldots \cdot n_{\nu_{\overline{N}_n}} \left(1 - n_{\nu_{\overline{N}_{n+1}}}\right) \cdot \ldots; \tag{4.46}$$

here $\sum_{\overline{\nu}} n_{\overline{\nu}}^Q + \overline{N}_n = N_n$, where N_n is the number of electrons in the system.

In the approximation adopted here formulas (3.25) and (3.26) are replaced by

$$\mathrm{Tr}(\widehat{W}\ln\widehat{W}) = \sum_n W_n \ln W_n =$$
$$\sum_Q W_Q \left\{ \ln W_Q + \sum_\nu [n_\nu \ln n_\nu + (1-n_\nu)\ln(1-n_\nu)] \right\},$$

$$\rho(q) = \sum_Q W_Q \left(\sum_{\overline{\nu}} n_{\overline{\nu}}^Q \left|\psi_{\overline{\nu}}^Q(q)\right|^2 + \sum_\nu n_\nu |\psi_\nu(q)|^2 \right). \tag{4.47}$$

Using formula (3.17) we can obtain the expression for the thermodynamic potential Ω in the one-particle approximation, in the sense explained above:

$$\Omega = \sum_Q W_Q E_Q + \theta \sum_Q W_Q \ln W_Q + \theta \sum_Q W_Q \sum_\nu [n_\nu \ln n_\nu + (1-n_\nu)\ln(1-n_\nu)] -$$
$$\mu \sum_Q W_Q \left[\sum_{\overline{\nu}} n_{\overline{\nu}}^Q + \sum_\nu n_\nu \right]. \tag{4.48}$$

Here E_Q is the energy of the configuration Q (see below).

We are looking for the minimum of Ω for arbitrary values of n_ν, W_Q (where $0 \le n_\nu \le 1$ and $\sum_Q W_Q = 1$), and also for functions $\psi_{\overline{\nu}}^Q(q), \psi_\nu(q)$, subject to the orthogonality and normalization conditions. Applying a variational principle in much the same way as we did in § 3.2, we obtain equations of Hartree-Fock type for each configuration Q and a separate equation for the average ion. Note that, as we shall see below, here the atomic potential of the average ion is obtained via an explicit averaging over the configurations Q.

In practice one usually resorts to an approximate description of exchange effects of Slater type for the temperature and density of matter (see § 3.3). This leads to a simplification of the functional Ω and of the corresponding equations obtained by the minimization of Ω. The expression for E_Q in the Hartree-Fock-Slater approximation has the form

$$E_Q = \sum_{\overline{\nu}} n_{\overline{\nu}}^Q \int \left(\psi_{\overline{\nu}}^Q(q)\right)^* \left[-\frac{1}{2}\Delta + V_a(\vec{r}) \right] \psi_{\overline{\nu}}^Q(q)\, dq +$$
$$\sum_\nu n_\nu \int \psi_\nu^*(q) \left[-\frac{1}{2}\Delta + V_a(\vec{r}) \right] \psi_\nu(q)\, dq + \frac{1}{2} \iint \frac{\rho^Q(q)\rho^Q(q')}{|\vec{r}-\vec{r}'|}\, dq\, dq' +$$
$$\int \varphi[\rho^Q(q), \theta]\, dq, \tag{4.49}$$

where $V_a(\vec{r})$ is the potential generated by nuclei,

$$\rho^Q(q) = \sum_{\overline{\nu}} n_{\overline{\nu}}^Q \left|\psi_{\overline{\nu}}^Q(q)\right|^2 + \sum_{\nu} n_\nu |\psi_\nu(q)|^2, \tag{4.50}$$

and $\varphi(\rho, \theta)$ is given by formula (3.51).

4.4.2 The Hartree-Fock-Slater equations of the ion method in the cell and plasma approximations

Further simplifications are achieved by passing to the cell or plasma models. First let us consider the cell model of matter for a dense high-temperature plasma. To describe a substance as a whole, the equations for E_Q and for the wave functions must contain the potential $V_{\mathrm{a}}(\vec{r})$ of the nuclei, corresponding to some position of the latter. To solve such a problem seems hardly possible, and so one usually considers the average potential near some nucleus, averaging it over the position of the other nuclei. If such an averaging is carried out directly in the HFS equations, then the problem becomes more complex due to the necessity of accounting in explicit form for the interaction of ions corresponding to different configurations and different charges of the ion core.

It is convenient to do the averaging before we apply the variational principle. Since in a plasma there is no distinguished direction, the resulting average atomic potentials will be spherically-symmetric. The corresponding nuclei can be enclosed in a sphere whose size is such that the cells with the selected ion configurations Q will be charge neutral. Thus, we will assume that the individual states of ions Q can be described by specifying charge neutral spherical cells, of different sizes, with volumes V_Q. Then the integration in (4.49) may be carried out only in the corresponding cells Q, and the result is summed over all cells.

From the condition of conservation of the average volume we have

$$\sum_Q W_Q V_Q = V = \frac{4}{3}\pi(1.388)^3 \frac{A}{\rho}, \tag{4.51}$$

where A is the atomic weight and ρ is the density of matter in g/cm^3. The requirement that the cells be charge neutral reads

$$\sum_{\overline{\nu}} n_{\overline{\nu}}^Q + \sum_{\nu} n_\nu = Z, \tag{4.52}$$

where Z is the charge of the nucleus. The charge neutrality condition (4.52) allows us to neglect the interaction of ions and instead of the potential of all nuclei, give only the potential of a single selected nucleus, i.e., $V_a(\vec{r}) = Z/r$, assuming that the total functional Ω can be represented as a sum over cells.

If we use the expression (4.48) for the functional Ω and the supplementary conditions

$$\sum_Q W_Q = 1, \quad \sum_{\overline{\nu}} n_{\overline{\nu}}^Q + \sum_\nu n_\nu = Z,$$

$$\int |\psi_{\overline{\nu}}^Q(q)|^2 dq = 1, \quad \int |\psi_\nu(q)|^2 \, dq = 1, \quad \sum_Q W_Q V_Q = V,$$

and then introduce the corresponding Lagrange multipliers $\mathcal{A}$, $\mathcal{B}_Q$, $\Lambda_{\overline{\nu}}^Q$, Λ_ν and $\mathcal{P}$, the variational principle reads

$$\delta\Bigg\{\sum_Q W_Q E_Q + \theta \sum_Q W_Q \ln W_Q + \theta \sum_Q W_Q \sum_\nu \left[n_\nu \ln n_\nu + (1-n_\nu) \ln(1-n_\nu)\right] -$$
$$\mu \sum_Q W_Q \left[\sum_{\overline{\nu}} n_{\overline{\nu}}^Q + \sum_\nu n_\nu\right] + \sum_Q \mathcal{B}_Q \left[\sum_{\overline{\nu}} n_{\overline{\nu}}^Q + \sum_\nu n_\nu - Z\right] +$$
$$\sum_Q \sum_{\overline{\nu}} \Lambda_{\overline{\nu}}^Q \left[\int \left|\psi_{\overline{\nu}}^Q(q)\right|^2 dq - 1\right] + \sum_\nu \Lambda_\nu \left[\int |\psi_\nu(q)|^2 \, dq - 1\right] +$$
$$\mathcal{A}\left[\sum_Q W_Q - 1\right] + \mathcal{P}\left[\sum_Q W_Q V_Q - V\right]\Bigg\} = 0.$$

Substituting here the expression (4.49) for E_Q and performing the variation, we obtain

$$\sum_Q \delta W_Q \Bigg\{ E_Q + \theta \ln W_Q + \theta \sum_\nu \left[n_\nu \ln n_\nu + (1 - n_\nu) \ln(1 - n_\nu)\right] +$$
$$\theta + \mathcal{A} + \mathcal{P} V_Q - \mu \left[\sum_{\overline{\nu}} n_{\overline{\nu}}^Q + \sum_\nu n_\nu\right]\Bigg\} +$$
$$\sum_Q W_Q \Bigg\{ \sum_\nu \delta n_\nu \left[\int \psi_\nu^*(q) \widehat{H}_Q \psi_\nu(q) \, dq + \theta \sum_\nu \ln \frac{n_\nu}{1 - n_\nu} - \mu - \frac{\mathcal{B}_Q}{W_Q}\right] +$$

$$\sum_\nu n_\nu \int \delta\psi_\nu^*(q)\widehat{H}_Q\psi_\nu(q)dq + \sum_{\overline{\nu}} n_{\overline{\nu}}^Q \int \delta\left(\psi_{\overline{\nu}}^Q(q)\right)^* \widehat{H}_Q\,\psi_{\overline{\nu}}^Q(q)\,dq +$$

$$\left.\sum_\nu n_\nu \int \psi_\nu^*(q)\widehat{H}_Q\delta\psi_\nu(q)\,dq + \sum_{\overline{\nu}} n_{\overline{\nu}}^Q \int \left(\psi_{\overline{\nu}}^Q(q)\right)^* \widehat{H}_Q\delta\psi_{\overline{\nu}}^Q(q)\,dq\right\} +$$

$$\sum_Q W_Q\delta V_Q\left\{\frac{\partial}{\partial V_Q}\left[E_Q + \theta\sum_\nu [n_\nu \ln n_\nu + (1-n_\nu)\ln(1-n_\nu)]\right] + \mathcal{A}\right\} +$$

$$\sum_{\overline{\nu}} \frac{\Lambda_{\overline{\nu}}^Q}{W_Q}\left[\int \delta\left(\psi_{\overline{\nu}}^Q(q)\right)^* \psi_{\overline{\nu}}^Q(q)dq + \int \left(\psi_{\overline{\nu}}^Q(q)\right)^* \delta\psi_{\overline{\nu}}^Q(q)dq\right] +$$

$$\sum_\nu \Lambda_\nu\left[\int \delta\psi_\nu^*(q)\psi(q)dq + \int \psi_\nu^*(q)\delta\psi(q)dq\right] = 0,$$

where

$$\widehat{H}_Q = -\frac{1}{2}\Delta + V_a(r) + \sum_{\overline{\nu}} n_{\overline{\nu}}^Q \int \frac{|\psi_{\overline{\nu}}^Q(q')|^2\,dq'}{|\vec{r}-\vec{r}\,'|} +$$

$$\sum_\nu n_\nu \int \frac{|\psi_\nu(q')|^2 dq'}{|\vec{r}-\vec{r}\,'|} - \frac{\partial\varphi(\rho^Q)}{\partial\rho^Q}. \quad (4.53)$$

Considering that the variations δn_ν, $\delta\psi_\nu$, $\delta\psi_\nu^*$, $\delta\psi_{\overline{\nu}}^Q$, $\delta\left(\psi_{\overline{\nu}}^Q\right)^*$, δW_Q and δV_Q are independent, we obtain the following conditions of minimum:

$$\widehat{H}_Q\psi_{\overline{\nu}}^Q(q) = -\frac{\Lambda_{\overline{\nu}}^Q}{W_Q n_{\overline{\nu}}^Q}\psi_{\overline{\nu}}^Q(q) \equiv \varepsilon_{\overline{\nu}}^Q\psi_{\overline{\nu}}^Q(q), \quad (4.54)$$

$$\widehat{H}_0\psi_\nu(q) = -\frac{\Lambda_\nu}{n_\nu}\psi_\nu(q) \equiv \varepsilon_\nu\psi_\nu(q), \quad (4.55)$$

$$n_\nu = \frac{1}{1+\exp\left(\dfrac{\varepsilon_\nu - \widetilde{\mu}}{\theta}\right)}, \quad (4.56)$$

$$W_Q = \exp\left(-\frac{E_Q - \mu\left(\sum n_{\overline{\nu}}^Q + \sum n_\nu\right) + \theta + \mathcal{A} + \mathcal{P}V_Q}{\theta}\right), \quad (4.57)$$

$$\frac{\partial}{\partial V_Q}\left(E_Q + \theta\sum_\nu [n_\nu \ln n_\nu + (1-n_\nu)\ln(1-n_\nu)]\right) + \mathcal{P} = 0, \quad (4.58)$$

where we used the notations

$$\widetilde{\mu} = \mu + \sum_Q \mathcal{B}_Q, \qquad \widehat{H}_0 = \sum_Q W_Q \widehat{H}_Q.$$

Since $\sum_{\overline{\nu}} n_{\overline{\nu}}^Q + \sum_\nu n_\nu = Z$, relation (4.57) yields

$$W_Q = C \exp\left(-\frac{E_Q + \mathcal{P}V_Q}{\theta} \right), \tag{4.59}$$

where the constant C is chosen from the normalization condition. From (4.58) it follows that

$$\mathcal{P}_{\mathrm{e}}^Q \equiv -\frac{\partial}{\partial V_Q} \left(E_Q + \theta \sum_\nu [n_\nu \ln n_\nu + (1 - n_\nu) \ln(1 - n_\nu)] \right) = \mathcal{P}. \tag{4.60}$$

The quantity $\mathcal{P}_{\mathrm{e}}^Q$ is the derivative of the free energy with respect to volume (taken with the minus sign), and according to the definition given in [17], it is the electron pressure in the cell Q. Therefore, condition (4.60) means that the electron pressures in all the cells Q must be equal to each other.

The cell approximation assumes a sufficiently high density of matter, at which violation of the charge neutrality of the cell leads to a highly disordered state of plasma with fast relaxation to the equilibrium state, in which every cell, the size of which is determined by the electron pressure produced in it, is charge neutral.

In the opposite case of low densities, for an ideal or weakly non-ideal plasma the requirement of charge neutrality of the cells may be too restrictive and in practice is not satisfied (for example, whenever negative ions are present). Under such circumstances, as a rule, one can neglect the interaction of cells, whose sizes are sufficiently large. Here we can assume that the free electrons are uniformly distributed in space with a constant density. In this case the cells of individual ion states may be considered of equal size. Further, let us note that in the case at hand charge neutrality may not hold in each cell, though for the plasma as a whole the charge neutrality condition is satisfied, i.e.,

$$\sum_Q W_Q \left(\sum_{\overline{\nu}} n_{\overline{\nu}}^Q + \sum_\nu n_\nu \right) = Z. \tag{4.61}$$

Application of the variational principle yields (in much the same way as above) the following system of Hartreee-Fock-Slater equations for the ion method in the plasma approximation:

$$\widehat{H}_Q \psi_{\overline{\nu}}^Q(q) = -\frac{\Lambda_{\overline{\nu}}^Q}{W_Q n_{\overline{\nu}}^Q} \psi_{\overline{\nu}}^Q(q) \equiv \varepsilon_{\overline{\nu}}^Q \psi_{\overline{\nu}}^Q(q), \tag{4.62}$$

$$\widehat{H}_0 \psi_\nu(q) = -\frac{\Lambda_\nu}{n_\nu} \psi_\nu(q) \equiv \varepsilon_\nu \psi_\nu(q), \tag{4.63}$$

$$n_\nu = \frac{1}{1 + \exp\left(\dfrac{\varepsilon_\nu - \mu}{\theta}\right)}, \tag{4.64}$$

$$W_Q = C \exp\left(-\frac{E_Q - \mu\left(\sum n_{\bar{\nu}}^Q + \sum n_\nu\right)}{\theta}\right). \tag{4.65}$$

Here the coefficient C and the chemical potential μ are determined from the normalization condition $\sum_Q W_Q = 1$ and the charge neutrality condition (4.61), respectively.

Thus, the cell and the plasma approximations correspond to the two limit cases of high and low densities of matter. Note that at high temperatures there exists a sufficiently wide range of intermediate densities for which the results of computations carried out in the two approximations are practically identical. This means that for high temperatures there is a smooth transition from the model (4.54)–(4.60) to the model (4.61)–(4.65) when the density decreases. For very small densities the plasma approximation (4.61)–(4.65) may become unapplicable because the condition of local thermodynamic equilibrium is violated (see § 5.7).

4.4.3 Wave functions and energy levels of ions in a plasma

Let us apply the average-atom model in the computation of wave functions and energy levels of ions in a plasma. As a rule, such computations rely on the one-electron wave functions obtained for free ions in the Hartree-Fock approximation. However, keeping in mind that for the calculation of Rosseland opacities it is necessary to take into account a huge number of ion states in the plasma, the computations can be simplified considerably by using the one-electron wave functions provided by the Hartree-Fock-Slater model of the average atom. Then, as numerical experiments have demonstrated, the physical accuracy is roughly the same as when the one-electron wave functions of the Hartree-Fock model are used.

It should be noted that when we used the HFS self-consistent potential $V(r)$ we have amended it so that $V(r) \geq 1/r$ for all $r < r_0$, where r_0 is the radius of the average atom cell [49].

Figure 4.6 compares the wave functions obtained via the Hartree-Fock-Slater model with those calculated via the Hartree-Fock model. The computations were carried out for an iron plasma at temperature $T = 20$ eV and with density $\rho = 10^{-4}$ g/cm^3. Under these conditions the average occupation numbers define the average configuration $1s^2 2s^2 2p^6 3s^{1.85} 3p^{4.12} 3d^{1.24}$, which is obtained by solving the Hartree-Fock-Slater equations for the average atom.

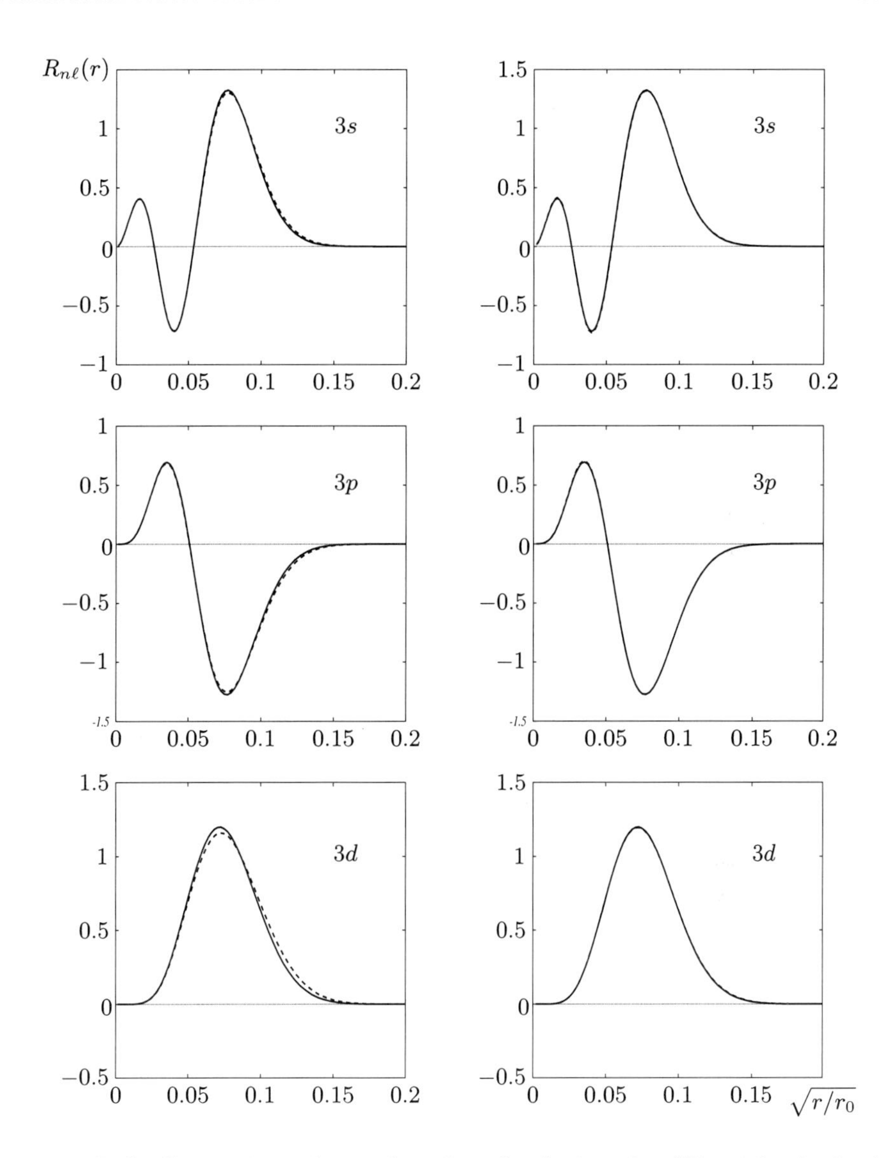

Figure 4.6: Left: Comparison of wave functions for the iron ion ($Z = 26$), obtained in the Hartree-Fock model for the configuration $1s^2 2s^2 2p^6 3s^2 3p^4 3d^1$ (solid curve) and in the Hartree-Fock-Slater model for the same configuration (dashed curve). Right: Comparison of wave functions in the Hartree-Fock model for the configuration $1s^2 2s^2 2p^6 3s^2 3p^4 3d^1$ (solid curve) and in the Hartree-Fock-Slater model for the average configuration $1s^2 2s^2 2p^6 3s^{1.85} 3p^{4.12} 3d^{1.24}$ (dashed curve). One can see that here the radial functions practically coincide

Tables 4.6–4.7 list the energy of terms, in atomic units, computed in the approximation of intermediate coupling scheme by using the wave functions obtained via the Hartree-Fock and the Hartree-Fock-Slater models (see Subsection 5.3.3). The letters $S, P, D, \ldots$ with a number in front of them indicate the dominating component of the term (the number corresponds to multiplicity). The computations were done for iron ($Z = 26$), and also for chromium ($Z = 24$). The experimental data were borrowed from the library of spectroscopic atomic data of the US National Institute of Standards and Technology (http://www.physics.nist.gov).

Table 4.6: Energies of terms of the configurations $3s^23p^63d^2$ and $3s^23p^63d4f$ for iron ($Z = 26$)

Term	J	experiment	HF	HFS
$Q = 3s^23p^63d^2$			$E_0 = -1250.436$	$E_0 = -1250.214$
$3F$	2	0.000	0.000	0.000
	3	0.005	0.005	0.005
	4	0.010	0.011	0.011
$1D$	2	0.080	0.097	0.088
$3P$	0	0.091	0.111	0.101
	1	0.093	0.113	0.103
	2	0.097	0.117	0.108
$1G$	4	0.132	0.148	0.134
$1S$	0	—	0.358	0.323
$Q = 3s^23p^63d4f$			$E_0 = -1247.402$	$E_0 = -1247.218$
$1G$	4	0.000	0.000	0.000
$3F$	2	0.000	0.008	0.007
	3	0.002	0.009	0.008
	4	0.006	0.008	0.007
$3G$	3	0.014	0.024	0.022
	4	0.029	0.027	0.026
	5	0.039	0.030	0.028
$1D$	2	0.018	0.022	0.021
$1F$	3	0.025	0.039	0.037
$3D$	1	0.027	0.035	0.032
	2	0.027	0.035	0.032
	3	0.031	0.034	0.032
$3P$	2	0.036	0.046	0.042
	1	0.038	0.048	0.044
	0	0.039	0.049	0.046
$1H$	5	0.046	0.066	0.061
$1P$	1	0.053	0.066	0.060
$3H$	4	—	0.012	0.012
	5	—	0.011	0.010
	6	—	0.016	0.015

As these tables and figures demonstrate, the approximation considered here, naturally, does not provide the accuracy needed in spectroscopic investigations, yet it is totally satisfactory for the computation of photon absorption coefficients and Rosseland mean opacities in a hot plasma.

Table 4.7: Energies of terms of the configuration $Q = 3s^2 3p^6 3d^4$ for chromium ($Z = 24$)

Term	J	experiment	HF	HFS
			$E_0 = -1041.056$	$E_0 = -1040.525$
$5D$	0	0.0000	0.0000	0.0000
	1	0.0003	0.0003	0.0002
	2	0.0008	0.0009	0.0007
	3	0.0016	0.0017	0.0015
	4	0.0026	0.0028	0.0025
$3P$	0	0.076	0.098	0.069
	1	0.078	0.100	0.070
	2	0.081	0.103	0.073
$3H$	4	0.079	0.090	0.064
	5	0.079	0.091	0.064
	6	0.080	0.092	0.065
$3F$	2	0.084	0.104	0.073
	3	0.084	0.104	0.074
	4	0.085	0.105	0.074
$3G$	3	0.094	0.114	0.082
	4	0.095	0.115	0.083
	5	0.096	0.116	0.083
$3D$	3	0.117	0.143	0.106
	2	0.117	0.148	0.106
	1	0.118	0.148	0.107
$1F$	3	0.187	0.207	0.148
$3F$	2	0.197	0.245	0.176
	3	0.197	0.245	0.175
	4	0.197	0.245	0.175
$3P$	0	0.222	0.247	0.177
	1	0.223	0.249	0.179
	2	0.225	0.250	0.180
$1I$	6	—	0.136	0.096
$1G$	4	—	0.143	0.101
$1D$	2	—	0.179	0.127
$1G$	4	—	0.277	0.198
$1S$	0	—	0.167	0.118
$1D$	2	—	0.374	0.269
$1S$	0	—	0.484	0.347

Part II

Radiative and thermodynamical properties of high-temperature dense plasma

Chapter 5

Interaction of radiation with matter

The higher the temperature of matter, the larger the role radiation plays in energy transfer processes. At very high temperatures (of the order of several million degrees), the main mechanism of energy redistribution is its transfer, effected by photons, because the energy flux carried by photons is considerably larger than the energy flux carried by other particles. If the photon free paths are small compared to the characteristic dimensions of the plasma, then the diffusion approximation holds, which leads to the *radiative heat conduction equation*. The conductivity coefficient in this equation depends in a complex nonlinear way on the temperature and density of matter as well as on the spectral photon absorption coefficients. The radiative heat conduction equation found wide applications in the description of various processes in high-temperature dense plasmas [234, 147, 67, 45, 130, 89].

When the photon free paths are comparable to the characteristic dimensions of the plasma, the diffusion approximation is not applicable, and in order to describe energy transfer processes it is necessary to solve the radiative transfer equation with the spectral dependence of the absorption coefficients of photons accounted for [50]. Furthermore, when part of the radiation escapes the plasma, the thermodynamic equilibrium condition may be violated, in which case, together with the photon transport, one has to account for the level kinetics of electrons [134].

To describe the interaction of radiation with matter we will use the radiative transfer equation, neglecting the diffraction and interference effects (see [67, 174]) and will consider the simplest methods for solving it.

5.1 Radiative heat conductivity of plasma

5.1.1 The radiative transfer equation

To study the energy transfer processes in matter we start from the photon *distribution function* $f(\vec{r}, \vec{\Omega}, \nu, t)$. The function $f(\vec{r}, \vec{\Omega}, \nu, t)$ measures the concentration of photons with direction $\vec{\Omega}$ and frequency ν, in a given point of space $\vec{r}$ and at time t. Therefore, the number of radiation quanta in the volume element dV, which contains the point $\vec{r}$, in the frequency interval $d\nu$ and propagating in the solid angle $d\Omega$ (it is assumed that $\vec{\Omega}$ is a unit vector) at time t is equal to

$$f(\vec{r}, \vec{\Omega}, \nu, t)\, dV\, d\Omega\, d\nu.$$

The spectral intensity of the radiation, i.e., the energy per unit of frequency transferred in a unit of time through the unit of surface in direction $\vec{\Omega}$ is given by

$$I_\nu = I_\nu(\vec{r}, \vec{\Omega}, t) = h\nu c f(\vec{r}, \vec{\Omega}, \nu, t). \tag{5.1}$$

The function $f(\vec{r}, \vec{\Omega}, \nu, t)$, depending on seven scalar variables, satisfies an integro-differential equation known as the *radiative transfer equation* [67, 174]:

$$\frac{\partial f}{\partial t} + c(\vec{\Omega}\,\mathrm{grad})f = N^* A^*_\nu - c(N_\mathrm{a}\sigma_\mathrm{a} + n\sigma_\mathrm{s})f + \\ cn\sigma_\mathrm{s} \iint K[(\vec{\Omega}, \vec{\Omega}'), \nu, \nu'] f(\vec{\Omega}', \nu')\, d\Omega'\, d\nu' \tag{5.2}$$

(under the integral sign we kept only the arguments of f with respect to which one integrates). The left-hand side of equation (5.2) represents the material (Lagrangian) derivative of the function f. If the photons would not interact with particles of matter, then this derivative would be equal to zero. The right-hand side of the equation represents the change in the number of photons due to emission and absorption by matter, as well as to scattering processes by particles of matter.

In the first right-hand side term, N^* denotes the concentration of particles of matter capable of emitting a photon of frequency ν; A^*_ν denotes the probability of emission of a photon with frequency ν in the direction $\vec{\Omega}$. The second term expresses the decrease in the number of photons, moving in direction $\vec{\Omega}$, due to elastic or inelastic interactions. The particles with which the photons interact during collisions are characterized by an effective scattering cross-section σ_s and an absorption cross-section σ_a, both depending on the frequency ν. Further, in equation (5.2) N_a [resp., n] is the concentration of particles of matter capable of absorbing [resp., scattering] photons. In all processes of interaction of radiation with matter a key role is played by the interaction with electrons, since the interaction cross-section is inverse proportional to the square of the particle mass.

Let us remark also that if the elementary cross-sections depend on the energy of particles, then the corresponding terms in (5.2) must be integrated with respect to the energy distribution function of these particles.

The last integral term in the right-hand side characterizes the increase of the number of photons, moving in direction $\vec{\Omega}$, due to photons that earlier moved in direction $\vec{\Omega}\,'$, and then were scattered at an angle of $(\vec{\Omega}, \vec{\Omega}\,')$ (i.e., the angle between the directions $\vec{\Omega}$ and $\vec{\Omega}\,'$), with the frequency changing from ν' to ν. The kernel of the integral, $K[(\vec{\Omega}, \vec{\Omega}\,'), \nu, \nu']$, is the probability of scattering by the angle $(\vec{\Omega}, \vec{\Omega}\,')$, which is determined by the interaction mechanism. Scattering of light by free electrons (Compton effect) is accompanied by a change in frequency for photon energies comparable with the electron's own energy mc^2, which is equal to 0.51 MeV. If, however, the energy of the photon is small compared to mc^2, then the frequency shift due to scattering is also small, and the cross-section tends to the *Thompson cross-section* $\sigma_0 = (8/3)\,\pi r_{\mathrm{e}}^2$, where $r_{\mathrm{e}} = e^2/(mc^2)$ is the classical radius of the electron. Scattering effects are studied in more detail in [199]. In what follows we will assume, as a rule, that we are dealing with pure Thompson scattering. In this case the kernel $K[(\vec{\Omega}, \vec{\Omega}\,'), \nu, \nu']$ of the integral degenerates to $K[(\vec{\Omega}, \vec{\Omega}\,')]$ and represents the probability of scattering by the angle $(\vec{\Omega}, \vec{\Omega}\,')$. As for the computations of the scattering cross-sections of the bound electrons, which belong to the discrete spectrum, here the scattering cross-section is much smaller than the corresponding absorption cross-section. For that reason we will assume that in (5.2) n stands for the concentration of free electrons and scattering effects will be accounted for in the nonrelativistic approximation.

Before we consider the effective absorption cross-section σ_{a} of the bound and free electrons and rewrite equation (5.2) to account for the approximations adopted here, we need to examine the first term $N^* A_\nu^*$ in the right-hand side of (5.2). If the photons would obey the classical statistics, the emission would be isotropic and its probability A_ν^* would not depend on the distribution function f. Since the photons actually obey the Bose statistics, the emission probability of photons of frequency ν depends on the number of photons already present in the phase-space cell under consideration. If one denotes by f the distribution function relative to the dimensionless unit volume of phase space $dV d\vec{p}/h^3$ (where $p = h\nu/c$, $d\vec{p} = dp_x dp_y dp_z$), the emission probability will be proportional to $1 + \mathrm{f}$, where f is the spectral density of photons with given direction of polarization. Therefore, $A_\nu^* = A_\nu(1 + \mathrm{f})$, where A_ν does not depend on f, and the distribution functions f and f are related as follows:

$$f\, dV\, d\Omega\, d\nu = 2\mathrm{f}\, \frac{dV\, d\vec{p}}{h^3}. \tag{5.3}$$

Since for photons with momentum p and direction of motion in the solid angle $d\Omega$ we have $d\vec{p} = p^2\, dp\, d\Omega$, this yields

$$\mathrm{f} = \frac{h^3}{2p^2}\, \frac{d\nu}{dp}\, f = \frac{c^3}{2\nu^2}\, f. \tag{5.4}$$

The scattering probability of photons in some cell of the phase space will also be proportional to $1 + \mathrm{f}$, but for scattering without energy change this effect drops from the transfer equation and can be neglected (see, e.g., [67]).

Substituting expression (5.4) in (5.2), we obtain the transfer equation for photons with the *induced emission* accounted for, in the case where scattering takes place without a change in energy :

$$\frac{1}{c}\frac{\partial f}{\partial t} + (\vec{\Omega}\,\mathrm{grad})f = \frac{N^*}{c}A_\nu\left(1 + \frac{c^3}{2\nu^2}f\right) - N_\mathrm{a}\sigma_\mathrm{a}f - n\sigma_\mathrm{s}f + n\sigma_\mathrm{s}\int K[(\vec{\Omega},\vec{\Omega}')]f(\vec{\Omega}')\,d\Omega'. \quad (5.5)$$

In practical computations the term $\frac{1}{c}\frac{\partial f}{\partial t}$ in (5.5) can usually be neglected (quasi-stationary case). Consequently, in the *nonrelativistic quasi-stationary* case the transfer equation can be recast in the form

$$(\vec{\Omega}\,\mathrm{grad})f = -N_\mathrm{a}\sigma_\mathrm{a}\left(1 - \frac{c^2}{2\nu^2}\frac{N^*}{N_\mathrm{a}}\frac{A_\nu}{\sigma_\mathrm{a}}\right)f + \frac{N^*}{c}A_\nu - n\sigma_\mathrm{s}f + n\sigma_\mathrm{s}\int K[(\vec{\Omega},\vec{\Omega}')]f(\vec{\Omega}')\,d\Omega'. \quad (5.6)$$

The quantity $\frac{c^2}{2\nu^2}\frac{N^*A_\nu}{N_\mathrm{a}\sigma_\mathrm{a}}$ represents the induced emission correction. The form (5.6) is convenient because the emission probability and the absorption cross-section are connected by the *principle of detailed balance* (for some peculiarities concerning strongly broadened lines the reader is referred to [235]). In the case when the matter can be assigned a determined temperature (i.e., when thermodynamic equilibrium is established among the particles of matter), this connection can be found with no difficulty. To that end it suffices to consider the isotropic case, when the radiation is in equilibrium with matter, so that $f = [f]$, where $[f]$ denotes the photon distribution function in the equilibrium state. Then, since

$$\int K[(\vec{\Omega},\vec{\Omega}')]\,d\Omega' = 1,$$

equation (5.6) yields

$$[f] = \frac{2\nu^2}{c^3}\frac{1}{\dfrac{2\nu^2}{c^2}\dfrac{N_\mathrm{a}\sigma_\mathrm{a}}{N^*A_\nu} - 1}.$$

According to the Boltzmann statistics,

$$\frac{N^*}{N_\mathrm{a}} = \frac{g^*}{g_\mathrm{a}}\exp\left(-h\nu/\theta\right),$$

where θ denotes the temperature of matter and g^*, g_a are the corresponding statistical weights of excited and absorbing states. Consequently, the formula for $[f]$ takes on the form

$$[f] = \frac{2\nu^2}{c^3} \frac{1}{\dfrac{2\nu^2}{c^2} \dfrac{\sigma_a}{A_\nu} \dfrac{g_a}{g^*} \exp(h\nu/\theta) - 1}.$$

On the other hand, by Planck's formula, the photon distribution function in the thermodynamic equilibrium state is

$$[f] = f_P(\nu) = \frac{2\nu^2}{c^3} \frac{1}{\exp(h\nu/\theta) - 1}, \tag{5.7}$$

whence

$$\frac{2\nu^2}{c^2} \frac{\sigma_a}{A_\nu} \frac{g_a}{g^*} = 1. \tag{5.8}$$

In the case of partial *local thermodynamic equilibrium*, when matter is characterized by a definite temperature θ, and the photon distribution function is not given by Planck's formula, the radiative transfer equation reads (using (5.7) and (5.8))

$$(\vec{\Omega}\,\mathrm{grad})f = -\left[N_a\sigma_a\left(1 - \exp(-h\nu/\theta)\right) + n\sigma_s\right] f + \frac{N^*}{c} A_\nu + n\sigma_s \int K[(\vec{\Omega}, \vec{\Omega}')] f(\Omega')\, d\Omega', \tag{5.9}$$

where

$$\frac{N^*}{c} A_\nu = N_a\sigma_a \frac{2\nu^2}{c^3} \exp(-h\nu/\theta) = N_a\sigma_a\left(1 - \exp(-h\nu/\theta)\right) f_P(\nu). \tag{5.10}$$

When the conditions of local thermodynamic equilibrium are not satisfied, in particular, when the radiation results in a nonequilibrium distribution of ions among excited states, equation (5.6) must be solved together with the system of equations governing the level kinetics. In this case the populations of the states N_a and N^* are determined by the relations between the rates of the collisional and radiative processes. Here an important circumstance is that the rates of the radiative processes depend on the local intensity of radiation I_ν (the photon distribution function f, see (5.1)). For that reason, in the calculation of the populations in a nonequilibrium plasma and of the distribution of its radiation, a certain consistency between these quantities must be ensured. Since, as a rule, the rates of the radiative processes are much larger that those of collisional processes, in a number of situations one can restrict ourselves to various quasi-stationary approximations, which simplifies the problem considerably [31, 134, 163]. In what follows, as a rule, we will assume that the conditions of local thermodynamic equilibrium are satisfied and, in addition, the radiation is nearly isotropic.

5.1.2 The diffusion approximation

If the angular distribution of photons is almost isotropic and the radiation mean free path may be considered to be small compared to the characteristic lengths, then to solve the problems of radiative gas dynamics one resorts to the diffusion approximation for the transfer equation (5.9). To this end it suffices to expand the distribution function f in spherical harmonics and then retain only the terms of first order in the cosines.

To derive the diffusion approximation we consider first the one-dimensional case, when the distribution function does not depend on y and z and (5.9) can be written as

$$\mu\frac{\partial f}{\partial x} = -\left[N_{\mathrm{a}}\sigma_{\mathrm{a}}\left(1-\exp\left(-h\nu/\theta\right)\right)+n\sigma_{s}\right]f + N_{\mathrm{a}}\sigma_{\mathrm{a}}\left(1-\exp\left(-h\nu/\theta\right)\right)f_{\mathrm{P}} + n\sigma_{\mathrm{s}}\int K[(\vec{\Omega},\vec{\Omega}')]f(\mu')\,d\Omega', \quad (5.11)$$

where $\mu = \cos\vartheta$, ϑ being the angle between the direction of the photon momentum and the x-axis. In the present case $d\Omega = \sin\vartheta\, d\vartheta\, d\varphi$. Let us introduce the *angular moments* of the distribution function:

$$f_k = \int \mu^k f(\mu)\,d\Omega = 2\pi\int_{-1}^{1}\mu^k f(\mu)\,d\mu, \quad k = 0,1,2.$$

The zeroth moment f_0 is simply the concentration of photons of frequency ν at the point with coordinate x. The first angular moment is equal, up to a constant factor, to the *flux* of photons of frequency ν in the direction of the x-axis:

$$j_x(\nu) = \int v_x f(\mu)\,d\Omega = \int c\mu f(\mu)\,d\Omega = cf_1. \quad (5.12)$$

Multiplying $j_x(\nu)$ by $h\nu$, we obtain the flux of radiative energy $F_x = h\nu cf_1$. The second angular moment gives the flux of the momentum, i.e., up to a constant factor, the pressure P exerted by photons on a surface perpendicular to the x-axis:

$$P = \int p_x v_x f(\mu)\,d\Omega = \int \frac{h\nu}{c}\mu\, c\mu\, f(\mu)\,d\Omega = h\nu f_2 \quad (5.13)$$

(here p_x and v_x are the projections of the momentum and the velocity on the x-axis).

Multiplying equation (5.11) by different powers of μ and integrating with respect to μ, we can obtain equations that connect angular moments of various orders. Indeed, let us multiply (5.11) by the first power of μ and integrate. We obtain

$$\frac{\partial f_2}{\partial x} = -\left[N_{\mathrm{a}}\sigma_{\mathrm{a}}\left(1-\exp\left(-h\nu/\theta\right)\right)+n\sigma_{\mathrm{s}}\right] f_1 + \\ n\sigma_{\mathrm{s}} \iint K[(\vec{\Omega},\vec{\Omega}')]\,\mu f(\nu')\,d\Omega\,d\Omega'. \quad (5.14)$$

In the right-hand side of (5.14) the integration must be carried out over all possible directions $\vec{\Omega}$ and $\vec{\Omega}'$. In the integration one can exploit the fact that the scattering probability $K[(\vec{\Omega},\vec{\Omega}')]$ depends only on the angle $(\vec{\Omega},\vec{\Omega}')$ between the directions $\vec{\Omega}$ and $\vec{\Omega}'$. Let $\mu^* = \cos(\vec{\Omega},\vec{\Omega}')$. Clearly, the result of the integration does not change if in the integrand in (5.14) we change the integration variables and for a fixed direction $\vec{\Omega}'$, instead of integrating with respect to the direction $\vec{\Omega}$, we integrate over all possible μ^* and the corresponding φ; here in the integrand the quantity μ (that is, the cosine of the angle between the direction $\vec{\Omega}$ and the x-axis) must be expressed through μ^* and μ'. After that we must integrate with respect to the direction $\vec{\Omega}'$. This yields

$$\iint K(\mu^*)\mu f(\mu')\,d\Omega\,d\Omega' = \int f(\mu')\left[\int \mu K(\mu^*)\,d\Omega^*\right] d\Omega'. \quad (5.15)$$

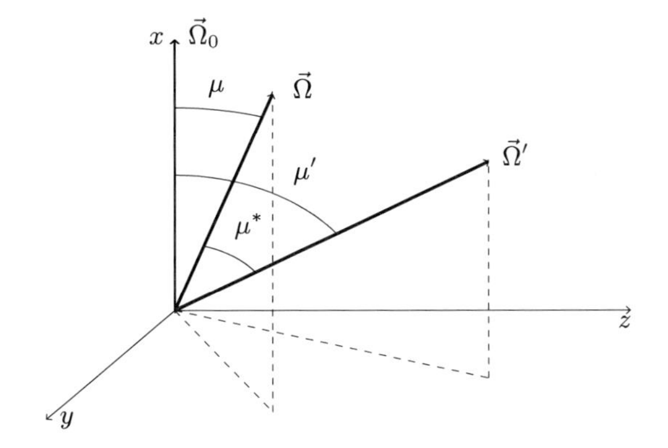

Figure 5.1: Coordinate axes

To integrate with respect to Ω^* it is convenient to choose a new system of coordinates (x', y', z'), in which the z'-axis points in the fixed direction $\vec{\Omega}'$. Then μ, i.e., the cosine of the angle between $\vec{\Omega}$ and the x-axis, can be written as the scalar product of the corresponding unit vectors $\vec{\Omega}$ and $\vec{\Omega}_0$:

$$\vec{\Omega} = (\sin\vartheta\,\cos\varphi,\ \sin\vartheta\,\sin\varphi,\ \cos\vartheta),$$

$$\vec{\Omega}_0 = (\sin\vartheta_0\,\cos\varphi_0,\ \sin\vartheta_0\,\sin\varphi_0,\ \cos\vartheta_0).$$

The angles ϑ, φ characterize the direction $\vec{\Omega}$ ($\cos\vartheta = \mu^*$); the x-axis points in the direction $\vec{\Omega}_0$ ($\cos\vartheta_0 = \mu'$, see Figure 5.1).

Therefore,

$$\mu = (\vec{\Omega} \cdot \vec{\Omega}_0) = \cos\vartheta\,\cos\vartheta_0 + \sin\vartheta\,\sin\vartheta_0\cos(\varphi - \varphi_0). \tag{5.16}$$

After integration over φ in the inner integral in (5.15), the second term of expression (5.16) for μ gives 0, because both $K(\mu^*)$ and $f(\mu')$ are independent of the angle φ. Further, since $\cos\vartheta\,\cos\vartheta_0 = \mu^*\mu'$, the integral in the right-hand side of (5.14) can be written as a product of two integrals:

$$\int f(\mu')\left[\int \mu K(\mu^*)\,d\Omega^*\right]d\Omega' = \int \mu' f(\mu')\,d\Omega' \int \mu^* K(\mu^*)\,d\Omega^*. \tag{5.17}$$

The first of these integrals is the first angular moment $f_1 = j_x(\nu)/c$, the second integral is the cosine of the scattering angle μ^*, averaged over the scattering indicatrix $K(\mu^*)$, i.e., the average cosine of the deviation angle ϑ of a particle due to scattering, which characterizes the anisotropy of scattering:

$$\int \mu^* K(\mu^*)\,d\Omega^* = \overline{\cos\vartheta}. \tag{5.18}$$

Substituting (5.17) in (5.14) we obtain an important relation connecting the first and second angular moments of the distribution function, i.e., the energy flux and the pressure:

$$\frac{\partial f_2}{\partial x} = -\left[N_{\rm a}\sigma_{\rm a}\left(1 - \exp\left(-h\nu/\theta\right)\right) + n\sigma_{\rm s}(1 - \overline{\cos\vartheta})\right] f_1. \tag{5.19}$$

As (5.19) shows, the role of the scattering is characterized by the quantity

$$\sigma_{\rm tr} = \sigma_{\rm s}(1 - \overline{\cos\vartheta}), \tag{5.20}$$

which is known as the *transport cross-section*. For isotropic scattering, as one can see from (5.18), $\overline{\cos\vartheta} = 0$ and $\sigma_{\rm tr} = \sigma_s$. If scattering is anisotropic, but symmetric, i.e., the scattering indicatrix is an even function of μ^*, then from (5.18) and (5.20) it follows that $\sigma_{\rm tr} = \sigma_{\rm s}$, as in the isotropic case.

Using (5.12) and (5.19), we obtain the following expression for the flux:

$$j_x(\nu) = -\ell(\nu)c\frac{\partial f_2}{\partial x}, \tag{5.21}$$

where

$$\ell(\nu) = \frac{1}{N_{\rm a}\sigma_{\rm a}\left(1 - \exp\left(-h\nu/\theta\right)\right) + n\sigma_{\rm s}(1 - \overline{\cos\vartheta})} \tag{5.22}$$

is the effective free path of photons with frequency ν, which is determined by the scattering cross-section, corrected for anisotropy, and the absorption cross-section, corrected for induced emission. In the nonrelativistic case the differential scattering cross-section is proportional to $1+\cos^2\vartheta$, and so according to (5.18) we will henceforth assume that $\overline{\cos\vartheta} = 0$ whenever scattering is accounted for.

It follows from (5.21) and (5.13) that the flux is proportional to the gradient of the photon pressure. Formula (5.21) resembles the elementary formula for particle diffusion in the one-dimensional case,

$$j = -\frac{\ell v}{3}\frac{df_0}{dx} \tag{5.23}$$

(where ℓ is the mean free path, v the velocity, and f_0 the concentration of particles).

Indeed, in the case when the distribution function has the form

$$f = \frac{1}{4\pi} f_0 + a\mu, \tag{5.24}$$

where a is a constant, the expressions (5.21) and (5.23) coincide, because

$$f_2 = 2\pi \int_{-1}^{1} f\,\mu^2\,d\mu = 2\pi \int_{-1}^{1} \left(\frac{f_0}{4\pi} + a\mu\right)\mu^2\,d\mu = \frac{1}{3} f_0. \tag{5.25}$$

Here it is natural to use, instead of the transfer equation, the *diffusion equation* (analogous to the heat conduction equation) and to regard the process of energy transfer by photons as *radiative heat conduction.*

If the distribution function f is not known, but one can assume that

$$f = \frac{1}{4\pi} f_0 + \delta, \tag{5.26}$$

and that f is almost isotropic ($|\delta| \ll f_0$), then in this very simple case we also have

$$f_2 = 2\pi \int_{-1}^{1} \left(\frac{f_0}{4\pi} + \delta\right)\mu^2\,d\mu \approx \frac{f_0}{2}\int_{-1}^{1} \mu^2\,d\mu = \frac{1}{3} f_0,$$

i.e., in accordance to (5.21),

$$j_x(\nu) = -\frac{\ell(\nu)c}{3}\frac{\partial f_0}{\partial x}. \tag{5.27}$$

Formula (5.27) expresses the *diffusion approximation* for the one-dimensional case. To generalize it to the three-dimensional case, it suffices to replace $\partial f_0/\partial x$ by grad f_0:

$$\vec{j}(\nu) = -\frac{\ell(\nu)c}{3}\,\mathrm{grad}\, f_0. \tag{5.28}$$

We can state that the diffusion approximation is applicable whenever in the spherical harmonics expansion of the angular part of the distribution function one can neglect all terms, except the first two. Notice that the simplest diffusion approximation with the coefficient 1/3, obtained by means of formula (5.25), can be refined considerably by resorting to the *quasi-diffusion approximation* [72, 174]. In the quasi-diffusion approximation the diffusion coefficients are calculated by taking into account the real distribution of the radiation fields.

In what follows, side by side with the notations h and ν, we shall also use $\hbar = h/(2\pi)$ and $\omega = 2\pi\nu$.

5.1.3 The Rosseland mean opacity

Under conditions of complete local thermodynamic equilibrium, when matter and radiation are characterized in each point of space by one and the same temperature, (5.7) yields for the photon concentration $f_0(\omega)$ the expression

$$f_0(\omega) = 4\pi f_{\mathrm{P}}(\omega) = 2f_{\mathrm{P}}(\nu) = \frac{\omega^2}{\pi^2 c^3}\,\frac{1}{\exp(\hbar\omega/\theta) - 1}.$$

According to (5.28), in the diffusion approximation the *flux of energy* transferred by photons with frequency ω is equal to

$$\vec{F}(\omega) = -\frac{\ell(\omega)c}{3}\,\mathrm{grad}\,u(\omega), \tag{5.29}$$

where

$$\ell(\omega) = \frac{1}{N_{\mathrm{a}}\sigma_{\mathrm{a}}\left(1 - \exp(-\hbar\omega/\theta)\right) + n\sigma_{\mathrm{s}}}, \tag{5.30}$$

$$u(\omega) = \hbar\,\omega\, f_0(\omega) = \frac{\hbar\omega^3}{\pi^2 c^3}\,\frac{1}{\exp(\hbar\omega/\theta) - 1} \tag{5.31}$$

is the spectral density of radiative energy.*) According to the Stefan-Boltzmann law,

$$u = \int_0^\infty u(\omega)\,d\omega = a\theta^4, \qquad a = \frac{\pi^2}{15c^3\hbar^3}. \tag{5.32}$$

The total flux of radiative energy

$$\vec{F} = \int_0^\infty \vec{F}(\omega)\,d\omega \tag{5.33}$$

*) $u(\omega) = \frac{4\pi}{c} B_\omega$, where B_ω is the Planckian: $B_\omega = \frac{\hbar\omega^3}{4\pi^3 c^2}\frac{1}{\exp(\hbar\omega/\theta) - 1}$.

is given in the diffusion approximation by an expression of the form

$$\vec{F} = -\frac{\ell c}{3}\,\mathrm{grad}\,u. \tag{5.34}$$

Combining formulas (5.29), (5.31) and (5.33), (5.34) one obtains the *Rosseland mean free path*

$$\ell = \frac{15}{4\pi^4}\int_0^\infty \ell(x)\,\frac{x^4 e^{-x}}{(1-e^{-x})^2}\,dx, \tag{5.35}$$

where $x = \hbar\omega/\theta$, and the spectral free path $\ell(x)$ for the energy of photons $\hbar\omega$ is given by formula (5.30).

In the astrophysics literature, to characterize radiative transfer properties of matter, together with the Rosseland mean free path ℓ, one uses the quantity

$$\kappa_{\mathrm{R}} = \frac{1}{\rho\ell},$$

called *opacity* or *Rosseland mean opacity* (ρ is the density of matter). The quantity κ_{R} is also referred to as the Rosseland mean absorption coefficient.

Although the flux of radiative energy (5.34), written as (see (5.32))

$$\vec{F} = -k\,\mathrm{grad}\,\theta,$$

is identical in form with the usual Fourier law of heat conduction, the equations describing energy transfer in a hot dense medium are highly nonlinear, because $k = (4/3)\,\ell c a\theta^3$, where $\ell = \ell(\rho,\theta)$ depends in a complicated way on temperature and density (see Figure (5.29)).

5.1.4 The Planck mean. Radiation of an optically thin layer

Let us examine the opposite limit case, when the photon free path is comparable with the characteristic dimensions of plasma and the radiative heat conduction approximation is not applicable. Ordinarily, in this case scattering effects can be neglected.

By (5.9), for a plane plasma layer in local thermodynamic equilibrium we have

$$\mu\,\frac{\partial f}{\partial x} = -\rho\kappa(\omega)\Big(f - f_{\mathrm{P}}(\omega)\Big), \tag{5.36}$$

where the spectral absorption coefficient with induced emission accounted for is given by

$$\kappa(\omega) = \frac{N_{\mathrm{a}}\sigma_{\mathrm{a}}(\omega)}{\rho}\,(1-\exp(-\hbar\omega/\theta))\,.$$

For a homogeneous plasma layer of thickness L, the solution of (5.36) in the absence of other external sources has the form (the x-axis is perpendicular to the layer):

$$f(x,\mu,\omega) = \begin{cases} f_{\rm P}(\omega)\,(1-\exp(-\kappa(\omega)\rho x/\mu)), & \text{if } 0 \le \mu \le 1, \\ f_{\rm P}(\omega)\,(1-\exp(\kappa(\omega)\rho(L-x)/\mu)), & \text{if } -1 \le \mu \le 0. \end{cases} \tag{5.37}$$

where $0 \le x \le L$.

The flux of radiation leaving the plasma layer at $x = L$ in the positive direction of the x-axis ($\mu > 0$) is given by the double integral

$$F^+ = \int d\omega \int_{\mu>0} d\Omega\; c\hbar\omega\,\mu f = 2\pi c\hbar \int_0^\infty d\omega \int_0^1 d\mu\,\omega\,\mu f.$$

In accordance with our assumption that the absorption coefficient $\kappa(\omega)$ is small, in (5.37) we replace the exponential by its series expansion and retain only the first two terms. This yields the following expression for the outgoing radiation (for a more accurate solution for a plane layer see Subsection 5.7.6):

$$F^+ = 2\pi\kappa_{\rm P}\rho L\,\frac{cu}{4\pi} = \kappa_{\rm P}\rho L\frac{ca\theta^4}{2},$$

where

$$\kappa_{\rm P} = \frac{\displaystyle\int \kappa(\omega)u(\omega)\,d\omega}{\displaystyle\int u(\omega)\,d\omega} = \frac{15}{\pi^4}\int_0^\infty \kappa(x)\,\frac{x^3e^{-x}}{1-e^{-x}}\,dx \tag{5.38}$$

is called the Planck mean absorption coefficient. Note that the expression for the flux F^+ can be recast as

$$F^+ = 2\kappa_{\rm P}\rho L\sigma T^4,$$

where $\sigma = ca/4 = 1.028\cdot 10^5$ W/(cm^2 eV^4) is the Stefan-Boltzmann constant.

5.2 Quantum-mechanical expressions for the effective photon absorption cross-sections

5.2.1 Absorption in spectral lines

The interaction of radiation with matter at high temperatures is determined by the behavior of electrons in the electromagnetic radiation field. To describe this behavior one usually applies the nonstationary perturbation theory to the electron–radiation-field system. It is assumed that in the absence of a field the electron is in some steady state belonging to the discrete or continuous spectrum.

In the nonrelativistic approximation the correction to the Hamiltonian is equal to the difference between the kinetic energy of the electron in the presence of the field and in its absence, respectively:

$$H' = \frac{1}{2m}\left(\widehat{\vec{p}} - \frac{e}{c}\vec{A}\right)^2 - \frac{1}{2m}\widehat{\vec{p}}^{\,2},$$

or

$$H' = -\frac{e}{mc}\widehat{\vec{p}}\cdot\vec{A} + \frac{e^2}{2mc^2}\vec{A}^2, \tag{5.39}$$

where $\widehat{\vec{p}} = -i\hbar\nabla$ is the electron momentum operator and $\vec{A}$ is the vector potential of the electromagnetic field.

The second term in (5.39) is essential only for high densities of radiation, when multi-photon processes become important. As a result, the ion–radiation-field interaction is approximately described by the operator

$$H' = -\frac{e}{mc}\sum_j \widehat{\vec{p}}_j \cdot \vec{A}_j,$$

where the sum is taken over all electrons of the ion under consideration and $\vec{A}_j$ is the vector potential of the radiation field acting on the j-th electron.

In the first approximation of perturbation theory the total probability of absorption of a photon per unit of time for the ion transition from state a to state b can be calculated by the formula (see, e.g., [67, 30, 123, 205]):

$$W_{ab} = \frac{1}{2}\sum_s \frac{\pi e^2}{\hbar\omega_{ba}^2 m^2}\, u(\omega_{ba}) \int d\Omega_k \left|\left\langle b\left|\sum_j e^{i\vec{k}\vec{r}_j}\left(\vec{u}^{(s)}\cdot\widehat{\vec{p}}_j\right)\right|a\right\rangle\right|^2 .$$

Here $u(\omega)\,d\omega$ is the energy density of radiation in the interval $(\omega, \omega + d\omega)$; $\omega_{ba} = (\varepsilon_b - \varepsilon_a)/\hbar$, where ε_a and ε_b are the energies of the ion in the initial and final state, respectively; $k = \omega_{ba}/c$ is the wave number of the photon; $d\Omega_k$ is the solid-angle element in the direction of the wave vector $\vec{k}$; $\vec{u}^{(s)}$ is the polarization unit vector of the photon ($s = 1, 2$); $\vec{r}_j$ is the position vector of the j-th electron of the ion.

The probability of absorption, per unit of time, of a photon with frequency in the interval $(\omega, \omega + d\omega)$ when the ion undergoes a transition from state a to state b is given by the expression

$$W(\omega)\,d\omega = W_{ab}J_{ab}(\omega)d\omega,$$

where $J_{ab}(\omega)$ denotes the profile of the absorption line ($\int J_{ab}(\omega)\,d\omega = 1$). Questions concerning the broadening of spectral lines are treated in § 5.4.

In practice it is convenient to pass from transition probabilities to effective cross-sections:

$$\sigma(\omega) = \frac{W(\omega)}{cu(\omega)/(\hbar\omega_{ba})} =$$

$$J_{ab}(\omega)\,\frac{\pi}{2}\alpha\, a_0^2\,\frac{1}{m\hbar\omega_{ba}}\sum_s\int d\Omega_k\left|\left\langle b\left|\sum_j e^{i\vec{k}\vec{r}_j}\,(\vec{u}^{(s)}\widehat{\vec{p}}_j)\right|a\right\rangle\right|^2 \qquad (5.40)$$

(here a_0 denotes the Bohr radius and α is the fine-structure constant).

In most cases expression (5.40) can be simplified considerably using the electric dipole approximation. For an electron in a bound state with principal quantum number n one has the estimate $r_j \sim n^2/Z$. Therefore, in (5.40) the argument of the exponential $\vec{k}\cdot\vec{r}_j \sim \omega\, r_j/c$ is usually small and for $\omega < Zc/n^2$ the exponential can be replaced by 1.

Passing to the dipole approximation in (5.40) and taking into account the relation between matrix elements of the momentum operator of the electron and matrix elements of the position-vector operator, we obtain the following expression for the absorption cross-section for the ion transition from state a to state b (integration over Ω_k for nonpolarized radiation gives the factor $8\pi/3$; the profile $J_{ab}(\omega)$ is normalized on the energy scale in atomic units):

$$\sigma(\omega) = 2\pi^2\alpha\, a_0^2 f_{ab} J_{ab}(\omega), \qquad (5.41)$$

where f_{ab} is the oscillator strength:

$$f_{ab} = \frac{2}{3}\,\frac{m\omega_{ba}}{\hbar}\left|\langle a|\sum_j \vec{r}_j|b\rangle\right|^2. \qquad (5.42)$$

The oscillator strength is a dimensionless quantity; its numerical values are smaller than unity (cf. the theorem on the sum of oscillator strengths [28]).

To derive expressions (5.41)–(5.42) for the oscillator strength we used the connection between the matrix elements of the momentum and position-vector operators: $\langle b|\widehat{\vec{p}}\,|a\rangle = -im\omega_{ba}\,\langle b|\vec{r}\,|a\rangle$. As shown, for example, in [28], in a Coulomb field Z/r one also has the relation

$$\langle b|\widehat{\vec{p}}\,|a\rangle = i\frac{Ze^2}{\omega_{ba}}\,\langle b\left|\frac{\vec{r}}{r^3}\right|a\rangle.$$

Thus, in addition to (5.42), one can obtain two other expressions for the oscillator strength if we replace the operator $\vec{r}_j$ by $\vec{p}_j/\omega_{ba}$, or by $Z\vec{r}_j/(\omega_{ba}^2\, r_j^3)$. If one would use the exact wave functions of the states a and b, then the three expressions would coincide. However, since in computations one uses approximate wave functions, the three different ways of writing the oscillator strengths may lead to different results. Usually the operators $\vec{r}$ and $\vec{p}/\omega_{ba}$ lead to identical results; when the operator $Z\vec{r}/(\omega_{ba}^2 r^3)$ is used, the calculation of oscillator strengths requires highly accurate wave functions for small values of r.

Now let us pass to the single-configuration approximation in (5.42). Suppose that for a given electron configuration Q there is defined the set of occupation numbers $\left\{N_{n\ell}^Q\right\}$ and the wave functions for the states $|a\rangle$ and $|b\rangle$ are linear combinations of determinants (3.22) (we use perturbation theory to account for the noncentral part of the electrostatic interaction of electrons and their spin-orbit interaction). In the absence of external fields (and neglecting the inner electric microfield) one can assume that the states $|a\rangle$ and $|b\rangle$ are characterized by some value of the momentum J and its projection on the z-axis, M. Since for a given configuration of the ion with fixed J and M there may be several states with different energies, we introduce an additional index γ, which specifies uniquely the state of the ion in the approximation adopted here. For the moment we will not specify the type of the vector coupling, allowing the types LS, jj, or the coupling of intermediate type (see § 5.3 below). We have

$$|a\rangle \equiv |\gamma JM\rangle = \sum_q C_{qM}^a \Psi_{qM}^a, \quad |b\rangle \equiv |\gamma' J'M'\rangle = \sum_{q'} C_{q'M'}^b \Psi_{q'M'}^b,$$

where Ψ_{qM}^a, $\Psi_{q'M'}^b$ are the one-determinant wave functions, constructed, for example, from one-electron functions of the form

$$\psi_\nu(q) = \psi_{n\ell m_l m_s}(\vec{r},\sigma) = \frac{1}{r} R_{n\ell}(r)\,(-1)^{m_\ell} Y_{\ell m_\ell}(\vartheta,\varphi)\chi_{\frac{1}{2}m_s}(\sigma).$$

Since the operator $\sum_j \vec{r}_j$ is a sum of operators, each of which acts on the coordinates of only one electron, the oscillator strength f_{ab} will be different from zero in the case when the final state of the ion differs from the initial state through the position of only one electron (the description of the states a and b uses the same one-electron wave functions), i.e.,

$$f_{ab} = \frac{2}{3}\omega_{ba} \sum_{qq'} \left(C_{q'M'}^b\right)^* C_{qM}^a |\vec{r}_{\alpha\beta}|^2,$$

$$\vec{r}_{\alpha\beta} = \int \psi_\alpha^*(q)\vec{r}\psi_\beta(q)\,dq, \tag{5.43}$$

where $\psi_\alpha(q)$, $\psi_\beta(q)$ are the electron wave functions before and after transition. Depending on the initial and final states of the electron, it is convenient to divide the photon absorption processes into three groups:

1) absorption in spectral lines, or bound–bound transitions;
2) photoionization, or bound–free transitions;
3) inverse bremsstrahlung, or free–free transitions.

The total cross-section of absorption in spectral lines (line absorption cross-section), denoted here by $\sigma_{\rm bb}(\omega)$, is obtained by averaging (5.41) with respect to the initial state $a = |Q\gamma JM\rangle$ and summing over the final states $b = |Q'\gamma' J'M'\rangle$:

$$\sigma_{\rm bb}(\omega) = 2\pi^2\alpha\, a_0^2 \sum_{ab} P_a\, f_{ab}\, J_{ab}(\omega), \tag{5.44}$$

where P_a is the probability of the initial state of the ion.

In the calculation of oscillator strengths, an error in the value of ω_{ba} has only a weak influence on the results of the computation, and so this quantity may be replaced by the difference of the average energies over the configurations Q and Q', which differ by a change in the state of a single electron. In this approximation $\omega_{ba} \approx E_{Q'} - E_Q$, where the energies averaged over configurations, E_Q and $E_{Q'}$, are determined by the sets of occupation numbers $\{N_{n\ell}^Q\}$ and $\{N_{n\ell}^{Q'}\}$. The probability of the state is $P_a \approx g_a P_Q / g_Q$, where g_a is the statistical weight of state a, P_Q is the probability of realization of configuration Q and g_Q is its statistical weight ($\sum_{a\in Q} g_a = g_Q$). Let us remark that such an approximation may turn to be rather crude if the splitting over energies inside subconfigurations is a quantity of the same order as or higher than the temperature of matter (see Subsection 5.5.3).

Since dipole transitions are possible only between configurations that differ through the state of a single electron, which changes its orbital momentum by one unit, summation over the final configurations Q' can be replaced by summation over the one-electron transitions $n\ell \to n'\ell'$. Suppose that in the initial state on the level $n\ell$ there are $N = N_{n\ell}^Q$ electrons, and that the corresponding number for the level $n'\ell'$ is $N' = N_{n'\ell'}^Q$. Then

$$g_Q = \binom{4\ell+2}{N}\binom{4\ell'+2}{N'},$$

where

$$\binom{G}{N} = \frac{G!}{(G-N)!N!}.$$

One can assume that the form of the line $J_{ab}(\omega)$ depends only on the configuration $\{N_{n\ell}^Q\}$ and the quantum numbers $n\ell$ and $n'\ell'$ of the initial and final state of the electron. This corresponds to a group of lines being replaced by one line of intensity equal to the sum of the intensities of all lines in the group (in the multiplet). Here we also use such an approximation; however, in § 5.5 it will be shown how to account for the multiplet structure of the spectrum by means of a certain effective profile.

Since (see [205])

$$\sum_{\gamma JM\gamma' J'M'} \left| \langle Q\gamma JM| \sum_j \vec{r}_j |Q'\gamma' J'M'\rangle \right|^2 = \binom{4\ell+1}{N-1}\binom{4\ell'+1}{N'} \sum_{mm_s m'm'_s} |\langle n\ell m m_s|\vec{r}\,|n'\ell' m' m'_s\rangle|^2 ,$$

we obtain the following expression for the line absorption cross-section [49, 191]:

$$\sigma_{\rm bb}(\omega) = 2\pi^2 \alpha\, a_0^2 \sum_Q P_Q \sum_{n\ell, n'\ell'} N_{n\ell}^Q \left[1 - \frac{N_{n'\ell'}^Q}{2(2\ell'+1)}\right] f_{n\ell,n'\ell'}^Q\, J_{n\ell,n'\ell'}^Q(\omega). \quad (5.45)$$

Here $J_{n\ell,n'\ell'}^Q(\omega)$ is the form of the absorption line for the transition $n\ell \to n'\ell'$, averaged over the states of configuration Q; $f_{n\ell,n'\ell'}^Q$ is the oscillator strength, averaged over the quantum numbers m, m_s of the initial states and summed over the quantum numbers m', m'_s of the final states:

$$f_{n\ell,n'\ell'}^Q = \frac{1}{2(2\ell+1)} \sum_{mm_s} \sum_{m'm'_s} \frac{2}{3} (E_{Q'} - E_Q) \left|\vec{r}_{n\ell m m_s, n'\ell' m' m'_s}\right|^2 , \quad (5.46)$$

$$\vec{r}_{n\ell m m_s, n'\ell' m' m'_s} = \int \psi_{n\ell m m_s}^*(q) \vec{r} \psi_{n'\ell' m' m'_s}(q)\, dq. \quad (5.47)$$

The sum over m, m_s and m', m'_s in (5.46) is calculated by means of the relation (see [28])

$$\sum_{m'm'_s} \left|\vec{r}_{n\ell m m_s, n'\ell' m' m'_s}\right|^2 = \begin{cases} \dfrac{\ell+\ell'+1}{2(2\ell+1)} (r_{n\ell,n'\ell'})^2, & \text{if} \quad \ell' = \ell \pm 1, \\ 0, & \text{if} \quad \ell' \neq \ell \pm 1, \end{cases}$$

where

$$r_{n\ell,n'\ell'} = \int_0^{r_0} R_{n\ell}(r) r R_{n'\ell'}(r)\, dr. \quad (5.48)$$

The difference of the average energies of configurations in (5.46) can be replaced by the corresponding difference in the average-atom approximation according to the Hartree-Fock-Slater model, which gives (see [204])

$$E_{Q'} - E_Q \approx \varepsilon_{n'\ell'} - \varepsilon_{n\ell}.$$

Therefore, for $\ell' = \ell \pm 1$

$$f^Q_{n\ell,n'\ell'} = \frac{2}{3}\left(\varepsilon_{n'\ell'} - \varepsilon_{n\ell}\right)\frac{\ell+\ell'+1}{2(2\ell+1)}\left(r_{n\ell,n'\ell'}\right)^2. \tag{5.49}$$

For high-Z plasmas relativistic effects may play essential role. In the first approximation, we replace the one-electron functions $\psi_{n\ell m m_s}$ by the functions $\psi_{n\ell jm}$, where j is the quantum number characterizing the total momentum of the electron (i.e., the sum of its orbital and spin momenta). Since in the calculation of $\sigma_{\rm bb}(\omega)$ it is particularly important to account for the position of the spectral line, the relativistic corrections to the energy will be calculated by means of solutions of the Dirac equations, while in the calculation of oscillator strengths for the wave functions $\psi_{n\ell jm}$ we will use the nonrelativistic approximation, in which

$$\psi_{n\ell jm}(q) = \sum_{m_\ell m_s} C^{jm}_{\ell m_\ell \frac{1}{2} m_s}\psi_{n\ell m_\ell m_s}(q), \tag{5.50}$$

where $C^{jm}_{\ell m_\ell \frac{1}{2} m_s}$ are the Clebsch-Gordan coefficients.

Thus, for high-Z plasmas in formula (5.45) the quantities

$$\varepsilon_{n\ell}, \quad N^Q_{n\ell}, \quad J^Q_{n\ell,n'\ell'}(\omega), \quad \frac{N^Q_{n'\ell'}}{2(2\ell'+1)}, \quad f^Q_{n\ell,n'\ell'}$$

must be replaced by, respectively,

$$\varepsilon_{n\ell j}, \quad N^Q_{n\ell j}, \quad J^Q_{n\ell j,n'\ell' j'}(\omega), \quad \frac{N^Q_{n'\ell' j'}}{2j'+1}, \quad f^Q_{n\ell j,n'\ell' j'}.$$

In this case we obtain for the oscillator strengths the expression

$$f^Q_{n\ell j,n'\ell' j'} = \frac{2}{3}\left(\varepsilon_{n'\ell' j'} - \varepsilon_{n\ell j}\right)\frac{1}{2j+1}\sum_{mm'}\left|\vec{r}_{n\ell jm,n'\ell' j'm'}\right|^2, \tag{5.51}$$

where

$$\vec{r}_{n\ell jm,n'\ell' j'm'} = \int \psi^*_{n\ell jm}\,\vec{r}\,\psi_{n'\ell' j'm'}\,dq. \tag{5.52}$$

As it follows from (5.50) and (5.52), the dipole matrix elements are expressed through Clebsch-Gordan coefficients. Using the Wigner-Eckhart theorem [220], one can carry out the summation in (5.51) over m, m'. This finally yields

$$f^Q_{n\ell j,n'\ell' j'} = \frac{1}{3}\left(\varepsilon_{n'\ell' j'} - \varepsilon_{n\ell j}\right)(2j'+1)(\ell+\ell'+1)W^2\left(\ell' j' \ell j; \frac{1}{2}1\right)(r_{n\ell,n'\ell'})^2,$$

where $W(\ell' j' \ell j; \frac{1}{2} 1)$ is the Racah coefficient.

We see that the line absorption cross-section for high-Z plasmas can be written in the form (compare with(5.45))

$$\sigma_{\mathrm{bb}}(\omega) = 2\pi^2 \alpha\, a_0^2 \sum_Q P_Q \sum_{n\ell j, n'\ell' j'} N^Q_{n\ell j} \left(1 - \frac{N^Q_{n'\ell' j'}}{2j'+1}\right) f^Q_{n\ell j, n'\ell' j'}\, J^Q_{n\ell j, n'\ell' j'}(\omega). \tag{5.53}$$

The main difficulty in calculating $\sigma_{\mathrm{bb}}(\omega)$ is that one has to account for a huge number of configurations Q, and in a number of cases for each such configuration one has to account for the splitting of levels (see Subsection 5.3.3). Simplifications are usually achieved by using various statistical methods to describe effectively groups of spectral lines that correspond to collections of configurations with close energies (see Subsections 5.5.3, 5.5.5), as well as by simplifying the quantities figuring in (5.53).

At high temperatures the hydrogen-like wave functions may be used [143]. This simplifies considerably the computation of dipole matrix elements (5.48). Substitution of the function

$$R_{n\ell}(r) = C_{n\ell} e^{-Z_{n\ell} r/n} \left(\frac{2Z_{n\ell} r}{n}\right)^{\ell+1} L^{2\ell+1}_{n-\ell-1}\left(\frac{2Z_{n\ell} r}{n}\right)$$

in formula (5.48) yields

$$r_{n\ell, n'\ell'} = \int_0^{x_0} e^{-x} P_N(x)\,dx \approx \int_0^{\infty} e^{-x} P_N(x)\,dx, \tag{5.54}$$

where $x = (Z_{n\ell}/n + Z_{n'\ell'}/n')\, r$, $x_0 = (Z_{n\ell}/n + Z_{n'\ell'}/n')\, r_0$, and $P_N(x)$ is a polynomial of degree $N = n + n'$.

Gauss quadrature using Laguerre polynomials allows one to calculate integrals of the form (5.54) exactly up to degree $N = 2s_{\max} - 1$, where $s_{\max}$ is the number of nodes in the quadrature formula (for $n \le 10$ one can take $s_{\max} = 12$). In this manner we obtain the formula

$$(r_{n\ell, n'\ell'})^2 = Z_{n\ell} Z_{n'\ell'} (\xi_{n\ell, n'\ell'})^2, \tag{5.55}$$

where

$$\xi_{n\ell, n'\ell'} = \frac{nn' \sum\limits_{s=1}^{s_{\max}} a_s x_s M_{n\ell}(x_{1s}) M_{n'\ell'}(x_{2s})}{(n' Z_{n\ell} + n Z_{n'\ell'})^2},$$

$$x_{1s} = \frac{2n' Z_{n\ell}}{n' Z_{n\ell} + n Z_{n'\ell'}} x_s, \quad x_{2s} = 2x_s - x_{1s},$$

and

$$M_{n\ell}(x) = \sqrt{\frac{(n-\ell-1)!}{(n+\ell)!}} x^{\ell+1} L_{n-\ell-1}^{2\ell+1}(x).$$

Here x_s and a_s are the nodes and respectively the weights in the Gauss quadrature formula for the integral (5.54) using the Laguerre polynomials $L_n^0(x)$.

Table 5.1: Dipole matrix elements $r_{n\ell,n'\ell'}$ for gold with temperature $T = 0.5$ keV and density $\rho = 1.93$ g/cm^3

$n\ell$	$n'\ell'$	formula (5.55)	WKB	HFS
1 0	2 1	0.016	0.016	0.016
1 0	3 1	0.006	0.006	0.006
2 0	3 1	0.029	0.037	0.037
2 0	4 1	0.012	0.016	0.016
2 1	3 0	0.009	0.013	0.015
2 1	3 2	0.058	0.059	0.060
2 1	4 0	0.001	0.005	0.006
2 1	4 2	0.026	0.023	0.024
3 0	4 1	0.037	0.068	0.068
3 1	4 0	0.021	0.045	0.047
3 1	4 2	0.066	0.093	0.093
3 2	4 1	0.019	0.031	0.035
3 2	4 3	0.146	0.149	0.151

Table 5.1 lists the dipole matrix elements $r_{n\ell,n'\ell'}$ for gold with temperature $T = 0.5$ keV and density $\rho = 1.93$ g/cm^3, calculated by means of numerical wave functions (HFS), the semiclassical wave functions (2.31) (WKB), and also by formula (5.55). The table shows that the hydrogen-like approximation (5.55) gives totally acceptable results for the basic transitions, though it fails to describe some transitions with small values of $r_{n\ell,n'\ell'}$ (see $r_{21,40}$). Note that the semiclassical wave functions yield a more accurate result than the hydrogen-like approximation for all matrix elements.

Thus, the hydrogen-like approximation (5.55) may prove too crude to describe details of the absorption spectrum when weak lines in the spectrum are important (see [142]).

5.2.2 Photoionization

We have considered transitions in which both the initial and the final state are bound states. Now let us address the case when the final state (and also both the initial and the final state) belong to the continuum. The photon absorption processes in such transitions are referred to respectively as photoabsorption (photoionization) and inverse bremsstrahlung. The cross-sections for such transitions can be derived from the formula for $\sigma_{\rm bb}(\omega)$ upon replacing the wave functions $R_{n'\ell'}(r)$ by the continuum wave functions $R_{\varepsilon'\ell'}(r)$, normalized according to the conditions given in § 2.3 (for a more accurate analysis of photoionization processes the reader is referred to, e.g., [229]).

Here the sum over the quantum numbers n' must be replaced by the integral over the energy ε'. Moreover, if in the consideration of photoionization processes one neglects broadening effects and the deviation of the occupancy of free electron states from the average one, then the expression for the photoionization cross-section in the nonrelativistic approximation takes on the form

$$\sigma_{\mathrm{bf}}(\omega) = 2\pi^2 \alpha\, a_0^2 \sum_Q P_Q \sum_{n\ell,\ell'} N_{n\ell}^Q \int_{\varepsilon_0}^{\infty} (1 - n_{\varepsilon'}) \delta(\omega - \varepsilon' + \varepsilon_{n\ell}^Q) f_{n\ell,\varepsilon'\ell'}^Q \, d\varepsilon', \tag{5.56}$$

$$f_{n\ell,\varepsilon'\ell'}^Q = \begin{cases} \dfrac{2}{3}(\varepsilon' - \varepsilon_{n\ell}) \dfrac{(\ell + \ell' + 1)}{2(2\ell + 1)} (r_{n\ell,\varepsilon'\ell'})^2, & \text{if} \quad \ell' = \ell \pm 1, \\ 0, & \text{if} \quad \ell' \neq \ell \pm 1. \end{cases} \tag{5.57}$$

For high-Z elements, if we introduce the relativistic corrections for the position of photoionization thresholds, formula (5.56) can be recast as

$$\sigma_{\mathrm{bf}}(\omega) = 2\pi^2 \alpha\, a_0^2 \sum_Q P_Q \sum_{n\ell j,\ell' j'} N_{n\ell j}^Q \int_{\varepsilon_0}^{\infty} (1 - n_{\varepsilon'}) \delta(\omega - \varepsilon' + \varepsilon_{n\ell j}^Q) f_{n\ell j,\varepsilon'\ell' j'}^Q \, d\varepsilon', \tag{5.58}$$

$$f_{n\ell j,\varepsilon'\ell' j'}^Q = \frac{1}{3}(\varepsilon' - \varepsilon_{n\ell j})(2j' + 1)(\ell + \ell' + 1) W^2 \left(\ell' j' \ell j; \frac{1}{2} 1\right) (r_{n\ell,\varepsilon'\ell'})^2. \tag{5.59}$$

Here

$$r_{n\ell,\varepsilon'\ell'} = \int_0^{r_0} R_{n\ell}(r) r R_{\varepsilon'\ell'}(r) \, dr, \tag{5.60}$$

n_ε is the distribution function of free electrons

$$n_\varepsilon = \frac{1}{1 + \exp\left(\dfrac{\varepsilon - \mu}{\theta}\right)}, \tag{5.61}$$

and ε_0 is the boundary of the continuum.

Let us examine a method for calculating the radial integrals (5.60) in the case when Coulomb wave functions are used. Obviously, the main contribution to the value of the integral $r_{n\ell,\varepsilon'\ell'}$ comes from the values of r in the region near the maximum of $R_{n\ell}(r)$. In accordance with the idea of the method of the trial potential by means of which the Coulomb functions $R_{n\ell}(r)$ are computed, in the domain where the function $R_{n\ell}(r)$ is notably different from zero the potential $V(r)$ can be replaced by the Coulomb potential with external screening

$$\widetilde{V}(r) = \frac{Z_{n\ell}}{r} - A_{n\ell}.$$

It follows that when one computes the integrals $r_{n\ell,\varepsilon'\ell'}$ in the essential integration domain $r_1 \le r \le r_2$ it is natural to use the radial wave functions $R_{\varepsilon'\ell'}(r)$, which are solutions of the Schrödinger equation for the potential $\widetilde{V}(r)$:

$$R_{\varepsilon'\ell'}(r) = C_{\varepsilon'\ell'}\, F_{\varepsilon'\ell'}(r) + D_{\varepsilon'\ell'}\, G_{\varepsilon'\ell'}(r) \quad (r_1 \le r \le r_2). \tag{5.62}$$

Here $F_{\varepsilon'\ell'}(r)$ is a particular solution of the Schrödinger equation with the potential $\widetilde{V}(r)$ that behaves like $r^{\ell'+1}$ when $r \to 0$, and $G_{\varepsilon'\ell'}(r)$ is another solution of the same equation. The constants $C_{\varepsilon'\ell'}$ and $D_{\varepsilon'\ell'}$ are to be found from the matching conditions in $r = r_1$ of the solution $R_{\varepsilon'\ell'}(r)$ given by (5.62) and the solution of the Schrödinger equation with the potential $V(r)$ in the interval $0 \le r \le r_1$. If the solution (5.62) is used starting with the value $r = 0$, then we must put $D_{\varepsilon'\ell'} = 0$. One may expect that for small values of r_1 the second term in (5.62) gives only a small contribution in the integral $r_{n\ell,\varepsilon'\ell'}$, especially for small values of the principal quantum number n. The validity of this assumption was verified by comparing the values of $r_{n\ell,\varepsilon'\ell'}$ obtained directly by means of the solutions $R_{n\ell}(r)$ and $R_{\varepsilon'\ell'}(r)$ of the Schrödinger equation with the values obtained by the method discussed here.

If one neglects the fact that the potential $V(r)$ differs from $\widetilde{V}(r)$ for $r \le r_1$ and one takes $D_{\varepsilon'\ell'} = 0$, then instead of (5.62) one has [147]:

$$R_{\varepsilon'\ell'}(r) = C_{\varepsilon'\ell'}\, F_{\varepsilon'\ell'}(r), \tag{5.63}$$

where for $\varepsilon' \ge A_{n\ell}$

$$F_{\varepsilon'\ell'}(r) = r^{\ell'+1}\, e^{-ikr}\, F\left(\frac{i\, Z_{n\ell}}{k} + \ell' + 1, 2\ell' + 2, 2ikr\right), \quad k = \sqrt{-2\varepsilon'}. \tag{5.64}$$

Using the hydrogen-like wave functions for the bound and free states we obtain

$$f_{n\ell,\varepsilon'\ell'} = f_{n\ell,\varepsilon'\ell\mp1} = \frac{2^{4\ell'+5}}{3(n-\ell-1)!(n+\ell)!}\left(\frac{n}{Z_{n\ell}}\right)^4 \frac{\ell_>}{2\ell+1} f_n(\lambda)\lambda^{\ell'+4} \times$$
$$\left[(n \pm \ell_>)(n-1\pm\ell_>)\varphi_{n-\ell'-2,\ell'}(\lambda) - \varphi_{n-\ell',\ell'}(\lambda)\right]^2 \prod_{s=1}^{\ell'} \left[(n^2-s^2)\lambda + s^2\right], \tag{5.65}$$

$$\lambda \equiv \lambda_{n\ell} = \frac{1}{\varepsilon' - \varepsilon_{n\ell}} \frac{Z_{n\ell}^2}{2n^2}, \quad \ell_> = \max(\ell, \ell');$$

$$f_n(\lambda) = \begin{cases} \dfrac{\exp\left(-4n\sqrt{\dfrac{\lambda}{1-\lambda}}\,\mathrm{arctg}\sqrt{\dfrac{1-\lambda}{\lambda}}\right)}{1-\exp\left(-2\pi n\sqrt{\dfrac{\lambda}{1-\lambda}}\right)}, & \text{if } \lambda < 1, \\ \exp(-4n), & \text{if } \lambda = 1, \\ \exp\left(-4n\sqrt{\dfrac{\lambda}{\lambda-1}}\,\mathrm{arcth}\sqrt{\dfrac{\lambda-1}{\lambda}}\right), & \text{if } \lambda > 1. \end{cases}$$

When $\ell = 1$, $\ell' = 0$ the product $\prod_{s=1}^{\ell'} \left[(n^2 - s^2)\lambda + s^2\right]$ must be replaced by 1. The functions $\varphi_{m\ell}(\lambda)$ are calculated by means of the recursion relations

$$\varphi_{m\ell}(\lambda) = -(m-1)(m+2\ell)\varphi_{m-2,\ell}(\lambda) + 2\big[2\lambda(m+\ell-n) - m - \ell\big]\varphi_{m-1,\ell}(\lambda),$$

where $m = 0, 1, 2, \ldots$, $\varphi_{0\ell}(\lambda) = 1$, $\varphi_{-1,\ell}(\lambda) = 0$.

Figure 5.2 ($T = 0$) compares experimental values of the photoionization cross-section in gold [243], obtained under normal conditions, with computational results in various approximations. The calculations were carried out for the temperature $T = 1$ eV and density $\rho = 0.1$ g/cm^3, since the photoionization cross-sections are practically independent of temperature in the considered range of photon energies $\hbar\omega > 0.1$ keV (to simplify notation, here and in what follows we will write ω instead $\hbar\omega$ in our graphs). For $T \sim 0$ there is also no dependence on density, since the cross-sections are expressed in cm^2/g. The calculations of the photoionization thresholds and effective charges were carried out by using the code THERMOS and the Hartree-Fock-Slater potential with relativistic corrections accounted for.

As the graphs demonstrate, the results obtained by using numerical wave functions agree well with the experimental values in the whole range of photon energies. As expected, for $T = 0$ the hydrogen-like approximation yields good results for all inner shells ($n \le 3$), which allows one to simplify considerably the calculations for large photon energies. For the temperature $T = 1$ keV the hydrogen-like approximation agrees with the numerical calculations for all the shells accounted for ($n \le 9$)

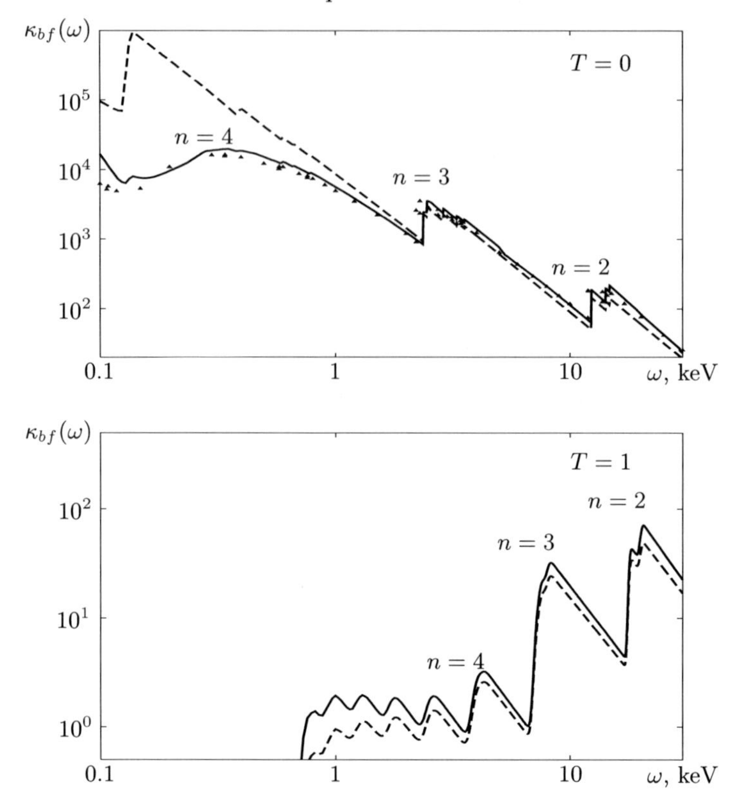

Figure 5.2: Dependence of the photoionization cross-section in gold (measured in cm^2/g) on the energy of photons in keV. The solid [resp., dashed] curve shows the results of calculations with the code THERMOS using numerical wave functions [resp., the hydrogen-like approximation (5.65)]. In the upper figure the small triangles indicate experimental values [243]

5.2.3 Inverse bremsstrahlung

When one calculates the inverse bremsstrahlung cross-sections one can neglect the deviation of the occupation numbers of the initial state from their mean values, i.e., one can replace P_Q, $\varepsilon_{n\ell}^Q$ and $N_{n\ell}^Q$ by 1, ε and $2(2\ell+1)n_\varepsilon$, respectively, and also neglect relativistic effects. In this approximation the cross-section, calculated per one atom cell, has the form

$$\sigma_{\rm ff}(\omega) = 2\pi^2\alpha\, a_0^2 \sum_{\ell} \sum_{\ell'=\ell\pm1} \int\limits_{\varepsilon_0}^{\infty}\int\limits_{\varepsilon_0}^{\infty} 2(2\ell+1)n_\varepsilon(1-n_{\varepsilon'})\delta(\omega+\varepsilon-\varepsilon') f_{\varepsilon\ell,\varepsilon'\ell'}\, d\varepsilon d\varepsilon'. \quad (5.66)$$

The summation over ℓ is carried out from $\ell = 0$ to $\ell = \ell_{\max}$. The maximal orbital momentum $\ell_{\max}$ is defined so that when $\ell > \ell_{\max}$ the domain of classical motion of the electron lies beyond the limits of the atom cell. To carry out the summation one can use the Gauss quadrature formulas, which rely on Chebyshev polynomials of a discrete variable [154]. To calculate the oscillator strengths it is convenient to use the matrix elements of the acceleration, which give the main contribution for small r:

$$f_{\varepsilon\ell,\varepsilon'\ell'} = \frac{\ell+\ell'+1}{3(2\ell+1)} \frac{1}{(\varepsilon'-\varepsilon)^3} \left[\int_0^{r_0} R_{\varepsilon\ell}(r) \frac{dV(r)}{dr} R_{\varepsilon'\ell'}(r)\, dr \right]^2 . \tag{5.67}$$

The inverse bremsstrahlung cross-section (5.66) may be alternatively written using the Gaunt generalized factor $g(\varepsilon, \varepsilon')$:

$$\sigma_{\text{ff}}(\omega) = \frac{16\pi\alpha a_0^2}{3\sqrt{3}} \frac{Z_0^2}{\omega^3} \int_0^{\infty} n_\varepsilon (1 - n_{\varepsilon'}) g(\varepsilon, \varepsilon')\, d\varepsilon, \tag{5.68}$$

where Z_0 is the average number of free electrons, $\varepsilon' = \varepsilon + \omega$. The Gaunt factor is chosen so that in the Kramers approximation $g(\varepsilon, \varepsilon') = 1$.

Calculations based on formula (5.66) are rather time consuming, so in practice one resorts to various approximations. For example, in the Born-Elwert approximation the function $g(\varepsilon, \varepsilon')$ can be written in the form

$$g(\varepsilon, \varepsilon') = \frac{\sqrt{3}}{\pi} \sqrt{\frac{\varepsilon'}{\varepsilon}} \ln \left(\frac{\sqrt{\varepsilon'} + \sqrt{\varepsilon}}{\sqrt{\varepsilon'} - \sqrt{\varepsilon}} \right) \frac{1 - \exp\left[-2\pi Z_0/\sqrt{2\varepsilon'}\right]}{1 - \exp\left[-2\pi Z_0/\sqrt{2\varepsilon}\right]}. \tag{5.69}$$

The inverse bremsstrahlung absorption cross-section (5.68) with the factor $g(\varepsilon, \varepsilon')$ in formula (5.69) differs from the commonly used Born-Elwert approximation [205] by accounting for degeneracy effects. Computational results for the inverse bremsstrahlung cross-section by formula (5.68) for an iron plasma in essentially different temperature and density ranges are shown in Figure 5.3. For comparison, results of computations carried out with the code STA [242] using numerical wave functions are shown as well. For temperature $T = 20$ eV and density $\rho = 10^{-4}$ g/cm^3 (rarefied plasma, see Figure 5.3a), in the calculation with numerical wave functions special methods of summation over the quantum number ℓ had to be used, because the number of terms in (5.66) is very large. For temperature $T = 600$ eV and density $\rho = 8$ g/cm^3 (dense hot plasma, see Figure 5.3b), the main difficulty is to integrate numerically with respect to the energy ε, because for large ε the integrands are rapidly oscillating functions.

As the figures show, in the region of not too large photon energies ($\omega/\theta < 5$) the results of the calculations using wave functions and those using the approximation (5.68), (5.69) practically coincide.

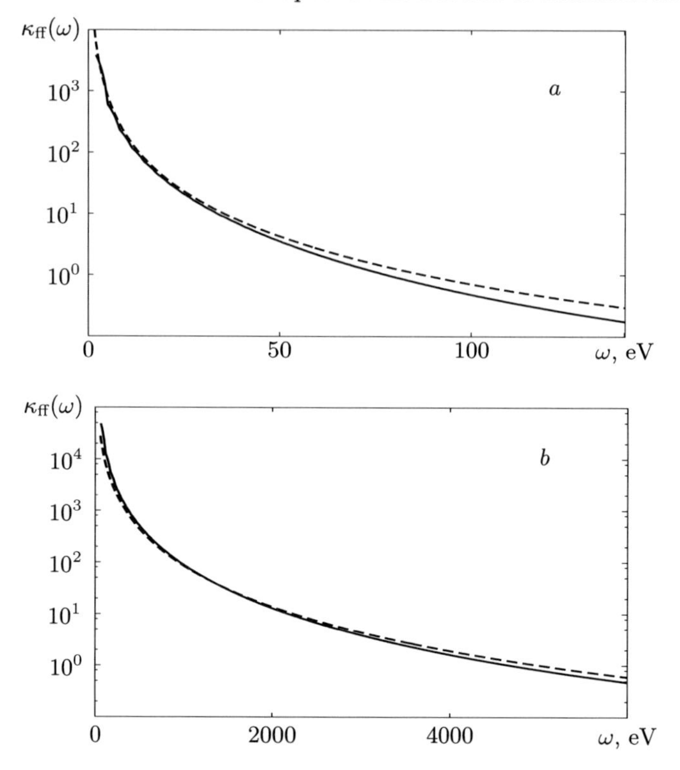

Figure 5.3: Dependence of the photoionization cross-section in iron (measured in cm^2/g) on the energy of photons, measured in eV. The continuous [resp., hatched] curve shows the results obtained using formula (5.68) [resp., the code STA]. The calculations were carried out for temperature $T = 20$ eV and density $\rho = 10^{-4}$ g/cm^3 [Figure (a)], and also for $T = 600$ eV and $\rho = 8$ g/cm^3 (b)

5.2.4 Compton scattering

Using the Klein-Nishina formula and results of [199], the following interpolation formula was derived for the Compton scattering cross-section [147]:

$$\sigma_s(\omega) = \frac{8\pi}{3}\alpha^4 a_0^2 Z_0 \, f(E,T), \tag{5.70}$$

where E is the photon energy, T the plasma temperature (both E and T are expressed in keV),

$$f(E,T) = \begin{cases} 1, & \text{if} \quad T < 2 \quad \text{or} \quad E < 0.25, \\ a(E)(T-2) + b(E), & \text{if} \quad T \geq 2 \quad \text{and} \quad E \geq 0.25; \end{cases}$$

$$a(E) = \begin{cases} 2.04 \cdot 10^{-5} E - 6.44 \cdot 10^{-4} \sqrt{E} + 5.13 \cdot 10^{-3}, & \text{if} \quad E < 210, \\ 3.04 \cdot 10^{-6} E - 1.455 \cdot 10^{-4} \sqrt{E} + 1.53 \cdot 10^{-3}, & \text{if} \quad E \geq 210; \end{cases}$$

$$b(E) = \begin{cases} 5.622 \cdot 10^{-4} E - 5.365 \cdot 10^{-2} \sqrt{E} + 1.1116, & \text{if} \quad E < 210, \\ 8.455 \cdot 10^{-4} E - 5.157 \cdot 10^{-2} \sqrt{E} + 1.0273, & \text{if} \quad E \geq 210. \end{cases}$$

5.2.5 The total absorption cross-section

The total absorption cross-section that accounts for induced emission and scattering processes is given by the sum

$$\sigma(\omega) = \left(1 - e^{-x}\right) \left[\sigma_{\text{bb}}(\omega) + \sigma_{\text{bf}}(\omega) + \sigma_{\text{ff}}(\omega)\right] + \sigma_{\text{s}}(\omega), \qquad x = \frac{\omega}{\theta}. \tag{5.71}$$

The absorption coefficient measured in cm^2/g (opacity) has the form

$$\kappa(\omega) = \frac{N_{\text{A}}}{A} \sigma(\omega), \tag{5.72}$$

where N_{A} is the Avogadro number and A the atomic weight.

The contribution of various processes to the Rosseland mean is illustrated in Table 5.2 for iron. As the table clearly shows, the lines determine the Rosseland mean free path for small and moderate densities; for very large densities inverse bremsstrahlung becomes dominant.

Table 5.2: Influence of various absorption processes on the Rosseland mean in an iron plasma with temperature $T = 100$ eV and different densities. The Rosseland means in cm^2/g are shown: κ_{R} accounts for all processes; $\kappa_{\text{R}}^{\text{con}}$ neglects the spectral lines; $\kappa_{\text{R}}^{\text{ff}}$ accounts only for inverse bremsstrahlung

ρ, g/cm^3	κ_{R}	$\kappa_{\text{R}}^{\text{con}}$	$\kappa_{\text{R}}^{\text{ff}}$
0.0001	9.19	1.57	0.0094
0.001	38.6	11.4	0.0773
0.01	261	89.4	0.641
0.1	1240	551	4.48
1	2820	1760	27.5
7.86	4130	3190	171
50	8500	8400	785
1000	21600	21600	20900
10000	29700	29600	28400

For a mixture of elements Z_i $(i=1,\cdots,N)$ with mass fractions m_i $(\sum m_i=1)$ we have

$$\kappa(\omega) = N_{\rm A} \sum_{i=1}^{N} \frac{m_i}{A_i} \sigma_i(\omega), \tag{5.73}$$

where A_i is the atomic weight and $\sigma_i(\omega)$ the absorption cross-section of the i-th element (see also [190, 181]).

5.3 Peculiarities of photon absorption in spectral lines

Among the processes of photon absorption by electrons listed in § 5.2, the most important from a theoretical as well as volume of computations point of view is the process of photon absorption in spectral lines. The reason is that in order to account for the spectral lines it is necessary, first, to describe in detail the structure of levels (terms) of a huge number of individual ion states in plasma and find the probability of realization of such states. Second, the structure of the terms gives only the position of the spectral lines, and to determine their intensities (oscillator strengths) it is necessary to construct the wave functions of all ion states in plasma. Finally, the most difficult task is the calculation of the form (profile) of the absorption lines, which determine the effects of the interaction of an ion with free electrons and other ions of the plasma, and with the radiation field, and also the Doppler effect. A large body of literature is devoted to these problems, including widely known monographs and textbooks (see, e.g., [75]). Our aim here is to construct algorithms that would yield satisfactory results even in the presence of hundreds of thousands of spectral lines. In this section we will consider the probability distributions of various ion states and the position of spectral lines. The calculations of line profiles will be treated in the next section.

5.3.1 Probability distribution of excited ion states

The probability distribution P_Q of excited ion states for a configuration Q with occupation numbers N_ν^Q (where ν is the set of quantum numbers determining a one-electron level) can be calculated starting from the Gibbs distribution. To simplify formulas, we will assume that the temperature is sufficiently high, so that the splitting with respect to terms may be considered small compared with the temperature. Moreover, we shall assume that the interaction with free electrons is accounted for in the calculation of the ion energy E_Q. In this approximation

$$P_Q = \frac{1}{\Omega} g_Q \exp\left(-\frac{E_Q - \mu N_Q}{\theta}\right),$$

where E_Q is the energy of the ion averaged over the configuration Q, μ is the chemical potential, $N_Q = \sum\limits_{\nu} N_\nu^Q$, and g_Q denotes the statistical weight of the configuration Q,

$$g_Q = \prod_{\nu} \binom{g_\nu}{N_\nu^Q} = \prod_{\nu} \frac{g_\nu!}{N_\nu^Q!(g_\nu - N_\nu^Q)!},$$

and finally Ω is the partition function over all ion configurations:

$$\Omega = \sum_Q g_Q \exp\left(-\frac{E_Q - \mu N_Q}{\theta}\right).$$

To get a preliminary estimate for the most probable configurations one can use the *binomial distribution* [45]. In this approximation the probabilities of level occupancy

$$p_\nu = \frac{1}{1 + \exp\left(\dfrac{\varepsilon_\nu - \mu}{\theta}\right)}$$

are assumed to be independent, and the total probability of realization of a configuration Q with given occupation numbers N_ν^Q is given by the binomial distribution

$$P_Q^{\mathrm{bin}} = \prod_{\nu} \binom{g_\nu}{N_\nu^Q} p_\nu^{N_\nu^Q} (1 - p_\nu)^{g_\nu - N_\nu^Q}. \tag{5.74}$$

For the Gibbs distribution it is convenient to use instead of P_Q the notation

$$P_{ks} = \frac{P_k}{\Omega_k} g_{ks} \exp\left(-\frac{E_{ks} - E_{k_0 s_0}}{\theta}\right), \tag{5.75}$$

where k is the ionization degree of the configuration Q ($k = 0, 1, \ldots, k_{\max} = Z$); s is the number of the excited state ($s = 0, 1, 2, \ldots$); k_0, s_0 are the ionization degree and the number of state of the most probable ion (the ion whose occupation numbers are the closest to the average values); P_{ks} is the probability of the ion being in the state s with ionization degree k ($P_k = \sum_s P_{ks}$), and Ω_k is the corresponding partition function:

$$\Omega_k = \sum_s g_{ks} \exp\left(-\frac{E_{ks} - E_{k_0 s_0}}{\theta}\right).$$

The probabilities P_k satisfy the normalization condition $\sum_k P_k = 1$.

In computations one usually takes into account only those configurations Q for which P_Q is larger than some prescribed value. This condition determines the range of variation of the occupation numbers N_ν^Q.

To calculate the probabilities P_{ks} it is necessary to find the connection between ionization and the chemical potential. Let us assume that the average-atom model gives the correct value of the ionization; then necessarily $\sum_k k P_k = Z_0$. Starting from (5.75), we obtain the following system of equations for the determination of P_k (see the derivation of the Saha equations in Subsection 6.2.2):

$$\begin{cases} \dfrac{P_{k+1}}{P_k} & = \zeta \dfrac{\Omega_{k+1}}{\Omega_k}, \\ \sum\limits_k P_k & = 1, \\ \sum\limits_k k P_k & = Z_0; \end{cases} \tag{5.76}$$

here $\zeta = \exp(-\mu/\theta)$ and μ is the chemical potential.

It is convenient to rewrite the last equation in (5.76) in the form

$$\zeta \frac{d \ln F}{d\zeta} = Z_0, \tag{5.77}$$

where

$$F = F(\zeta) = \zeta^{k_0} f_0(\zeta), \quad f_m(\zeta) = \sum_{k=0}^{k_{\max}} \zeta^{k-k_0} k^m \Omega_k.$$

Equation (5.77) is solved by Newton's method:

$$\zeta^{(p+1)} = \zeta^{(p)} \left. \frac{f_0 f_2 + f_0^2 Z_0 - f_1^2}{f_0 f_2 + f_0 f_1 - f_1^2} \right|_{\zeta=\zeta^{(p)}}, \quad p = 0, 1, \ldots, \tag{5.78}$$

where the initial value $\zeta^{(0)}$ is calculated by means of the average-atom model. The iteration scheme (5.78) is analogous to the scheme used to find the ion concentrations in the Saha model (6.55). Let us remark that for the calculation of $f_m(\zeta)$ in the summation over k one can start with $k = k_0$, and subsequently increase or decrease k till the corresponding states become of low probability.

As an example let us consider the probability distribution of excited ion states in gold, computed for various temperatures (see Figure 5.4). It is evident that for temperature T =1 keV the binomial distribution and the Gibss distribution differ only slightly from one another, whereas for T = 50 eV the difference is considerable.

5.3.2 Position of spectral lines

To calculate the center of a group of absorption spectral lines for a one-electron transition $\alpha \to \beta$, let us consider two configurations $Q = \{N_\nu^Q\}$ and $Q' = \{N_\nu^{Q'}\}$. Configuration Q' differs from Q by the position of one electron:

$$N_\nu^{Q'} = \begin{cases} N_\nu^Q, & \text{if } \nu \neq \alpha, \beta, \\ N_\nu^Q - 1, & \text{if } \nu = \alpha, \\ N_\nu^Q + 1, & \text{if } \nu = \beta. \end{cases}$$

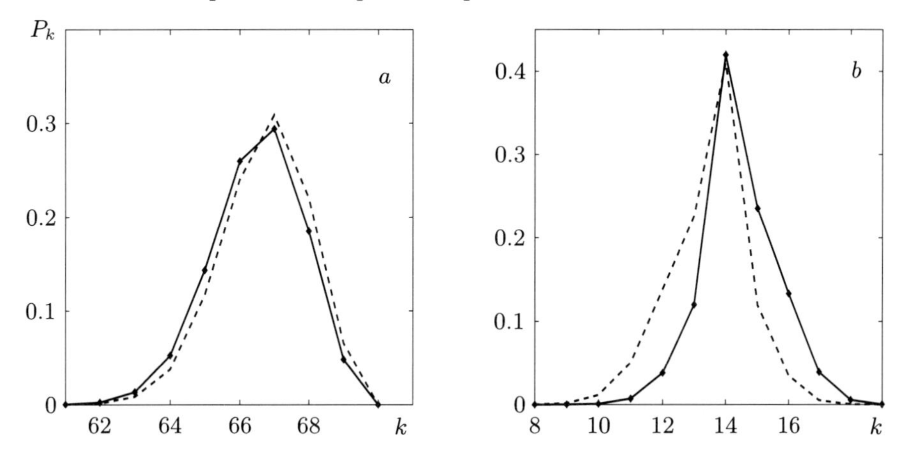

Figure 5.4: Dependence of the probability distribution of excited ion states on the ionization degree k for a gold plasma with density $\rho = 0.1$ g/cm^3 at temperatures: (a) $T = 1$ keV, (b) $T = 50$ eV. The continuous [resp., hatched] line represents the probability distribution obtained by using the Gibbs distribution (5.75) [resp., the binomial distribution (5.74)]

In the nonrelativistic approach the center of the group of lines is approximately determined by the difference between the energies of the configuration Q and Q' (see formula (3.39)):

$$E_{Q'} - E_Q = I_\alpha - I_\beta + \sum_\nu N_\nu^Q (H_{\nu\beta} - H_{\nu\alpha}) + (H_{\alpha\alpha} - H_{\alpha\beta}). \tag{5.79}$$

Calculating E_Q by means of the wave functions obtained, say, in the Hartree-Fock-Slater model, one can set, in the simplest approximation,

$$\omega_{\alpha\beta} = E_{Q'} - E_Q. \tag{5.80}$$

When hydrogen-like wave functions are used, if we put $\alpha = n\ell$, $\beta = n'\ell'$, $\nu = \bar{n}\bar{\ell}$ and retain only the first terms in the expansion of E_Q in Slater integrals, we obtain the expression (see Subsection 3.2.3):

$$I_{n\ell} = \int_0^\infty R_{n\ell}(r) \left(-\frac{1}{2} \frac{d^2}{dr^2} + \frac{\ell(\ell+1)}{2r^2} - \frac{Z}{r} \right) R_{n\ell}(r) dr \approx \frac{Z_{n\ell}(Z_{n\ell} - 2Z)}{2n^2},$$

$$H_{n\ell,\bar{n}\bar{\ell}} \approx F^{(0)}_{n\ell,\bar{n}\bar{\ell}} = \int_0^\infty\int_0^\infty \frac{1}{r_>} R^2_{n\ell}(r_1) R^2_{\bar{n}\bar{\ell}}(r_2) dr_1 dr_2 \approx$$

$$\frac{Z_{n\ell}}{n^2} - \frac{Z_{n\ell}}{2n^2\bar{n}} \sum_{s=1}^{s_{\max}} a_s \frac{M^2_{n\ell}(u_s)}{u_s} \sum_{s'=1}^{s_{\max}} a_{s'} u_{s'} \frac{M^2_{\bar{n}\bar{\ell}}(v_{ss'})}{v_{ss'}},$$

where

$$u_s = \frac{\bar{n} Z_{n\ell}}{\bar{n} Z_{n\ell} + n Z_{\bar{n}\bar{\ell}}} x_s, \qquad v_{ss'} = x_{s'} + \frac{n Z_{\bar{n}\bar{\ell}}}{\bar{n} Z_{n\ell} + n Z_{\bar{n}\bar{\ell}}} x_s,$$

$$M_{n\ell}(x) = \sqrt{\frac{(n-\ell-1)!}{(n+\ell)!}}\, x^{\ell+1} L^{2\ell+1}_{n-\ell-1}(x).$$

Relativistic corrections can be incorporated by setting

$$\omega_{n\ell j, n'\ell' j'} = \omega_{n\ell, n'\ell'} + \varepsilon_{n'\ell' j'} - \varepsilon_{n'\ell'} - \varepsilon_{n\ell j} + \varepsilon_{n\ell}. \tag{5.81}$$

For the oscillator strengths the formulas (5.49) and (5.55) remain valid.

5.3.3 Atom wave functions and addition of momenta

In the central-field approximation, to each configuration Q there corresponds a set of states that differ from one another through the orientation of the orbital and spin momenta of the electrons. Assigning all these states to one and the same energy level of the ion (atom) is possible only as long as we neglect the spin-orbit interaction as well as that part of the electrostatic interaction between electrons that is not accounted for in the central-field approximation.

The joint consideration of such interactions is a rather difficult task, so one usually resorts to perturbation theory, assuming that one of the interactions is much weaker than the other. In the resulting approximation the ion wave function can be constructed by specifying a scheme for addition of momenta, based on some type of vector coupling between them. When one speaks about different types of coupling it is implied that the electrostatic and spin-orbit interactions that are not accounted for in the self-consistent potential may be interpreted as couplings of various types between the orbital momentum vectors $\vec{\ell}$ and the spin momentum vectors $\vec{s}$ of the electrons of the ion.

The principal part of the energy of spin-orbit interaction is given by the expression

$$V_{\text{so}} = -\frac{\alpha^2}{2}\frac{1}{r}\frac{dV}{dr}\vec{\ell}\,\vec{s}, \tag{5.82}$$

where $V = V(r)$ is the potential in which the electron moves. As a rule, for low-Z atoms the spin-orbit interaction is considerably weaker than the electrostatic one, and so the former is neglected due to the presence of the small parameter α^2 in (5.82). In such cases one uses the approximation of the LS-coupling; here the orbital and spin momenta are coupled (added) independently, and the total momentum of the ion is $\vec{J} = \vec{L} + \vec{S}$.

When Z is increased, the relative role played by the spin-orbit interaction grows rapidly, and it no longer makes sense to speak about the orbital and spin momenta of the electron as separate entities. Rather, it only makes sense to speak about the total momentum of each electron, $\vec{j} = \vec{\ell} + \vec{s}$. In this case one usually resorts to the jj-coupling approximation, in which one first couples the momenta $\vec{\ell}_i$ and $\vec{s}_i$ in the total momentum $\vec{j}_i$, and then all these individual momenta are coupled in the total momentum of the ion, $\vec{J} = \sum \vec{j}_i$. For a detailed description of the corresponding methods the reader is referred to the books [48, 102, 156, 37, 221, 195, 212].

For multi-charged, multi-electron ions the LS- or jj-coupling approximation, as a rule, is not applicable, since the spin-orbit interaction and the electrostatic interaction not accounted for in the self-consistent potentials are quantities of the same order of magnitude (especially for $p - d$ or $d - f$ transitions inside a shell). In this case it is necessary to resort to an approximation based on an intermediate-type coupling scheme and account simultaneously for the spin-orbit interaction and the noncentral part of the Coulomb interaction. In many cases it is necessary to incorporate the interaction between configurations and thus use a multi-configuration approximation [195]. We shall use a method for the construction of ion wave functions [48] which, despite being relatively demanding from the computational point of view, is sufficiently transparent.

Let us consider a coupling of intermediate type that uses the one-electron functions in the representation $n\ell jm$. For a given configuration $Q = \{N_{n\ell}\}$, we will take into account all its subconfigurations $q = \{N_{n\ell jm}\}$ with occupation numbers $N_{n\ell jm}$, which can take the values 0 or 1, so that $\sum_{jm} N_{n\ell jm} = N_{n\ell}$. For each subconfiguration q one can construct the one-determinant function

$$\Psi_{qM} = \frac{1}{\sqrt{N!}} \det ||\psi_i(\vec{r}_k, \sigma_k)||, \tag{5.83}$$

where $k = 1, 2, \ldots, N$ (N is the number of electrons: $N = \sum N_{n\ell}$), and to each i we associate the quantum numbers $n\ell jm$, which determine the state of an electron with the wave function

$$\psi_{n\ell jm}(\vec{r}, \sigma) = \frac{1}{r} R_{n\ell}(r) \sum_{m_\ell, m_s} C^{jm}_{\ell m_\ell \frac{1}{2} m_s} (-1)^{m_\ell} Y_{\ell m_\ell}(\vartheta, \varphi)\, \chi_{\frac{1}{2} m_s}(\sigma). \tag{5.84}$$

The radial functions $R_{n\ell}(r)$ can be obtained by solving the Hartee-Fock-Slater equation, in which case the functions $\psi_{n\ell jm}$ will be eigenfunctions of the Hamiltonian

$$\widehat{H}_0(\vec{r}) = -\frac{1}{2}\Delta - V(r), \tag{5.85}$$

where $V(r)$ is the Hartree-Fock-Slater self-consistent potential.

For a multi-electron ion the Hamiltonian $\widehat{H}$ is not the sum of the Hamiltonians (5.85) for all electrons, because (5.85) accounts only for the principal part of the interaction, which can be described by the central self-consistent potential. The remaining, noncentral part and other interactions (in particular, the spin-orbit one) represent corrections to the sum of the expressions (5.85) over all electrons, $\sum_i \widehat{H}_0(\vec{r}_i)$, and can be incorporated by perturbation theory. For the sake of definiteness we will consider only the main corrections, setting (see [156])

$$\widehat{H} = \sum_i \left[-\frac{1}{2}\Delta_i - \frac{Z}{r_i}\right] + \sum_{i<j} \frac{1}{|\vec{r}_i - \vec{r}_j|} - \frac{\alpha^2}{2}\sum_i \frac{1}{r}\frac{dV}{dr}\vec{\ell}_i\,\vec{s}_i, \tag{5.86}$$

where in addition to the electrostatic interactions of electrons with the nucleus and between themselves we have included the main part of their spin-orbit interaction.

The one-determinant function Ψ_{qM} is an eigenfunction of the operator of the projection of the total angular momentum on the z-axis, the magnitude of which is given by

$$M = \sum_{n,\ell,j,m} mN_{n\ell jm}.$$

We shall seek the wave function of the ion, which is a joint eigenfunction of the operators $\widehat{J}_z$, $\widehat{\vec{J}^2}$ and $\widehat{H}$, in the first order of perturbation theory. It is known that in our case it suffices that the sought-for function $\Psi_{\gamma JM}$ be an eigenfunction of the operator $\widehat{H}$, because then it also is an eigenfunction of the operators $\widehat{\vec{J}^2}$ and $\widehat{J}_z$ (see [48]; the supplementary index γ was introduced to distinguish between the solutions with the same J and M). Let us represent the function $\Psi_{\gamma JM}$ as a linear combination of functions (5.83):

$$\Psi_{\gamma JM} = \sum_q C_{qM}^{\gamma J}\Psi_{qM}. \tag{5.87}$$

The coefficients of this decomposition are found by solving the secular equation

$$\sum_{q'} C_{q'M}^{\gamma J}\left(\langle q|\widehat{H}|q'\rangle - E\right) = 0, \tag{5.88}$$

where each eigenvalue $E = E_{\gamma JM}$ and each eigenvector $\left\{C_{qM}^{\gamma J}\right\}$ correspond to specified values of γ, J and M. The condition that the energy E is independent of M allows one to classify the states and determine the quantities J and M for each solution. If for certain J and M there are several such vectors, then the index γ is used to distinguish between them.

The matrix $\langle q|\widehat{H}|q'\rangle$ is calculated as follows (see [48]). For diagonal elements, i.e., for $q \equiv q'$, we have

$$\langle q|\widehat{H}|q\rangle = \sum_{\alpha} N_\alpha (I_\alpha + \Delta E_\alpha) + \frac{1}{2}\sum_{\alpha,\beta} N_\alpha N_\beta H_{\alpha\beta}, \tag{5.89}$$

where

$$\alpha = \{n\ell jm\}, \qquad \beta = \{n'\ell' j'm'\},$$

$$\Delta E_\alpha = \frac{\zeta_{n_\alpha \ell_\alpha}}{2}\left[j_\alpha(j_\alpha+1) - \ell_\alpha(\ell_\alpha+1) - \frac{3}{4}\right],$$

$$\zeta_{n\ell} = -\frac{\alpha^2}{2}\int\limits_0^{r_0} \frac{1}{r}\frac{dV}{dr} R_{n\ell}^2(r)\,dr.$$

A two-electron matrix element is given by the expression

$$H_{\alpha\beta} = \left\langle\alpha\beta\left|\frac{1}{r_{12}}\right|\alpha\beta\right\rangle - \left\langle\alpha\beta\left|\frac{1}{r_{12}}\right|\beta\alpha\right\rangle,$$

where

$$\left\langle\alpha\beta\left|\frac{1}{r_{12}}\right|\alpha\beta\right\rangle = \sum_{\sigma\sigma'}\iint \frac{\psi_\alpha^*(\vec r_1,\sigma)\psi_\beta^*(\vec r_2,\sigma')\psi_\alpha(\vec r_1,\sigma)\psi_\beta(\vec r_2,\sigma')}{|\vec r_1 - \vec r_2|}\,d\vec r_1\,d\vec r_2,$$

$$\left\langle\alpha\beta\left|\frac{1}{r_{12}}\right|\beta\alpha\right\rangle = \sum_{\sigma\sigma'}\iint \frac{\psi_\alpha^*(\vec r_1,\sigma)\psi_\beta^*(\vec r_2,\sigma')\psi_\alpha(\vec r_2,\sigma')\psi_\beta(\vec r_1,\sigma)}{|\vec r_1 - \vec r_2|}\,d\vec r_1\,d\vec r_2$$

(in the last expression the nonzero terms correspond to the states α and β with the same direction of the spin). If the subconfiguration q' is obtained from the subconfiguration q by changing the state of a single electron (say, in configuration q that electron occupies the state α, while in q' it occupies the state γ), then

$$\langle q'|\widehat{H}|q\rangle = (-1)^P \sum_\beta N_\beta\left(\left\langle\alpha\beta\left|\frac{1}{r_{12}}\right|\gamma\beta\right\rangle - \left\langle\alpha\beta\left|\frac{1}{r_{12}}\right|\beta\gamma\right\rangle\right),$$

where P is the number of permutations needed to obtain subconfinguration q from subconfiguration q' (see [48]). If one chooses some enumeration of all possible states of electrons for which $\alpha < \beta$, $\gamma < \delta$, then P will be equal to the sum of all occupied states lying between α and β, as well as between γ and δ.

In the case when two electrons change state ($\alpha \to \gamma,\ \beta \to \delta$),

$$\langle q'|\widehat{H}|q\rangle = (-1)^{P}\left(\langle\alpha\beta\left|\frac{1}{r_{12}}\right|\gamma\delta\rangle - \langle\alpha\beta\left|\frac{1}{r_{12}}\right|\delta\gamma\rangle\right),$$

In the remaining cases, when more than two electrons change states, $\langle q'|\widehat{H}|q\rangle = 0$.

In the calculation of the matrix elements $\langle\alpha\beta\left|\frac{1}{r_{12}}\right|\gamma\delta\rangle$ by using the one-electron wave functions (5.84) one can integrate with respect to the angular variables and sum with respect to the spin variable σ. This yields

$$\langle\alpha\beta\left|\frac{1}{r_{12}}\right|\gamma\delta\rangle = \sum_{m_s,m_s'}\ \sum_{\bar{m}_\alpha,\bar{m}_\beta,\bar{m}_\gamma,\bar{m}_\delta} \delta_{m_\alpha-m_\gamma,m_\delta-m_\beta}(-1)^{m_\alpha+m_\delta-m_s-m_s'}\times$$
$$C^{j_\alpha m_\alpha}_{\ell_\alpha\bar{m}_\alpha\,\frac{1}{2}m_s} C^{j_\beta m_\beta}_{\ell_\beta\bar{m}_\beta\,\frac{1}{2}m_s'} C^{j_\gamma m_\gamma}_{\ell_\gamma\bar{m}_\gamma\,\frac{1}{2}m_s} C^{j_\delta m_\delta}_{\ell_\delta\bar{m}_\delta\,\frac{1}{2}m_s'}\times$$
$$\sum_k D^k_{\ell_\alpha,m_\alpha-m_s;\ell_\gamma,m_\gamma-m_s} D^k_{\ell_\beta,m_\beta-m_s';\ell_\delta,m_\delta-m_s'} R^{(k)}_{\alpha\beta\gamma\delta},$$

where

$$D^k_{\ell\tilde{m},\ell'\tilde{m}'} = (-1)^{\tilde{m}-\tilde{m}'}\sqrt{(2\ell+1)(2\ell'+1)}\times$$
$$\begin{pmatrix}\ell & k & \ell'\\ 0 & 0 & 0\end{pmatrix}\begin{pmatrix}\ell & k & \ell'\\ -\tilde{m} & \tilde{m}-\tilde{m}' & \tilde{m}'\end{pmatrix},$$

$$R^{(k)}_{\alpha\beta\gamma\delta} = \iint \frac{r_<^k}{r_>^{k+1}} R_{n_\alpha\ell_\alpha}(r_1)R_{n_\beta\ell_\beta}(r_2)R_{n_\gamma\ell_\gamma}(r_1)R_{n_\delta\ell_\delta}(r_2)\,dr_1\,dr_2. \tag{5.90}$$

The calculation of the eigenvectors and eigenfunctions can be done by means of the standard schemes, for example, by Jacobi's method [16, 226].

EXAMPLE. Let us examine in more detail the case of two p-electrons in the one-configuration approximation, assuming a coupling of intermediate type, i.e., considering that the corrections due to the spin-orbit interaction and the noncentral part of the electrostatic interaction are of the same order. Two p-electrons may occupy 6 different states with different quantum numbers j, m ($j = 1/2$, $m = \pm 1/2$; $j = 3/2, m = \pm 1/2, \pm 3/2$). The one-electron wave functions of these states are given by formula (5.85). The one-determinant function of a subconfiguration q whose states $\nu_1 = j_1 m_1$ and $\nu_2 = j_2 m_2$ are occupied has the form

$$\Psi_{qM} = \frac{1}{\sqrt{2}}\begin{vmatrix}\psi_{\nu_1}(\vec{r}_1,\sigma_1) & \psi_{\nu_1}(\vec{r}_2,\sigma_2)\\ \psi_{\nu_2}(\vec{r}_1,\sigma_1) & \psi_{\nu_2}(\vec{r}_2,\sigma_2)\end{vmatrix}$$

(here we neglect the interaction between unoccupied and occupied shells).

Since $M = m_1 + m_2$, then, for example, for $M = 2$ one has two different one-determinant functions with $j_1 = 3/2, m_1 = 3/2;\ j_2 = 3/2,\ m_2 = 1/2$ and $j_1 = 3/2, m_1 = 3/2;\ j_2 = 1/2,\ m_2 = 1/2$. In what follows q will be specified by the occupancy of the state $j_i m_i$, i.e., $q = 3/2, +3/2;\ 3/2, +1/2$ means that $N_{\frac{3}{2}\frac{3}{2}} = 1$, $N_{\frac{3}{2}\frac{1}{2}} = 1$, and all the remaining $N_{j_i m_i} = 0$. For $M = 2$ the matrix $< q|H|q' >$ has the form

q / q'	$\frac{3}{2}+\frac{3}{2};\frac{3}{2}+\frac{1}{2}$	$\frac{3}{2}+\frac{3}{2};\frac{1}{2}+\frac{1}{2}$
$\frac{3}{2}+\frac{3}{2};\frac{3}{2}+\frac{1}{2}$	$F_0 - 3F_2 + \zeta$	$-\sqrt{8}F_2$
$\frac{3}{2}+\frac{3}{2};\frac{1}{2}+\frac{1}{2}$	$-\sqrt{8}F_2$	$F_0 - F_2 - \zeta/2$

Here

$$\zeta = -\frac{\alpha^2}{2}\int \frac{V'(r)}{r} R_{n\ell}^2(r)\,dr, \quad F_0 = R^{(0)}_{n\ell n\ell, n\ell n\ell}, \quad F_2 = \frac{1}{25} R^{(2)}_{n\ell n\ell, n\ell n\ell},$$

where $R^{(k)}_{n\ell n\ell, n\ell n\ell}$ are the Slater integrals (5.90) for the shell with principal quantum number n and orbital number $\ell = 1$. We remind the reader that we are calculating the energy for only two p-electrons. The diagonalization of this matrix yields two levels:

$$\left.\begin{matrix} {}^1D_2' \\ {}^3P_2' \end{matrix}\right\} = F_0 - 2F_2 + \zeta/4 \pm \sqrt{9F_2^2 - \frac{3}{2}F_2\zeta + \frac{9}{16}\zeta^2}$$

(we use the traditional notation, indicating the dominating term).

For $M = 1$ we have

q / q'	$\frac{3}{2}+\frac{3}{2};\frac{3}{2}-\frac{1}{2}$	$\frac{3}{2}+\frac{3}{2};\frac{1}{2}-\frac{1}{2}$	$\frac{3}{2}+\frac{1}{2};\frac{1}{2}+\frac{1}{2}$
$\frac{3}{2}+\frac{3}{2};\frac{3}{2}-\frac{1}{2}$	$F_0 - 3F_2 + \zeta$	$-\sqrt{2}F_2$	$-\sqrt{6}F_2$
$\frac{3}{2}+\frac{3}{2};\frac{1}{2}-\frac{1}{2}$	$-\sqrt{2}F_2$	$F_0 - 4F_2 - \zeta/2$	$\sqrt{3}F_2$
$\frac{3}{2}+\frac{1}{2};\frac{1}{2}+\frac{1}{2}$	$-\sqrt{6}F_2$	$\sqrt{3}F_2$	$F_0 - 2F_2 - \zeta/2$

The diagonalization of this matrix gives three levels; two of them coincide with the ones obtained above, while the third has the form

$${}^3P_1' = F_0 - 5F_2 + \zeta/2.$$

Finally, for $M = 0$ we have

q' \ q	$\frac{3}{2}+\frac{3}{2};\frac{3}{2}-\frac{3}{2}$	$\frac{3}{2}+\frac{1}{2};\frac{3}{2}-\frac{1}{2}$	$\frac{3}{2}+\frac{1}{2};\frac{1}{2}-\frac{1}{2}$	$\frac{3}{2}-\frac{1}{2};\frac{1}{2}+\frac{1}{2}$	$\frac{1}{2}+\frac{1}{2};\frac{1}{2}-\frac{1}{2}$
$\frac{3}{2}+\frac{3}{2};\frac{3}{2}-\frac{3}{2}$	$F_0+F_2+\zeta$	$-4F_2$	$-\sqrt{2}F_2$	$-\sqrt{2}F_2$	$5F_2$
$\frac{3}{2}+\frac{1}{2};\frac{3}{2}-\frac{1}{2}$	$-4F_2$	$F_0+F_2+\zeta$	$-\sqrt{2}F_2$	$-\sqrt{2}F_2$	$-5F_2$
$\frac{3}{2}+\frac{1}{2};\frac{1}{2}-\frac{1}{2}$	$-\sqrt{2}F_2$	$-\sqrt{2}F_2$	$F_0-3F_2+\zeta/2$	$2F_2$	0
$\frac{3}{2}-\frac{1}{2};\frac{1}{2}+\frac{1}{2}$	$-\sqrt{2}F_2$	$-\sqrt{2}F_2$	$2F_2$	$F_0-3F_2+\zeta/2$	0
$\frac{1}{2}+\frac{1}{2};\frac{1}{2}-\frac{1}{2}$	$5F_2$	$-5F_2$	0	0	$F_0-2\zeta$

The diagonalization of this matrix yields, in addition to the levels already obtained, the following levels:

$$\left.\begin{array}{c}{}^1S_0'\\{}^3P_0'\end{array}\right\} = F_0 + \frac{5}{2}F_2 - \zeta/2 \pm \sqrt{\frac{225}{4}F_2^2 + \frac{15}{2}F_2\zeta + \frac{9}{4}\zeta^2}.$$

Table 5.3 shows numerical results for various configurations of np^2 ions of aluminum, iron, molybdenum and gold, obtained by direct diagonalization of the matrix considered here (the results coincide with those of calculations employing the analytic formulas). The term energies are given in atomic units and are measured from the minimal energy for the configuration. The values of the Slater integrals and of the parameters ζ, calculated according to the Hartree-Fock-Slater model for the corresponding configurations, are displayed in Table 5.4.

Table 5.3: Term energies, in atomic units, for various configurations of np^2 ions of aluminum, iron, molybdenum and gold

term	J	Al	Fe	Mo	Au
${}^3P'$	0	0.0000	0.0000	0.0000	0.0000
	1	0.0084	0.0410	0.0695	0.5300
	2	0.0230	0.0849	0.1153	0.6037
${}^1D'$	2	0.2183	0.2324	0.2411	1.1892
${}^1S'$	0	0.5212	0.4640	0.3974	1.3566

Table 5.4: The Slater integrals and the parameters ζ, in atomic units, for the above configurations of np^2 ions of aluminum, iron, molybdenum and gold (see Table 5.3)

Element	n	Configuration (lower shells are fully occupied)	$R^0_{n1n1,n1n1}$	$R^2_{n1n1,n1n1}$	ζ
Al	2	$2s^2\ 2p^2$	1.79271	0.84045	0.00151
Fe	3	$3s^2\ 3p^2$	1.25282	0.63666	0.00551
Mo	4	$4s^2\ 4p^2$	0.80900	0.43070	0.07596
Au	5	$5s^2\ 5p^2$	0.90334	0.49448	0.40953

As one sees in the tables, the magnitude of the multiplet splitting reaches 35 eV for gold, which is considerably higher than the width of the spectral line (see further in § 5.4). The effects of multiplet splitting are particularly important when the main contributors to the Rosseland mean free path are the $p-d$ and $d-f$ transitions of multielectron ions (see § 5.5). In this case the number of components of the multiplet reaches several hundreds or even thousands.

5.4 Shape of spectral lines

Since an ion can be in an excited state only for a finite time interval τ, according to the well-known uncertainty principle the energy of the ion in such a state is defined up to an error of $\Delta E \sim h/\tau$. Consequently, a spectral line cannot be monochromatic; rather, it is a set of lines, i.e., it has a certain shape. The probability of absorption of a photon with frequency in the interval ω, $\omega + d\omega$, per unit of time, when the ion changes from a state a to a state b is equal to

$$W(\omega)\,d\omega = W_{ab}\,J(\omega)\,d\omega, \tag{5.91}$$

where W_{ab} is the total transition probability (per unit of time) from state a to state b, i.e.,

$$\int W(\omega)\,d\omega = W_{ab},$$

which is equivalent to the normalization $\int J(\omega)\,d\omega = 1$. The expression for the transition probability W_{ab} is given in Subsection 5.2.1. Here we are interested only in the relative distribution of the intensity $J(\omega)$.

In the framework of classical electrodynamics the intensity distribution of a spectral line is described by the Lorentz dispersion formula

$$J(\omega) = \frac{1}{\pi}\,\frac{\gamma/2}{(\omega-\omega_0)^2 + (\gamma/2)^2} \tag{5.92}$$

(here ω_0 is the position of the center of the spectral line and γ is called the *line width*). The quantum-mechanical consideration of the interaction of a free atom with an electromagnetic field also leads to formula (5.92). In that approach the quantity γ is called the *natural* (or *radiation*) *width* of the line. In the general case, when the profile of the line is not Lorentzian, the term line width signifies the distance between the points of the contour in which the intensity is equal to half the maximal value.

According to quantum theory, the natural width γ of the line in the transition from state a to state b is equal to the sum of the widths of the initial and final states: $\gamma = \gamma_a + \gamma_b$, where $\gamma_a = 1/\tau_a$, τ_a is the mean lifetime of the atom in state a; $\gamma_b = 1/\tau_b$, τ_b is the mean life time of the atom in state b. In addition to the natural broadening, there are other effects that result in the broadening of spectral lines.

First, collisions of electrons (and also of ions) with the ion under consideration reduces the ion's lifetime in the given excited state.

Second, electrical fields in a plasma split and shift energy levels. Consequently, instead of a single line with definite frequency one observes many lines with frequencies close to one another, emitted by different ions. This effect is called Stark broadening.

Further, the chaotic thermal motion of the radiating ions results in a frequency shift of the spectral lines emitted by a moving ion. This is known as the Doppler broadening of lines.

Moreover, the autoionization process that accompanies photon emission may also contribute to line broadening (Auger effect, see Subsection 5.4.7). As a rule, for a dense hot plasma this effect may be neglected.

At thermodynamic equilibrium the shape of the absorption line coincides with the shape of the emission line. Due to the fact that spectral lines are not strictly monochromatic, but occupy some frequency band, bound–bound transitions may exert a considerable influence on the Rosseland mean free path of photons, because the number of spectral lines for high-Z elements may be huge.

5.4.1 Doppler effect

The simplest among the effects leading to broadening of spectral lines is the Doppler effect.

Let $J(\omega)\,d\omega$ be the intensity distribution when the Doppler effect is neglected. If an ion emits a spectral line with frequency ω' and moves with velocity v in the direction of the observer, then what one sees in the fixed coordinate system is a spectral line with frequency $\omega = \omega'(1 + v/c)$. This means that the intensity distribution of the spectral line for the given values ω' and v will be $J(\omega') = \delta(\omega - \omega'(1 + v/c))\,d\omega$.

Let $W(v)$ be the distribution function (normalized to 1) of ions according to the projections of the velocities on a distinguished direction. Then the probability that an ion whose velocity in the direction of the line of sight lies in the interval $(v, v + dv)$ will emit a line with frequency in the interval $(\omega', \omega' + d\omega')$ is equal to the product $J(\omega')\,d\omega'\,W(v)\,dv$. We see that intensity distribution in the fixed coordinate system with the Doppler effect accounted for has the form

$$\Phi(\omega)\,d\omega = d\omega \int \delta\left(\omega - \omega'\left(1 + \frac{v}{c}\right)\right) J(\omega')\,W(v)\,d\omega'\,dv.$$

Integrating with respect to ω' and putting $(1 + v/c)^{-1} \approx 1 - v/c$, we obtain

$$\Phi(\omega) \approx \int J\left(\omega - \omega\frac{v}{c}\right) W(v)\,dv.$$

In the case when the velocity distribution of ions is Maxwellian, i.e.,

$$W(v)\,dv = \sqrt{\frac{M}{2\pi\theta}} \exp\left(-\frac{Mv^2}{2\theta}\right) dv, \tag{5.93}$$

where $M = 1836\,A$ is the ion mass and A the atomic weight, we have

$$\Phi(\omega) = \frac{1}{\sqrt{\pi}D} \int\limits_0^\infty J\,(\omega - s)\;e^{-s^2/D^2}\,ds, \tag{5.94}$$

$$D = \frac{\omega}{c}\sqrt{\frac{2\theta}{M}}.$$

5.4.2 Electron broadening in the impact approximation

The effect of the interaction of a radiating ion with other ions and electrons on the shape of spectral lines was studied in a large number of works (see, e.g., [18, 112, 85, 75, 114, 177, 178, 84]). Here we shall confine ourselves to the consideration of two limit cases, the first of which corresponds to the interaction of slow particles with a radiating atom (quasi-static broadening), while the second deals with the interaction of fast particles (impact broadening).

In the first case the interaction changes little over the duration of the photon absorption process under consideration, i.e., it is quasi-static and leads mainly to a line shift. In this case the line intensity distribution $J(\omega)\,d\omega$ is equal to the probability that the position of the line will lie in the interval ω, $\omega + d\omega$ when the interaction with perturbing particles is accounted for. In the temperature and density ranges we are interested in, the quasi-static theory describes well the interaction with ions, since the mean lifetime of excited states of an ion is considerably shorter than the time a ion requires to fly a distance of the order of the mean distance between ions. Ion broadening accounted for in the framework of the quasi-static approximation will be considered in Subsection 5.4.6.

Electron broadening is usually considered in the impact approximation, because electron velocities are much greater than ion velocities. An elementary treatment of the impact theory was first given by Lorentz. According to Lorentz, when the perturbing particles collide with a radiating atom they completely terminate the radiation process, as a result of which radiation is no longer monochromatic.

Suppose that an atom emits radiation of frequency ω_0 in the time interval from 0 to T. Then the amplitude distribution, which can be obtained by expansion in a Fourier integral, will be equal to

$$J(\omega, T) \sim \int_0^T e^{i(\omega-\omega_0)t}\, dt = \left.\frac{e^{i(\omega-\omega_0)t} - 1}{i(\omega - \omega_0)}\right|_{t=T}.$$

If one could observe a single radiation act, then the intensity distribution according to frequencies would be proportional to $|J(\omega, T)|^2 d\omega$. However, the observed line is a mixture of radiation emmited by many atoms, which radiate during different time intervals, the mean value of which τ is given by $\tau = 1/\nu_c$, where ν_c is the collision frequency. The probability that a given atom radiates in the time interval T is $\tau^{-1}e^{-T/\tau}$. It follows that the intensity distribution of the observed line is

$$J(\omega) = \frac{1}{\tau}\int_0^\infty |J(\omega, T)|^2\, e^{-T/\tau}\, dT \sim \frac{1}{(\omega - \omega_0)^2 + \nu_c^2}.$$

Thus, we obtained the Lorentz dispersion curve, whose total width at half of the maximum of the intensity is $2\nu_c$.

The Lorentz profile may be obtained by examining the radiation processes in a more rigorous way. Let us consider a quantum-mechanical system consisting of an ion, free electrons and a radiation field. The total Hamiltonian of this system can be written as a sum

$$\mathcal{H} = H_{\mathrm{i}} + H_{\mathrm{e}} + H_{\mathrm{r}} + H_{\mathrm{ie}} + H_{\mathrm{ir}} + H_{\mathrm{er}}, \tag{5.95}$$

where H_{i} is the Hamiltonian of the considered ion in the field $\vec{F}$ generated by other ions, H_{e} the Hamiltonian of the free electrons and H_{r} the Hamiltonian of the radiation field; H_{ie}, H_{ir} and H_{er} describe the interaction of the ion with the free electrons, of the ion with the radiation field, and of the free electrons with the radiation field, respectively. The term H_{er} has only a weak influence on the shape of spectral lines and will be neglected in what follows, as is usually done in the study of line broadening.

5.4.3 The nondegenerate case

Let us denote by $\psi_n^{(i)}$ and ε_n the eigenfunctions and energy eigenvalues of the ion Hamiltonian H_{i}. Suppose that at the initial time $t = 0$ the ion is in one of the steady states $\psi_{n_1}^{(i)}$. We will study the absorption of photons with frequencies close to some resonance frequency ω_0, corresponding to the transition $n_1 \to n_2$. The levels n_1 and n_2 are assumed to be nondegenerate.

To solve the Schrödinger equation

$$i\hbar \frac{\partial \Psi}{\partial t} = \mathcal{H}\Psi,$$

where Ψ is the wave function of the system, we resort to perturbation theory. To this end it is convenient to represent the Hamiltonian (5.95) in the form

$$\mathcal{H} = H_0 + H,$$

where

$$\begin{aligned} H_0 &= H_{\rm i} + H_{\rm e} + \langle n_1|H_{\rm ie}|n_1\rangle + H_{\rm r}, \\ H &= H_{\rm ir} + H_{\rm ie} - \langle n_1|H_{\rm ie}|n_1\rangle, \end{aligned} \tag{5.96}$$

note that $H \ll H_0$ [86, 147].

The eigenfunctions of the Hamiltonian H_0 have the form

$$\psi^{(0)}_{n\lambda\nu} = \psi^{(\rm i)}_n \psi^{(\rm e)}_\lambda \psi^{(\rm r)}_\nu,$$

where $\psi^{(\rm i)}_n$, $\psi^{(\rm e)}_\lambda$, $\psi^{(\rm r)}_\nu$ are eigenfunctions of the Hamiltonians $H_{\rm i}$, $H_{\rm e} + \langle n_1|H_{\rm ie}|n_1\rangle$ and $H_{\rm r}$, respectively:

$$H_{\rm i}\psi^{(\rm i)}_n = \varepsilon_n \psi^{(\rm i)}_n, \tag{5.97}$$

$$\left[H_{\rm e} + \langle n_1|H_{\rm ie}|n_1\rangle\right] \psi^{(\rm e)}_\lambda = \varepsilon_\lambda\, \psi^{(\rm e)}_\lambda, \tag{5.98}$$

$$H_{\rm r}\psi^{(\rm r)}_\nu = \varepsilon_\nu \psi^{(\rm r)}_\nu. \tag{5.99}$$

Therefore,

$$H_0\psi^{(0)}_{n\lambda\nu} = \varepsilon_{n\lambda\nu}\psi^{(0)}_{n\lambda\nu}, \qquad \varepsilon_{n\lambda\nu} = \varepsilon_n + \varepsilon_\lambda + \varepsilon_\nu.$$

In accordance with nonstationary perturbation theory, let us expand the wave function of the system, $\Psi(t)$, with respect to the set of functions $\psi_{n\lambda\nu}(t) = \psi^{(0)}_{n\lambda\nu} \exp\left(-\frac{it}{\hbar}\,\varepsilon_{n\lambda\nu}\right)$:

$$\Psi(t) = \sum_{n\lambda\nu} a_{n\lambda\nu}(t)\,\psi_{n\lambda\nu}(t).$$

The coefficients $a_{n\lambda\nu}(t)$ of the expansion satisfy the equations

$$i\hbar\dot{a}_{n\lambda\nu} = \sum_{n'\lambda'\nu'} \langle n\lambda\nu|H|n'\lambda'\nu'\rangle \exp\left[\frac{it}{\hbar}\left(\varepsilon_{n\lambda\nu} - \varepsilon_{n'\lambda'\nu'}\right)\right] a_{n'\lambda'\nu'}(t). \tag{5.100}$$

By (5.96), the matrix elements $\langle n\lambda\nu|H|n'\lambda'\nu'\rangle$ have the form

$$\begin{aligned} \langle n\lambda\nu|H|n'\lambda'\nu'\rangle = {} & \langle n\nu|H_{\rm ir}|n'\nu'\rangle \cdot \delta_{\lambda\lambda'} + \\ & \langle n\lambda|H_{\rm ie}|n'\lambda'\rangle \cdot \delta_{\nu\nu'} - \langle n_1\lambda|H_{\rm ie}|n_1\lambda'\rangle \cdot \delta_{nn'}\,\delta_{\nu\nu'} \end{aligned} \tag{5.101}$$

(here we used the fact that the systems of functions $\psi_n^{(\mathrm{i})}$, $\psi_\lambda^{(\mathrm{e})}$, $\psi_\nu^{(\mathrm{r})}$ are orthonormal). The system of equations (5.100) will be solved with the initial conditions

$$a_{n\lambda\nu}(0) = \begin{cases} 1, & \text{if } \ n\lambda\nu = n_1\lambda_1\nu_1, \\ 0, & \text{if } \ n\lambda\nu \neq n_1\lambda_1\nu_1. \end{cases}$$

Here we will neglect a number of terms in the right-hand sides of the equations, motivated by the following qualitative description of the behavior of the coefficients $a_{n\lambda\nu}(t)$. We shall assume that for values of t that are not too large, as long as $|a_{n_1\lambda_1\nu_1}| \gg |a_{n\lambda\nu}|$, the functions $|a_{n_1\lambda_1\nu_1}(t)|$ decrease, while the functions $|a_{n\lambda\nu}(t)|$ with $n\lambda\nu \neq n_1\lambda_1\nu_1$ grow due to the transitions from the initial state $n_1\lambda_1\nu_1$.

Ordinarily, the transition probability due to the interaction with free electrons is large only when the energy of the ion does not change significantly as a result of these interactions — inelastic collisions with large transfer of energy have low probability. Since the energy level ε_{n_1} is nondegenerate, the most important contributions are due to the transitions corresponding to the matrix elements $\langle n_1\lambda|H_{\mathrm{ie}}|n_1\lambda_1\rangle$. But according to (5.101), in the expression for the matrix element $\langle n\lambda\nu|H|n_1\lambda_1\nu_1\rangle$ the matrix elements $\langle n_1\lambda|H_{\mathrm{ie}}|n_1\lambda_1\rangle$ cancel out. It follows that for moderately large values of t the probability of transition from the initial state is determined mainly by the interaction with radiation (the first term in (5.101)), and the fastest growing among the functions $a_{n\lambda\nu}(t)$ are the functions $a_{n_2\lambda_1\nu_2}(t)$ corresponding to the resonant absorption of photons with frequency $\omega_0 = (\varepsilon_{n_2} - \varepsilon_{n_1})/\hbar$.

Thus, over a sufficiently long time interval the functions $|a_{n_1\lambda_1\nu_1}(t)|$ and $|a_{n_2\lambda_1\nu_2}(t)|$ will be larger than all the remaining functions $|a_{n\lambda\nu}(t)|$. Consequently, in that time interval transitions between states that do not include the states $n_1\lambda_1\nu_1$ and $n_2\lambda_1\nu_2$ can be neglected. This leads to the following system of equations:

$$\begin{cases} i\hbar\dot{a}_{n_1\lambda_1\nu_1}(t) = \sum\limits_{n\lambda\nu}\langle n_1\lambda_1\nu_1|H|n\lambda\nu\rangle \exp\left[\frac{it}{\hbar}\left(\varepsilon_{n_1\lambda_1\nu_1} - \varepsilon_{n\lambda\nu}\right)\right] a_{n\lambda\nu}(t), \\[2ex] i\hbar\dot{a}_{n_2\lambda_1\nu_2}(t) = \sum\limits_{n\lambda\nu}\langle n_2\lambda_1\nu_2|H|n\lambda\nu\rangle \exp\left[\frac{it}{\hbar}\left(\varepsilon_{n_2\lambda_1\nu_2} - \varepsilon_{n\lambda\nu}\right)\right] a_{n\lambda\nu}(t), \\[2ex] i\hbar\dot{a}_{n\lambda\nu}(t) = \langle n\lambda\nu|H|n_1\lambda_1\nu_1\rangle \exp\left[\frac{it}{\hbar}\left(\varepsilon_{n\lambda\nu} - \varepsilon_{n_1\lambda_1\nu_1}\right)\right] a_{n_1\lambda_1\nu_1}(t) + \\[2ex] \qquad \langle n\lambda\nu|H|n_2\lambda_1\nu_2\rangle \exp\left[\frac{it}{\hbar}\left(\varepsilon_{n\lambda\nu} - \varepsilon_{n_2\lambda_1\nu_2}\right)\right] a_{n_2\lambda_1\nu_2}(t). \end{cases} \tag{5.102}$$

In the last of these equations $n\lambda\nu \neq n_1\lambda_1\nu_1$, $n\lambda\nu \neq n_2\lambda_1\nu_2$ and there is no summation over different values of ν_2, the reason being that we neglect the matrix elements where the energy conservation law is strongly violated, i.e., the difference $\varepsilon_{n\lambda\nu} - \varepsilon_{n_2\lambda_1\nu_1}$ is large (we neglect the matrix elements $\langle n\nu_1|H_{\text{ir}}|n_2\nu_2\rangle$ and $\langle n_1\lambda_1|H_{\text{ie}}|n\lambda_1\rangle$ with $n \neq n_1$, as well as $\langle n_2\lambda_1|H_{\text{ie}}|n\lambda_1\rangle$ and $\langle n_1\nu_1|H_{\text{ir}}|n\nu_2\rangle$ with $n \neq n_2$). For the sake of convenience, let us introduce the notation

$$a_{n_1\lambda_1\nu_1} = a_0, \quad \varepsilon_{n_1\lambda_1\nu_1} = \varepsilon_0,$$

$$a_{n_2\lambda_1\nu_2} = a_\alpha, \quad \varepsilon_{n_2\lambda_1\nu_2} = \varepsilon_\alpha,$$

$$a_{n\lambda\nu} = a_k, \quad \varepsilon_{n\lambda\nu} = \varepsilon_k \quad (n\lambda\nu \neq n_1\lambda_1\nu_1,\ n\lambda\nu \neq n_2\lambda_1\nu_2).$$

The form of the representation of $\psi_\lambda^{(e)}$ was chosen so that ε_0 does not change when λ_1 does, and the energy of the level ε_α may change weakly with λ_1 due to the interaction with free electrons.

To carry out the calculations we resort to a method described in W. Heitler's book [86], Ch. IV, § 16 and in the paper [85]. Following this method, we pass from the functions $a_k(t)$, $a_\alpha(t)$ to the functions $u_k(\varepsilon)$, $u_\alpha(\varepsilon)$ by the Fourier transformation, and then letting $t \to \infty$ we obtain from the system of equations (5.102) the relations

$$\begin{cases} a_0(\infty) = 0, \\ a_\alpha(\infty) = \dfrac{u_\alpha(\varepsilon_\alpha)}{\varepsilon_\alpha - \varepsilon_0 + \frac{1}{2}i\hbar\Gamma(\varepsilon_\alpha)}, \\ a_k(\infty) = \dfrac{u_k(\varepsilon_k)}{\varepsilon_k - \varepsilon_0 + \frac{1}{2}i\hbar\Gamma(\varepsilon_k)}. \end{cases} \tag{5.103}$$

The functions $u_\alpha(\varepsilon)$, $u_k(\varepsilon)$ and $\Gamma(\varepsilon)$ satisfy the system of equations

$$\begin{cases} u_\alpha(\varepsilon) = H_{\alpha 0} + H_{\alpha\alpha}\zeta(\varepsilon - \varepsilon_\alpha)u_\alpha(\varepsilon) + \sum\limits_k H_{\alpha k}\zeta(\varepsilon - \varepsilon_k)u_k(\varepsilon), \\ u_k(\varepsilon) = H_{k0} + H_{k\alpha}\zeta(\varepsilon - \varepsilon_\alpha)u_\alpha(\varepsilon), \\ \dfrac{\hbar}{2}\Gamma(\varepsilon) = i\sum\limits_\alpha H_{0\alpha}\zeta(\varepsilon - \varepsilon_\alpha)u_\alpha(\varepsilon) + i\sum\limits_k H_{0k}\zeta(\varepsilon - \varepsilon_k)u_k(\varepsilon), \end{cases} \tag{5.104}$$

where $\zeta(x)$ is the singular function

$$\zeta(x) = \mathcal{P}\frac{1}{x} - i\pi\delta(x) = \lim_{\substack{\sigma\to 0 \\ (\sigma>0)}} \frac{1}{x + i\sigma}.$$

The symbol $\mathcal{P}\displaystyle\int_{-\infty}^{\infty} \frac{1}{x} f(x)\,dx$ denotes the integral in the sense of principal value. In what follows we shall use the fact that $x\zeta(x) = 1$.

The first two equations of system (5.104) yield

$$u_\alpha(\varepsilon) = \frac{H_{\alpha 0} + \sum_k H_{\alpha k} H_{k0} \zeta(\varepsilon - \varepsilon_k)}{1 - \zeta(\varepsilon - \varepsilon_\alpha)\left[H_{\alpha\alpha} + \sum_k |H_{\alpha k}|^2 \zeta(\varepsilon - \varepsilon_k)\right]} =$$

$$\frac{(\varepsilon - \varepsilon_\alpha)\left[H_{\alpha 0} + \sum_k H_{\alpha k} H_{k0} \zeta(\varepsilon - \varepsilon_k)\right]}{(\varepsilon - \varepsilon_\alpha) - \left[H_{\alpha\alpha} + \sum_k |H_{\alpha k}|^2 \zeta(\varepsilon - \varepsilon_k)\right]}.$$

This expression shows that $u_\alpha(\varepsilon_\alpha) = 0$, since for $\varepsilon = \varepsilon_\alpha$ the denominator is different from zero. Indeed, for $\varepsilon = \varepsilon_\alpha$ the imaginary part of the denominator is equal to

$$\pi \sum_k |H_{\alpha k}|^2 \delta(\varepsilon_\alpha - \varepsilon_k) > 0.$$

Therefore, by (5.103), $a_\alpha(\infty) = 0$. The expression obtained for $u_\alpha(\varepsilon)$ can be simplified since the term $H_{\alpha 0}$ is of first order with respect to the magnitude of the perturbation, while the sum $\sum_k H_{\alpha k} H_{k0} \zeta(\varepsilon - \varepsilon_k)$ is of second order (in the sum with respect to k the singularity of the function $\zeta(\varepsilon - \varepsilon_k)$ is inessential, because this summation presumes an integration over a continuous spectrum of energies ε_k). Based on the same considerations, in the denominator of the expression for $u_\alpha(\varepsilon)$ one can neglect $\mathrm{Re} \sum_k |H_{\alpha k}|^2 \zeta(\varepsilon - \varepsilon_k)$ in comparison with $H_{\alpha\alpha}$. This finally yields

$$u_\alpha(\varepsilon) = \frac{(\varepsilon - \varepsilon_\alpha) H_{\alpha 0}}{(\varepsilon - \varepsilon_\alpha) - \left[H_{\alpha\alpha} - i\pi \sum_k |H_{\alpha k}|^2 \delta(\varepsilon - \varepsilon_k)\right]}.$$

For $u_k(\varepsilon)$ we derive from (5.104) the expression

$$u_k(\varepsilon) = H_{k0} + \frac{H_{k\alpha} H_{\alpha 0}}{(\varepsilon - \varepsilon_\alpha) - \left[H_{\alpha\alpha} - i\pi \sum_k |H_{\alpha k}|^2 \delta(\varepsilon - \varepsilon_k)\right]} =$$

$$H_{k0} + \frac{H_{k\alpha} H_{\alpha 0}}{(\varepsilon - \varepsilon_\alpha - H_{\alpha\alpha}) + \frac{1}{2} i\hbar\gamma(\varepsilon)},$$

where

$$\gamma(\varepsilon) = \frac{2\pi}{\hbar} \sum_k |H_{\alpha k}|^2 \delta(\varepsilon - \varepsilon_k).$$

Substituting these expressions for $u_\alpha(\varepsilon)$ and $u_k(\varepsilon)$ in the third equation of system (5.104) we obtain

$$\frac{1}{2}\hbar\Gamma(\varepsilon) = i\sum_\alpha \frac{|H_{0\alpha}|^2}{(\varepsilon - \varepsilon_\alpha - H_{\alpha\alpha}) + \frac{1}{2}i\hbar\gamma(\varepsilon)} + i\sum_k |H_{0k}|^2\zeta(\varepsilon - \varepsilon_k) +$$

$$i\sum_\alpha \frac{H_{0k}H_{k\alpha}H_{\alpha 0}\zeta(\varepsilon - \varepsilon_k)}{(\varepsilon - \varepsilon_\alpha - H_{\alpha\alpha}) + \frac{1}{2}i\hbar\gamma(\varepsilon)} \simeq$$

$$i\sum_\alpha \frac{|H_{0\alpha}|^2}{(\varepsilon - \varepsilon_\alpha - H_{\alpha\alpha}) + \frac{1}{2}i\hbar\gamma(\varepsilon)} + i\sum_k |H_{0k}|^2\zeta(\varepsilon - \varepsilon_k).$$

Since the quantity $\hbar\gamma(\varepsilon)$ is small, applying the limit relation

$$\zeta(x) = \lim_{\substack{\sigma\to 0\\ (\sigma>0)}} \frac{1}{x + i\sigma},$$

one obtains

$$\frac{1}{2}\hbar\Gamma(\varepsilon) = i\sum_\alpha |H_{0\alpha}|^2\zeta(\varepsilon - \varepsilon_\alpha - H_{\alpha\alpha}) + i\sum_k |H_{0k}|^2\zeta(\varepsilon - \varepsilon_k).$$

The form of the matrix elements allows us to conclude that $H_{\alpha\alpha}$ does not change when α varies (assuming the state n_2 fixed). Let us denote $H_{\alpha\alpha} = \Delta$. The sum $\sum_\alpha |H_{0\alpha}|^2\zeta(\varepsilon - \varepsilon_\alpha - \Delta)$ is a slowly varying function of the variable ε. Therefore,

$$\sum_\alpha |H_{0\alpha}|^2\zeta(\varepsilon - \varepsilon_\alpha - \Delta) \simeq \sum_\alpha |H_{0\alpha}|^2\zeta(\varepsilon - \varepsilon_\alpha)$$

whence

$$\frac{1}{2}\hbar\Gamma(\varepsilon) \simeq i\sum_\alpha |H_{0\alpha}|^2\zeta(\varepsilon - \varepsilon_\alpha) + i\sum_k |H_{0k}|^2\zeta(\varepsilon - \varepsilon_k).$$

It follows that

$$\operatorname{Re}\Gamma(\varepsilon) = \frac{2\pi}{\hbar}\left[\sum_\alpha |H_{0\alpha}|^2\delta(\varepsilon - \varepsilon_\alpha) + \sum_k |H_{0k}|^2\delta(\varepsilon - \varepsilon_k)\right],$$

$$\operatorname{Im}\Gamma(\varepsilon) = \frac{2}{\hbar}\mathcal{P}\left[\sum_\alpha \frac{|H_{0\alpha}|^2}{\varepsilon - \varepsilon_\alpha} + \sum_k \frac{|H_{0k}|^2}{\varepsilon - \varepsilon_k}\right].$$

Let us find the probability of photon absorption in the frequency interval $\omega, \omega + d\omega$ for the transition of the ion from the state n_1 to the state n_2. Directly, that is, by using $a_\alpha(\infty)$, this is not possible, because $a_\alpha(\infty) = 0$. For this reason we will proceed as follows. Consider those values of k for which $H_{\alpha k} \neq 0$. It turns out that for these values of k the matrix elements $H_{0k} = 0$. Indeed, by (5.101), $H_{\alpha k} \neq 0$ for states k such that

(1) $\lambda = \lambda_1$, and the number of photons in the state k differs by the number of the photons in the state α by one, and so $H_{0k} = 0$;

(2) $\nu = \nu_2$, but for such states $H_{0k} = 0$, because we have neglected the matrix elements $\langle n_1\lambda_1|H_{\mathrm{i}\ell}|n\lambda_1\rangle$ with $n \neq n_1$ and $\langle n_1\nu_1|H_{\mathrm{ir}}|n\nu_2\rangle$ with $n \neq n_2$.

Therefore, for values of k for which $H_{\alpha k} \neq 0$, the quantity $|a_k(\infty)|^2$ may be regarded as the conditional probability of the following transition: as a result of absorbing a photon, the system passes first to the state α, and then to the considered state $k = k_\alpha$ (transition through an intermediate state).

Let us calculate the probability $J(\varepsilon, \varepsilon')\, d\varepsilon\, d\varepsilon'$ that the energy ε_α lies in the interval $\varepsilon, \varepsilon + d\varepsilon$ and the energy ε_{k_α} lies in the interval $\varepsilon', \varepsilon' + d\varepsilon'$.

Obviously,

$$J(\varepsilon, \varepsilon')\, d\varepsilon\, d\varepsilon' = \sum_{\substack{\alpha \\ (\varepsilon \le \varepsilon_\alpha \le \varepsilon + d\varepsilon)}} \sum_{\substack{k_\alpha \\ (\varepsilon' \le \varepsilon_{k_\alpha} \le \varepsilon' + d\varepsilon')}} |a_{k_\alpha}(\infty)|^2 =$$

$$\sum_{\alpha, k_\alpha} \frac{|H_{k_\alpha \alpha} H_{\alpha 0}|^2}{|\varepsilon_{k_\alpha} - \varepsilon_0 + \frac{1}{2} i\hbar\Gamma(\varepsilon_{k_\alpha})|^2\, |\varepsilon_{k_\alpha} - \varepsilon_\alpha - \Delta + \frac{1}{2} i\hbar\gamma(\varepsilon_{k_\alpha})|^2} \simeq$$

$$\frac{\sum_{\alpha, k_\alpha} |H_{k_\alpha \alpha} H_{\alpha 0}|^2}{|\varepsilon' - \varepsilon_0 + \frac{1}{2} i\hbar\Gamma(\varepsilon')|^2\, |\varepsilon' - \varepsilon - \Delta + \frac{1}{2} i\hbar\gamma(\varepsilon')|^2}.$$

The sum figuring in the numerator of the expression for $J(\varepsilon, \varepsilon')\, d\varepsilon\, d\varepsilon'$ may be written in the form $A(\varepsilon, \varepsilon')\, d\varepsilon\, d\varepsilon'$, where $A(\varepsilon, \varepsilon')$ is a continuous function.

Indeed,

$$\sum_{\substack{\alpha \\ (\varepsilon \le \varepsilon_\alpha \le \varepsilon + d\varepsilon)}} \sum_{\substack{k_\alpha \\ (\varepsilon' \le \varepsilon_{k_\alpha} \le \varepsilon' + d\varepsilon')}} |H_{k_\alpha \alpha} H_{\alpha 0}|^2 \simeq$$

$$d\varepsilon\, d\varepsilon' \sum_\alpha \sum_{k_\alpha} |H_{k_\alpha \alpha} H_{\alpha 0}|^2\, \delta(\varepsilon - \varepsilon_\alpha)\delta(\varepsilon' - \varepsilon_{k_\alpha}).$$

Therefore,

$$J(\varepsilon, \varepsilon') = \frac{A(\varepsilon, \varepsilon')}{|\varepsilon' - \varepsilon_0 + \frac{1}{2}\, i\hbar\Gamma(\varepsilon')|^2\, |\varepsilon' - \varepsilon - \Delta + \frac{1}{2}\, i\hbar\gamma(\varepsilon')|^2}.$$

The function $J(\varepsilon, \varepsilon')$ has a sharp maximum in the region $\varepsilon = \varepsilon' = \varepsilon_0$, because the quantities Δ, $\hbar\Gamma$, $\hbar\gamma$ are small. Near this maximum

$$J(\varepsilon, \varepsilon') \simeq \frac{A_0}{|\varepsilon' - \varepsilon_0 + \frac{1}{2}i\hbar\Gamma|^2 \, |\varepsilon' - \varepsilon - \Delta + \frac{1}{2}i\hbar\gamma|^2},$$

where

$$A_0 = A(\varepsilon_0, \varepsilon_0), \quad \Gamma = \Gamma(\varepsilon_0), \quad \gamma = \gamma(\varepsilon_0).$$

The probability that the energy ε_α lies in the interval $\varepsilon, \varepsilon + d\varepsilon$ is then given by

$$J(\varepsilon) = \int_{-\infty}^{\infty} J(\varepsilon, \varepsilon')\, d\varepsilon' \sim \frac{1}{(\varepsilon_0 - \varepsilon - \Delta + \frac{1}{2}\hbar\, \mathrm{Im}\,\Gamma)^2 + [\frac{1}{2}\hbar(\gamma + \mathrm{Re}\,\Gamma)]^2}.$$

Compared to Δ, the term $\frac{1}{2}\hbar\, \mathrm{Im}\,\Gamma$ can be neglected, being a quantity of the next order of smallness with respect to the perturbation H, and we obtain

$$J(\varepsilon) \sim \frac{1}{(\varepsilon_0 - \varepsilon - \Delta)^2 + [\frac{1}{2}\hbar(\gamma + \mathrm{Re}\,\Gamma)]^2}.$$

Further, since $\varepsilon_0 - \varepsilon = \hbar\omega + \varepsilon_{n_1} - \varepsilon_{n_2}$ ($\varepsilon = \varepsilon_\alpha$ and ω is the frequency of the absorbed photon in the transition from the state with energy ε_0 to the state with energy ε_α), the probability of absorption of a photon with frequency in the interval $\omega, \omega + d\omega$ is equal to

$$J(\omega)\, d\omega \sim \frac{1}{(\omega - \omega_0)^2 + (\Gamma_0/2)^2}, \tag{5.105}$$

where

$$\omega_0 = \frac{1}{\hbar}(\varepsilon_{n_2} - \varepsilon_{n_1} + \Delta), \quad \Gamma_0 = \gamma + \mathrm{Re}\,\Gamma.$$

Thus, we arrived at the Lorentz formula with a frequency shift $\Delta/\hbar$ and width Γ_0. The quantity

$$\Gamma_0 = \frac{2\pi}{\hbar}\left(\sum_k |H_{\alpha k}|^2 \delta(\varepsilon_0 - \varepsilon_k) + \sum_\alpha |H_{0\alpha}|^2 \delta(\varepsilon_0 - \varepsilon_\alpha) + \sum_k |H_{0k}|^2 \delta(\varepsilon_0 - \varepsilon_k)\right) \tag{5.106}$$

represents the total probability of transition, per unit of time, from the initial state $n_1\lambda_1\nu_1$ to the final state $n_2\lambda_1\nu_2$, under the action of the perturbation H.

According to the definition of Δ, the position of the spectral line ω_0 is given by

$$\omega_0 = \frac{1}{\hbar}(\varepsilon'_{n_2} - \varepsilon'_{n_1}), \tag{5.107}$$

where

$$\varepsilon'_{n_2} = \varepsilon_{n_2} + \langle n_2\lambda_1 | H_{\mathrm{ie}} | n_2\lambda_1\rangle, \qquad \varepsilon'_{n_1} = \varepsilon_{n_1} + \langle n_1\lambda_1 | H_{\mathrm{ie}} | n_1\lambda_1\rangle$$

are the energies of the ion in the states n_2 and n_1 with the corrections due to the interaction of the ion with free electrons.

5.4.4 Accounting for degeneracy

To clarify how degeneracy affects the shape of spectral lines, let us consider the case where the top level consists of a set of levels $\varepsilon_{n_{2s}}$, $(s = 1, 2, \ldots)$ that are close to one another, and the frequencies of the other transitions, $\omega_{\ell m} = (\varepsilon_\ell - \varepsilon_m)/\hbar$ differ sharply from the frequencies $\omega_{0s} = (\varepsilon_{n_{2s}} - \varepsilon_{n_1})/\hbar$ for $\ell \neq n_2$, $m \neq n_1$.

In this case we obtain a system of equations similar to (5.102), where now instead of n_2 we have n_{2s} and in the right-hand side one sums over s. Let

$$a_{n_{2s}\lambda_1\nu_2} = a_{\alpha s}, \quad \varepsilon_{n_{2s}\lambda_1\nu_2} = \varepsilon_{\alpha s}.$$

Then, similarly to (5.103), when $t \to \infty$ we obtain

$$\begin{cases} a_0(\infty) = 0, \\ a_{\alpha s}(\infty) = \dfrac{u_{\alpha s}(\varepsilon_{\alpha s})}{\varepsilon_{\alpha s} - \varepsilon_0 + \frac{1}{2} i\hbar\Gamma(\varepsilon_{\alpha s})}, \\ a_k(\infty) = \dfrac{u_k(\varepsilon_k)}{\varepsilon_k - \varepsilon_0 + \frac{1}{2} i\hbar\Gamma(\varepsilon_k)}. \end{cases}$$

The functions $u_{\alpha s}(\varepsilon)$, $u_k(\varepsilon)$ and $\Gamma(\varepsilon)$ satisfy the system of equations

$$\begin{cases} u_{\alpha s}(\varepsilon) = H_{\alpha s,0} + \sum\limits_{s'} H_{\alpha s,\alpha s'}\zeta(\varepsilon - \varepsilon_{\alpha s'})u_{\alpha s'}(\varepsilon) + \sum\limits_{k} H_{\alpha s,k}\zeta(\varepsilon - \varepsilon_k)u_k(\varepsilon), \\ u_k(\varepsilon) = H_{k0} + \sum\limits_{s'} H_{k,\alpha s'}\zeta(\varepsilon - \varepsilon_{\alpha s'})u_{\alpha s'}(\varepsilon), \\ \dfrac{\hbar}{2}\Gamma(\varepsilon) = i\sum\limits_{\alpha s'} H_{0,\alpha s'}\zeta(\varepsilon - \varepsilon_{\alpha s'})u_{\alpha s'}(\varepsilon) + i\sum\limits_{k'} H_{0k'}\zeta(\varepsilon - \varepsilon_{k'})u_{k'}(\varepsilon), \end{cases} \tag{5.108}$$

whence

$$u_{\alpha s} = H_{\alpha s,0} + \sum_{s'} H_{\alpha s,\alpha s'}\zeta(\varepsilon - \varepsilon_{\alpha s'})u_{\alpha s'} + \\ \sum_{k,s'} H_{\alpha s,k}H_{k,\alpha s'}\zeta(\varepsilon - \varepsilon_k)\zeta(\varepsilon - \varepsilon_{\alpha s'})u_{\alpha s'} + \sum_k H_{\alpha s,k}\zeta(\varepsilon - \varepsilon_k)H_{k0}. \tag{5.109}$$

As we did when we solved (5.104), the last term in (5.109) can be neglected. To solve the resulting system, we replace $u_{\alpha s}$ by the functions

$$c_{\alpha s} = \zeta(\varepsilon - \varepsilon_{\alpha s})u_{\alpha s},$$

which in turn will be, for each fixed value of α, solutions of the finite system of equations

$$(\varepsilon - \varepsilon_{\alpha s})c_{\alpha s} = H_{\alpha s,0} + \sum_{s'} c_{\alpha s'}\Big[H_{\alpha s,\alpha s'} + \sum_k H_{\alpha s,k}H_{k,\alpha s'}\zeta(\varepsilon - \varepsilon_k)\Big]. \tag{5.110}$$

Let $\lambda_\alpha^{(p)}$ and $\{b_{\alpha s}^{(p)}\}$ be the eigenvalues and the eigenvectors of the homogeneous system associated to the system (5.110):

$$(\lambda_\alpha^{(p)} - \varepsilon_{\alpha s})b_{\alpha s}^{(p)} = \sum_{s'} b_{\alpha s'}^{(p)} \Big[H_{\alpha s,\alpha s'} + \sum_k H_{\alpha s,k} H_{k,\alpha s'} \zeta(\varepsilon - \varepsilon_k)\Big]. \tag{5.111}$$

It is readily verified that the solution of (5.110) has the form

$$c_{\alpha s} = \sum_p \frac{b_{\alpha s}^{(p)}}{\varepsilon - \lambda_\alpha^{(p)}}, \tag{5.112}$$

where the eigenvectors $\{b_{\alpha s}^{(p)}\}$ are normalized so that

$$\sum_p b_{\alpha s}^{(p)} = H_{\alpha s,0}. \tag{5.113}$$

The eigenvalues $\lambda_\alpha^{(p)}$ have negative imaginary parts. Indeed, from (5.111) it follows that

$$(\lambda_\alpha^{(p)} - \bar{\lambda}_\alpha^{(p)}) \sum_s |b_{\alpha s}^{(p)}|^2 = \sum_{ss'} b_{\alpha s'}^{(p)} \bar{b}_{\alpha s}^{(p)} \sum_k H_{\alpha s,k} H_{k,\alpha s'} \big[\zeta(\varepsilon - \varepsilon_k) - \bar{\zeta}(\varepsilon - \varepsilon_k)\big],$$

(the overbar denotes complex conjugation), whence

$$\operatorname{Im} \lambda_\alpha^{(p)} = -\frac{\pi}{\sum_s |b_{\alpha s}^{(p)}|^2} \sum_k \delta(\varepsilon - \varepsilon_k) \left|\sum_s H_{k,\alpha s} b_{\alpha s}^{(p)}\right|^2 < 0.$$

Note also that, since the perturbation is small, $\operatorname{Re} \lambda_\alpha^{(p)} \approx \varepsilon_{\alpha p}$ (i.e., $\varepsilon_{\alpha s}$ at $s = p$); accordingly, henceforth we will replace $\operatorname{Re} \lambda_\alpha^{(p)}$ by $\varepsilon_{\alpha p}$.

Since the sum $\sum_k H_{\alpha s,k} H_{k,\alpha s'} \zeta(\varepsilon - \varepsilon_k)$ is a slowly varying function of ε, it follows from (5.111) that the eigenvectors $\{b_{\alpha s}^{(p)}\}$ and eigenvalues $\lambda_\alpha^{(p)}$ will depend only weakly on ε (i.e., if ε changes by a quantity of the order of the width of the spectral line, $\{b_{\alpha s}^{(p)}\}$ and $\lambda_\alpha^{(p)}$ may be assumed to be constants). Using (5.112), equations (5.108) yield

$$u_k(\varepsilon) = H_{k0} + \sum_s H_{k,\alpha s} \sum_p \frac{b_{\alpha s}^{(p)}}{\varepsilon - \lambda_\alpha^{(p)}},$$

whence

$$a_k(\infty) = \frac{1}{\varepsilon - \varepsilon_0 + \frac{1}{2} i\hbar\Gamma(\varepsilon)} \left[H_{k0} + \sum_s H_{k,\alpha s} \sum_p \frac{b_{\alpha s}^{(p)}}{\varepsilon - \lambda_\alpha^{(p)}}\right]\Bigg|_{\varepsilon=\varepsilon_k}$$

and

$$\frac{\hbar}{2}\Gamma(\varepsilon) = i\sum_{\alpha s} H_{0,\alpha s}\sum_p \frac{b_{\alpha s}^{(p)}}{\varepsilon - \lambda_\alpha^{(p)}} + i\sum_k H_{0k}\zeta(\varepsilon-\varepsilon_k)\left[H_{k0} + \sum_s H_{k,\alpha s}\sum_p \frac{b_{\alpha s}^{(p)}}{\varepsilon - \lambda_\alpha^{(p)}}\right] \approx i\sum_k |H_{0k}|^2\zeta(\varepsilon-\varepsilon_k) + i\sum_{\alpha s} H_{0,\alpha s}\sum_p \frac{b_{\alpha s}^{(p)}}{\varepsilon - \lambda_\alpha^{(p)}}.$$

Since $\zeta(x) = \lim\limits_{\substack{\sigma\to 0\\ (\sigma>0)}} \dfrac{1}{x+i\sigma}$, we have $\dfrac{1}{\varepsilon - \lambda_\alpha^{(p)}} \approx \zeta(\varepsilon - \varepsilon_{\alpha p})$, and so

$$\frac{\hbar}{2}\Gamma(\varepsilon) \approx i\sum_k |H_{0k}|^2\zeta(\varepsilon-\varepsilon_k) + i\sum_p\sum_{\alpha s} H_{0,\alpha s} b_{\alpha s}^{(p)}\zeta(\varepsilon - \varepsilon_{\alpha p}).$$

The second term in the right-hand side depends weakly on ε, and consequently it changes only slightly if in the function $\zeta(\varepsilon - \varepsilon_{\alpha p})$ we replace ε by the close value $\varepsilon + \varepsilon_{\alpha p} - \varepsilon_{\alpha s}$. Using (5.113), we obtain

$$\frac{\hbar}{2}\Gamma(\varepsilon) \approx i\sum_k |H_{0k}|^2\zeta(\varepsilon-\varepsilon_k) + i\sum_{\alpha s} |H_{0,\alpha s}|^2\zeta(\varepsilon - \varepsilon_{\alpha s}). \tag{5.114}$$

It can be easily shown that if for some value $k = k_\alpha$ we have $H_{k,\alpha s} \neq 0$, then $H_{0k} = 0$. Therefore, the quantity $|a_{k_\alpha}(\infty)|^2$ can be interpreted as the conditional probability that the system, after absorbing a photon with frequency ω_α, such that $\varepsilon_{\alpha s} = \varepsilon_0 + \hbar\omega_{0s} - \hbar\omega_\alpha$, will pass to the state k_α.

The probability of interest here, namely, the probability of absorption of a photon with frequency in the interval $\omega, \omega + d\omega$, can be calculated by means of the formula

$$J(\omega)\,d\omega = \sum_{\omega<\omega_\alpha<\omega+d\omega}\ \sum_{k_\alpha} |a_{k_\alpha}(\infty)|^2.$$

In the present case the line intensity $J(\omega)$ is not represented just by a sum of Lorentz contours corresponding to transitions to each of the upper levels n_{2s}; rather, it also contains interference terms. Nevertheless, the intensity $J(\omega)$ can be represented approximately as a Lorentz contour if we use the following approximate equality (see (5.113)):

$$\sum_p \frac{b_{\alpha s}^{(p)}}{\varepsilon - \lambda_\alpha^{(p)}} \approx \frac{1}{\varepsilon - \lambda_\alpha}\sum_p b_{\alpha s}^{(p)} = \frac{H_{\alpha s,0}}{\varepsilon - \lambda_\alpha},$$

where λ_α is some average of the values $\lambda_\alpha^{(p)}$. From the secular equation for $\lambda_\alpha^{(p)}$ corresponding to (5.111) we conclude that the arithmetic mean of the values $\lambda_\alpha^{(p)}$ is given by the formula

$$\lambda_\alpha = \Big\langle \varepsilon_{\alpha s} + H_{\alpha s,\alpha s} + \sum_k |H_{\alpha s,k}|^2 \zeta(\varepsilon - \varepsilon_k) \Big\rangle_s$$

(here the brackets $\langle \cdots \rangle_s$ denote the arithmetic mean value over the states s). We conclude that

$$\mathrm{Re}\,\lambda_\alpha \approx \langle \varepsilon_{\alpha s} \rangle_s, \quad \mathrm{Im}\,\lambda_\alpha = -\pi \Big\langle \sum_k |H_{\alpha s,k}|^2 \delta(\varepsilon - \varepsilon_k) \Big\rangle_s = -\frac{\hbar}{2} \langle \gamma_{\alpha s} \rangle_s .$$

In this approximation

$$a_{k_\alpha}(\infty) = \frac{\sum_s H_{k,\alpha s} H_{\alpha s,0}}{\left[\varepsilon_k - \varepsilon_0 + \frac{1}{2} i\hbar\Gamma(\varepsilon_k)\right] \left[\varepsilon_k - \langle \varepsilon_{\alpha s} \rangle_s + \frac{1}{2} i\hbar \langle \gamma_{\alpha s} \rangle_s \right]}$$

and we obtain the following formula for $J(\omega)$:

$$J(\omega) \sim \frac{1}{\big(\omega - \langle \omega_{0s} \rangle_s\big)^2 + \big(\langle \Gamma_{0s} \rangle_s / 2\big)^2}, \tag{5.115}$$

where

$$\Gamma_{0s} = \frac{2\pi}{\hbar} \Bigg\{ \sum_k |H_{\alpha s,k}|^2 \delta(\varepsilon_{\alpha s} - \varepsilon_k) + \sum_{\alpha s} |H_{0,\alpha s}|^2 \delta(\varepsilon_0 - \varepsilon_{\alpha s}) + \sum_k |H_{0k}|^2 \delta(\varepsilon_0 - \varepsilon_k) \Bigg\}, \tag{5.116}$$

which is given by an expression similar to (5.106), is the total probability of the transition, per unit of time, from the initial state $n_1\lambda_1\nu_1$ to the final state $n_{2s}\lambda_1\nu_2$, under the action of the perturbation H. The quantity ω_{0s}, which determines the position of the spectral line, can be represented in the form

$$\omega_{0s} = \frac{1}{\hbar}(\varepsilon'_{n_{2s}} - \varepsilon'_{n_1}),$$

where

$$\varepsilon'_{n_{2s}} = \varepsilon_{n_{2s}} + \langle n_{2s}\lambda_1 | H_{ie} | n_{2s}\lambda_1 \rangle, \quad \varepsilon'_{n_1} = \varepsilon_{n_1} + \langle n_1\lambda_1 | H_{ie} | n_1\lambda_1 \rangle$$

are the energies of the ion in the states n_{2s} and n_1, respectively.

In view of result obtained above, one should expect that for transitions between groups of levels that are close to one another, the intensity distribution $J(\omega)$ will also be described, in some approximation, by the Lorentz formula, in which now the line width is equal to the sum of the mean widths for the upper and lower groups of levels.

Substituting expression (5.101) for the matrix elements in (5.116), we obtain the following expression for the line width (we assume that averaging has been carried out):

$$\gamma \equiv \Gamma_0 = \gamma_{n_1}^{(r)} + \gamma_{n_2}^{(r)} + \gamma_{n_1}^{(e)} + \gamma_{n_2}^{(e)} + \gamma_{n_1 n_2}^{(u)}.$$

Here

$$\gamma_n^{(r)} = \frac{2\pi}{\hbar} \sum_{n'\nu} \left| \langle n\nu_1 | H_{\mathrm{ir}} | n'\nu \rangle \right|^2 \delta(\varepsilon_n + \varepsilon_{\nu_1} - \varepsilon_{n'} - \varepsilon_\nu) \tag{5.117}$$

is the natural (radiation) width of the energy level of the ion in the state n, which is equal to the probability of transition, per unit of time, from the state n due to the interaction with the electromagnetic field;

$$\gamma_n^{(e)} = \frac{2\pi}{\hbar} \sum_{n' \neq n} \sum_{\lambda} \left| \langle n\lambda_1 | H_{\mathrm{ie}} | n'\lambda \rangle \right|^2 \delta(\varepsilon_n + \varepsilon_{\lambda_1} - \varepsilon_{n'} - \varepsilon_\lambda) \tag{5.118}$$

is the broadening due to the interaction with the free electrons; and

$$\gamma_{n_1 n_2}^{(u)} = \frac{2\pi}{\hbar} \sum_{\lambda} \left| \langle n_2\lambda_1 | H_{\mathrm{ie}} | n_2\lambda \rangle - \langle n_1\lambda_1 | H_{\mathrm{ie}} | n_1\lambda \rangle \right|^2 \delta(\varepsilon_{\lambda_1} - \varepsilon_\lambda) \tag{5.119}$$

is, in the terminology of [112], the universal broadening, which is a result of elastic collisions.

5.4.5 Methods for calculating radiation and electron broadening

In the dipole approximation the natural width of the line (5.117) for the state $n\ell$ (i.e., the reciprocal of the lifetime of this state, see Subsection 5.7.4) equals

$$\gamma_{n\ell}^{(r)} = 2\pi^2 \alpha \sum_{n'\ell' \neq n\ell} \left(1 - \frac{N_{n'\ell'}}{2(2\ell'+1)}\right) F_{n\ell,n'\ell'}(|\varepsilon_{n\ell} - \varepsilon_{n'\ell'}|/\hbar), \tag{5.120}$$

where

$$F_{n\ell,n'\ell'}(\omega) = \begin{cases} \left(\dfrac{\hbar\omega^3}{\pi^2 c^2} + 4\pi B_\omega\right) \dfrac{2\ell'+1}{2\ell+1} \dfrac{f_{n'\ell',n\ell}}{\hbar\omega}, & \text{if } \varepsilon_{n'\ell'} < \varepsilon_{n\ell}, \\ 4\pi B_\omega \dfrac{f_{n\ell,n'\ell'}}{\hbar\omega}, & \text{if } \varepsilon_{n'\ell'} > \varepsilon_{n\ell}, \end{cases}$$

$$B_\omega = \frac{\hbar\omega^3}{4\pi^3 c^2} \frac{1}{e^x - 1},$$

$\omega = |\varepsilon_{n\ell} - \varepsilon_{n'\ell'}|/\hbar$, $x = \hbar\omega/\theta$, $\alpha = 1/c = 1/137.036$, and the factor $\left(1 - \frac{N_{n'\ell'}}{2(2\ell'+1)}\right)$ accounts for the occupancy of the level $n'\ell'$ (see the derivation of (5.45)). The quantity $\gamma_{n\ell}^{(\mathrm{r})}$ is expressed in atomic units.

The sum for $\varepsilon_{n'\ell'} < \varepsilon_{n\ell}$ accounts for induced and spontaneous emission, while for $\varepsilon_{n'\ell'} > \varepsilon_{n\ell}$ it accounts for absorption. In the latter case one should, generally speaking, account for the states belonging to the continuum, but due to the presence of the factor $1/(e^x - 1)$ those states have practically no contribution to the natural width.

To calculate the electron widths $\gamma_{n_1}^{(\mathrm{e})}$, $\gamma_{n_2}^{(\mathrm{e})}$ and $\gamma_{n_1 n_2}^{(\mathrm{u})}$ it is necessary to construct wave functions of the ion-free electron system, adopting some coupling scheme of momenta. Width calculations then become rather tedious, despite the fact that the influence of errors on widths is sufficiently weak, as we shall see below. For that reason, we shall use the one-electron approximation, i.e., we shall assume that a spectral line is connected with a change of state of only one electron of the ion. To obtain the line width connected with the transitions of the electron in question, it suffices that, in formulas (5.118)–(5.119), in the summation over various states of the ion, we restrict ourselves to those states in which our electron changes its position. After we obtain the one-electron widths, we must average them over the quantum numbers of the electron within a group of one-electron energy levels that are close to one another.

To calculate the widths of spectral lines using formulas (5.118)–(5.119) we must know the eigenfunctions $\psi_n^{(\mathrm{i})}$, $\psi_\lambda^{(\mathrm{e})}$, $\psi_\nu^{(\mathrm{r})}$. The photon wave functions are well known (see, e.g., [86]). The calculation of the electron wave functions for the bound and continuum states is considered in §§ 2.2, 2.3.

In the calculation of the matrix elements $\langle n\lambda_1 | H_{\mathrm{ie}} | n'\lambda\rangle$ that appear in the expressions for the widths $\gamma_n^{(\mathrm{e})}$ and $\gamma_{n_1 n_2}^{(\mathrm{u})}$ we may assume, with little loss of accuracy, that the potentials V_λ and V_{λ_1} that act on the continuum electrons are equal. Indeed, let us put $V_{\lambda_1} = V_\lambda$ and calculate the matrix elements $\langle n\lambda_1 | H_{\mathrm{ie}} | n'\lambda\rangle$. Obviously, the Hamiltonian H_{ie} can be written as a sum of Hamiltonians, each of which acts on the coordinates of a single electron in the ion under consideration:

$$H_{\mathrm{ie}} = \sum_s \frac{1}{|\vec{r}_s - \vec{r}_p|},$$

where $\vec{r}_s$ is the position vector of the s-th electron of the ion and $\vec{r}_p$ is the position vector of the perturbing electron. For $V_\lambda = V_{\lambda_1}$, the one-electron wave functions corresponding to the states λ and λ_1 are orthogonal. Therefore, in the one-electron approximation the matrix element $\langle n\lambda_1 | H_{\mathrm{ie}} | n'\lambda\rangle$ will be different from zero if and only if the initial state differs from the final state only through the position of one electron of the ion and the state of one of the perturbing electrons. If the wave functions and the energies of these two electrons in the initial state $n\lambda_1$ [resp., final state $n'\lambda$] are $\psi_k(\vec{r})$, ε_k [resp., $\psi_{k'}(\vec{r})$, $\varepsilon_{k'}$] for the electron of the ion and $\varphi_\mu(\vec{r}\,')$, ε_μ [resp., $\varphi_{\mu'}(\vec{r}\,')$, $\varepsilon_{\mu'}$] for the perturbing electron, then the matrix element $\langle n\lambda_1 | H_{\mathrm{ie}} | n'\lambda\rangle$ takes on the form

$$\langle n\lambda_1|H_{\rm ie}|n'\lambda\rangle = \int \psi_k^*(\vec{r})\,\varphi_\mu^*(\vec{r}\,')\,\frac{1}{|\vec{r}-\vec{r}\,'|}\,\psi_{k'}(\vec{r})\,\varphi_{\mu'}(\vec{r}\,')\,d\vec{r}\,d\vec{r}\,'. \tag{5.121}$$

Summing with respect to n' and λ in (5.118) and applying the Pauli principle, we obtain

$$\gamma_n^{(\rm e)} = \sum_{k'}(1-n_{k'})\,\gamma_{kk'}^{(\rm e)}, \tag{5.122}$$

where

$$\gamma_{kk'}^{(\rm e)} = 2\pi\sum_{\mu,\mu'} n_\mu(1-n_{\mu'})\left|\int \psi_k^*(\vec{r})\varphi_\mu^*(\vec{r}\,')\frac{1}{|\vec{r}-\vec{r}\,'|}\psi_{k'}(\vec{r})\varphi_{\mu'}(\vec{r}\,')\,d\vec{r}\,d\vec{r}\,'\right|^2 \times \delta(\Delta\varepsilon_{kk'}+\varepsilon_\mu-\varepsilon_{\mu'}); \tag{5.123}$$

here n_k and n_μ are the occupation coefficients of the electron states, equal to 1 for the occupied states and to 0 for the unoccupied ones and $\Delta\varepsilon_{kk'}$ is the difference between the initial and final energies of the ion when one of the electrons passes from state k to state k'. In what follows, instead of $\Delta\varepsilon_{kk'}$ we will use its approximation $\varepsilon_k-\varepsilon_{k'}$, since this has practically no influence on the value of $\gamma_n^{(\rm e)}$.

According to its meaning, the quantity $\gamma_{kk'}^{(\rm e)}$ is the probability of transition of a bound-state electron from state k to state k' per unit of time due to the interaction with the free electrons.

In much the same way we obtain for $\gamma_{n_1n_2}^{(\rm u)}$ the expression

$$\gamma_{n_1n_2}^{(\rm u)} = \gamma_{kk'}^{(\rm u)} = 2\pi\sum_{\mu,\mu'} n_\mu(1-n_{\mu'})\times \left|\int \left[|\psi_{k'}(\vec{r})|^2-|\psi_k(\vec{r})|^2\right]\frac{\varphi_\mu^*(\vec{r}\,')\,\varphi_{\mu'}(\vec{r}\,')}{|\vec{r}-\vec{r}\,'|}\,d\vec{r}\,d\vec{r}\,'\right|^2 \delta(\varepsilon_\mu-\varepsilon_{\mu'}). \tag{5.124}$$

A sufficiently good approximation in calculations of line widths for high temperatures is the one in which $\gamma_{kk'}^{(\rm e)}$ and $\gamma_{kk'}^{(\rm u)}$ are calculated for the average ion configuration, neglecting the action of electrical fields of the other ions and the relativistic splitting of lines. As the wave functions of bound-state electrons one uses the functions

$$\psi_{n\ell m}(\vec{r}) = \frac{1}{r}\,R_{n\ell}(r)\,(-1)^m Y_{\ell m}(\vartheta,\varphi),$$

while for the continuum electrons one takes

$$\psi_{\varepsilon\ell m}(\vec{r}) = \frac{1}{r}\,R_{\varepsilon\ell}(r)\,(-1)^m Y_{\ell m}(\vartheta,\varphi),$$

where $R_{n\ell}(r)$, $R_{\varepsilon\ell}(r)$ are solutions of the Schrödinger equation with the Hartree-Fock-Slater self-consistent potential.

Let us replace summation with respect to the free-electron states μ and μ' by integration with respect to ε and ε'. When we count the number of states of continuum electrons μ' we need to keep in mind that in a transition between states the spin of a free electron is conserved. Therefore, if for k and k' we use the quantum numbers $n\ell m$ and $n'\ell'm'$, we arrive at the following expression for $\gamma_{kk'}^{(e)}$:

$$\gamma_{n\ell m,n'\ell'm'}^{(e)} = 4\pi e^{-\eta} \sum_{\bar\ell\bar m} \sum_{\bar\ell'\bar m'} \int (1-n_\varepsilon)(1-n_{\varepsilon'}) e^{-\varepsilon/\theta}\, d\varepsilon \times$$

$$\left| \int R_{n\ell}(r)\, Y_{\ell m}^*(\vartheta,\varphi)\, R_{\varepsilon\bar\ell}(r')\, Y_{\bar\ell\bar m}^*(\vartheta',\varphi') \times \right.$$

$$\left. \frac{1}{|\vec r - \vec r\,'|} R_{n'\ell'}(r')\, Y_{\ell'm'}(\vartheta,\varphi)\, R_{\varepsilon'\bar\ell'}(r')\, Y_{\bar\ell'\bar m'}(\vartheta',\varphi')\, dr\, d\Omega\, dr'\, d\Omega' \right|^2, \quad (5.125)$$

where $n\ell m \neq n'\ell'm'$, the integration with respect to $\vec r$ and $\vec r\,'$ is carried over the atom cell for $\varepsilon' = \varepsilon + \varepsilon_{n\ell} - \varepsilon_{n'\ell'}$, $n_\varepsilon = 1/[1+\exp(\varepsilon/\theta+\eta)]$ and η is the reduced chemical potential: $\eta = -\mu/\theta$.

In much the same way we obtain for $\gamma_{kk'}^{(u)}$ the expression

$$\gamma_{n\ell m,n'\ell'm'}^{(u)} = 4\pi e^{-\eta} \sum_{\bar\ell\bar m} \sum_{\bar\ell'\bar m'} \int (1-n_\varepsilon)^2 e^{-\varepsilon/\theta}\, d\varepsilon \times$$

$$\left| \int \left[|R_{n'\ell'}(r)\, Y_{\ell'm'}(\vartheta,\varphi)|^2 - |R_{n\ell}(r)\, Y_{\ell m}(\vartheta,\varphi)|^2 \right] \times \right.$$

$$\left. \frac{1}{|\vec r - \vec r\,'|} R_{\varepsilon\bar\ell}(r')\, Y_{\bar\ell\bar m}^*(\vartheta',\varphi')\, R_{\varepsilon\bar\ell'}(r')\, Y_{\bar\ell'\bar m'}(\vartheta',\varphi')\, dr\, d\Omega\, dr'\, d\Omega' \right|^2. \quad (5.126)$$

To obtain the average widths $\gamma_{n\ell}^{(e)}$ we carry out the summation in (5.122) with respect to the quantum numbers n', ℓ', m' and average over m, replacing $1-n_{k'}$ by its mean value $1 - N_{n'\ell'}/(2(2\ell'+1))$. After that we average the universal widths over m and m'. This finally yields

$$\gamma_{n\ell}^{(e)} = \frac{1}{2\ell+1} \sum_m \sum_{n'\ell'm'} \left(1 - \frac{N_{n'\ell'}}{2(2\ell'+1)}\right) \gamma_{n\ell m,n'\ell'm'}^{(e)} \quad (n\ell m \neq n'\ell'm') \quad (5.127)$$

and

$$\gamma^{(\mathrm{u})}_{n\ell,n'\ell'} = \frac{1}{(2\ell+1)(2\ell'+1)} \sum_{m} \sum_{m'} \gamma^{(u)}_{n\ell m,n'\ell' m'}. \tag{5.128}$$

The total width for the transition $n\ell \to n'\ell'$ will be equal to

$$\gamma_{n\ell,n'\ell'} = \gamma^{(\mathrm{r})}_{n\ell} + \gamma^{(\mathrm{r})}_{n'\ell'} + \gamma^{(\mathrm{e})}_{n\ell} + \gamma^{(\mathrm{e})}_{n'\ell'} + \gamma^{(\mathrm{u})}_{n\ell,n'\ell'}. \tag{5.129}$$

To calculate the electron widths (5.125) and (5.126) we shall use the expansion of $1/|\vec{r}-\vec{r}\,'|$ in spherical harmonics,

$$\frac{1}{|\vec{r}-\vec{r}\,'|} = \sum_{s,\tilde{m}} \frac{4\pi}{2s+1} \frac{r^s_<}{r^{s+1}_>} Y_{s\tilde{m}}(\vartheta,\varphi) Y^*_{s\tilde{m}}(\vartheta',\varphi') \tag{5.130}$$

and express the integrals over the angular variables in terms of the Wigner $3j$-symbols. This yields

$$\begin{aligned}\gamma^{(\mathrm{e})}_{n\ell m,n'\ell' m'} = 4\pi e^{-\eta} \sum_{\bar{\ell}\bar{\ell}'} \sum_{\bar{m}\bar{m}'} (2\ell+1)(2\ell'+1)(2\bar{\ell}+1)(2\bar{\ell}'+1) \times \\ \int (1-n_\varepsilon)(1-n_{\varepsilon'}) e^{-\varepsilon/\theta}\, d\varepsilon \times \\ \left| \sum_{s,\tilde{m}} \begin{pmatrix} \ell & \ell' & s \\ 0 & 0 & 0 \end{pmatrix} \begin{pmatrix} \bar{\ell} & \bar{\ell}' & s \\ 0 & 0 & 0 \end{pmatrix} \begin{pmatrix} \ell & \ell' & s \\ -m & m' & \tilde{m} \end{pmatrix} \begin{pmatrix} \bar{\ell} & \bar{\ell}' & s \\ -\bar{m} & \bar{m}' & -\tilde{m} \end{pmatrix} R^{(s)}_{n\ell n'\ell'\varepsilon\bar{\ell}\varepsilon'\bar{\ell}'} \right|^2,\end{aligned} \tag{5.131}$$

$$\varepsilon' = \varepsilon + \varepsilon_{n\ell} - \varepsilon_{n'\ell'},$$

$$\begin{aligned}\gamma^{(\mathrm{u})}_{n\ell m,n'\ell' m'} = 4\pi e^{-\eta} \sum_{\bar{\ell}\bar{\ell}'} \sum_{\bar{m}\bar{m}'} \int (1-n_\varepsilon)^2 e^{-\varepsilon/\theta}\, d\varepsilon \times \\ \left| \sum_{s\tilde{m}} \sqrt{(2\bar{\ell}+1)(2\bar{\ell}'+1)} \begin{pmatrix} \bar{\ell} & \bar{\ell}' & s \\ 0 & 0 & 0 \end{pmatrix} \begin{pmatrix} \bar{\ell} & \bar{\ell}' & s \\ -\bar{m} & \bar{m}' & -\tilde{m} \end{pmatrix} \times \right. \\ \left[(2\ell'+1)\, R^{(s)}_{n'\ell' n'\ell'\varepsilon\bar{\ell}\varepsilon'\bar{\ell}'} (-1)^{m'} \begin{pmatrix} \ell' & \ell' & s \\ 0 & 0 & 0 \end{pmatrix} \begin{pmatrix} \ell' & \ell' & s \\ -m' & m' & \tilde{m} \end{pmatrix} - \right. \\ \left. \left. (2\ell+1)\, R^{(s)}_{n\ell n\ell\varepsilon\bar{\ell}\varepsilon'\bar{\ell}'} (-1)^{m} \begin{pmatrix} \ell & \ell & s \\ 0 & 0 & 0 \end{pmatrix} \begin{pmatrix} \ell & \ell & s \\ -m & m & \tilde{m} \end{pmatrix} \right] \right|^2.\end{aligned} \tag{5.132}$$

Here

$$R^{(s)}_{n\ell n'\ell'\varepsilon\bar{\ell}\varepsilon'\bar{\ell}'} = \int R_{n\ell}(r)\, R_{\varepsilon\bar{\ell}}(r') \frac{r^s_<}{r^{s+1}_>} R_{n'\ell'}(r)\, R_{\varepsilon'\bar{\ell}'}(r')\, dr\, dr'.$$

Summing with respect to m, m' and using the orthogonality relations for the Wigner coefficients (see, e.g., [220, 230]), one can show that for an arbitrary function $f(s,\tilde{m})$ the following relation holds:

$$\sum_{mm'}\left|\sum_{s\tilde{m}}\begin{pmatrix}\ell & \ell' & s\\ -m & m' & \tilde{m}\end{pmatrix} f(s,\tilde{m})\right|^2 =$$

$$\sum_{ss'}\sum_{\tilde{m}\tilde{m}'} f(s,\tilde{m})\, f(s',\tilde{m}') \sum_{mm'}\begin{pmatrix}\ell & \ell' & s\\ -m & m' & -\tilde{m}\end{pmatrix}\begin{pmatrix}\ell & \ell' & s'\\ -m & m' & -\tilde{m}'\end{pmatrix} =$$

$$\sum_{s\tilde{m}}\frac{1}{2s+1}\, f^2(s,\tilde{m})\{\ell\,\ell'\,s\}. \quad (5.133)$$

Using (5.133), we have

$$\gamma^{(e)}_{n\ell m,n'\ell'm'} = (2\ell+1)(2\ell'+1)\sum_{s,\tilde{m}}\frac{1}{2s+1}\begin{pmatrix}\ell & \ell' & s\\ 0 & 0 & 0\end{pmatrix}^2\begin{pmatrix}\ell & \ell' & s\\ -m & m' & \tilde{m}\end{pmatrix}^2\gamma^{(e)}_{n\ell,n'\ell',s}, \quad (5.134)$$

where

$$\gamma^{(e)}_{n\ell,n'\ell',s} = 4\pi e^{-\eta}\sum_{\bar{\ell}\bar{\ell}'}(2\bar{\ell}+1)(2\bar{\ell}'+1)\begin{pmatrix}\bar{\ell} & \bar{\ell}' & s\\ 0 & 0 & 0\end{pmatrix}^2 \times$$

$$\int(1-n_\varepsilon)(1-n_{\varepsilon'})e^{-\varepsilon/\theta}\left[R^{(s)}_{n\ell n'\ell'\varepsilon\bar{\ell}\varepsilon'\bar{\ell}'}\right]^2 d\varepsilon,$$

$$\varepsilon' = \varepsilon + \varepsilon_{n\ell} - \varepsilon_{n'\ell'}.$$

Similarly,

$$\gamma^{(u)}_{n\ell m,n'\ell'm'} = \gamma^{(e)}_{n'\ell'm',n'\ell'm'} + \gamma^{(e)}_{n\ell m,n\ell m} - 2(2\ell+1)(2\ell'+1)\times$$

$$\sum_{s,\tilde{m}}\frac{1}{2s+1}\begin{pmatrix}\ell' & \ell' & s\\ 0 & 0 & 0\end{pmatrix}\begin{pmatrix}\ell & \ell & s\\ 0 & 0 & 0\end{pmatrix}\times$$

$$(-1)^{m+m'}\begin{pmatrix}\ell' & \ell' & s\\ -m' & m' & \tilde{m}\end{pmatrix}\begin{pmatrix}\ell & \ell & s\\ -m & m & \tilde{m}\end{pmatrix}\gamma^{(u)}_{n\ell,n'\ell',s}, \quad (5.135)$$

where

$$\gamma^{(u)}_{n\ell,n'\ell',s} = 4\pi e^{-\eta}\sum_{\bar{\ell}\bar{\ell}'}(2\bar{\ell}+1)(2\bar{\ell}'+1)\begin{pmatrix}\bar{\ell} & \bar{\ell}' & s\\ 0 & 0 & 0\end{pmatrix}^2 \times$$

$$\int(1-n_\varepsilon)^2 e^{-\varepsilon/\theta}\, R^{(s)}_{n'\ell'n'\ell'\varepsilon\bar{\ell}\varepsilon\bar{\ell}'}\, R^{(s)}_{n\ell n\ell\varepsilon\bar{\ell}\varepsilon\bar{\ell}'}\, d\varepsilon.$$

Table 5.5 lists the values of the radiation width $\gamma^{(r)}_{n\ell}$, of the electron widths $\gamma^{(e)}_{n\ell}$ and $\gamma^{(u)}_{n\ell,n'\ell'}$, and of their sum $\gamma_{n\ell,n'\ell'}$, for a gold plasma with density ρ=0.1 g/cm^3 and at various temperatures (see formulas (5.120), (5.127), (5.128) and (5.129)). For illustration purposes, we have chosen transitions from the levels $n\ell = 31$, 41, 43. As the table shows, the Lorentz widths of spectral lines are of the order of fractions of eV.

Table 5.5: The radiation width $\gamma^{(r)}_{n\ell}$ and the electron widths $\gamma^{(e)}_{n\ell}$, $\gamma^{(u)}_{n\ell,n'\ell'}$ of levels, in eV, and the total widths $\gamma_{n\ell,n'\ell'}$ in a gold plasma at density 0.1 g/cm^3 and various temperatures (the notation 1.45−13, say, means $1.45{\cdot}10^{-13}$)

T, eV	$n\ell$	$n'\ell'$	$\gamma^{(r)}_{n\ell}$	$\gamma^{(r)}_{n'\ell'}$	$\gamma^{(e)}_{n\ell}$	$\gamma^{(e)}_{n'\ell'}$	$\gamma^{(u)}_{n\ell,n'\ell'}$	$\gamma_{n\ell,n'\ell'}$
100	31	32	1.45−13	1.01−11	6.55−14	5.51−14	1.47−06	1.47−06
	31	40	1.45−13	1.44−06	6.55−14	1.20−03	2.36−04	1.44−03
	31	42	1.45−13	2.82−05	6.55−14	2.72−04	2.20−04	5.20−04
	31	50	1.45−13	5.29−05	6.55−14	1.30−02	7.89−04	1.38−02
	31	52	1.45−13	2.12−04	6.55−14	2.43−02	8.50−04	2.53−02
	31	60	1.45−13	2.52−04	6.55−14	2.79−02	1.48−03	2.97−02
	31	62	1.45−13	2.38−04	6.55−14	3.15−02	1.61−03	3.33−02
	41	42	5.60−06	2.82−05	3.05−04	2.72−04	4.72−07	6.12−04
	41	50	5.60−06	5.29−05	3.05−04	1.30−02	4.54−04	1.38−02
	41	52	5.60−06	2.12−04	3.05−04	2.43−02	5.34−04	2.53−02
	41	60	5.60−06	2.52−04	3.05−04	2.79−02	1.54−03	3.00−02
	41	62	5.60−06	2.38−04	3.05−04	3.15−02	1.77−03	3.38−02
	43	52	5.49−05	2.12−04	1.88−04	2.43−02	5.77−04	2.53−02
	43	54	5.49−05	8.64−04	1.88−04	5.01−03	1.04−03	7.15−03
	43	62	5.49−05	2.38−04	1.88−04	3.15−02	1.85−03	3.38−02
	43	64	5.49−05	5.73−04	1.88−04	2.92−02	2.55−03	3.26−02
1000	31	32	4.13−03	7.15−03	8.66−04	3.52−04	1.68−05	1.25−02
	31	40	4.13−03	2.11−02	8.66−04	2.32−03	4.39−04	2.89−02
	31	42	4.13−03	5.45−02	8.66−04	1.75−03	2.98−04	6.16−02
	31	50	4.13−03	2.04−02	8.66−04	3.92−03	1.32−03	3.06−02
	31	52	4.13−03	4.34−02	8.66−04	3.44−03	1.17−03	5.30−02
	31	60	4.13−03	1.62−02	8.66−04	8.35−03	2.29−03	3.19−02
	31	62	4.13−03	3.09−02	8.66−04	7.88−03	2.16−03	4.60−02
	41	42	2.82−02	5.45−02	2.25−03	1.75−03	1.38−05	8.67−02
	41	50	2.82−02	2.04−02	2.25−03	3.92−03	5.99−04	5.53−02
	41	52	2.82−02	4.34−02	2.25−03	3.44−03	4.67−04	7.78−02
	41	60	2.82−02	1.62−02	2.25−03	8.35−03	1.79−03	5.68−02
	41	62	2.82−02	3.09−02	2.25−03	7.88−03	1.61−03	7.09−02
	43	52	1.10−01	4.34−02	7.22−04	3.44−03	1.07−03	1.58−01
	43	54	1.10−01	5.25−02	7.22−04	9.20−04	5.29−04	1.64−01
	43	62	1.10−01	3.09−02	7.22−04	7.88−03	2.62−03	1.52−01
	43	64	1.10−01	3.47−02	7.22−04	4.22−03	2.01−03	1.51−01

5.4.6 Ion broadening

The starting assumption of the quasi-static theory of ion broadening is that the splitting of a spectral line in separate components and the position of these components in the absorption spectrum depends on the electric field F, whose probability distribution, $W(F)$, is assumed to be known. The total profile of the spectral line is obtained by averaging the set of profiles of components with the distribution function of their shifts in dependence of the field [38].

Therefore, in order to calculate the profile we must first find the distribution function of the ion microfield $W(F)$ and the Stark frequency-shift $\Delta\omega = \Delta\omega(F)$ for each component of the spectral line in each of the ions.

Here we shall consider only temperatures and densities for which the effective radius of the ion core is considerably smaller than the radius of the atom cell. In these circumstances one can assume, with a high degree of accuracy, that the electric field generated by ions is homogeneous within the limits of the ion core under consideration. Let us consider a system of N ions of charge Z_0, whose positions are specified by position vectors $\vec{r}_1, \vec{r}_2, \ldots, \vec{r}_N$. The intensity of the field generated by these ions in the origin of coordinates where the ion under study is located is equal (in atomic units) to

$$\vec{F} = -Z_0 \sum_{i=1}^{N} \frac{\vec{r}_i}{r_i^3}.$$

For a given position of the ions, the probability that the field intensity will lie in the interval $\vec{F}, \vec{F} + d\vec{F}$ is given by the expression

$$Q(\vec{r}_1, \vec{r}_2, \ldots, \vec{r}_N, \vec{F})\, d\vec{F} = \delta\left(\vec{F} + Z_0 \sum_{i=1}^{N} \frac{\vec{r}_i}{r_i^3}\right) d\vec{F}.$$

Here $\delta(\vec{x})$ is Dirac's delta-function. According to the Boltzmann statistical distribution, the probability of a given position $\vec{r}_1, \vec{r}_2, \ldots, \vec{r}_N$ of the ions is proportional to

$$\exp\left[-\frac{1}{\theta}\, U(\vec{r}_1, \vec{r}_2, \ldots, \vec{r}_N)\right] d\vec{r}_1 d\vec{r}_2 \ldots d\vec{r}_N,$$

where $U(\vec{r}_1, \vec{r}_2, \ldots, \vec{r}_N)$ is the potential energy of the system of N ions. Therefore, the probability that the field intensity will lie in the interval $\vec{F}, \vec{F} + d\vec{F}$ can be found by averaging over all positions of the ions:

$$p(\vec{F})\, d\vec{F} = \frac{\displaystyle\int \exp\left[-\frac{1}{\theta}\, U(\vec{r}_1, \vec{r}_2, \ldots, \vec{r}_N)\right] \cdot \delta\left(\vec{F} + Z_0 \sum_{i=1}^{N} \frac{\vec{r}_i}{r_i^3}\right) d\vec{r}_1 d\vec{r}_2 \ldots d\vec{r}_N}{\displaystyle\int \exp\left[-\frac{1}{\theta}\, U(\vec{r}_1, \vec{r}_2, \ldots, \vec{r}_N)\right] d\vec{r}_1 d\vec{r}_2 \ldots d\vec{r}_N}\, d\vec{F}.$$

To simplify the calculations, let us expand the delta-function in plane waves:

$$\delta\left(\vec{F}+Z_0\sum_{i=1}^{N}\frac{\vec{r}_i}{r_i^3}\right)=\frac{1}{(2\pi)^3}\int\exp\left[i\left(\vec{F}+Z_0\sum_{i=1}^{N}\frac{\vec{r}_i}{r_i^3}\right)\vec{\ell}\right]d\vec{\ell}.$$

Then

$$p(\vec{F})=\frac{1}{(2\pi)^3}\int T(\vec{\ell})\exp(i\vec{F}\vec{\ell})\,d\vec{\ell},\tag{5.136}$$

where

$$T(\vec{\ell})=\frac{\int\exp\left[-\frac{1}{\theta}U(\vec{r_1},\vec{r_2},\ldots,\vec{r_N})+i\vec{\ell}Z_0\sum_{i=1}^{N}\frac{\vec{r}_i}{r_i^3}\right]d\vec{r}_1d\vec{r}_2\ldots d\vec{r}_N}{\int\exp\left[-\frac{1}{\theta}U(\vec{r}_1,\vec{r}_2,\ldots,\vec{r}_N)\right]d\vec{r}_1d\vec{r}_2\ldots d\vec{r}_N}.\tag{5.137}$$

Since the function $U(\vec{r}_1,\vec{r}_2,\ldots,\vec{r}_N)$ does not depend on the choice of the coordinate system, but depends only on the mutual distances between ions, $T(\vec{\ell})$ is actually a function of $\ell=|\vec{\ell}|$. Accordingly, $p(\vec{F})=p(F)$, where $F=|\vec{F}|$.

Let $W(F)\,dF$ denote the probability that the absolute value of the electric field lies in the interval $F, F+dF$. Then, by (5.136),

$$W(F)=4\pi F^2p(F)=\frac{2F}{\pi}\int_0^\infty\ell\sin(\ell F)\cdot T(\ell)\,d\ell.\tag{5.138}$$

Starting with formulas (5.136), (5.137) and applying the Bohm-Pines separation of electrostatic interactions into long-distance and short-distance interactions one can obtain various approximate expressions for $W(F)$. For a rarefied plasma the Boltzmann factor $\exp\left[-U(\vec{r}_1,\vec{r}_2,\ldots,\vec{r}_N)/\theta\right]$, which accounts for the nonuniformity of the ion distribution, can be replaced by 1 since in the present situation the distribution of the ions in space can indeed be assumed to be homogeneous. This leads to the Holtsmark distribution for $W(F)$. This distribution is clearly the limit case obtained by letting $\theta\to\infty$ while keeping $\rho=\text{const}$, or by letting $\rho\to 0$ at fixed temperature. The Holtsmark distribution is obtained by using formulas (5.137), (5.138) and replacing the Boltzmann factor by 1:

$$T(\ell)=\lim_{N\to\infty}\frac{\int\exp\left(i\vec{\ell}Z_0\sum_{i=1}^{N}\frac{\vec{r}_i}{r_i^3}\right)d\vec{r}_1d\vec{r}_2\ldots d\vec{r}_N}{\int d\vec{r}_1d\vec{r}_2\ldots d\vec{r}_N}=$$

$$\lim_{N\to\infty}\left[\frac{1}{V}\int\exp\left(i\vec{\ell}Z_0\frac{\vec{r}}{r^3}\right)d\vec{r}\right]^N;$$

here $N = nV$, V is the volume of the system of ions and n is the concentration of ions in atomic units ($n = 1/v$, where $v = \frac{4}{3}\pi {r_0}^3$ is the volume of the atom cell). For fixed values of n the above limit is easily calculated if one observes that the integral

$$\int \left[1 - \exp\left(\frac{i\vec{\ell}\vec{r}}{r^3} Z_0\right)\right] d\vec{r}$$

over an infinite volume converges.

Indeed,

$$T(\ell) = \lim_{V\to\infty} \left[1 - \frac{1}{V}\int_V \left(1 - \exp\frac{i\vec{\ell}\vec{r}}{r^3} Z_0\right) d\vec{r}\right]^{nV} =$$

$$\exp\left[-n \int_{-1}^{1}\int_{0}^{\infty} 2\pi r^2\left(1 - \exp\frac{i\ell\mu}{r^2} Z_0\right) d\mu dr\right] =$$

$$\exp\left[-4\pi n \int_0^{\infty}\left(1 - \frac{\sin(\ell Z_0/r^2)}{(\ell Z_0/r^2)}\right) r^2 dr\right] =$$

$$\exp\left[-2\pi n(\ell Z_0)^{3/2} \int_0^{\infty}\left(1 - \frac{\sin x}{x}\right)\frac{dx}{x^{5/2}}\right].$$

The integral appearing in the argument of the exponential reduces via integration by parts to the Fresnel integral [10]. Consequently, $T(\ell)$ is given by

$$T(\ell) = \exp\left[-\frac{4}{15}(2\pi\ell Z_0)^{3/2} n\right]. \tag{5.139}$$

Substituting (5.139) in expression (5.138) for $W(F)$ we obtain the *Holtsmark distribution*

$$W(F) = \frac{2}{\pi F}\int_0^{\infty} x \sin x \cdot \exp\left[-\frac{4}{15}(2\pi x Z_0/F)^{3/2} n\right] dx. \tag{5.140}$$

In the case of high densities, for small values of the amplitude of oscillations of the central ion about the equilibrium position (i.e., when this amplitude is considerably smaller than the radius of the ion sphere), the *simple harmonic-oscillator approximation* is applicable. To obtain the corresponding distribution, let us calculate the force that arises when the ion is displaced at a distance r from its equilibrium position. We can assume that the external charges do not act on the ion. Then the ion is acted upon by the force of attraction corresponding to a charge of magnitude

$$q = -Z_0\left(\frac{r}{r_0}\right)^3.$$

Therefore, the absolute magnitude of the intensity of the field acting on the ion is

$$F = \frac{|q|}{r^2} = \frac{Z_0}{r_0^3}\, r,$$

and the potential energy of the ion is

$$U(r) = Z_0 \int_0^r F(r)\, dr = \frac{1}{2} F^2 r_0^3.$$

By the Boltzmann statistics, the probability that the ion will be located in the domain $r, r+dr$ is proportional to

$$\exp\left[-\frac{1}{\theta} U(r)\right] r^2\, dr \sim \exp\left[-\frac{F^2 r_0^3}{2\theta}\right] F^2\, dF.$$

Thus, in the approximation considered here

$$W(F) \sim \exp\left[-\frac{F^2 r_0^3}{2\theta}\right] F^2. \tag{5.141}$$

A more precise distribution of the ion microfield, which is valid in a wide range of temperatures and densities, can be derived by resorting to a model of perturbing independent particles and results of computer modelling [91]. These results are approximated by the following simple dependence [194]:

$$W(F) = \frac{H(u)}{F_0}, \quad H(u) = \frac{2u}{\pi} \int_0^\infty x \sin(ux) \cdot \exp\left[-\frac{x^{3/2}}{\left(1 + \frac{\Gamma}{\sqrt{x}}\right)^{1.2876}}\right] dx, \tag{5.142}$$

where $u = F/F_0$, $F_0 = 2\pi Z_0 \left(4n/15\right)^{2/3}$ is the normal Holtsmark intensity of the field and Γ is a nonideality parameter. As nonideality parameter one can take the corresponding parameter for the mutual interaction between ions in the average-atom approximation: $\Gamma = Z_0^2/(\theta r_0)$. The distribution (5.142) goes over into the Holtsmark distribution when $\Gamma \to 0$, and into the simple harmonic-oscillator approximation when $\Gamma \to \infty$.

Figure 5.5 displays the distribution functions $W(F)$ of the ion microfield obtained in various approximations, as functions of the variable $u = F/F_0$, for different values of the dimensionless parameter Γ. This parameter can vary in wide limits; for example, for gold at $T = 1$ keV, $\rho = 1$ g/cm^3 the parameter $\Gamma \sim 7$; for iron at $T = 20$ eV, $\rho = 10^{-4}$ g/cm^3 — $\Gamma \sim 0.7$; and for hydrogen at $T = 1$ eV, $\rho = 10^{-9}$ g/cm^3 — $\Gamma \sim 0.02$. As Figure 5.5 clearly shows, the approximation (5.142) gives good results throughout the entire domain of parameters, practically coinciding with (5.140) for $\Gamma < 0.03$ and with (5.141) for $\Gamma > 3$.

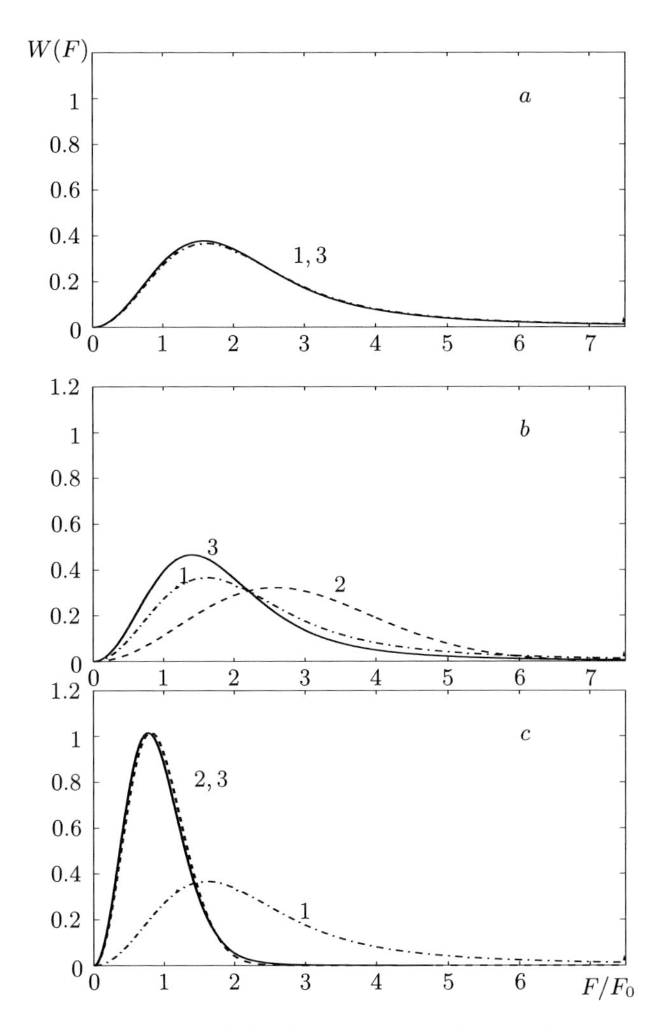

Figure 5.5: The distribution $W(F)$ of the ion microfield as function of $u = F/F_0$ for $\Gamma = 0.03$ (a), $\Gamma = 0.3$ (b) and $\Gamma = 3$ (c): (1) the Holtsmark distribution (5.140), (2) the simple harmonic oscillators approximation (5.141), (3) formula (5.142), continuous line

Including in the considerations even a constant electric filed complicates substantially the task of determining the energy levels of a multi-electron ion [48]. With the more modest aim of deriving the influence of the Stark effect on the Rosseland mean free path, we shall account for this effect in the simplest one-electron approximation.

In a given constant electric field $\vec{F}$, the one-electron energy levels and wave functions of the electrons can be found by perturbation theory. As the unperturbed Hamiltonian we take $H_0 = -\frac{1}{2}\Delta - V(r)$, where the effective central potential $V(r)$ is the self-consistent potential of the relativistic Hartree-Fock-Slater model, corresponding to the most probable state of the ion and free electrons. The perturbation has the form

$$H' = \vec{F} \cdot \vec{r}.$$

As the unperturbed wave functions, corresponding to the total angular momentum of the electron motion $\vec{j} = \vec{\ell} + \vec{s}$, we take

$$\psi_{n\ell jm} = \sum_{m_\ell, m_s} C^{j\,m}_{\ell m_\ell\,\frac{1}{2} m_s}\, \psi_{n\ell m_\ell m_s}.$$

Then the matrix of the operator H_0 in this approximation is diagonal and its elements are independent of m:

$$\langle n\,\ell\,j\,m\,|\,H_0\,|\,n'\ell'j'm'\rangle = \varepsilon_{n\ell j}\delta_{nn'}\delta_{\ell\ell'}\delta_{jj'}\delta_{mm'}.$$

Let us calculate the matrix elements of the operator $\vec{F} \cdot \vec{r}$, taking the z-axis to be directed along the field: $\vec{F} \cdot \vec{r} = Fz$. Since $Fz = \sqrt{4\pi/3}\, FrY_{10}(\theta, \varphi)$ is an irreducible rank-1 tensor operator that commutes with the spin operator, the Wigner-Eckart theorem gives [220]

$$\langle n\,\ell\,j\,m\,|\,Fz\,|\,n'\,\ell'\,j'\,m'\rangle = (-1)^{j-m} \begin{pmatrix} j & 1 & j' \\ -m & 0 & m' \end{pmatrix} (-1)^{\frac{3}{2}-\ell'+j} \times$$
$$\sqrt{(2j+1)(2j'+1)}\, W(\ell\,j\,\ell'\,j'\,;\,\frac{1}{2}\,1) \cdot (n\,\ell\,\|\,Fz\,\|\,n'\,\ell'), \quad (5.143)$$

where the reduced matrix element

$$(n\,\ell\,\|\,Fz\,\|\,n'\,\ell') = (-1)^{(\ell+\ell'+1)/2} \sqrt{\frac{\ell'+\ell+1}{2}}\, F\, r_{n\ell, n'\ell'} \quad (\ell' = \ell \pm 1).$$

Since the Wigner $3jm$-symbol $\begin{pmatrix} j & 1 & j' \\ -m & 0 & m' \end{pmatrix}$ is different from zero only when $m' = m$, this shows that, besides the principal quantum number n, the energy levels are characterized by the quantum number m. The eigenfunctions of the Hamiltonian $H = H_0 + H'$ can be written as linear combinations of functions $\psi_{n\ell jm}$ with fixed values of n and m:

$$\psi^{(s)}_{nm} = \sum_{\ell j} c^{(s)}_{n\ell jm} \psi_{n\ell jm},$$

where s is the number of the eigenvalue $\varepsilon^{(s)}_{nm}$. The coefficients $c^{(s)}_{n\ell jm}$ are determined by solving the secular equation

$$(\varepsilon^{(s)}_{nm} - \varepsilon_{n\ell j}) c^{(s)}_{n\ell jm} = \sum_{\ell' j'} \langle n\,\ell\,j\,m\,|\,Fz\,|\,n\,\ell'\,j'\,m\rangle c^{(s)}_{n\ell' j' m}.$$

To calculate the line absorption cross-sections, instead of the transitions $n\ell j \to n'\ell' j'$ in formula (5.53) we must account for the transitions $nms \to n'm's'$ ($m' = m$, $m \pm 1$), setting

$$\sigma_{\rm bb}(\omega) = 2\pi^2 \alpha a_0^2 \int_0^\infty dF\, W(F) \times \sum_Q P_Q \sum_{nms} \sum_{n'm's'} N_{nms}^Q \left(1 - N_{n'm's'}^Q\right) f_{nms,n'm's'} J_{nms,n'm's'}(\omega), \quad (5.144)$$

where

$$f_{nms,n'm's'} = \frac{2}{3}\left(\varepsilon_{n'm'}^{(s')} - \varepsilon_{nm}^{(s)}\right) \left| \int (\psi_{n'm'}^{(s')})^* \vec{r} \psi_{nm}^{(s)} d\vec{r} \right|^2$$

and $J_{nms,n'm's'}(\omega)$ is the corresponding line profile.

In those cases where the influence of ion broadening is considerably stronger than other broadening effects the calculations can be simplified. For example, in the case of the linear Stark effect, if we take for the position of the center of the components of the spectral line

$$\omega_{nms,n'm's'}(F) = \omega_{nms,n'm's'}(0) + \widetilde{C}_{nms,n'm's'} \cdot F$$

and consider the shape of each component to be a δ-function, then one can carry out the integration with respect to the field F in (5.144). This yields

$$\sigma_{\rm bb}(\omega) \approx 2\pi^2 \alpha a_0^2 \sum_Q P_Q \sum_{nms} \sum_{n'm's'} N_{nms}^Q \left(1 - N_{n'm's'}^Q\right) \times \tilde{f}_{nms,n'm's'} \frac{W\left[(\omega - \omega_0)/\widetilde{C}_{nms,n'm's'}\right]}{\widetilde{C}_{nms,n'm's'}}, \quad (5.145)$$

where $\omega_0 = \omega_{nms,n'm's'}(0)$ and the quantities $\tilde{f}_{nms,n'm's'}$, $\widetilde{C}_{nms,n'm's'}$ are calculated for the most probable field F. A similar formula can be derived for the case of the quadratic Stark effect.

Computations based on formulas (5.144) and (5.145) are rather tedious, although in many cases the influence of the Stark broadening on the Rosseland mean free path is of about 1% [147]. In table calculations, ion broadening is usually accounted for in a simpler way [189]. Let us assume that only two levels are so close to one another that it is necessary to account for their splitting. This leads to an eigenvalue problem for a 2×2 matrix. Since the role played by the Stark effect is the largest for upper levels, where the influence of relativistic corrections is small, in the calculation of the Stark splitting one can restrict ourselves to the classification of levels according to the quantum numbers $n\ell$, considering that the shifts $\varepsilon_{n\ell j}$ for $j = \ell \pm 1/2$ are equal.

Consider the shift due to the Stark effect, taking into account a level $\varepsilon_{n'\ell'}$ that is closest to the level $\varepsilon_{n\ell}$, taking, for instance, $n' = n$ and $\ell' = \ell - 1$. Write $\varepsilon_{n\ell} = \varepsilon_0 + \Delta/2$, $\varepsilon_{n,\ell-1} = \varepsilon_0 - \Delta/2$, where $\varepsilon_0 = (\varepsilon_{n\ell} + \varepsilon_{n,\ell-1})/2$, $\Delta = \varepsilon_{n\ell} - \varepsilon_{n,\ell-1}$. Then the‘ energy levels ε with respect to ε_0 in a constant electric field of intensity F is found from the equation

$$\begin{vmatrix} \dfrac{\Delta}{2} - \varepsilon & V_{12} \\ V_{21} & -\dfrac{\Delta}{2} - \varepsilon \end{vmatrix} = 0, \tag{5.146}$$

where $V_{12} = V_{21} = \langle n\,\ell\,m\,|Fz|\,n'\ell'm'\rangle$ is the matrix element of the electron–electric field interaction (see (5.143)):

$$\langle n\,\ell\,m\,|Fz|\,n'\ell'm'\rangle = \begin{cases} \delta_{mm'}\sqrt{\dfrac{(\ell+1)^2 - m^2}{(2\ell+3)(2\ell+1)}}\,F\,r_{n\ell,n'\ell'} & \text{if } \ell' = \ell+1, \\ \delta_{mm'}\sqrt{\dfrac{\ell^2 - m^2}{(2\ell+1)(2\ell-1)}}\,F\,r_{n\ell,n'\ell'} & \text{if } \ell' = \ell-1; \end{cases}$$

$\langle n\,\ell\,m\,|Fz|\,n'\ell'm'\rangle = 0$ if $\ell' \neq \ell \pm 1$.

Equation (5.146) yields $\varepsilon = \sqrt{(\Delta/2)^2 + V_{12}V_{21}}$, and the shift of the level $\varepsilon_{n\ell}$ in the electric field F is

$$\Delta\varepsilon = \varepsilon - \frac{\Delta}{2} = \sqrt{\left(\frac{\varepsilon_{n\ell} - \varepsilon_{n,\ell-1}}{2}\right)^2 + V_{12}^2} - \frac{\varepsilon_{n\ell} - \varepsilon_{n,\ell-1}}{2}. \tag{5.147}$$

Let us average the quantity $\Delta\varepsilon$ over the direction of the field $\vec{F}$; this is equivalent to averaging over the projection of the orbital momentum m on the direction of the filed. Assuming that the dependence on the field is essentially quadratic $\left((\Delta/2)^2 \gg V_{12}^2\right)$, the averaged value can be obtained if in (5.147) we replace V_{12}^2 by the quantity

$$\langle V_{12}^2\rangle \equiv v_{n\ell,n'\ell'}^2 = \frac{1}{2\ell_<}\sum_{m=-\ell_<}^{\ell_<} |\langle n\ell m|Fz|n'\ell' m\rangle|^2 = \frac{\ell_> F^2 r_{n\ell,n'\ell'}^2}{3(2\ell_< + 1)}$$

$$(\ell_< = \min\{\ell,\ell'\}, \quad \ell_> = \max\{\ell,\ell'\}).$$

Considering two nearest-neighbor levels $n, \ell - 1$ and $n, \ell + 1$ of the level $n\ell$ (in the case $\ell = 0$ or $\ell = n - 1$ — one such nearest-neighbor level), we obtain for the average shift in the field F the value

$$d_{n\ell} = \Delta_{n\ell}^{n,\ell-1} + \Delta_{n\ell}^{n,\ell+1},$$

where

$$\Delta_{n\ell}^{n'\ell'} = \sqrt{\left(\frac{\varepsilon_> - \varepsilon_<}{2}\right)^2 + v_{n\ell,n'\ell'}^2} - \frac{\varepsilon_> - \varepsilon_<}{2}$$

$$(\varepsilon_< = \min\{\varepsilon_{n\ell}, \varepsilon_{n'\ell'}\}, \quad \varepsilon_> = \max\{\varepsilon_{n\ell}, \varepsilon_{n'\ell'}\}).$$

In the calculation of the line absorption coefficients the influence of ions will be accounted for in the form of an additional Lorentz broadening (see [189]):

$$\gamma_{n\ell,n'\ell'}^{(\mathrm{st})} = d_{n\ell} + d_{n'\ell'}, \tag{5.148}$$

where the average Stark shifts of the levels $d_{n\ell}$ and $d_{n'\ell'}$ are calculated in the maximally probable field F with respect to the distribution (5.142) (see Figure 5.5). Generally speaking, the influence of ion broadening, manifested by an additional Lorentz width (5.148), is usually overestimated. Note that if the additional broadening is accounted for as a Doppler broadening, then the effects of the ion broadening, as a rule, are underestimated.

5.4.7 The Voigt profile

If the Doppler effect is superposed on the Lorentz profile of the spectral line by means of formula (5.94), we arrive at the Voigt line profile [205]:

$$J_{\alpha\beta}(\omega) = \frac{1}{\sqrt{\pi}D} K\left(\frac{\omega - \omega_{\alpha\beta}}{D}, \frac{\gamma}{D}\right). \tag{5.149}$$

Here

$$K(x,y) = \frac{y}{\pi} \int_{-\infty}^{\infty} \frac{\exp(-s^2)}{(x-s)^2 + y^2}\, ds, \tag{5.150}$$

$\omega_{\alpha\beta}$ is the position of the line center for the transition $\alpha \to \beta$ $(\varepsilon_\alpha < \varepsilon_\beta)$, $\gamma = \gamma_{\alpha\beta}$ is the Lorentz line width, and D is the Doppler broadening parameter:

$$D = \frac{\omega_{\alpha\beta}}{c} \sqrt{\frac{2\theta}{1836A}}. \tag{5.151}$$

The Lorentz width is provided by the formula (see (5.127))

$$\gamma_{\alpha\beta} = \gamma_{n\ell,n'\ell'} = \gamma_{n\ell}^{(\mathrm{r})} + \gamma_{n'\ell'}^{(\mathrm{r})} + \gamma_{n\ell}^{(\mathrm{e})} + \gamma_{n'\ell'}^{(\mathrm{e})} + \gamma_{n\ell,n'\ell'}^{(\mathrm{u})} + \gamma_{n\ell,n'\ell'}^{(\mathrm{st})}. \tag{5.152}$$

To calculate the Voigt integral one can use the scheme described in [154]. Due to the presence of the factor $\exp(-s^2)$, the most essential domain for the integration with respect to s in (5.150) corresponds to $|s| < 1$. In that domain the function $y/[(x-s)^2 + y^2]$ is, for each fixed value of x, a sufficiently smooth function of s, provided that y is relatively large. It follows that, for $y > 1$, in order to calculate $K(x, y)$ one can use Gauss-type quadrature formulas based on Hermite polynomials:

$$K(x, y) \approx \frac{1}{\pi} \sum_{j=1}^{n} \lambda_j \frac{y}{(x - s_j)^2 + y^2}. \tag{5.153}$$

For small values of y the function $y/[(x - s)^2 + y^2]$ has a sharp maximum in the point $x = s$, as a result of which the quadrature formula (5.153) fails when x is small. To circumvent this, it is desirable to transform the expression for $K(x, y)$ to a form that is more suitable for applying Gauss-type quadrature formulas. Namely, we have

$$K(x, y) = \frac{1}{\pi} \operatorname{Im} \int_{-\infty}^{\infty} \frac{e^{-s^2}}{x - s - iy} ds.$$

If we now use the Cauchy theorem to replace integration along the real axis by integration along a line parallel to it, setting $s = ai + t$ ($a > 0, -\infty < t < \infty$), we obtain

$$K(x, y) = \frac{1}{\pi} \operatorname{Im} \int_{-\infty}^{\infty} \frac{e^{-(ai+t)^2}}{(x - t) - i(a + y)} dt =$$

$$\frac{1}{\pi} e^{a^2} \int_{-\infty}^{\infty} \frac{e^{-t^2} [(a + y) \cos\, 2at - (x - t) \sin\, 2at]}{(x - t)^2 + (a + y)^2} dt. \tag{5.154}$$

Thanks to this transformation, the function $1/[(x - s)^2 + y^2]$, which has a sharp maximum, is replaced by the function $1/[(x - t)^2 + (a + y)^2]$, which in the domain essential for integration can be sufficiently well approximated by polynomials of relatively small degree. However, this procedure introduces a supplementary, oscillating factor. It is natural to take $a = 1$, and then in the domain essential for integration the function $(a + y) \cos\, 2at - (x - t) \sin\, 2at$ becomes sufficiently smooth as well. Now instead of (5.153) we obtain

$$K(x, y) \approx \frac{e^{a^2}}{\pi} \sum_{j=1}^{n} \lambda_j \frac{(a + y) \cos 2as_j - (x - s_j) \sin 2as_j}{(x - s_j)^2 + (a + y)^2}. \tag{5.155}$$

Formula (5.155) gives good results for arbitrary values of x and y even for a small number of quadrature points [154].

5.4.8 Line profiles of a hydrogen plasma in a strong magnetic field

An external magnetic field, in addition to the electric fields of the ions and electrons of a plasma, exerts a strong influence on the shape of spectral lines. Let us examine the influence of a magnetic field on the example of a hydrogen (deuterium) plasma [140, 53]. Such a problem arises, in particular, in the investigation of radiative transfer processes in low-temperature edge plasma in a tokamak. For example, in a deuterium plasma in the divertor of the Alcator C-Mod tokamak, with temperature $T \sim 1$ eV, electron density $n_e \sim 10^{15}$ cm^{-3} and magnetic field $B \sim 8$ T, radiation in spectral lines plays a determining role in the energy transfer process, and this transfer depends in an essential manner on the shapes of lines [173].

In electric and magnetic fields the degeneracy is removed and the spectral lines corresponding to transitions between levels with quantum numbers $\bar{n}$ and n split into separate components. The shift of each component and its intensity depend on the value of the magnetic field $\vec{B}$, the value of the quasi-static ion microfield $\vec{F}$ and the angle between the vectors $\vec{F}$ and $\vec{B}$. Moreover, the intensity of a component depends on the direction in which the radiation propagates, more precisely, on the angles between the wave vector $\vec{k}$ and the vectors $\vec{F}$ and $\vec{B}$. The total profile of the line for the transition $\bar{n} \to n$ can be calculated by averaging over the magnitude and direction of the quasi-static ion microfield (in what follows the dependence on the magnetic field $\vec{B}$ is not indicated explicitly, though it is implied):

$$\Phi_{\bar{n}n}(\omega) = \frac{1}{4\pi} \int W(F) \sum_{\overline{\nu}\nu} G_{\overline{\nu}\nu}(\vec{F}) \phi_{\overline{\nu}\nu}(\omega, \vec{F}) d\vec{F}. \tag{5.156}$$

Here $W(F)$ is the probability distribution function of the electric ion microfield, $G_{\overline{\nu}\nu}(\vec{F})$ and $\phi_{\overline{\nu}\nu}(\omega, \vec{F})$ are the relative intensity and respectively the profile of the component $\overline{\nu} \to \nu$ of the transition $\bar{n} \to n$ ($\overline{\nu}$ and ν designate collections of quantum numbers that specify the initial and final states of each component).

The profile of each component is determined by the Doppler effect and the interaction of the atom with the free electrons (the natural broadening is here negligibly small):

$$\phi_{\overline{\nu}\nu}(\omega, \vec{F}) = \frac{1}{\sqrt{\pi} D_{\overline{\nu}\nu}} \int \varphi_{\overline{\nu}\nu}(\omega - \omega_{\overline{\nu}\nu} - s, \vec{F}) e^{-(s/D_{\overline{\nu}\nu})^2} \, ds,$$

where $D_{\overline{\nu}\nu} = (\omega_{\overline{\nu}\nu}/c)(2T/M)^{1/2}$ is the Doppler broadening parameter, M is the atom mass, and $\omega_{\overline{\nu}\nu}$ is the position of the center of the component $\overline{\nu} \to \nu$. For the profile $\varphi_{\overline{\nu}\nu}$ one uses the approximation proposed by Seaton for the calculation of the broadening of Stark components of a hydrogen plasma by electrons [201]. In our case the splitting of lines is due not only to the Stark effect, but also to the Zeeman effect, which imposes a specific character to the method for calculating component profiles and requires some modification of the theory [201].

The relative intensities of the components $\overline{\nu} \to \nu$ in the dipole approximation are given by the expression

$$G_{\overline{\nu}\nu} = \frac{\left|\vec{u}^{(s)} \langle \nu | \vec{r} | \overline{\nu} \rangle\right|^2}{\sum\limits_{\overline{\nu}\nu} \left|\vec{u}^{(s)} \langle \nu | \vec{r} | \overline{\nu} \rangle\right|^2}, \tag{5.157}$$

where $\vec{u}^{(\mathrm{s})}$ is the unit polarization vector of the photon. For unpolarized light

$$G_{\bar{\nu}\nu} = \frac{\omega_{\bar{n}n}}{n^2 f_{n\bar{n}}} \sum_{s=1,2} \left|\vec{u}^{(\mathrm{s})} \langle \nu | \vec{r} | \bar{\nu} \rangle \right|^2, \tag{5.158}$$

where $f_{n\bar{n}}$ is the total oscillator strength of the transition $n \to \bar{n}$.

The lines $\bar{n} \to n$ will be regarded as isolated; for this to be possible it is necessary that their characteristic width be smaller than the distance between the nearest levels, which imposes restrictions on the magnitude of the magnetic field and on the electron density:

$$B < c/\bar{n}^4, \quad n_{\mathrm{e}} < 0.05/\bar{n}^{15/2}. \tag{5.159}$$

Let us remark that the straight-line approximation of classical trajectories for perturbing electrons, which is customary in the theory of spectral-line broadening, does not lead to additional constraints on the magnitude of the magnetic field. Indeed, that condition requires that the Larmor frequency be smaller than the plasma frequency, or $B < (8\pi c^2 n_{\mathrm{e}})^{1/2}$. But as one can readily verify, the last inequality is always satisfied if the constraints (5.159) hold. The spin-orbit interaction can also be neglected, since the fine-structure splitting at temperature $T \sim 1$ eV is smaller than the Doppler width of levels.

As Epstein has shown in the framework of Bohr's theory (see [35, 40]), the joint influence of the magnetic and electric fields on the orbit of an electron in the state with principal quantum number n and orbital number ℓ can be described, in the first approximation with respect to the field, as a uniform and independent precession of the vectors $\frac{3}{2}n\vec{\ell} \mp \vec{r}_{\mathrm{a}}$ with angular velocities $\vec{\omega}_1$ and $\vec{\omega}_2$:

$$\vec{\omega}_{1,2} = \frac{1}{2c}\vec{B} \mp \frac{3}{2}n\vec{F}$$

(here $\vec{\ell}$ is the angular momentum of the electron and $\vec{r}_{\mathrm{a}}$ is its position vector, averaged over the orbital motion.) In the same approximation one can obtain the corrections to the energy of the electron. The quantum-mechanical approach yields — in the first approximation of perturbation theory — the same result [53]. The Hamiltonian $\widehat{H}$ is written as the sum $\widehat{H} = \widehat{H}_0 + \widehat{H}'$ of the unperturbed Hamiltonian $\widehat{H}_0 = -\Delta/2 - 1/r$ and the perturbation $\widehat{H}' = (1/2c)\,\vec{B}\cdot\vec{\ell} + \vec{F}\cdot\vec{r}$, which in the subspace of states with given n can be represented in the form

$$\widehat{H}' = \frac{1}{2c}\vec{B}\cdot\vec{\ell} - \frac{3}{2}n\vec{F}\cdot\vec{A} = \vec{\omega}_1\cdot\vec{I}_1 + \vec{\omega}_2\cdot\vec{I}_2.$$

Here $\vec{I}_{1,2} = \dfrac{1}{2}\left(\vec{\ell} \pm \vec{A}\right)$ and $\vec{A}$ is the Runge-Lenz vector, which in the indicated subspace satisfies the relation $\vec{A} = -2\vec{r}/(3n)$ (see, e.g., [53]).

The operators $\vec{I}_{1,2}$ commute with $\widehat{H}_0$ and obey the usual commutation relations for the angular momentum operator. It follows that $I_1^2 = I_2^2 = j(j+1)$, where j is determined by the number of possible states, i.e., $2(2j+1)^2 = 2n^2$ and $j = (n-1)/2$; moreover, the projections of $\vec{I}_1$ on the axis $\vec{\omega}_1$ and of $\vec{I}_2$ on the axis $\vec{\omega}_2$ (denoted here by n' and n'', respectively) may take the $2j+1$ integer or half-integer values $-j,\ -j+1,\ \ldots,\ j-1,\ j$.

In the first order of perturbation theory

$$\varepsilon_{nn'n''m_s} = -\frac{1}{2n^2} + \omega_1 n' + \omega_2 n'' + \frac{1}{c} B m_s, \tag{5.160}$$

where $m_s = \pm\frac{1}{2}$ is the projection of the spin on the axis directed along the magnetic field.

The corresponding wave function $\psi_{nn'n''}$ can be written as a linear combination of wave functions in parabolic coordinates with the z-axis directed along the electric field $\vec{F}$:

$$\psi_{nn'n''} = \sum_{i_1=-j}^{j} \sum_{i_2=-j}^{j} d^j_{n'i_1}(\alpha_1) d^j_{n''i_2}(\alpha_2)\, \psi_{ni_1i_2}, \tag{5.161}$$

where $d^j_{kk'}(\alpha) = D^j_{kk'}(0,\alpha,0)$ is the Wigner function corresponding to the rotation by angle α around the z-axis (see [220]) and $\psi_{ni_1i_2} \equiv \psi_{n_1n_2m}$ are wave functions in parabolic coordinates. In the present case these functions are conveniently characterized by the quantum numbers i_1 and i_2, representing the projections of the operators $\vec{I}_1$ and $\vec{I}_2$ on the z-axis; these numbers are connected with the ordinary parabolic quantum numbers n_1, n_2 and the magnetic quantum number m $(n_1 + n_2 + |m| + 1 = n)$ by the following relations:

$$i_1 + i_2 = m, \quad i_1 - i_2 = n_2 - n_1.$$

Further, α_1 and α_2 are the angles between the z-axis (i.e., the vector $\vec{F}$) and the vectors $\vec{\omega}_1$ and $\vec{\omega}_2$, respectively:

$$\cos\alpha_1 = \frac{\frac{1}{2c} B\cos\vartheta - \frac{3}{2} nF}{\omega_1}, \qquad \cos\alpha_2 = \frac{\frac{1}{2c} B\cos\vartheta + \frac{3}{2} nF}{\omega_2},$$

where ϑ is the angle between $\vec{F}$ and $\vec{B}$.

The shift of the component $\bar{\nu} \to \nu$ relative to the center of the line (ν denotes the set of quantum numbers $nn'n''$) can be calculated with the help of (5.160) to be

$$\omega_{\bar{\nu}\nu} - \omega_{\bar{n}n} = \bar{\omega}_1 \bar{n}' + \bar{\omega}_2 \bar{n}'' - \omega_1 n' - \omega_2 n'',$$

where $\bar{\omega}_1$ and $\bar{\omega}_2$ are the angular velocities of precession for the upper level $\bar{n}$.

The dipole matrix elements $\langle \nu | \vec{r} | \bar{\nu} \rangle$ with the wave functions (5.161) can be represented as linear combinations of matrix elements calculated in parabolic coordinates. Let $u_x^{(s)}$, $u_y^{(s)}$, $u_z^{(s)}$ be the Cartesian coordinates of the unit polarization vector $\vec{u}^{(s)}$ in a system of coordinates with the z-axis directed along $\vec{F}$ and the x-axis lying in the plane of the vectors $\vec{F}$ and $\vec{B}$. Then

$$\vec{u}^{(s)}\langle\overline{\nu}|\vec{r}|\nu\rangle = \sum_{a=x,y,z} u_a^{(s)}\langle\overline{\nu}|a|\nu\rangle.$$

Here

$$\langle\overline{\nu}|a|\nu\rangle \equiv \langle\bar{n}\bar{n}'\bar{n}''\,|a|\,nn'n''\rangle =$$

$$\sum_{i_1,i_2=-j}^{j}\sum_{\bar{i}_1,\bar{i}_2=-\bar{j}}^{\bar{j}} d^j_{n'i_1}(\alpha_1)\,d^j_{n''i_2}(\alpha_2)\times$$

$$d^{\bar{j}}_{\bar{n}'\bar{i}_1}(\bar{\alpha}_1)\,d^{\bar{j}}_{\bar{n}''\bar{i}_2}(\bar{\alpha}_2)\langle\bar{n}_1\bar{n}_2\bar{m}|a|n_1n_2m\rangle, \quad (5.162)$$

where the matrix elements $\langle\bar{n}_1\bar{n}_2\bar{m}|a|n_1n_2m\rangle$ in parabolic coordinates for $a = x, y, z$ are calculated by means of Gordon's formulas [28].

The Wigner functions $d^j_{kk'}(\alpha)$ in (5.161) and (5.162) can be expressed in terms of Jacobi polynomials [220]:

$$d^j_{kk'}(\alpha) = \xi_{kk'}\left[\frac{s!(s+\mu+\nu)!}{(s+\mu)!(s+\nu)!}\right]^{1/2}\left(\sin\frac{\alpha}{2}\right)^{\mu}\left(\cos\frac{\alpha}{2}\right)^{\nu} P_s^{(\mu,\nu)}(\cos\alpha),$$

where $\mu = |k-k'|$, $\nu = |k+k'|$, $s = j-(\mu+\nu)/2$, $\xi_{kk'} = 1$ for $k' \geq k$ and $\xi_{kk'} = (-1)^{k'-k}$ for $k' < k$. The Jacobi polynomials $P_s^{(\mu,\nu)}$ are conveniently calculated by means of recursion relations [154].

In the case of polarized light we direct one of the polarization vectors, $\vec{u}^{(1)}$, along the normal to the plane of the vectors $\vec{k}$ and $\vec{B}$, and take the other, $\vec{u}^{(2)}$, to lie in this plane and be orthogonal to $\vec{k}$. It is convenient to carry out the averaging with respect to the direction $\vec{u}^{(s)}$ in (5.156) in a system of coordinates attached to the magnetic field. Let φ denote the angle between the projections of the vectors $\vec{k}$ and $\vec{F}$ in the plane orthogonal to $\vec{B}$. Then

$$u_x^{(1)} = \cos\vartheta\sin\varphi, \quad u_y^{(1)} = \cos\varphi, \quad u_z^{(1)} = \sin\vartheta\sin\varphi,$$

$$u_x^{(2)} = -\cos\beta\cos\vartheta\cos\varphi - \sin\beta\sin\vartheta, \quad u_y^{(2)} = \cos\beta\sin\varphi,$$

$$u_z^{(2)} = -\cos\beta\sin\vartheta\cos\varphi + \sin\beta\cos\vartheta,$$

where β is the angle between $\vec{k}$ and $\vec{B}$.

Using (5.156), (5.157) and assuming that the profiles of individual components are independent of the angle φ, we finally obtain for the transition $\overline{n} \to n$:

$$\Phi_{\bar{n}n}(\omega) = \frac{\omega_{\bar{n}n}}{4n^2 f_{n\bar{n}}}\int_0^{\infty} dF W(F)\sum_{\nu\overline{\nu}}\int_0^{\pi} d\vartheta\sin\vartheta\,\phi_{\overline{\nu}\nu}(\omega,\vartheta,F)\times$$

$$\left\{2\sin^2\beta(x_{\nu\overline{\nu}}\sin\vartheta - z_{\nu\overline{\nu}}\cos\vartheta)^2 + (1+\cos^2\beta)\times\right.$$

$$\left.\left[(x_{\nu\overline{\nu}}\cos\vartheta + z_{\nu\overline{\nu}}\sin\vartheta)^2 + (y_{\nu\overline{\nu}})^2\right]\right\}, \quad (5.163)$$

where the matrix elements $x_{\nu\overline{\nu}}$, $y_{\nu\overline{\nu}}$, $z_{\nu\overline{\nu}}$ are given by (5.162).

Line shapes have been calculated for a deuterium plasma with temperature $T = 1$ eV and electron density $n_e = 3 \cdot 10^{15}$ cm^{-3}. Under these conditions the influence of the magnetic field, the Stark effect and the Doppler effect on the lower lines of the Lyman and Balmer series are of the same order (see Table 5.6), which requires that these effects be computed simultaneously.

Table 5.6: Broadening parameters of lines (in eV) in a deuterium plasma with $T = 1$ eV, $n_e = 3 \cdot 10^{15}$ cm^{-3}, $B = 8$ T

Line	Electron width	Doppler parameter	Stark shift	Shift in a magnetic field
Ly_α	$4.1 \cdot 10^{-5}$	$3.3 \cdot 10^{-4}$	$1.2 \cdot 10^{-4}$	$4.6 \cdot 10^{-4}$
Ly_β	$1,7 \cdot 10^{-4}$	$3.7 \cdot 10^{-4}$	$1.9 \cdot 10^{-4}$	$4.6 \cdot 10^{-4}$
Ly_γ	$4.7 \cdot 10^{-4}$	$4.2 \cdot 10^{-4}$	$2.5 \cdot 10^{-4}$	$4.6 \cdot 10^{-4}$
D_α	$1.7 \cdot 10^{-4}$	$6.2 \cdot 10^{-5}$	$1.9 \cdot 10^{-4}$	$4.6 \cdot 10^{-4}$

Figure 5.6 shows the profiles of the spectral lines Ly_α $(2 \to 1)$ and D_α $(3 \to 2)$, computed by means of formula (5.163). The computations were done in the absence of a magnetic field as well as in the presence of a magnetic field of strength $B = 8$ T directed parallel to the wave vector $\vec{k}$, perpendicular to $\vec{k}$, and at an angle of 45° with $\vec{k}$. The graphs show that the magnetic field exerts a strong influence on the shape of the spectral lines, which is confirmed experimentally [3, 161].

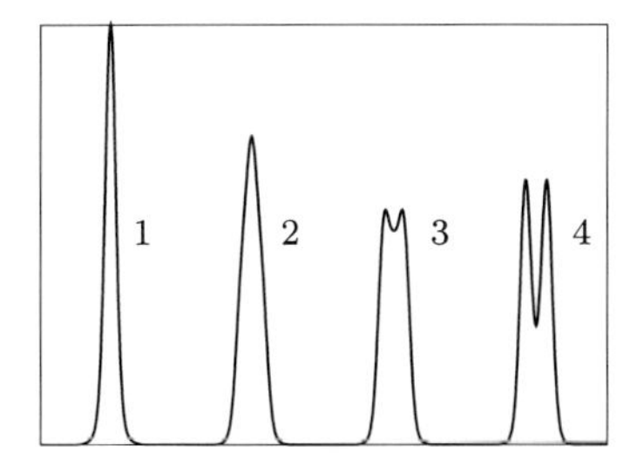

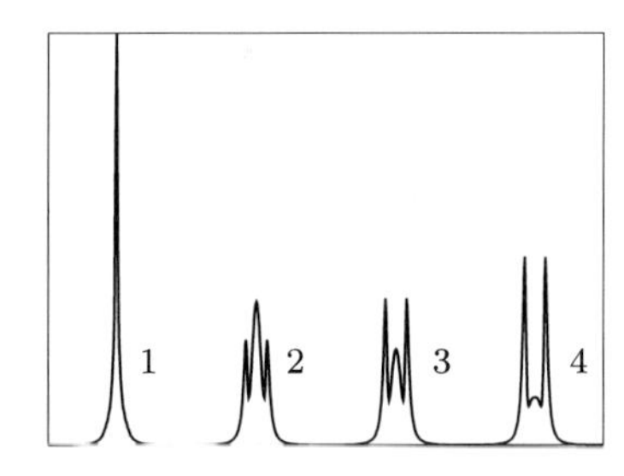

Figure 5.6: Profiles of the lines Ly_α (left) and D_α (right) with no magnetic field (1) and in the presence of a magnetic field $B = 8$ T directed perpendicularly (2), at an angle of 45° (3) and parallel (4) to the wave vector $\vec{k}$

5.5 Statistical method for line-group accounting

The computation of line absorption cross-sections using formula (5.44) requires a considerable amount of computer time. A main reason for this is the presence of a huge number spectral lines in a dense plasma. This is connected with the realization of a large number of states of various ions which, having a small concentration, may contribute to the absorption coefficient because the value of the cross-section near the center of the line is large. In addition, there are considerable computational difficulties due to the fact that it is necessary to account for various effects that lead to line splitting and broadening in plasma. For example, accounting in a sufficiently accurate manner for line splitting and broadening due to the electric fields of neighboring ions results increase of several tens of times in computing time.

To deal with this it is necessary to considerably simplify the computations. This can be done by grouping spectral lines and describing the resulting groups by means of some effective profile. Clearly, the application of such a description for groups of nonoverlapping lines can lead to considerable errors and to distortion of the absorption spectrum. To avoid this, it is necessary to analyze how line splitting and broadening mechanisms affect the photon absorption coefficients.

5.5.1 Shift and broadening parameters of spectral lines in plasma

Let us examine the most important effects leading to line splitting and broadening in a hot dense plasma. First of all, a large number of lines arise because identical one-electron transitions (see (5.53)) take place in fields of ions that differ from one another through their configuration $Q = \left\{N_{n\ell j}^{Q}\right\}$ and, connected with this, through the effective charge of the ion core, external screening, etc. Moreover, for a given configuration Q an essential role is played by the multiplet splitting of levels according to the total momentum of the ion's electrons.

In the examination of the effects that influence the shape and width of a line, it is necessary to bring into picture, in addition to the natural broadening, the broadening due to the interaction of the ion with free electrons, with other ions, and also to the Doppler and other effects (for example, the Auger effect). All the line splitting and broadening effects listed above lead to line overlapping and to blurring of details of the spectrum. To characterize the degree to which lines may overlap, let us consider a gold plasma ($Z = 79$) with temperature $T = 1$ keV and several different values of the density: $\rho = 0.1$ g/cm^3, $\rho = 1$ g/cm^3 and $\rho = 10$ g/cm^3.

Table 5.7 lists for these cases the energy levels ε_α and the average occupation numbers of electron states, N_α ($\alpha \equiv n\ell j$). Moreover, Figure 5.7 below shows the graphs of the ion probability distribution P_k with respect to the degree of ionization k.

Table 5.7: Energy levels (in eV) and the average occupation numbers in a gold plasma with temperature $T = 1$ keV and densities $\rho = 0.1$ g/cm^3 ($Z_0 = 66.4$), $\rho = 1$ g/cm^3 ($Z_0 = 60.9$), $\rho = 10$ g/cm^3 ($Z_0 = 52.1$). The computations took into account shells with $n \leq 9$

			$\rho = 0.1$ g/cm^3		$\rho = 1$ g/cm^3		$\rho = 10$ g/cm^3	
n	ℓ	j	$\lvert\varepsilon_{n\ell j}\rvert$	$N_{n\ell j}$	$\lvert\varepsilon_{n\ell j}\rvert$	$N_{n\ell j}$	$\lvert\varepsilon_{n\ell j}\rvert$	$N_{n\ell j}$
1	0	1/2	86396	2.0000	85269	2.0000	83850	2.0000
2	0	1/2	19648	2.0000	18619	2.0000	17299	2.0000
2	1	1/2	19153	2.0000	18100	2.0000	16759	2.0000
2	1	3/2	17261	3.9988	16224	3.9996	14897	3.9998
3	0	1/2	7919	0.4419	7108	1.0628	6001	1.5157
3	1	1/2	7745	0.3848	6911	0.9645	5782	1.4309
3	1	3/2	7229	0.4983	6420	1.4525	5317	2.4497
3	2	3/2	7011	0.4106	6147	1.2114	4991	2.1312
3	2	5/2	6889	0.5520	6034	1.6769	4886	3.0395
4	0	1/2	4129	0.0128	3514	0.0606	2644	0.1972
4	1	1/2	4048	0.0118	3422	0.0555	2541	0.1796
4	1	3/2	3882	0.0200	3277	0.0963	2419	0.3213
4	2	3/2	3791	0.0183	3161	0.0860	2277	0.2817
4	2	5/2	3746	0.0262	3122	0.1242	2246	0.4104
4	3	5/2	3686	0.0247	3027	0.1131	2106	0.3603
4	3	7/2	3665	0.0322	3010	0.1484	2093	0.4745
5	0	1/2	2508	0.0025	2030	0.0141	1330	0.0572
5	1	1/2	2468	0.0024	1985	0.0135	1280	0.0544
5	1	3/2	2389	0.0045	1919	0.0252	1229	0.1036
5	2	3/2	2344	0.0043	1861	0.0238	1158	0.0967
5	2	5/2	2322	0.0063	1843	0.0351	1145	0.1431
5	3	5/2	2293	0.0061	1796	0.0335	1077	0.1339
5	3	7/2	2283	0.0081	1788	0.0444	1071	0.1776
5	4	7/2	2264	0.0080	1754	0.0429	1018	0.1686
5	4	9/2	2258	0.0099	1749	0.0533	1015	0.2102
6	0	1/2	1659	0.0011	1268	0.0661	686.6	0.0345
6	1	1/2	1637	0.0011	1243	0.0064	658.8	0.0296
6	1	3/2	1593	0.0020	1207	0.0124	633.5	0.0578
6	2	3/2	1568	0.0020	1175	0.0121	694.0	0.0556
6	2	5/2	1555	0.0029	1165	0.0179	587.1	0.0828
6	3	5/2	1539	0.0029	1139	0.0174	548.7	0.0798
6	3	7/2	1533	0.0038	1134	0.0231	545.7	0.1060
6	4	7/2	1522	0.0038	1116	0.0227	515.4	0.1029
6	4	9/2	1519	0.0047	1113	0.0283	513.8	0.1284
6	5	9/2	1515	0.0047	1103	0.0280	493.0	0.1258
6	5	11/2	1513	0.0056	1102	0.0336	492.0	0.1508

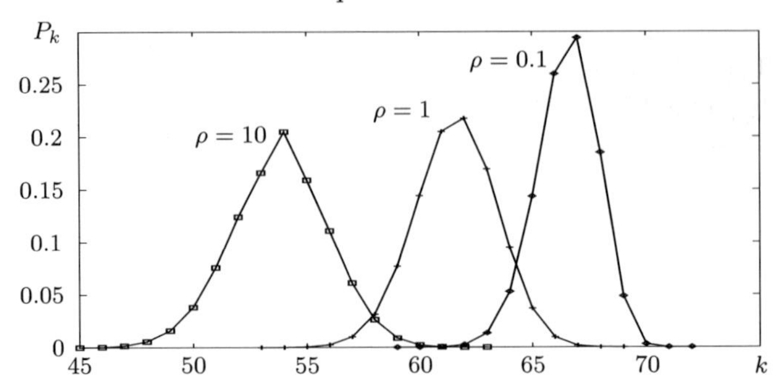

Figure 5.7: Ion probability distribution with respect to the ionization degree k for a gold plasma with temperature $T = 1$ keV and densities $\rho = 0.1$ g/cm^3, $\rho = 1$ g/cm^3 and $\rho = 10$ g/cm^3

Knowledge of the average occupation numbers allows one to estimate the number of spectral lines corresponding to a given one-electron transition $n\ell j \to n'\ell'j'$. To do the estimation we will assume that with sufficiently large probability in the plasma there are realized configurations Q with occupation numbers of electron states N_α^Q such that

$$N_\alpha - \delta N_\alpha \le N_\alpha^Q \le N_\alpha + \delta N_\alpha, \tag{5.164}$$

where δN_α is the mean square dispersion of the occupation numbers N_α^Q relative to their mean values N_α.

To calculate δN_α we will use the binomial distribution (5.74), which yields

$$\delta N_\alpha = \left[g_\alpha N_\alpha \left(1 - \frac{N_\alpha}{g_\alpha} \right) \right]^{1/2} \tag{5.165}$$

(here g_α denotes the statistical weight of the level α). Since the number of distinct ion states $\mathcal{N}$ (neglecting splitting effects) correspond approximately to the number of configurations considered, we obtain the estimate

$$\mathcal{N} \approx \prod_\alpha (2 \cdot [\delta N_\alpha + 0.9] + 1), \tag{5.166}$$

where $[x]$ denotes the integer part of x. Generally speaking, the number of spectral lines is obviously considerably higher than $\mathcal{N}$, in particular due to multiplet splitting of levels. In the case $Z = 79$, $T = 1$ keV, $\rho = 1$ g/cm^3 the estimate (5.166) yields $\mathcal{N} \sim 10^5$, and when splitting effects are incorporated one has $\mathcal{N} \sim 10^8$.

Next let us examine the typical line shifts and broadening parameters for one of the one-electron transitions $n\ell j \to n'\ell'j'$, with $n = 3$, $\ell = 1$, $j = 3/2$, $n' = 4$, $\ell' = 2$, $j' = 5/2$. The results of the computation of the various relevant quantities are shown in Table 5.8. As one can see, a change in the ion charge by the amount

Table 5.8: Typical values of the line shift and broadening (in eV) for the transition $3p_{3/2} \to 4d_{5/2}$ in gold at temperature $T = 1$ keV

Effect	$\rho = 0.1$ g/cm^3	$\rho = 1$ g/cm^3	$\rho = 10$ g/cm^3
1. Change of ion charge	50–200	45–200	40–150
2. Change of occupation of outer shells	0.1–6	0.2–8	0.3–12
3. Multiplet splitting with respect to total momentum	0–28	0–28	0–28
4. Natural broadening	0.06	0.04	0.03
5. Doppler effect	0.4	0.4	0.4
6. Electron broadening	0.07	0.1	0.3
7. Ion broadening	0.006	0.04	0.2

$\Delta Z \sim 1$–5 results in a shift of lines and also of photoionization thresholds by $\simeq$ 40–200 eV (see the first row in Table 5.8).

The magnitude of the shift can be easily estimated using the hydrogen-like approximation for the energy levels:

$$\varepsilon_{n\ell} = -\frac{Z_{n\ell}^2}{2n^2} + A_{n\ell},$$

whence

$$\Delta\varepsilon_{n\ell} \approx \frac{Z_{n\ell}\,\Delta Z_{n\ell}}{n^2}, \tag{5.167}$$

i.e., for instance, $\Delta\varepsilon_{n\ell} \approx 120\,\Delta Z_{n\ell}$ eV for $n = 4$ (see Table 2.3, which lists effective charges $Z_{n\ell}$ for the average ion).

The influence of the external screening is much weaker, the corresponding shift being only of $\simeq 0.1$–12 eV. A similar shift $\simeq 0$–28 eV is obtained for the value of the level splitting with respect to the total momentum (see row 3 in the table). The total dispersion of the positions of lines in ions with identical charge is $\Delta E \simeq 20$–30 eV.

If the number of lines is sufficiently large, so that the condition $\mathcal{N}\gamma \gg \Delta E$ is satisfied, where γ is the typical line width for the given transition, then the groups of spectral lines corresponding to the given energy interval ΔE will overlap. For density $\rho = 0.1$ g/cm^3 one has $\mathcal{N} \sim 10^3$, while for densities $\rho = 1$ g/cm^3 and $\rho = 10$ g/cm^3 one has $\mathcal{N} \sim 10^5$ and $\mathcal{N} \sim 10^7$, respectively. As one can see from Table 5.8, the condition of line overlapping for the transition in question is practically satisfied already for a density $\rho = 1$ g/cm^3. For larger densities, in particular for $\rho = 10$ g/cm^3, the lines overlap completely, which changes the situation cardinally compared with the case of absorption in a single spectral line.

It is interesting to note that until relatively recently researchers thought that spectral lines have no effect on the Rosseland opacity. For instance, in the monograph [67], in his analysis of the influence of various processes on the radiative heat conductivity of matter, the author argues as follows:

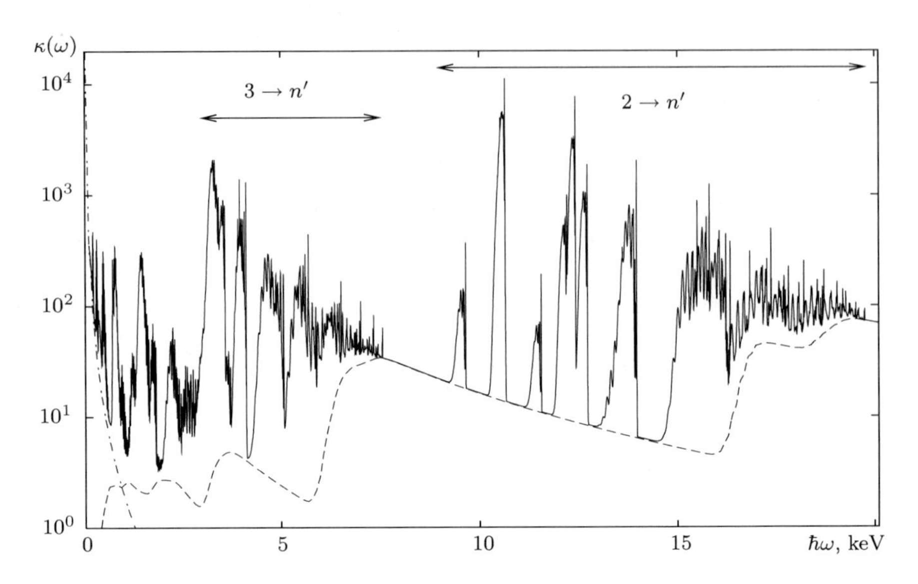

Figure 5.8: Contribution of various processes to the spectral absorption coefficient for a gold plasma ($Z = 79$, $T = 1$ keV, $\rho = 0.1$ g/cm^3). The solid curve corresponds to the total absorption coefficient $\kappa(\omega)$; also shown are the contributions of photoionization (dashed curve), bremsstrahlung (dashed-dotted curve), and the domains of influence of the main transitions $n \to n'$

"We did not include among the processes considered those for which both the initial and the final state are bound states. This refers to absorption, which results in the presence of spectral lines. Generally speaking, this process cannot be always neglected. However, in the special case of radiative heat conductivity it practically has no importance due to the specific features of the Rosseland mean. If we are dealing with a thin line, then the opacity will be very large in a very narrow frequency interval. But according to Rosseland, what is being averaged over frequency is not the opacity, but the reciprocal quantity, i.e., the free path. Within the limits of the spectral line the free path is very small, but so is the frequency interval. Even if one considers that the free path within the limits of the line is zero, that would only mean that from the Rosseland integral we must exclude the frequency interval corresponding to the line width. In the limit of infinitely thin lines, their influence on the Rosseland mean is identically equal to zero, regardless of how strong the absorption in the spectral lines is. From this point of view, the natural and Doppler width of lines are negligible. Absorption in lines may manifest itself in the radiative heat conductivity only due to the broadening of lines at high densities as a result of collisions. However, quantitative estimates show that, in fact, in stellar conditions this effect can be neglected. If the density is so high that the lines become very broad due to collisions, then inverse bremsstrahlung plays a more essential role than all transitions from bound states. In view of the above argument, in the consideration of radiative heat conductivity we will consider only the three processes listed above."

As a matter of fact, in many cases absorption in spectral lines in a hot plasma is the most important process in the description of radiative heat conductivity. For example, accounting for spectral lines in gold with temperature $T = 1$ keV and density $\rho = 0.1$ g/cm^3 leads to a decrease of about ten times in the Rosseland mean free path (see Figure 5.8 above).

It is precisely the line groups that determine the absorption spectrum (see Figure 5.9*a* in the natural scale, in contrast to the logarithmic scale used in Figure 5.8) and the emissivity (Figure 5.9*b*), exerting a strong influence on the magnitude of the free paths of photons (Figure 5.9*c*).

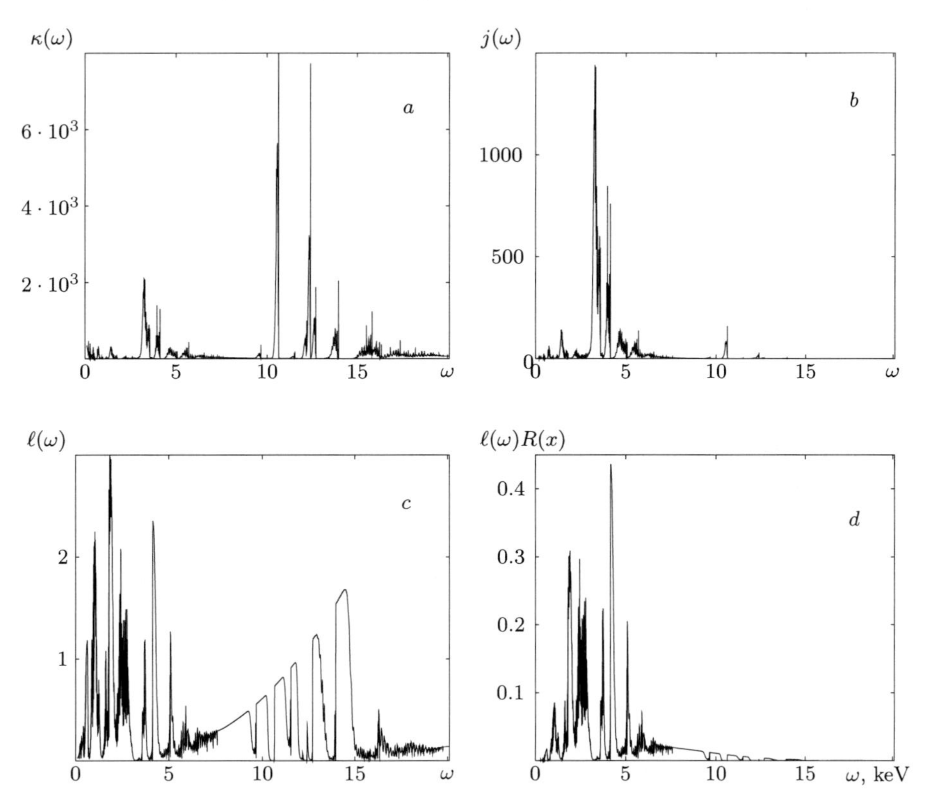

Figure 5.9: Dependence of the spectral characteristics of a gold plasma in the interval of photon energies $0 < \hbar\omega < 20T$, at temperature $T = 1$ keV and density $\rho = 0.1$ g/cm^3 on the photon energy in keV: (*a*) absorption coefficient, in cm^2/g (opacity); (*b*) emissivity in TW/(cm^3·eV·steradian); (*c*) free path, in cm; (*d*) free path times the Rosseland weight function

The emissivity $j(\omega)$ was calculated in accordance with Kirchhoff's law (see Subsection 5.1.3):

$$j(\omega) = \rho\,\kappa(\omega)\,B_\omega = \frac{\hbar\,\omega^3 \rho\,\kappa(\omega)}{4\pi^3 c^2}\,\frac{e^{-x}}{1-e^{-x}}, \quad \text{where } x = \frac{\hbar\,\omega}{T}. \tag{5.168}$$

Note that for the calculation of the Rosseland mean free path the most important domain is $T < \hbar\omega < 10T$ (see Figure 5.9d, which shows the subintegral curve $\ell(\omega)R(x)$, $R(x) = \dfrac{15}{4\pi^4}\dfrac{x^4 e^{-x}}{(1-e^{-x})^2}$).

Fluctuations of the occupation numbers lead to a broadening of the domain of influence of the spectral lines corresponding to one and the same one-electron transition. This factor is most prominent for high densities (in the case of gold, for example, for densities $\rho \geq 0.01$ g/cm^3). For smaller densities the multiplet structure of the spectrum (see Subsection 5.5.3 below) becomes the dominating factor. Next to it, especially for low-Z elements in a highly ionized state, when the number of possible ion states is not large, the Stark effect, which determines the broadening of spectral lines, may prove important [75, 189].

5.5.2 Fluctuations of occupation numbers in a dense hot plasma

A detailed description of the level structure of each ion in a hot plasma requires enormous computational resources. In practice it is convenient to group together levels that are close to one another, thereby creating a transition arrays from one group of levels to another.

In particular, such numerical algorithms using an unresolved transition array (UTA) method are constructed in a natural manner if, for example, one does not do a detailed accounting with respect to the quantum numbers ℓ, j and assumes that different configurations of ions differ only through the total number of electrons on the shell with principal quantum number n. In this way one has to deal with a relatively small number of configurations $\{N_n^Q\}$, with occupation numbers $N_n = N_n^Q = 0, 1, 2, \ldots, 2n^2$.

Once the numbers N_n^Q are given, the occupation numbers of the corresponding averaged representative of this group of configurations are given by the formula

$$\bar{N}_{n\ell j}^Q = \frac{2j+1}{2n^2}\,N_n^Q. \tag{5.169}$$

Similarly, one can carry out a more detailed configuration accounting by setting, for example,

$$\bar{N}_{n\ell j}^Q = \frac{2j+1}{2(2\ell+1)}\,N_{n\ell}^Q. \tag{5.170}$$

and sorting the configurations $\{N_{n\ell}^Q\}$.

In such approach it is convenient to start with quantities obtained in the average-atom approximation, regarding the different configurations as fluctuations

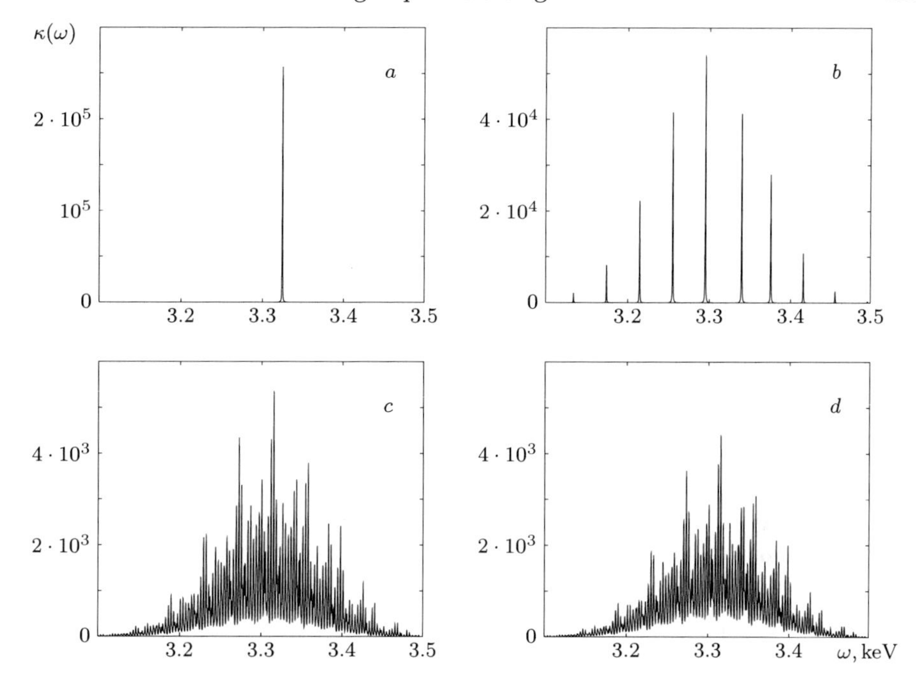

Figure 5.10: Line absorption coefficient (without accounting for multiplet splitting) for the transition $3p\frac{3}{2} \to 4d\frac{5}{2}$ in a gold plasma at temperature $T = 1$ keV and density $\rho = 1$ g/cm^3, for different numbers of ion configurations accounted for: (a) one configuration, (b) 24 configurations, corresponding to various degrees of ionization, (c) about 6000 configuration, (d) more than 70000 configurations

of the occupation numbers about their average values. It is clear that one needs to take into considerations only the shells that are not filled, more specifically, those among them for which the occupation numbers are not too small. In practice it suffices to take into account 2 or 3 shells, and the amount of details may be decreased with the growth of the principal quantum number.

Figure 5.10 shows the absorption coefficients for the transition $3p_{3/2} \to 4d_{5/2}$ at $T = 1$ keV, $\rho = 1$ g/cm^3, computed by formula (5.53). In the first case (Figure 5.10a) one considers only one configuration, corresponding to the average ion. Figure 5.10b accounts for the main configurations, corresponding to various degrees of ionization. For Figure 5.10c about 6000 configurations, with different numbers $\{N_{n\ell}^Q\}$, were accounted for. Finally, Figure 5.10d accounts for practically all possible configurations $\{N_{n\ell j}^Q\}$ (more than 70000). As a result of accounting for an increasing number of configurations, the total absorption coefficient of the lines $3p\frac{3}{2} \to 4d\frac{5}{2}$ fills out more and more densely the spectrum interval 3.1–3.5 keV, approaching the smooth curve obtained by Moszkowski's method, described below in Subsection 5.5.3 (see Figure 5.11).

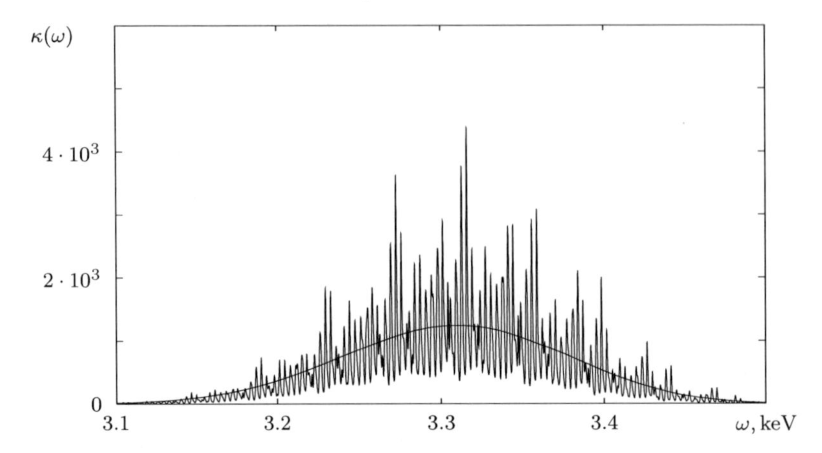

Figure 5.11: Line absorption coefficient for the transition $3p_{\frac{3}{2}} \to 4d_{\frac{5}{2}}$ at temperature $T = 1$ keV and density $\rho = 1$ g/cm^3 for a detailed configuration accounting without multiplet splitting and with multiplet splitting accounted for by Moszkowski's method (smooth curve)

5.5.3 Statistical description of overlapping multiplets

Multiplet splitting leads to a further increase in the number of lines, which is particularly important for multielectron ions of high-Z elements. Such ions may exhibit states with a large number of partially occupied shells (including the d and f shells), which results in a huge number of multiplet components [166]. In a number of cases, to account for the term structure one needs to use the intermediate-coupling approximation [71, 93, 132]. This makes the computations considerably more difficult, and they can be done only on supercomputers.

To calculate the averaged coefficients — the Rosseland and Planck mean free paths, the group constants, — it hardly makes sense to give a minute description of fine details of the absorption spectrum. In this case we may regard the multiplet-splitting effects as an additional broadening of the lines and calculate the combined profile of the multiplet on the basis of some statistical distribution of its components [22, 26, 138].

Assuming that in the one-configuration approximation the ion wave functions can be constructed as linear combinations of Slater determinants (5.84), we obtain the following expression for the average of the ion energy over the configuration Q:

$$E_Q = \frac{\sum\limits_{\gamma JM} g(\gamma JM)\,\langle\gamma JM|\widehat{H}|\gamma JM\rangle}{\sum\limits_{\gamma JM} g(\gamma JM)} \equiv \langle\gamma JM|\widehat{H}|\gamma JM\rangle_Q, \tag{5.171}$$

where $\widehat{H}$ is the Hamiltonian of the system of electrons of the ion, $g(\gamma JM)$ is the statistical weight of the state γJM and $\langle \cdots \rangle_Q$ denotes averaging over all possible states of configuration Q.

Next, by a know relation from probability theory, the dispersion of the energy of an electron configuration is given by

$$\sigma_Q^2 = \langle E_{\gamma JM}^2 \rangle_Q - \langle E_{\gamma JM} \rangle_Q^2, \tag{5.172}$$

where $E_{\gamma JM} = \langle \gamma JM | \widehat{H} | \gamma JM \rangle$.

In much the same manner one can define the dispersion of the energy of the transition between two configurations Q and Q':

$$\sigma_{Q \to Q'}^2 = \langle (E_{\gamma' J' M'} - E_{\gamma JM})^2 \rangle_{Q,Q'} - (\langle E_{\gamma' J' M'} - E_{\gamma JM} \rangle_{Q,Q'})^2, \tag{5.173}$$

where $\langle \cdots \rangle_{Q,Q'}$ denotes the averaging with the weights of the corresponding components of the multiplet.

Explicit calculations based on formulas (5.171)–(5.173) were carried out in [23, 24, 25], where analytic formulas for the dispersion (5.173) were obtained in the approximation of a different type of coupling. However, for practical opacity computations such expressions are too tedious, and so it is more convenient to confine ourselves to a simpler approximation, whose accuracy is comparable to that in [25]. If one neglects the interaction between electron shells, then to calculate the dispersion of energies of a transition between arbitrary configurations one can use the sum of the corresponding squared dispersions of transitions for two-electron configurations.

Let us consider the one-electron transition $\alpha \to \beta$ from configuration Q to configuration Q', where $Q = \alpha^N$ and $Q' = \alpha^{N-1}\beta$, assuming that the electron in the state β does not influence the state α. Then, as it was shown on a number of examples in [138] by Moszkowski, for such a transition the dispersion can be calculated by means of the formula

$$\sigma_{Q \to Q'}^2 \simeq \frac{(N-1)(g_\alpha - N)}{g_\alpha - 2} \sigma_{\alpha\alpha \to \alpha\beta}^2, \tag{5.174}$$

where $\sigma_{\alpha\alpha \to \alpha\beta}^2$ is the dispersion for the transition $\alpha^2 \to \alpha\beta$ and g_α is the statistical weight of the state α.

Formula (5.174) can be derived by a simple argument. First, we shall assume a quadratic dependence of the dispersion of the multiplet splitting on the number of electrons N. Next, we observe that the dispersion is equal to zero for $N = 1$ and $N = g_\alpha$; also, if $N = 2$, then $\sigma_{Q \to Q'}^2$ must coincide with $\sigma_{\alpha\alpha \to \alpha\beta}^2$. Note that in this approach the dispersion attains its maximum for $N \sim g_\alpha/2$, as it should be.

Accordingly, for the transition $\alpha^{N_1}\beta^{N_2} \to \alpha^{N_1-1}\beta^{N_2+1}$ we have

$$\sigma_{Q \to Q'}^2 \simeq \frac{(N_1-1)(g_\alpha - N_1)}{g_\alpha - 2} \sigma_{\alpha\alpha \to \alpha\beta}^2 + \frac{N_2(g_\beta - N_2 - 1)}{g_\beta - 2} \sigma_{\alpha\beta \to \beta\beta}^2. \tag{5.175}$$

Summing the expressions for the dispersions of the transitions $\alpha\nu^N \to \beta\nu^N$,

$$\sigma^2_{\alpha\nu^N\to\beta\nu^N} \simeq \frac{N(g_\nu - N)}{g_\nu - 1}\,\sigma^2_{\alpha\nu\to\beta\nu}, \tag{5.176}$$

and adding the expression (5.175) to the result, we conclude that in the general case the mean square dispersion of the components of the multiplet $Q \to Q'$ of the one-electron transition $\alpha \to \beta$ is given by

$$\sigma^2_{\alpha\beta} \simeq \frac{(N^Q_\alpha - 1)(g_\alpha - N^Q_\alpha)}{g_\alpha - 2}\sigma^2_{\alpha^2\to\alpha\beta} + \frac{N^Q_\beta(g_\beta - N^Q_\beta - 1)}{g_\beta - 2}\sigma^2_{\alpha\beta\to\beta^2} + \sum_{\nu\neq\alpha,\beta}\frac{N^Q_\nu(g_\nu - N^Q_\nu)}{g_\nu - 1}\sigma^2_{\alpha\nu\to\beta\nu}, \tag{5.177}$$

where the two-electron dispersions $\sigma_{\alpha\nu\to\beta\nu}$ are calculated in the approximation provided by some coupling scheme. In practice it is convenient to use the jj- or LS-coupling, since for these choices the energy of two electrons can be calculated by means of simple analytic formulas.

Expression (5.177) is convenient for practical computations, because the calculation of the dispersion of the transition $\alpha \to \beta$ for a given configuration Q reduces to that of dispersions of transitions in two-electron configurations, which in the approximation adopted here are independent of the concrete states of ions. This allows one, based on two-electron dispersions calculated previously, to determine the dispersion for any multi-electron configuration with a minimum number of numerical operations.

Let us consider, for example, the situation when the two-electron dispersions are treated in the approximation of the jj-coupling:

$$\sigma^2_{\alpha\nu\to\beta\nu} = \langle\Delta E^2_{\alpha\nu\to\beta\nu}\rangle - \langle\Delta E_{\alpha\nu\to\beta\nu}\rangle^2,$$

where the parameters $\langle\Delta E^m_{\alpha\nu\to\beta\nu}\rangle$ for $m = 1, 2$ are calculated using the shifts of ion energies due to a change in the state of one of the electrons:

$$\langle\Delta E^m_{\alpha\nu\to\beta\nu}\rangle = \frac{\sum_{J,J'}(2J+1)(2J'+1)|\Delta E_{\alpha\nu\to\beta\nu}(J,J')|^m}{\sum_{J,J'}(2J+1)(2J'+1)}.$$

Here

$$\Delta E_{\alpha\nu\to\beta\nu}(J,J') = E(j_\beta, j_\nu, J') - E(j_\alpha, j_\nu, J),$$

$$|j_\nu - j_\alpha| \le J \le j_\nu + j_\alpha, \quad |j_\nu - j_\beta| \le J' \le j_\nu + j_\beta, \quad |J - J'| \le 1.$$

To calculate $E(j_1, j_2, J)$ we assume that the one-electron wave functions remain unchanged when the electron states change; what changes is only the total momentum J and, correspondingly, the multi-electron wave function, i.e., the linear combination of determinants (5.177). In this approximation, the energy of two electrons with total momentum J is determined by means of the Slater integrals $F^s_{\alpha\nu}$ and $G^s_{\alpha\nu}$ (see [156] and also the analogous formulas (3.39), (3.44) for the nonrelativistic case):

$$E(j_\alpha, j_\nu, J) = I_\alpha + I_\nu + \sum_{s=|j_\nu - j_\alpha|}^{j_\nu + j_\alpha} (a_s F^s_{\alpha\nu} - b_s G^s_{\alpha\nu}), \tag{5.178}$$

where

$$a_s = (-1)^{j_\alpha + j_\nu - J} \sqrt{(2j_\alpha + 1)(2j_\nu + 1)} W(j_\alpha j_\nu j_\alpha j_\nu; Js) C^{j_\alpha \frac{1}{2}}_{j_\alpha \frac{1}{2}\ s0} C^{j_\nu \frac{1}{2}}_{j_\nu \frac{1}{2}\ s0},$$

$$b_s = (2j_\alpha + 1) W(j_\alpha j_\nu j_\nu j_\alpha; Js) \left(C^{j_\nu \frac{1}{2}}_{j_\alpha \frac{1}{2}\ s0} \right)^2 \left[1 + (-1)^{l_\alpha + l_\nu + s} \right] /2,$$

(in the case $\alpha = \nu$ the coefficient $b_s = 0$).

To illustrate the use of formula (5.177), we give here results of the computed transition arrays $4s \to 4p$ for the ions Ni II, Mo XVI and Pr XXXIII (see figures 5.12–5.14 below). The computations were carried out for the configuration $3d^8\,4s$, for various temperatures and densities (see Table 5.9). As the figures show, formula (5.177) describes well the distribution of overlapping clusters of lines. Moreover, when the density increases the profiles of such clusters, computed with the term structure accounted for, approach the profile given by dispersion (5.177). Note that the computing time required for the detailed calculation is several orders larger that in the case of Moszkowski's method.

Table 5.9: Lorentz widths γ in eV for the lines of the transition $3d^8\,4s \to 3d^8\,4p$ in the ions Ni II, Mo XVI and Pr XXXIII for different values of the temperature T (measured in eV) and density ρ (measured in g/cm^3). Also shown is the average-ion charge Z_0

	ρ	10^{-3}	10^{-2}	10^{-1}	1
Ni II	T	2	2.5	3	
	γ	0.0058	0.042	0.6	
	Z_0	1.18	1.06	0.82	
Mo XVI	T		65	75	115
	γ		0.013	0.12	0.74
	Z_0		15.8	14.6	15.36
Pr XXXIII	T		200	250	350
	γ		0.0051	0.031	0.29
	Z_0		32.9	32.9	32.4

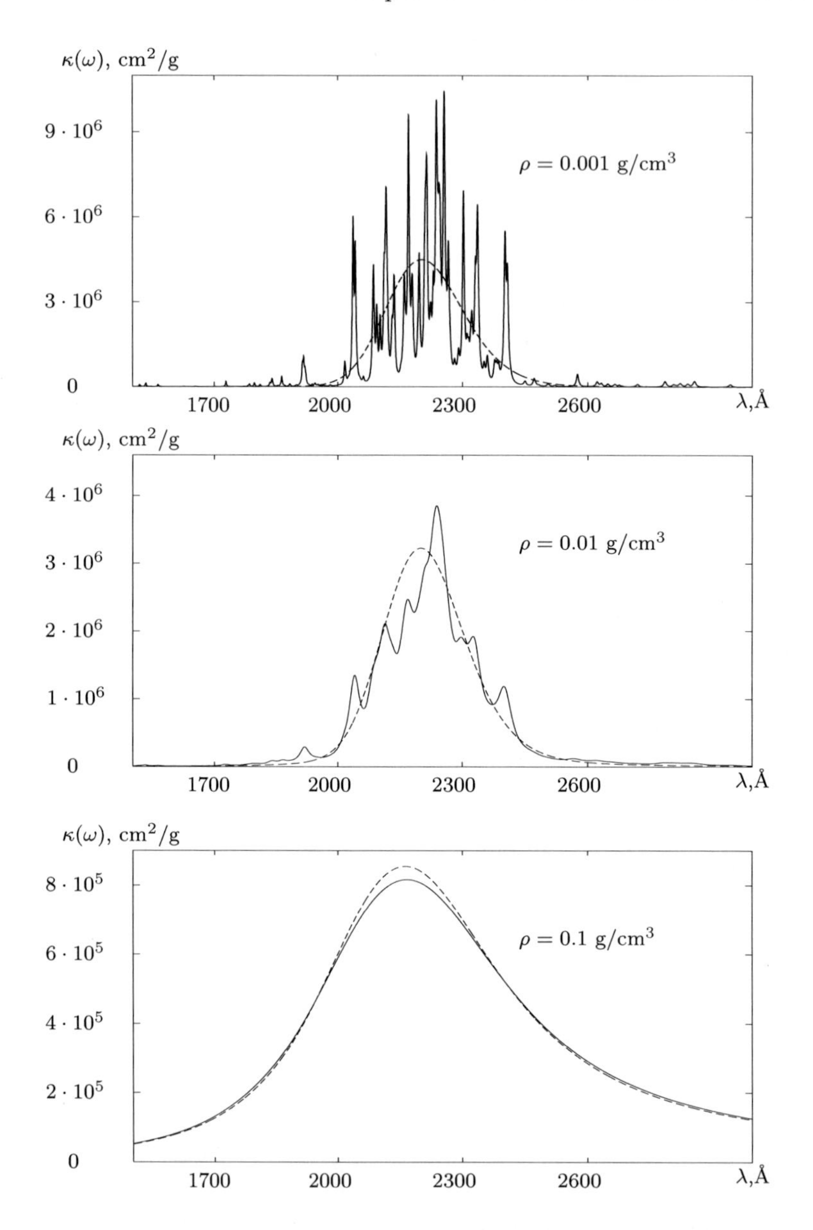

Figure 5.12: Spectral absorption coefficients for the transition $3d^8\ 4s \rightarrow 3d^8\ 4p$ in the Ni II ion for different densities ρ, with the spin-orbit interaction accounted for via a coupling of intermediate type (continuous curves) and computed by Moszkowski's method (dashed curves)

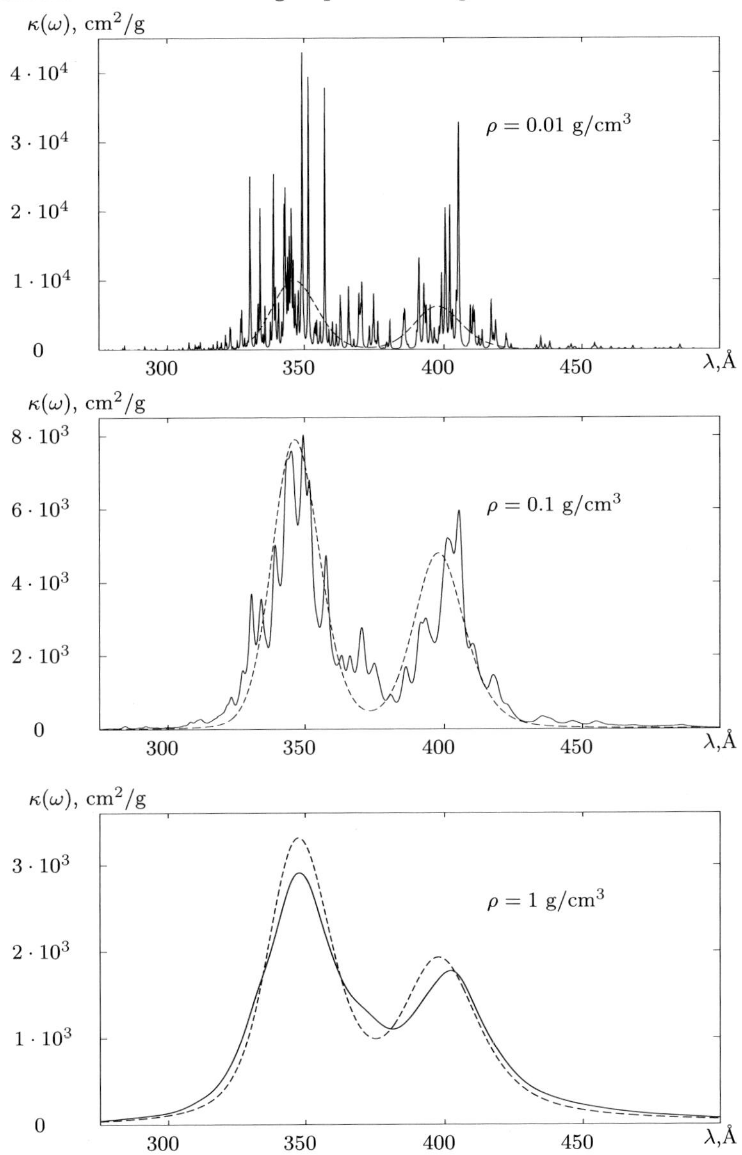

Figure 5.13: Same as in Figure 5.12, but for the Mo XVI ion

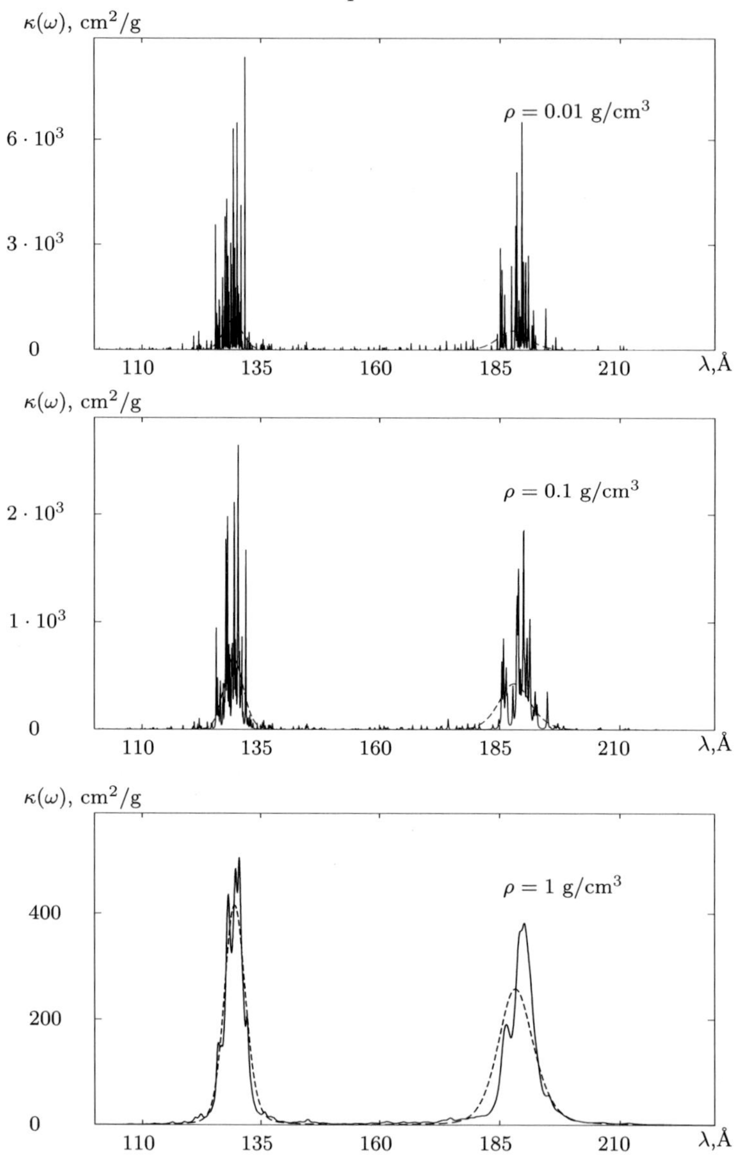

Figure 5.14: Same as in Figure 5.12, but for the Pr XXXIII ion

For a comparison with the more accurate method found in [25], Table 5.10 lists the dispersions for several transitions in a tungsten plasma, computed with the wave functions of the average ion for $T = 0.5$ keV, $\rho = 0.5$ g/cm^3 (in this case the average charge of the ion is $Z_0 = 46$). If one recalls that the charge of the ion

W^{48} in [25] differs from the charge of the average ion, in the potential of which the wave functions were calculated, one can consider that the approximation (5.177) yields a satisfactory accuracy.

Table 5.10: Dispersions of several transitions for a tungsten plasma. The computations using formula (5.177) were carried out with the wave functions of the average ion at temperature $T = 0.5$ keV and density $\rho = 0.5$ g/cm^3

Transition	$\sigma_{\alpha\beta}$ according to (5.177)	$\sigma_{\alpha\beta}$ according to ref. [25]
$3d^3_{3/2}d^6_{5/2} \to 3d^2_{3/2}d^6_{5/2}4p_{1/2}$	6.50	8.14
$3d^3_{3/2}d^6_{5/2} \to 3d^3_{3/2}d^5_{5/2}4p_{3/2}$	6.11	5.18
$3d^4_{3/2}d^5_{5/2} \to 3d^4_{3/2}d^4_{5/2}4p_{3/2}$	6.13	6.43
$3d^4_{3/2}d^5_{5/2} \to 3d^3_{3/2}d^5_{5/2}4p_{3/2}$	6.68	6.47

The influence of the multiplet splitting effects on the absorption coefficients is illustrated on Figure 5.15 below (see also Table 5.11), which shows some results of computations for an iron plasma, carried out in various approximations with the codes THERMOS, STA and OPAL. The codes OPAL and STA, in contrast to THERMOS, use a parametric potential (see, e.g., [180]). Absorption in spectral lines is accounted for either in the approximation of a coupling scheme of intermediate type with a detailed accounting for terms (OPAL), or in the statistical approximation [26], which is more precise than Moszkowski's method, but requires a considerably larger volume of computations (STA).

As one can see from Figure 5.15 and Table 5.11, the multiplet splitting determines the radiative heat conductivity of iron at temperatures $T < 0.1$ keV and for densities $\rho \sim 10^{-4}$ g/cm^3 (see also Figure 5.16, which displays the spectral absorption coefficients for an iron plasma at $T = 20$ eV, $\rho = 10^{-4}$ g/cm^3).

The redistribution of spectral lines due to multiplet splitting modifies the absorption spectrum in essential manner, and this effect becomes even more important for low densities (see Figure 5.17 for a xenon plasma at temperature $T =$ 15 eV and density $\rho = 10^{-6}$ g/cm^3). As the density increases—in particular, for densities $\rho > 0.01$ g/cm^3, the influence of the term structure becomes less notable, because for such densities the determining role is played by the fluctuations of the occupation numbers.

The practical importance of accounting for the effects of multiplet splitting in the computation of absorption coefficients is particularly transparent in as-

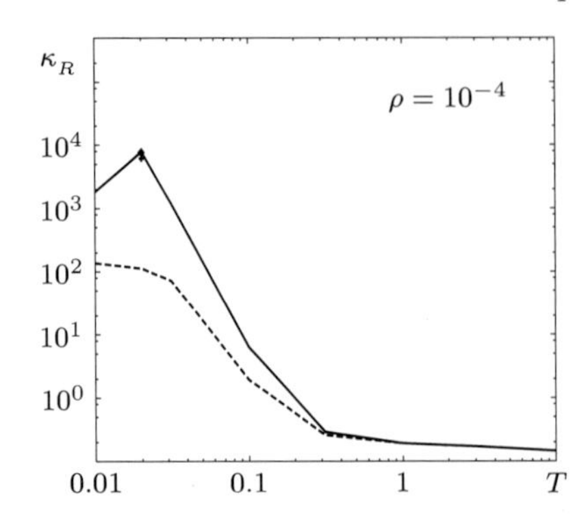

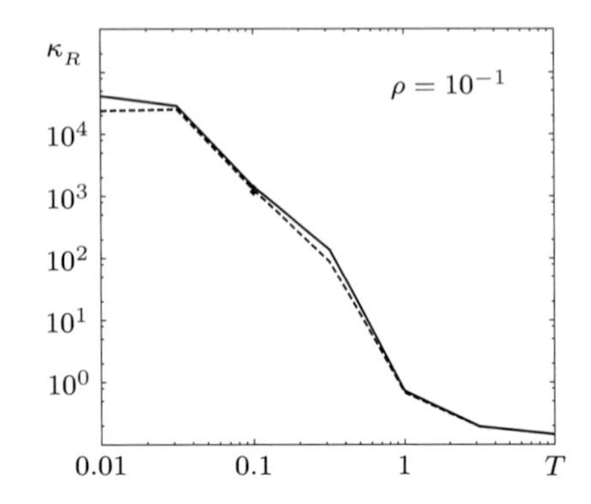

Figure 5.15: Dependence of the Rosseland mean opacity κ_R for iron on the temperature T in keV for densities $\rho = 10^{-4}$ g/cm^3 and $\rho = 10^{-1}$ g/cm^3, calculated with the code THERMOS. The dashed [resp., continuous] curve represents the values calculated without the multiplet structure of the levels [resp., with the multiplet structure accounted for in the Moszkowski approximation]. Also shown are the values obtained with the codes OPAL and STA for $T = 0.02$ keV, $\rho = 10^{-4}$ g/cm^3 and for $T = 0.1$ keV, $\rho = 0.1$ g/cm^3 (they practically lie on the continuous curve, see Table 5.11)

Table 5.11: Rosseland mean opacity κ_R for an iron plasma

T, eV	ρ, g/cm^3	Code	κ_R, cm^2/g
20	10^{-4}	THERMOS	$7.87 \cdot 10^3$
		STA	$7.68 \cdot 10^3$
		OPAL	$6.11 \cdot 10^3$
100	10^{-1}	THERMOS	$1.43 \cdot 10^3$
		STA	$1.38 \cdot 10^3$
		OPAL	$1.18 \cdot 10^3$

trophysical applications. As it turned out, the heaviest component of the stellar atmospheres of cepheids, iron plasma, despite having very small concentration, determines their opacity [209]. Highly accurate computations carried out with the OPAL code at the Livermore National Laboratory, USA, [182] and the Sandia Laboratory experiments which confirm these computations, allowed researchers to explain the puzzling behavior of cepheids in the Herzsprung-Russell diagram [209].

Therefore, for low densities of matter, in many cases, in addition to a detailed accounting of configurations, a detailed accounting of terms is necessary, and satisfactory results are obtained only when a coupling scheme of intermediate type is used.

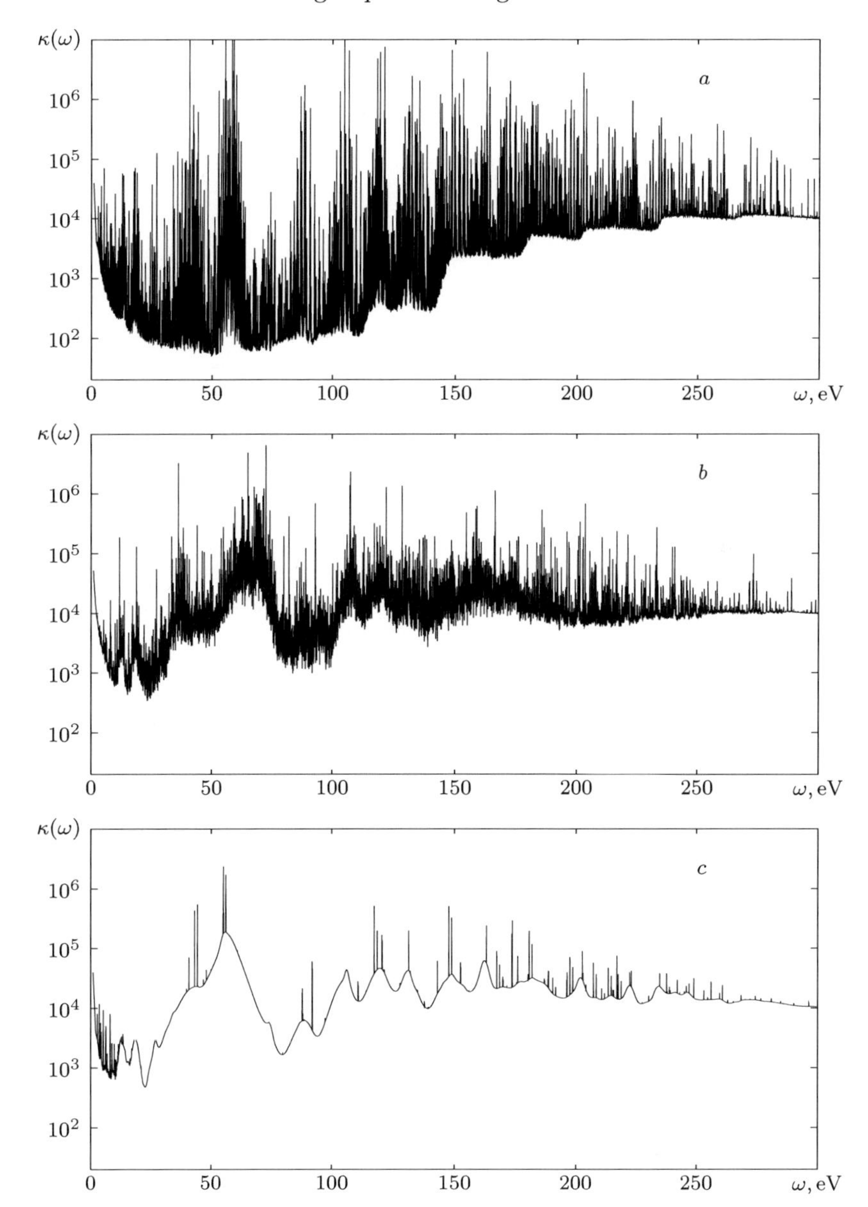

Figure 5.16: Spectral absorption coefficients of iron at temperature $T = 20$ eV and density $\rho = 10^{-4}$ g/cm^3: (*a*) values computed without accounting for the multiplet structure of levels; (*b*) computed with the multiplet structure of levels accounted for in the approximation of an intermediate-type coupling, using the code OPAL [182], (*c*) computed with the multiplet structure of levels accounted for in the Moszkowski approximation, using the code THERMOS

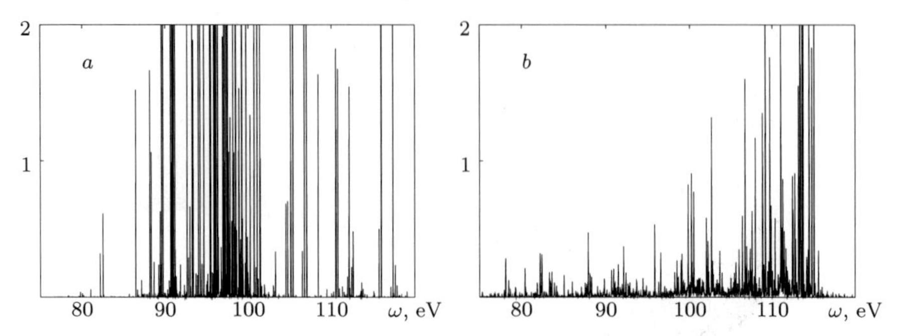

Figure 5.17: Absorption coefficients (in 10^5 cm^2/g) in various approximations : (a) configurations are accounted for by using perturbation theory without calculation of the multiplet structure; (b) the multiplet structure of levels is accounted for in the approximation of an intermediate coupling scheme. The computations were carried out for a xenon plasma ($Z = 54$) at temperature $T = 15$ eV and density $\rho = 10^{-6}$ g/cm^3

5.5.4 Effective profile for a group of lines

In a sufficiently dense plasma it is possible that the conditions for the overlapping of lines on characteristic energy intervals, which are determined by the magnitude of the multiplet splitting and the line shift due to the change of the occupation numbers in different configurations Q, are realized. To account for the influence of various configurations, let us first combine the contribution of the broadening effects (5.151), (5.152) and of the multiplet splitting (5.177) in the form of a profile $\Phi^Q_{\alpha\beta}(\varepsilon - \varepsilon^Q_{\alpha\beta})$, where $\varepsilon^Q_{\alpha\beta}$ is the energy of the transition $\alpha \to \beta$, averaged over all states of the configuration Q (including averaging with respect to the total momentum J). Using $\Phi^Q_{\alpha\beta}(\varepsilon - \varepsilon^Q_{\alpha\beta})$, the absorption cross-section corresponding to the one-electron transition $\alpha \to \beta$ can be written as

$$\sigma^{\alpha\beta}_{\text{bb}}(\varepsilon) = C_0 \sum_Q P_Q N^Q_\alpha \left(1 - \frac{N^Q_\beta}{g_\beta}\right) f^Q_{\alpha\beta} \Phi^Q_{\alpha\beta}(\varepsilon - \varepsilon^Q_{\alpha\beta}), \tag{5.179}$$

where the oscillator strength $f^Q_{\alpha\beta}$ is also averaged over the states of configuration Q, and $C_0 = 2\pi^2 \alpha a_0^2 = 4 \cdot 10^{-18}$ cm^2.

In the simplest case of total overlapping of lines one can sum over the configurations Q, relying on the average-atom approximation:

$$\sigma^{\alpha\beta}_{\text{bb}}(\varepsilon) = C_0 N_\alpha \left(1 - \frac{N_\beta}{g_\beta}\right) f_{\alpha\beta} J_{\alpha\beta}(\varepsilon). \tag{5.180}$$

Here $J_{\alpha\beta}(\varepsilon)$ is some combined profile for the set of overlapped lines of the transition $\alpha \to \beta$ and $f_{\alpha\beta}$ is the corresponding oscillator strength for the average ion with occupation numbers N_α.

Let us calculate $J_{\alpha\beta}(\varepsilon)$ in the case where all the lines of the transition $\alpha \to \beta$ are grouped near one maximum [56]. We make the approximate assumption that the profiles of the lines of one and the same transition are independent of the configuration Q, differing only through the position of the line's center. We shall also assume that the wave functions of the ion under consideration differ only slightly from the wave functions of the average atom. Then for the transition energies $\varepsilon^Q_{\alpha\beta}$ and the oscillator strengths $f^Q_{\alpha\beta}$, corresponding to the average values over the configuration Q, we have

$$\varepsilon^Q_{\alpha\beta} \simeq \varepsilon_{\alpha\beta} + \Delta\varepsilon_Q = \varepsilon_{\alpha\beta} + \sum_q \delta N_q G_q^{\alpha\beta}, \tag{5.181}$$

$$f^Q_{\alpha\beta} \approx \frac{2}{3}\varepsilon^Q_{\alpha\beta} \left| \left(\sum_j \vec{r}_j \right)^Q_{\alpha\beta} \right|^2 \approx \frac{2}{3}\varepsilon^Q_{\alpha\beta} |\vec{r}_{\alpha\beta}|^2 = \frac{\varepsilon^Q_{\alpha\beta}}{\varepsilon_{\alpha\beta}} f_{\alpha\beta}, \tag{5.182}$$

where $\delta N_q = N_q^Q - N_q$, $G_q^{\alpha\beta} \equiv G_q = H_{\beta q} - H_{\alpha q}$, H_{kq} is the matrix element of the pair interaction of electrons in the states k and q, calculated from the wave functions of the average ion (see (3.39)); $\varepsilon_{\alpha\beta}$ and $f_{\alpha\beta}$ are the energy and respectively the oscillator strength for the transition $\alpha \to \beta$, corresponding to the average ion.

Substituting (5.181) and (5.182) in (5.179), we have

$$\sigma^{\alpha\beta}_{\text{bb}}(\varepsilon) \simeq C_0 \frac{f_{\alpha\beta}}{\varepsilon_{\alpha\beta}} g_\alpha \sum_Q P_Q(\varepsilon_{\alpha\beta}+\Delta\varepsilon_Q)(n_\alpha+\delta n_\alpha)(1-n_\beta-\delta n_\beta)\Phi_{\alpha\beta}(\varepsilon-\varepsilon_{\alpha\beta}-\Delta\varepsilon_Q), \tag{5.183}$$

where $n_\alpha = N_\alpha / g_\alpha$, $\delta n_\alpha = \delta N_\alpha / g_\alpha$.

To continue the calculation, it is convenient to represent the function $\Phi_{\alpha\beta}$ as a Fourier integral:

$$\Phi_{\alpha\beta}(\varepsilon - \varepsilon_{\alpha\beta} - \Delta\varepsilon_Q) = \frac{1}{\sqrt{2\pi}} \int_{\infty}^{\infty} \phi(s) \exp[is(\varepsilon_{\alpha\beta} + \Delta\varepsilon_Q - \varepsilon)]ds. \tag{5.184}$$

This yields a formula of the form (5.180):

$$\sigma^{\alpha\beta}_{\text{bb}}(\varepsilon) = C_0 g_\alpha n_\alpha (1 - n_\beta) f_{\alpha\beta} J_{\alpha\beta}(\varepsilon), \tag{5.185}$$

where

$$J_{\alpha\beta}(\varepsilon) = \frac{1}{\sqrt{2\pi}} \int_{-\infty}^{\infty} \phi(s) \exp[is(\varepsilon - \varepsilon_{\alpha\beta})] \sum_Q P_Q \exp[-is\Delta\varepsilon_Q] \times$$
$$\left[1 + \frac{\Delta\varepsilon_Q}{\varepsilon_{\alpha\beta}}\right] \left[1 + \frac{\delta n_\alpha}{n_\alpha} - \frac{\delta n_\beta}{1 - n_\beta} - \frac{\delta n_\alpha \delta n_\beta}{n_\alpha(1 - n_\beta)}\right] ds. \tag{5.186}$$

The expression (5.186) describes the distribution of the population of spectral lines corresponding to the transition $\alpha \to \beta$. If we introduce the functions

$$\mathcal{F}(s) = \sum_Q P_Q \exp[-is\Delta\varepsilon_Q] \left[1 + \frac{\Delta\varepsilon_Q}{\varepsilon_{\alpha\beta}}\right]\left[1 + \frac{\delta n_\alpha}{n_\alpha} - \frac{\delta n_\beta}{1-n_\beta} - \frac{\delta n_\alpha \delta n_\beta}{n_\alpha(1-n_\beta)}\right], \tag{5.187}$$

$$F(\xi) = \frac{1}{\sqrt{2\pi}} \int_{-\infty}^{\infty} \mathcal{F}(s) \exp(-is\xi)\, ds, \tag{5.188}$$

and recall that $\int_{-\infty}^{\infty} \exp[-i\xi(s-s')]\, d\xi = 2\pi\delta(s-s')$, then the profile (5.186) can be recast as

$$J_{\alpha\beta}(\varepsilon) = \int_{-\infty}^{\infty} \Phi_{\alpha\beta}(\varepsilon - \varepsilon_{\alpha\beta} - \xi) F(\xi) d\xi. \tag{5.189}$$

Thus, in order to calculate $J_{\alpha\beta}(\varepsilon)$ it suffices to determine the function $\mathcal{F}(s)$. To this end let us substitute expression (5.181) in (5.187) and expand the exponential in a series. Then, up to terms $(\delta n)^2$, we have

$$\begin{aligned}\mathcal{F}(s) = \sum_Q P_Q \Bigg\{ 1 &+ \frac{\delta n_\alpha}{n_\alpha} - \frac{\delta n_\beta}{1-n_\beta} + \\ &\sum_q g_q \frac{\delta n_q G_q}{\varepsilon_{\alpha\beta}} \left[1 + \frac{\delta n_\alpha}{n_\alpha} - \frac{\delta n_\beta}{1-n_\beta} - \frac{\delta n_\alpha \delta n_\beta}{n_\alpha(1-n_\beta)}\right] + \\ &is \sum_q g_q \delta n_q G_q \left[1 + \frac{\delta n_\alpha}{n_\alpha} - \frac{\delta n_\beta}{1-n_\beta} + \sum_p g_p \frac{\delta n_p G_p}{\varepsilon_{\alpha\beta}}\right] + \\ &\frac{(is)^2}{2} \sum_{p,q} g_q g_p \delta n_q \delta n_p G_q G_p \Bigg\}.\end{aligned}$$

Henceforth we will denote averaging with respect to the binomial distribution (5.74) with brackets $\langle \cdots \rangle$. Under the no-correlation assumption (i.e., in the case when $\langle \delta n_p \delta n_q \rangle = 0$ whenever $p \neq q$), we have

$$\begin{aligned}\mathcal{F}(s) \cong 1 &+ \frac{\langle \delta n_\alpha^2 \rangle}{\varepsilon_{\alpha\beta} n_\alpha} g_\alpha G_\alpha - \frac{\langle \delta n_\beta^2 \rangle}{\varepsilon_{\alpha\beta}(1-n_\beta)} g_\beta G_\beta + \\ &is \left[\frac{\langle \delta n_\alpha^2 \rangle}{n_\alpha} g_\alpha G_\alpha - \frac{\langle \delta n_\beta^2 \rangle}{(1-n_\beta)} g_\beta G_\beta + \sum_q g_q^2 \frac{\langle \delta n_q^2 \rangle G_q^2}{\varepsilon_{\alpha\beta}}\right] + \\ &\frac{(is)^2}{2} \sum_q g_q^2 \langle \delta n_q^2 \rangle G_q^2.\end{aligned} \tag{5.190}$$

By a property of the binomial distribution,

$$\langle \delta n_q^2 \rangle = \frac{n_q(1-n_q)}{g_q}, \tag{5.191}$$

and so

$$\mathcal{F}(s) = \left(1 + \frac{E_1}{\varepsilon_{\alpha\beta}}\right)\left[1 + \frac{E_1 + E_2/\varepsilon_{\alpha\beta}}{1 + E_1/\varepsilon_{\alpha\beta}} is + \frac{E_2}{1 + E_1/\varepsilon_{\alpha\beta}} \frac{(is)^2}{2}\right], \tag{5.192}$$

where

$$E_1 = G_\alpha(1-n_\alpha) - G_\beta n_\beta, \qquad E_2 = \sum_q g_q n_q (1-n_q) G_q^2. \tag{5.193}$$

Therefore, in terms of two known moments of the distribution (5.192), the function $F(\xi)$ (see (5.188)) can be approximated by a Gaussian

$$F(\xi) \simeq \frac{1 + E_1/\varepsilon_{\alpha\beta}}{\sqrt{2\pi}\sigma} \exp\left[-\frac{(\xi - \Delta)^2}{2\sigma^2}\right], \tag{5.194}$$

where

$$\Delta = \frac{E_1 + E_2/\varepsilon_{\alpha\beta}}{1 + E_1/\varepsilon_{\alpha\beta}}, \qquad \sigma^2 = \frac{E_2}{1 + E_1/\varepsilon_{\alpha\beta}}. \tag{5.195}$$

Thus, the profile of the group of lines is given by the expression

$$J_{\alpha\beta}(\varepsilon) = \frac{1 + E_1/\varepsilon_{\alpha\beta}}{\sqrt{2\pi}\,\sigma} \int_{-\infty}^{\infty} \Phi_{\alpha\beta}(\varepsilon - \varepsilon_{\alpha\beta} - \xi) \exp\left[-\frac{(\xi - \Delta)^2}{2\sigma^2}\right] d\xi. \tag{5.196}$$

To derive formula (5.196) we have used the expression (5.181) for the transition energy and the binomial distribution (5.74) for the probability of different configurations. Furthermore, we have assumed that the lines are substantially overlapping and that the fluctuations of the occupation numbers are small compared with the statistical weight of levels, i.e., $\delta n_\alpha = \delta N_\alpha / g_\alpha \ll 1$. Both these conditions are well satisfied for high-Z elements at temperatures when roughly half of the electrons are ionized. Note that if the binomial distribution is applicable, then, as follows from (5.191), the condition that the fluctuations be small is also satisfied.

In most practical problems the individual profiles $\Phi_{\alpha\beta}(\varepsilon - \varepsilon_{\alpha\beta})$ of spectral lines are described by the Voigt function (see (5.149)):

$$\Phi_{\alpha\beta}(\varepsilon - \varepsilon_{\alpha\beta}) = \frac{1}{\sqrt{\pi}D} K\left(\frac{\varepsilon - \varepsilon_{\alpha\beta}}{D}, \frac{\gamma}{D}\right),$$

where γ is the Lorentz width of the line and D is the Doppler broadening parameter. The convolution of the Voigt function with the Gauss distribution (5.194) according to formula (5.196) yields again the Voigt function, but with different parameters:

$$J_{\alpha\beta}(\varepsilon) = \frac{1 + E_1/\varepsilon_{\alpha\beta}}{\sqrt{\pi(D^2 + 2\sigma^2)}} K \left(\frac{\varepsilon - \varepsilon_{\alpha\beta} - \Delta}{\sqrt{D^2 + 2\sigma^2}}, \frac{\gamma}{\sqrt{D^2 + 2\sigma^2}} \right). \quad (5.197)$$

Therefore, in the approximation adopted here the total profile of the set of Voigt lines is also described by the Voigt function. In the case where the density of lines is large, when the width of an individual line γ can be neglected in comparison with their mean square dispersion σ, putting $\Phi_{\alpha\beta}(\varepsilon - \varepsilon_{\alpha\beta}) = \delta(\varepsilon - \varepsilon_{\alpha\beta})$ and assuming that the conditions $E_1/\varepsilon_{\alpha\beta} \ll 1$, $E_2/\varepsilon_{\alpha\beta} \ll 1$, are satisfied, we obtain for the profile (5.196) the Gaussian curve

$$J_{\alpha\beta}(\varepsilon) = \frac{1}{\sqrt{2\pi E_2}} \exp\left(-\frac{(\varepsilon - \varepsilon_{\alpha\beta} - E_1)^2}{2E_2} \right). \quad (5.198)$$

Expression (5.198) was used by Stein, Shalitin and Ron in [210], where some other parameters E_1 and E_2 were obtained.

Formula (5.180) gives one smooth envelope for the entire group of lines of a one-electron transition $\alpha \to \beta$. Such an approximation is rather crude, because in many cases the profile of overlapped lines corresponding to a one-electron transition has a more complex shape and is given by a function with a certain number of maxima. These maxima arise due to the presence of states of the group of ions that differ considerably through their transition energies. To sharpen formula (5.180) it seems natural to group together the lines near such maxima and describe the entire population of lines of the considered one-electron transition as a set of several effective profiles (5.196).

Such a generalization is carried out in the papers [19, 55]. In particular, in [55] the authors use the fact that the fluctuations of the occupation numbers on the lower shells "shift" the lines more notably than the fluctuations on the upper levels. It is precisely such configurations, in which the occupancy of the lowest shells is affected by ionization changes, that will determine the main groups of lines. Hence, it is appropriate to single out a small number of shells, for which it is desirable that the summation in (5.179) over their occupation numbers be done explicitly. Suppose, for example, that μ is such a shell (the generalization to the case of several shells is straightforward). In (5.179) the summation over the shells $q \neq \mu$ can be carried out approximately up to terms $(\delta n)^2$, by analogy with the way we proceed above.

In this manner we obtain

$$\sigma_{\mathrm{bb}}^{\alpha\beta}(\varepsilon) = C_0 \frac{f_{\alpha\beta}}{\varepsilon_{\alpha\beta}} \sum_{\widetilde{N}_\mu = 0}^{g_\mu} P_\mu\, \varepsilon_{\alpha\beta}^{\mu}\, N_\alpha^{\mu} \left(1 - \frac{N_\beta^{\mu}}{g_\beta} \right) J_{\alpha\beta}^{\mu}(\varepsilon). \quad (5.199)$$

Here P_μ is the probability that on the shell μ there are $\widetilde{N}_\mu$ electrons ($\widetilde{N}_\mu = 0, 1, 2, \dots g_\mu$), $\varepsilon_{\alpha\beta}^{\mu} = \varepsilon_{\alpha\beta} + (\widetilde{N}_\mu - N_\mu) G_\mu^{\alpha\beta}$, N_μ is the average number of electrons on the shell μ,

$$N_q^{\mu} = \begin{cases} \widetilde{N}_{\mu}, & \text{if } q = \mu, \\ N_q, & \text{if } q \neq \mu, \end{cases}$$

$$J_{\alpha\beta}^{\mu}(\varepsilon) = \frac{1 + E_1^{\mu}/\varepsilon_{\alpha\beta}^{\mu}}{\sqrt{2\pi}\sigma_{\mu}} \int_{-\infty}^{\infty} \Phi_{\alpha\beta}(\varepsilon - \varepsilon_{\alpha\beta}^{\mu} - \xi) \exp\left[-\frac{(\xi - \Delta_{\mu})^2}{2\sigma_{\mu}^2}\right] d\xi, \tag{5.200}$$

$$\Delta_{\mu} = \frac{E_1^{\mu} + E_2^{\mu}/\varepsilon_{\alpha\beta}^{\mu}}{1 + E_1^{\mu}/\varepsilon_{\alpha\beta}^{\mu}}, \qquad \sigma_{\mu}^2 = \frac{E_2^{\mu}}{1 + E_1^{\mu}/\varepsilon_{\alpha\beta}^{\mu}},$$

$$E_1^{\mu} = \begin{cases} -G_{\beta}^{\alpha\beta} N_{\beta}/g_{\beta} & \text{if } \alpha = \mu, \beta \neq \mu, \\ G_{\alpha}^{\alpha\beta} (1 - N_{\alpha}/g_{\alpha}) & \text{if } \alpha \neq \mu, \beta = \mu, \\ G_{\alpha}^{\alpha\beta} (1 - N_{\alpha}/g_{\alpha}) - G_{\beta}^{\alpha\beta} N_{\beta}/g_{\beta} & \text{if } \alpha \neq \mu, \beta \neq \mu, \end{cases}$$

$$E_2^{\mu} = \sum_{q \neq \mu} N_q \left(1 - \frac{N_q}{g_q}\right) \left(G_q^{\alpha\beta}\right)^2.$$

Using in (5.180) the Voigt profiles for $\Phi_{\alpha\beta}$ we obtain the following expression for the effective profile:

$$J_{\alpha\beta}^{\mu}(\varepsilon) = \frac{1 + E_1^{\mu}/\varepsilon_{\alpha\beta}^{\mu}}{\sqrt{\pi(D^2 + 2\sigma_{\mu}^2)}} K\left(\frac{\varepsilon - \varepsilon_{\alpha\beta}^{\mu} - \Delta_{\mu}}{\sqrt{D^2 + 2\sigma_{\mu}^2}}, \frac{\gamma_{\alpha\beta}^{\mu}}{\sqrt{D^2 + 2\sigma_{\mu}^2}}\right). \tag{5.201}$$

Here $\gamma_{\alpha\beta}^{\mu}$ is the characteristic Lorentz width of the transition $\alpha \to \beta$ for the group of lines corresponding to the configuration $\{\widetilde{N}_{\mu}\}$. Note that instead of a shell μ one could take some group of states with close energies.

To illustrate our discussion we show in Figure 5.18 the results of computations for the transition $3p_{\frac{3}{2}} \to 4d_{\frac{5}{2}}$ in gold (for $T = 1$ keV, $\rho = 1$ g/cm^3 and $\rho = 0.1$ g/cm^3) according to formula (5.53), with a detailed accounting for configurations (with the multiplet structure accounted for in the Moszkowski approximation), using formula (5.180) (we refer to this approximation as the *effective* method) and using formula (5.199). One can see that for density $\rho = 1$ g/cm^3 the effective method (5.180) yields results that are close to those obtained by the detailed computation. By contrast, for $\rho = 0.1$ g/cm^3 formula (5.180) does not render well the results of the detailed computation, but the results provided by the sharpened formula (5.199) — the dotted curve in Figure 5.18 — practically coincide with those of the detailed computation.

5.5.5 Statistical description of the photoionization process

An approximation similar to the approximation (5.180), but designed to account for spectral lines, can be obtained also for photoionization processes [57]. To that

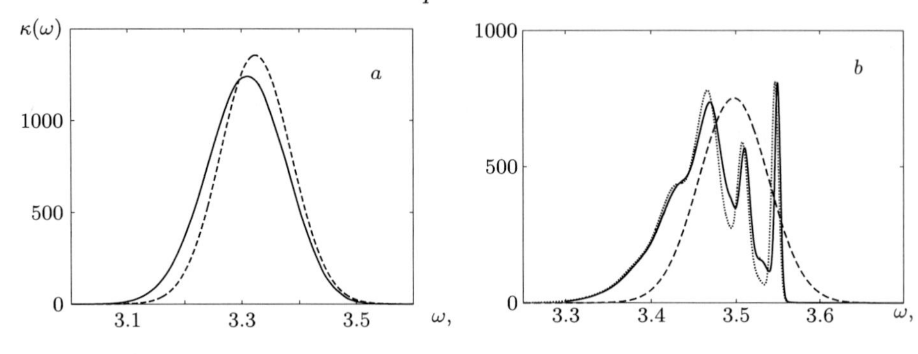

Figure 5.18: The line absorption coefficient for the transition $3p_{\frac{3}{2}} \to 4d_{\frac{5}{2}}$ in gold with detailed accounting for configurations, with the multiplet structure accounted for in the Moszkowski approximation (solid curves) and obtained via the effective method (dashed curves): (*a*) $T = 1$ keV, $\rho = 1$ g/cm^3; (*b*) $T = 1$ keV, $\rho = 0.1$ g/cm^3. The dashed curves with one maximum represent the result of the computation based on formula (5.180); the dotted curve with three maxima represents the result of the computation based on formula (5.199)

end let us write the photoionization cross-section of the one-electron process $\varepsilon_\alpha + \varepsilon \to \varepsilon'$ in the form

$$\sigma_{\text{bf}}^{\alpha\varepsilon'}(\varepsilon) = C_0 \sum_Q \sum_{\ell'} P_Q N_\alpha^Q [1 - n(\varepsilon')] f_{\alpha\varepsilon'\ell'}^Q \Phi_\alpha(\varepsilon), \tag{5.202}$$

where the function $\Phi_\alpha(\varepsilon)$ defines the contribution of various broadening processes, $n(\varepsilon) = 1/\left[1 + \exp\left((\varepsilon - \mu)/\theta\right)\right]$ and $C_0 = 4 \cdot 10^{-18}$ cm^2 (see (5.179)). Using approximations analogous to (5.181), (5.182), and also the binomial distribution, we have

$$\sigma_{\text{bf}}^{\alpha\varepsilon'}(\varepsilon) = C_0 f_{\alpha\varepsilon'\ell'} N_\alpha [1 - n(\varepsilon')] J_\alpha(\varepsilon), \tag{5.203}$$

where the profile of all photoionization thresholds of the process $\alpha \to \varepsilon'\ell'$ is described by an expression similar to (5.196):

$$J_\alpha(\varepsilon) = \frac{1 + E_1/(\varepsilon' - \varepsilon_\alpha)}{\sqrt{2\pi}\,\sigma} \int \Phi_\alpha(\varepsilon - \varepsilon_\alpha - \xi) \exp\left[-\frac{(\xi - \Delta)^2}{2\sigma^2}\right] d\xi. \tag{5.204}$$

Here

$$\Delta = \frac{E_1 + E_2/\varepsilon}{1 + E_1/\varepsilon}, \qquad \sigma^2 = \frac{E_2}{1 + E_1/\varepsilon}, \qquad \varepsilon = \varepsilon' - \varepsilon_\alpha,$$

$$E_1 = H_{\alpha\alpha}(1 - n_\alpha) - n(\varepsilon') w(\varepsilon') H_{\varepsilon'\ell',\varepsilon'\ell'},$$

$$E_2 = \sum_q g_q n_q (1 - n_q) \left[H_{\alpha q} - \sqrt{w(\varepsilon')} H_{\varepsilon'\ell',q}\right]^2.$$

The density of free electron states $w(\varepsilon)$ is introduced in order that the normalization of the continuum wave functions adopted in § 2.3 be preserved during the calculation of the matrix elements $H_{\alpha\beta}$.

To calculate the total photoionization cross-section from the level α we must integrate expression (5.203) over all final states ε' (the summation over the orbital momentum ℓ' is carried out in the resultant formula):

$$\sigma_{\mathrm{bf}}^{\alpha}(\varepsilon) = \int_0^\infty \sigma_{\mathrm{bf}}^{\alpha\varepsilon'}(\varepsilon) d\varepsilon'. \tag{5.205}$$

Since $J_\alpha(\varepsilon)$ has a sharp maximum for the value $\varepsilon' = \varepsilon^* = \varepsilon + \varepsilon_\alpha - E_1$ and the functions $f_{\alpha\varepsilon'}$ and $n(\varepsilon')$ change only slightly in the essential part of the integration domain, we have

$$\sigma_{\mathrm{bf}}^{\alpha}(\varepsilon) \cong C_0 \sum_{\ell'} f_{\alpha\varepsilon^*\ell'} N_\alpha [1 - n(\varepsilon^*)] \int_0^\infty J_\alpha(\varepsilon)\, d\varepsilon'. \tag{5.206}$$

Putting, as is customary, $\Phi(\varepsilon) = \delta(\varepsilon)$, and using the conditions $E_1/(\varepsilon' - \varepsilon_\alpha) \ll 1$, $E_2/(\varepsilon' - \varepsilon_\alpha) \ll 1$, which are satisfied almost always, the summation yields

$$\sigma_{\mathrm{bf}}^{\alpha}(\varepsilon) \cong C_0 \sum_{\ell'} f_{\alpha\varepsilon^*\ell'} N_\alpha [1 - n(\varepsilon^*)] \left[1 + \mathrm{erf}\left(\frac{\varepsilon^*}{\sqrt{2E_2}}\right)\right], \tag{5.207}$$

where

$$\mathrm{erf}(x) = \frac{2}{\sqrt{\pi}} \int_0^x \exp(-y^2) dy.$$

The statistical methods (5.180), (5.199) and (5.207) allow one, in a number of situations, to reduce the computing time without a major loss of accuracy.

5.6 Computational results for Rosseland mean paths and spectral photon-absorption coefficients

5.6.1 Comparison of the statistical method with detailed computation

The spectral absorption coefficients at temperature $T = 1$ keV and density $\rho = 0.1$ g/cm^3 for gold, calculated by means of the effective method (5.180), as well as by a detailed accounting of more than 70000 configurations using formula (5.179), are shown in Figure 5.19. One can see that the effective method renders well the behavior of the absorption coefficient in the mean, coarsening, of course, the details of the spectrum. The mean characteristics are rendered far more accurately (see Table 5.12).

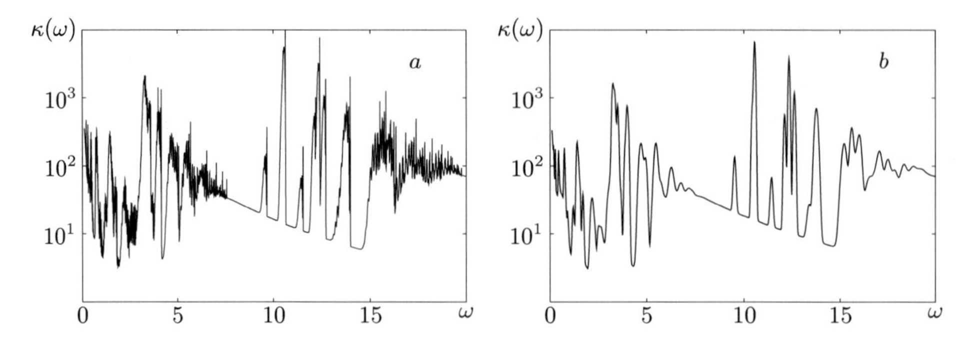

Figure 5.19: Absorption coefficient, measured in cm^2/g, for a gold plasma with temperature $T = 1$ keV and density $\rho = 0.1$ g/cm^3: (a) for a detailed configuration accounting ($\ell = 0.422$ cm), (b) obtained by the effective method ($\ell = 0.510$ cm)

We must emphasize that as the plasma density increases, accounting for configurations becomes considerably more complicated, because the fluctuation width of the distribution over the degrees of ionization increases (see Figure 5.7). However, as has to be expected, the accuracy of the description of the absorption spectrum by means of the effective method, as a rule, increases.

The absorption coefficient for $T = 1$ keV and $\rho = 1$ g/cm^3 is shown in Figure 5.20. The detailed computation required to account for more than 10^6 configurations in order that the absorption coefficient reached saturation and approached the results of the computation using the effective method (the detailed

Table 5.12: Absorption coefficients in a gold plasma: Rosseland mean (κ_R) and Planck mean (κ_P), in cm^2/g. The computations were carried out with a detailed accounting for configurations and by using the effective method (5.180)

		κ_R		κ_P	
T	ρ	Detailed	Effective	Detailed	Effective
0.1	0.1	3710	4760	7806	7723
	1	5927	7112	9288	9346
	10	8274	7668	10301	10050
0.316	0.1	678	674	1507	1473
	1	1218	1290	2410	2467
	10	1555	1764	3156	3489
1	0.1	23.7	19.6	157	149
	1	137	130	449	438
	10	274	282	791	805
3.16	0.1	0.36	0.37	2.63	2.79
	1	2.00	1.61	18.8	17.9
	10	14.6	10.9	66.8	65.7

computation required roughly a 1000 times more time). Compared with the case $T = 1$ keV, $\rho = 0.1$ g/cm^3, here the lines are considerably wider and the result produced by the effective method is much closer to that of the detailed calculation: $\ell_{\text{det}} = 0.0073$ cm, $\ell_{\text{eff}} = 0.0077$ cm.

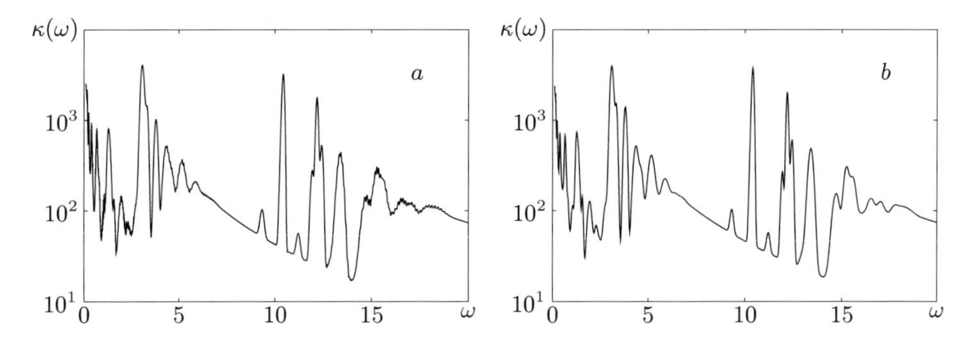

Figure 5.20: Absorption coefficient, measured in cm^2/g, for a gold plasma with temperature $T = 1$ keV and density $\rho = 1$ g/cm^3: (a) for a detailed configuration accounting ($\ell = 0.0073$ cm), (b) obtained by the effective method ($\ell = 0.0077$ cm)

Therefore, computations based on the effective method, on the one hand, give an acceptable approximate result (especially for mean values), and on the other hand, they yield a limit estimate of the absorption coefficient in spectral lines in the presence of a very large influence of the splitting and broadening effects (in this connection see [192]). Let us mention that the differences between the results of the detailed computation and the effective method seen in Table 5.12 are connected mainly with the different way in which these methods account for photoionization effects.

To illustrate the application of the effective method (5.207), we give here results of the computation of the photoionization cross-sections for $T = 3$ keV, $\rho = 1.9$ g/cm^3 for gold and for $T = 20$ eV, $\rho = 10^{-4}$ g/cm^3 for iron. For gold the shift E_1 is not very important (being of several atomic units), while the dispersion (the quantity E_2 in formula (5.204)) leads to a considerable smoothing of the behavior of the photoionization cross-section near the threshold. This is clearly seen in Figure 5.21, which shows the photoionization cross-section obtained with a detailed configuration accounting (768 ion states were accounted for), as well as by means of the effective method, for the level $n = 2$, $\ell = 0$, $j = 1/2$.

As one can see in Figure 5.21, the calculation of the photoionization cross-section without accounting for configurations gives a rather crude result, whereas the effective method agrees well with the detailed computation. Moreover, by increasing the number of configurations considered in the detailed computation and accounting for the multiplet structure of the spectrum and other broadening and splitting effects one can apparently reach an even better agreement.

Figure 5.21 makes clear also that accounting for configurations leads to quali-

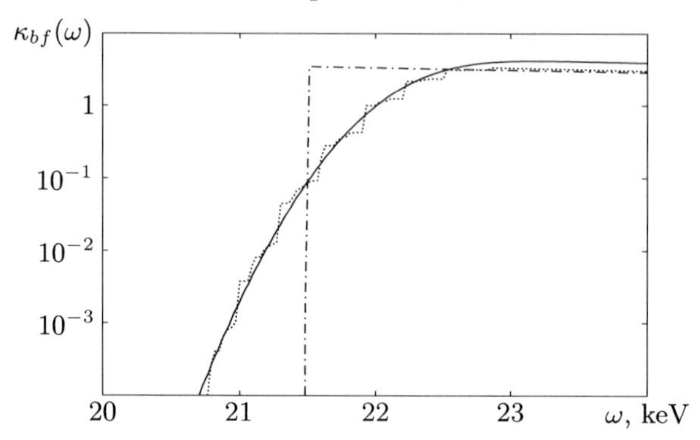

Figure 5.21: The absorption coefficient (in cm^2/g) under photoionization of the level $2s_{1/2}$ in a gold plasma with temperature $T = 3$ keV and density $\rho = 1.9$ $\mathrm{g/cm}^3$, calculated with various methods of accounting for configurations: the hatched-dotted line with one step was obtained in the average-atom approximation, the dotted line with steps—by the detailed account of ion configurations, and the continuous line—by the effective method

tative changes in the absorption spectrum compared with the average-atom model. Note also the quantitative changes in the average characteristics, in particular, accounting for fluctuations of photoionization thresholds may result in a change of $\sim 20\%$ in the Rosseland free path (spectral lines where not included in the computations). At lower temperatures the divergence of the results obtained in the average-atom approximation from those furnished by the detailed computation becomes even more notable.

Figure 5.22 displays the results of the computation of photoionization cross-section for iron at temperature $T = 20$ eV and density $\rho = 10^{-4}$ $\mathrm{g/cm}^3$, calculated by means of a detailed configuration accounting according to formula (5.202), as well as by means of the effective method (5.207). Here the result obtained by means of the average-atom approximation differs by a factor of 2 from that furnished by the detailed computation, whereas the effective method (5.207) agrees well with the detailed computation.

The results of computations based on formula (5.199) are compared with experimental data [228] for a holmium plasma ($Z = 67$) in Figure 5.23. The figure shows the spectral transmission coefficient, which has calculated by means of the formula

$$T(\omega) = e^{-\kappa(\omega)\rho L}, \tag{5.208}$$

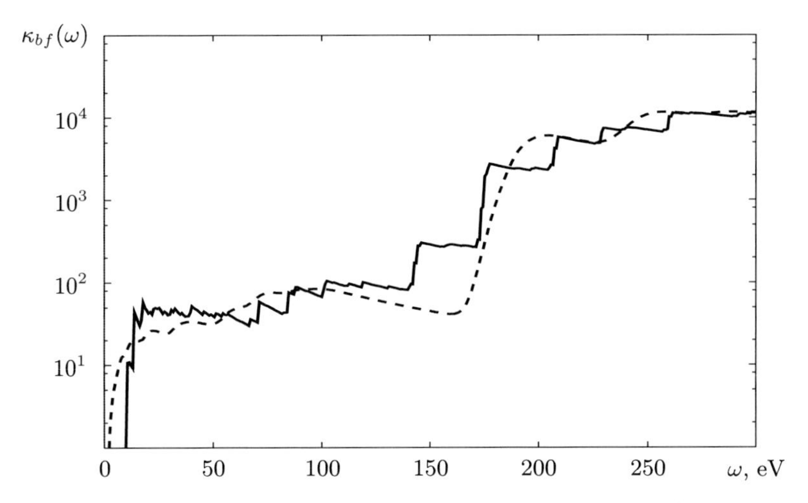

Figure 5.22: The absorption coefficient (measured in cm^2/g) under photoionization of an iron plasma at temperature $T = 20$ eV and density $\rho = 10^{-4}$ g/cm^3, calculated with various ways of accounting for configurations: detailed approach (solid curve); effective accounting (dashed curve)

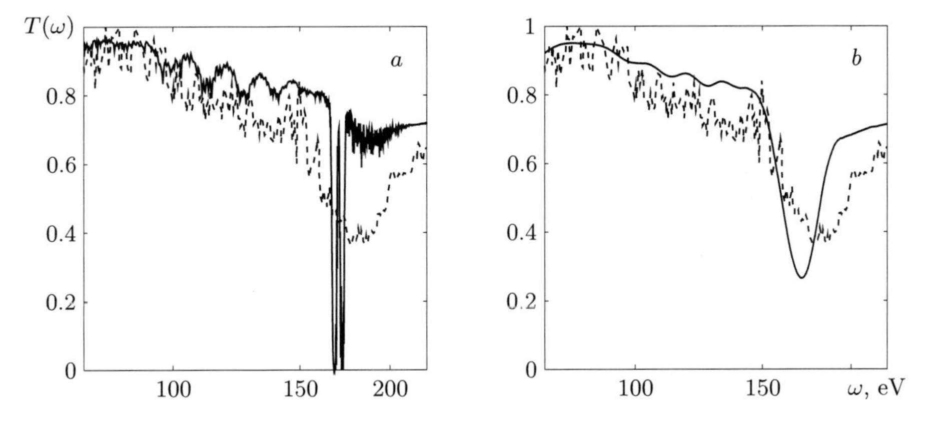

Figure 5.23: Transmission coefficient in a holmium plasma at temperature $T = 20$ eV and density $\rho = 0.03$ g/cm^3 (the thickness of the plasma layer is $L = 5 \cdot 10^{-4}$ cm) as a function of the photon energy in eV (solid curves): (a) neglecting the multiplet structure; (b) with the multiplet structure of the spectrum accounted for in the Moszkowski approximation and with the fluctuations of the occupation numbers accounted for by means of formula (5.199). Experimental data [228] (dashed curve) are also provided for each of the two cases (a) and (b)

where $\kappa(\omega)$ is the spectral absorption coefficient, ρ is the plasma density, and L is the thickness of the plasma layer (for the holmium plasma $L = 5 \cdot 10^{-4}$ cm). One can see that computations based on formula (5.199) yield satisfactory agreement with the experiment. At the same time, computations that neglect the multiplet structure of the spectrum do not agree with the experimental results in the domain of strong lines.

5.6.2 Dependence of the absorption coefficients on the element number, temperature and density of the plasma

The computation of absorption coefficients and free paths of photons (spectral, as well as Rosseland means) are carried out by using various codes (programs). Brief information about such codes and the approximations adopted in them can be found in papers and reports of international workshops on *opacity* computations [247, 241, 242]. Therein there are also given computational results for a number of elements and mixtures and comparisons with experimental data (see [179, 227]). Let us remark that the successes achieved to this date in the development of experimental methods as well as the utilization of multiprocessor supercomputers allow one to make major advances in the theoretical description of radiative properties of matter (this is demonstrated by a whole series of recently published works; see, e.g., [172, 207, 208, 228, 1, 126, 185, 51, 39, 84, 11, 69], as well as the special issue *Radiative Properties of Hot Dense Matter* of the Journal of Quantitative Spectroscopy and Radiative Transfer, vol. 81, no. 1–4, 2003.

Based on the codes developed, databases (libraries) of photon absorption coefficients were and continue to be created (see Subsection 6.5.6). Two well-known such data bases are OPLIB (Lawrence Livermore National Laboratory) and SESAME (Los Alamos National Laboratory), both in the US. A similar data base and the code THERMOS[*)] were developed at the M. V. Keldysh Institute of Applied Mathematics in Moscow. THERMOS makes it possible to describe radiative and thermodynamic properties of various substances and mixtures in a wide range of temperatures and densities. The reader can find the models and algorithms implemented in the code THERMOS in this book.

The computation of detailed tables imposes special requirements on the codes used for that purpose, namely, the codes must be not only precise, but also sufficiently fast, in order to be able to carry out the computation of a large number of cases with respect to temperature and density of matter. To ensure precision it is

---

[*)] The name THERMOS stands for:

T	—	Thomas–Fermi model
H	—	Hartree, Hartree–Fock, Hartree–Fock–Slater self-consistent field models
E	—	Equation of state
R	—	Rosseland mean free path
M	—	Mixture of elements
O	—	Opacity calculations
S	—	Storage of data

necessary to accurately compute each of the individual processes discussed in the present chapter, since any of them may prove to be determining factor in some range of temperatures and densities. Along with the approximations employed, an important role is played by the choice of code parameters, such as the number of shells accounted for, the amount of ions accounted for, the computational meshes, etc.

Table 5.13: Influence of the number of shells accounted for, $n_{\max}$, on the absorption coefficients in a gold plasma with density $\rho = 0.1$ g/cm^3 and temperature $T =$ 1 keV (also shown is the corresponding average charge of the ion, Z_0)

$n_{\max}$	Z_0	κ_R, cm^2/g	κ_P, cm^2/g
6	66.47	19.4	155
8	66.41	23.1	157
9	66.39	23.7	157
10	66.36	24.3	157
12	66.28	24.5	157
14	66.20	24.5	157

Since the most computationally-demanding process is accounting for absorption in spectral lines, the main parameter that determines the time required for computations is the number of electron shells accounted for. Note that actually the amount of ions (configurations) accounted for has a larger influence on the computing time; however, the effective methods considered in § 5.5, allow one to simplify the calculation process in the most complex situations. Table 5.13 lists results of computations of the Rosseland mean free path in gold with different numbers n of shells accounted for, where $n \le n_{\max}$, $n_{\max} = 6 \div 14$. The computing times for a detailed configuration accounting with $n_{\max} = 6$ and $n_{\max} = 14$ differ by a factor of 15. This clearly demonstrates that, in order to achieve a high efficiency and accuracy of codes, it is necessary to enrich the computational procedures by various physical models and methods. The universality of the codes allows one to compute in a wide range of temperatures and densities, and, by comparing the results obtained by different methods, to ensure their reliability.

The values of the averaged absorption coefficients in the state of local thermodynamic equilibrium are determined by the temperature, density and composition of matter. In the region of high temperatures and relatively small densities, when matter is almost fully ionized, the determining role is played by the scattering processes (5.62). Under such conditions the following simple formula for the Rosseland mean free path, measured in cm, holds:

$$\ell = 2.5 \frac{A}{\rho Z_0}, \qquad (5.209)$$

where Z_0 is the average charge of the ion, A is its atomic weight, and ρ is the density in g/cm^3.

If for high temperatures one increases the plasma density, then inverse bremsstrahlung processes become the dominant factor. For a fully ionized dense plasma the corresponding formula for the Rosseland free path in the Kramers approximation (5.228) is (see [234]):

$$\ell = 70\,\frac{A^2 T^{7/2}}{\rho^2 Z_0^3}, \tag{5.210}$$

where T is the temperature in keV.

Formulas (5.209), (5.210) are valid when the influence of the electrons in bound states can be neglected. As a rule, the dependence of the absorption coefficients on temperature and density has a more complex character. Usually a considerable influence on absorption coefficients is exerted by spectral lines, and the photoionization processes are also rather important.

A comparison of the photoionization cross-sections obtained by means of the codes THERMOS, OPAL and STA for an iron plasma at temperature $T = 20$ eV and density $\rho = 10^{-4}$ g/cm^3 is given in Figure 5.24. As one can see, the more detailed accounting for the photoionization thresholds made in the OPAL code compared with the THERMOS and STA codes (in OPAL the energy levels are calculated in the approximation of an intermediate coupling scheme) results in a smoothing of the photoionization thresholds.

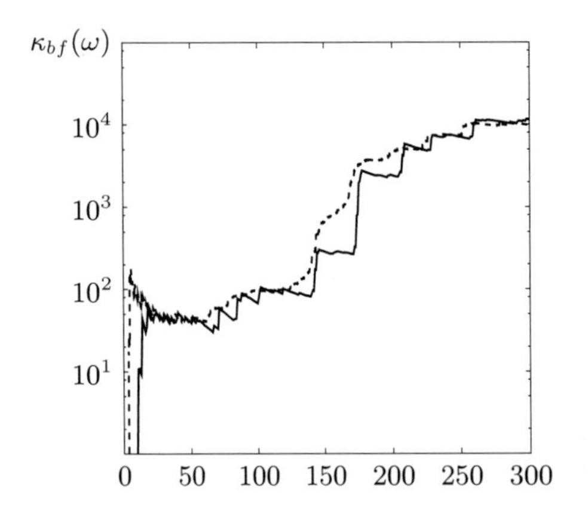

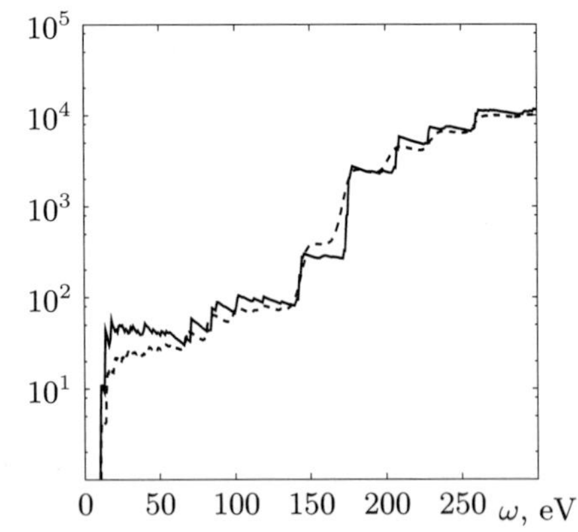

Figure 5.24: The absorption coefficient (measured in cm^2/g) under photoionization in an iron plasma at temperature $T = 20$ eV and density $\rho = 10^{-4}$ g/cm^3 with detailed configuration accounting, computed by the THERMOS code (solid curve), the OPAL code (dashed curve—left figure) and the STA code (dashed curve—right figure)

The spectral dependence of the absorption coefficients $\kappa(\omega)$ with all processes considered in § 5.2 accounted for, is illustrated in Figure 5.25 for molybdenum ($Z = 42$) with density $\rho = 0.1$ g/cm^3. As the figure demonstrates, the spectral dependence of the absorption coefficient for molybdenum in the indicated range is determined by the processes of photoionization and absorption in the spectral lines. The influence of these processes on the Rosseland mean free path depends in turn of the role they play near the maximum of the Rosseland weight function. Note that the most important contribution to the Rosseland mean free path comes from the domain with minimal opacity (see Figure 5.26, which shows the subintegral curves for the computation of Rosseland mean free paths for molybdenum).

The role of the spectral lines is clearly seen in Figure 5.27, which compares the absorption coefficients with, as well as without the spectral lines accounted for, for the relatively light element aluminum ($Z = 13$) and for gold ($Z = 79$). It is evident that the influence of bound-bound transitions depends on the occupation numbers of the electron shells. As an analysis reveals, the domain where the lines exert maximal influence correspond to an ionization such that on the first nonfilled shell one finds only roughly half of the electrons (see also Table 5.14). Thus, for example, in the case of gold with $T = 1$ keV, $\rho = 1$ g/cm^3, the level with number $n = 3$ hosts about half of the maximum possible number of electrons.

Let us mention that in a substantial range of temperatures and densities the spectral lines fill densely certain parts of the spectrum and their profiles overlap essentially due to fluctuations of the occupation numbers and multiplet splitting effects. As a consequence, the results of the calculation of Rosseland free paths are usually little sensitive to the magnitude of electron width and the method used to account for the ion broadening (Stark effect). In particular, when the electron widths are doubled in this domain, the mean free path changes only by a few percent.

Table 5.14: Average ion charge Z_0 of gold at different temperatures and densities

ρ, g/cm^3	T, keV			
	0.1	0.316	1.0	3.16
10^{-4}	32.655	52.906	69.522	77.000
10^{-3}	29.688	51.034	69.021	76.998
10^{-2}	26.357	48.648	68.573	76.996
10^{-1}	22.282	43.920	66.400	76.809
10^{0}	17.868	36.813	60.877	75.601
10^{1}	16.095	30.414	52.124	71.329
10^{2}	19.119	29.098	45.644	65.198
10^{3}	43.001	43.547	47.811	58.950
10^{4}	61.000	61.000	61.052	63.067

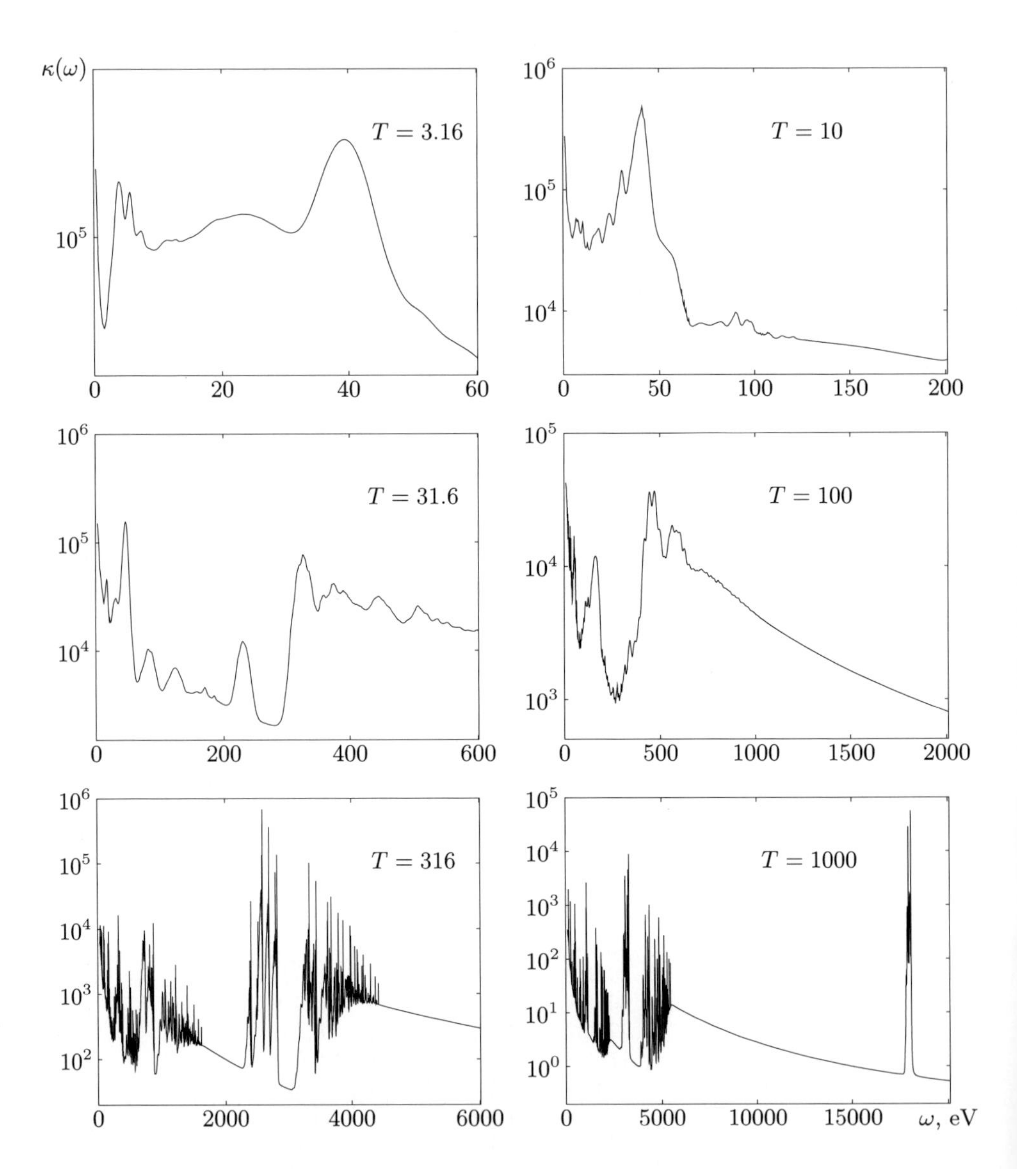

Figure 5.25: Absorption coefficients (in cm^2/g) in a molybdenum plasma for the energy interval $0 < \hbar\omega < 20T$, for different temperatures T in eV and density $\rho = 0.1$ g/cm^3

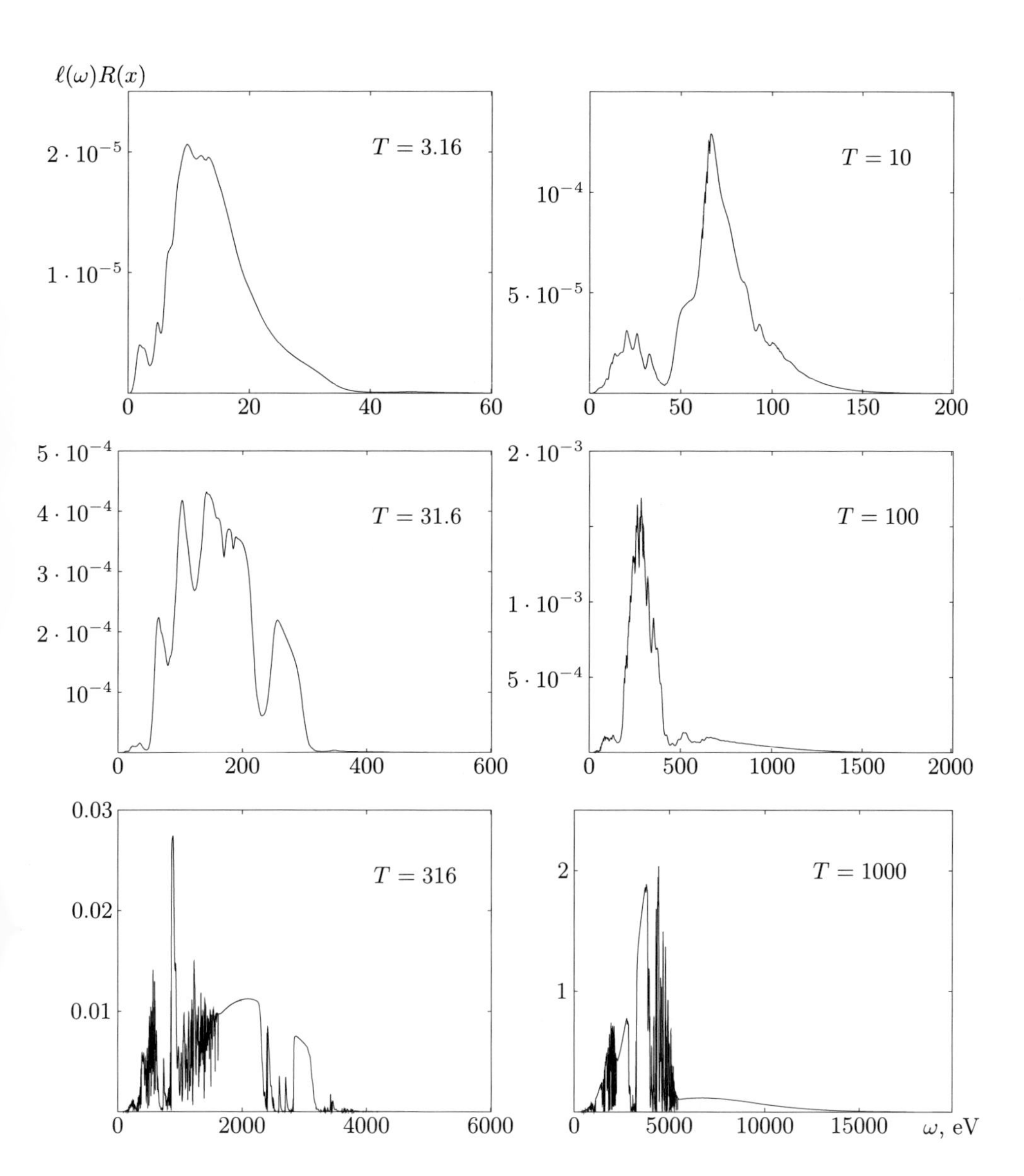

Figure 5.26: Photon free path $\ell(\omega)$ (in cm), multiplied by the Rosseland weight function $R(x)$, $x = \hbar\omega/T$, in a molybdenum plasma for the energy interval $0 < \hbar\omega < 20T$, for different temperatures T in eV and density $\rho = 0.1$ g/cm^3 (compare with Figure 5.25)

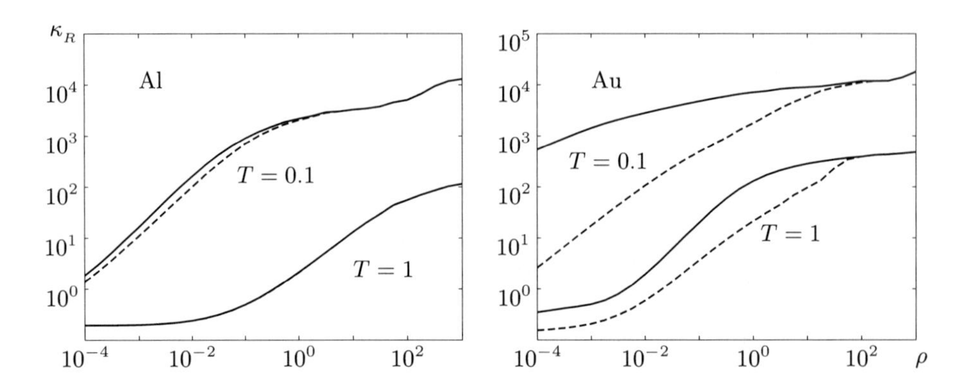

Figure 5.27: Dependence of the absorption coefficients (opacity $\kappa_{\mathrm{R}} = 1/(\rho\ell)$) on the density ρ (in g/cm^2), in an aluminum plasma and a gold plasma, with the spectral lines accounted for (solid curves) and not accounted for (dashed curves), at the temperatures $T = 0.1$ keV and $T = 1$ keV

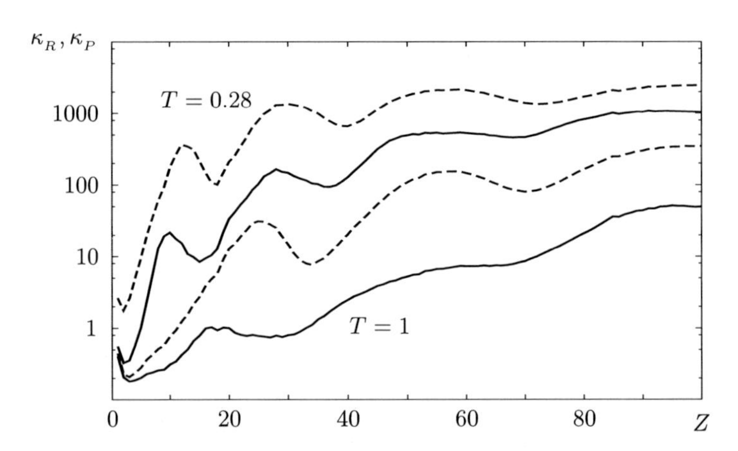

Figure 5.28: The Rosseland mean (solid curves) and Planck mean (dashed curves) absorption coefficients, in cm^2/g, as functions of the element number Z, for $T =$ 1 keV, $\rho = 0.1$ g/cm^3 and $T = 0.28$ keV, $\rho = 0.05$ g/cm^3

The dependence of the absorption coefficients on the element number is illustrated in Figure 5.28, which displays the Rosseland and Planck mean absorption coefficients for two cases with respect to the temperature and density ($T = 1$ keV, $\rho = 0.1$ g/cm^3 and $T = 0.28$ keV, $\rho = 0.05$ g/cm^3).

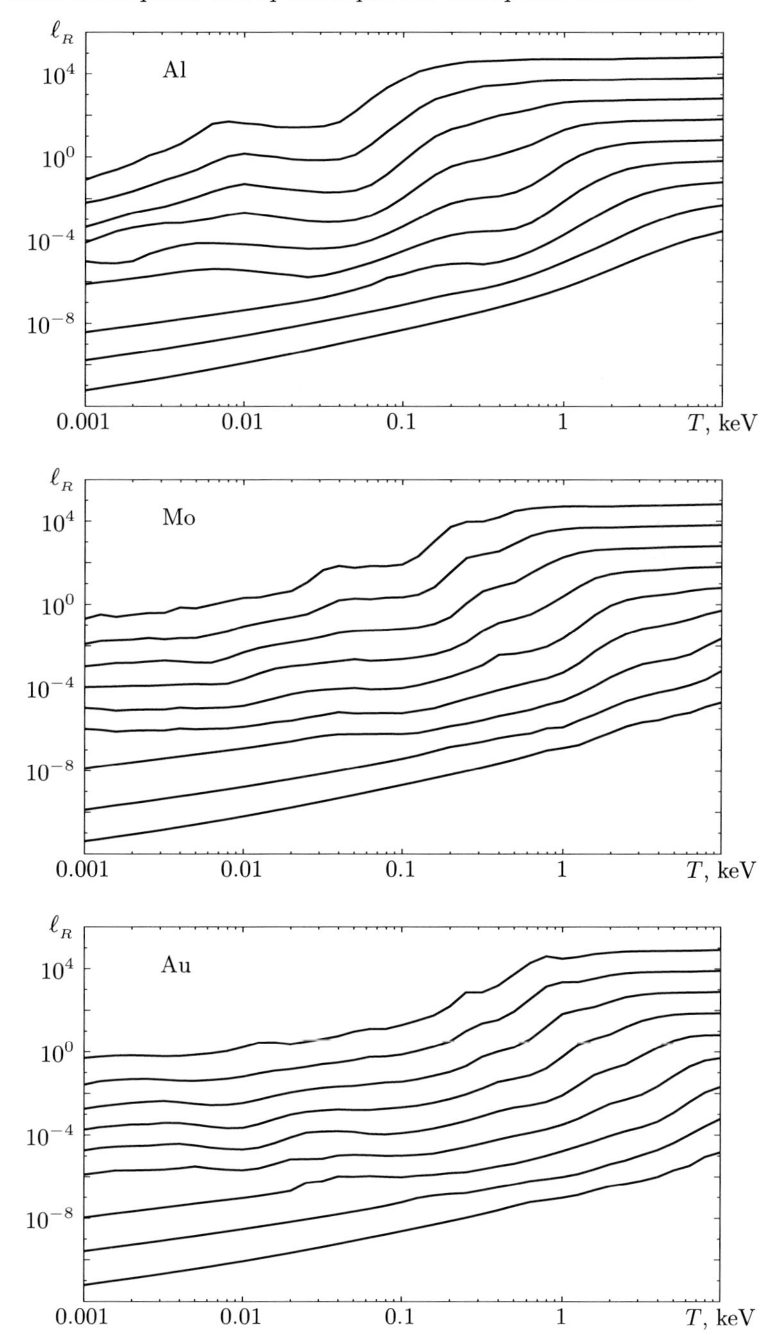

Figure 5.29: Dependence of the Rosseland mean free path ℓ_{R} (in cm) for aluminum, molybdenum, and gold plasmas on temperature T (in keV). The curves are calculated for different densities $\rho = 10^k$ g/cm^3, where $k = -4, -3, -2, -1, 0, 1, 2, 3, 4$

Figure 5.29 on the preceding page shows the Rosseland mean free paths for aluminum ($Z = 13$), molybdenum ($Z = 42$) and gold ($Z = 79$) plasmas as functions of the temperature T (in keV) for densities ρ from 10^{-4} g/cm^3 up to 10^4 g/cm^3; the density changes by one order of magnitude from one curve to the next (the free path decreases when the density increases). The character of the dependence of the absorption coefficient $\kappa = 1/(\rho\ell)$ on temperature and density depends in turn on the degree of ionization of the electron shells. At beginning of the ionization of each shell the coefficient decreases, but later, when the shell becomes almost half ionized, the absorption coefficient may increase notably due to the increasing influence of spectral lines.

A comparison of the computational results for Rosseland mean free paths at density $\rho = 0.1$ g/cm^3 in aluminum ($Z = 13$), copper ($Z = 29$) and europium ($Z = 63$) carried out with the code THERMOS and the SESAME database [245], published in [215] (in graphic form) was made in [162]. This comparison has shown that at temperatures $T > 100$ eV the differences between the two compared results are very small (the graphs are indistinguishable). At lower temperatures, where the main contribution to the Rosseland integral is due to photoionization, some differences are observed [162].

Experimental values of the absorption coefficient κ_R in aluminum and iron for density $\rho \simeq 1$ g/cm^3 and temperatures from 0.5 keV to 1.25 keV are published in [12, 13]. Tables 5.15–5.16 compare those data with results of computations done in the Thomas-Fermi model [147, 150] and in the Hartree-Fock-Slater model (using the code THERMOS), and also with results of computations using the codes LEDCOP, OPAL, HOPE and STA [241].

Table 5.15: Rosseland mean absorption coefficient κ_R, in cm^2/g, for aluminum

T, keV	0.5	0.75	1.0	1.25
Thomas-Fermi	40.5	9.09	2.09	0.83
THERMOS	46.1	10.3	2.14	0.83
OPAL	59.3		2.13	
HOPE	53.1		1.92	
STA	52.5		2.08	
Experiment [12, 13]	51 ± 8	13 ± 2	2.9 ± 0.4	1.1 ± 0.2

Table 5.16: Rosseland mean absorption coefficient κ_R, in cm^2/g, for iron

T, keV	0.5	0.75	1.0	1.25
Thomas-Fermi	62.5	7.81	2.70	1.59
THERMOS	79.6	8.37	3.14	2.29
LEDCOP	74.8		2.79	
OPAL	84.2		3.48	
HOPE	78.2		3.56	
STA	71.8		2.74	
Experiment [12, 13]	82 ± 12	7.8 ± 1	2.3 ± 0.4	1.3 ± 0.2

Table 5.17: The absorption coefficient κ_R in cm^2/g for gold

T	ρ	[34]	[162]	THERMOS detailed	THERMOS effective
0.316	19.3	1731	2116	2013	1901
0.316	1.93	1323	1433	1276	1447
0.75	19.3	526	552	545	624
0.75	1.93	260	329	312	374
1	19.3	245	308	287	326
1	1.93	104	167	145	145

Note that there is reasonably good agreement with experimental data for $T \leq 0.75$ keV within the limits of experimental errors, whereas for temperatures higher than 1 keV there are notable differences. On the other hand, the results of the computations done with the different codes agree well with one another. Since the data in [12] were obtained by processing computations using certain codes for radiative gas dynamics, further analysis and processing of those data by incorporating new results are apparently necessary.

For a dense hot plasma of heavy elements only few theoretical values of the absorption coefficients are known and those are rarely found in the literature. The most complete and accurate data on opacity computations are provided by the SESAME database at the Los Alamos National Laboratory in the US, though they are not readily accessible to all researchers. Among the published results there are data for gold [139, 33, 34], lead [32] and europium [215, 88]. Table 5.17 compares results obtained by using the code THERMOS with the data for gold published in [34].

Also shown in the table are data given in [162], obtained earlier without accounting for the multiplet structure of the spectrum, but using a variant of the effective method (5.199). One can see that the various approximations considered yield close results (differences are smaller than 20 %), except for the case $T = 1$ keV, $\rho = 1.93$ g/cm^3, for which the THERMOS results are 1.5 larger than the data in [34].

5.6.3 Spectral absorption coefficients

Figures 5.30–5.31 compare computed and experimental values of the transmission function for a germanium plasma ($Z = 32$, $T = 76$ eV, $\rho = 0.05$ g/cm^3) and an iron plasma with a 19.8 % NaF admixture ($T = 59$ eV, $\rho = 0.013$ g/cm^3). The thickness of the plasma layer is $L = 0.0032$ cm for germanium and $L = 0.03$ cm for the iron with the NaF admixture. Satisfactory agreement between the THERMOS results and experimental data is observed.

Next, figures 5.32 and 5.33 compare results obtained by means of the code THERMOS (with detailed accounting for configurations and accounting for the term structure in the Moszkowski approximation) with results provided by the OPAL and STA codes [241, 242]. As indicated earlier (see Subsection 5.5.5), the codes THERMOS, OPAL and STA employ different models and approximations in the calculation of energy levels of ions and of wave functions, as well as in that of cross-sections of radiation-matter interaction. Comparisons of the different methods allow one to draw conclusions about the degree of uncertainty of the data and on their reliability.

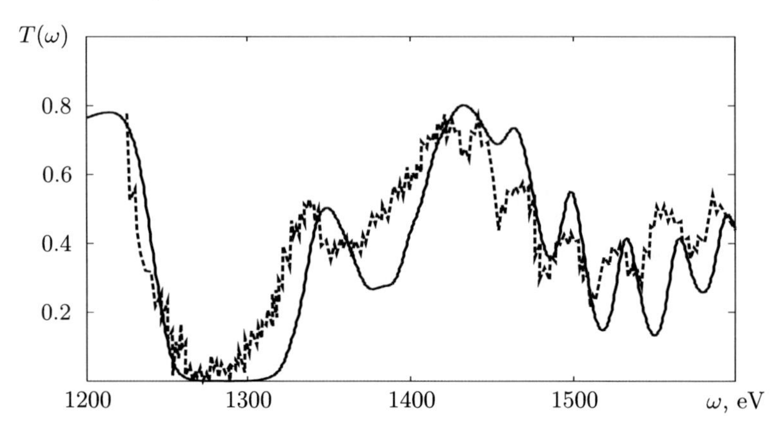

Figure 5.30: Transmission of a germanium plasma at temperature $T = 76$ eV and density $\rho = 0.05$ g/cm^3 as a function of the photon energy (in eV), obtained by means of the code THERMOS (solid curve) and from experimental data [66] (dashed curve)

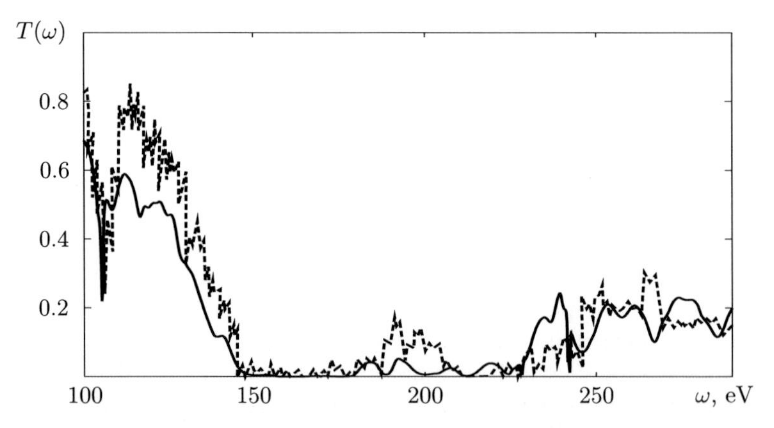

Figure 5.31: Transmission of an iron plasma with an NaF admixture at temperature $T = 59$ eV and density $\rho = 0.013$ g/cm^3 as a function of the photon energy (in eV), obtained by means of the code THERMOS (solid curve) and from experimental data [207] (dashed curve)

Modern information technologies allow researchers to create and amass databases of spectral and mean photon absorption coefficients for various elements and mixtures (cf. Subsection 5.6.5). Such databases are necessary for solving many problems of radiative gas dynamics and in the design of new technologies and devices.

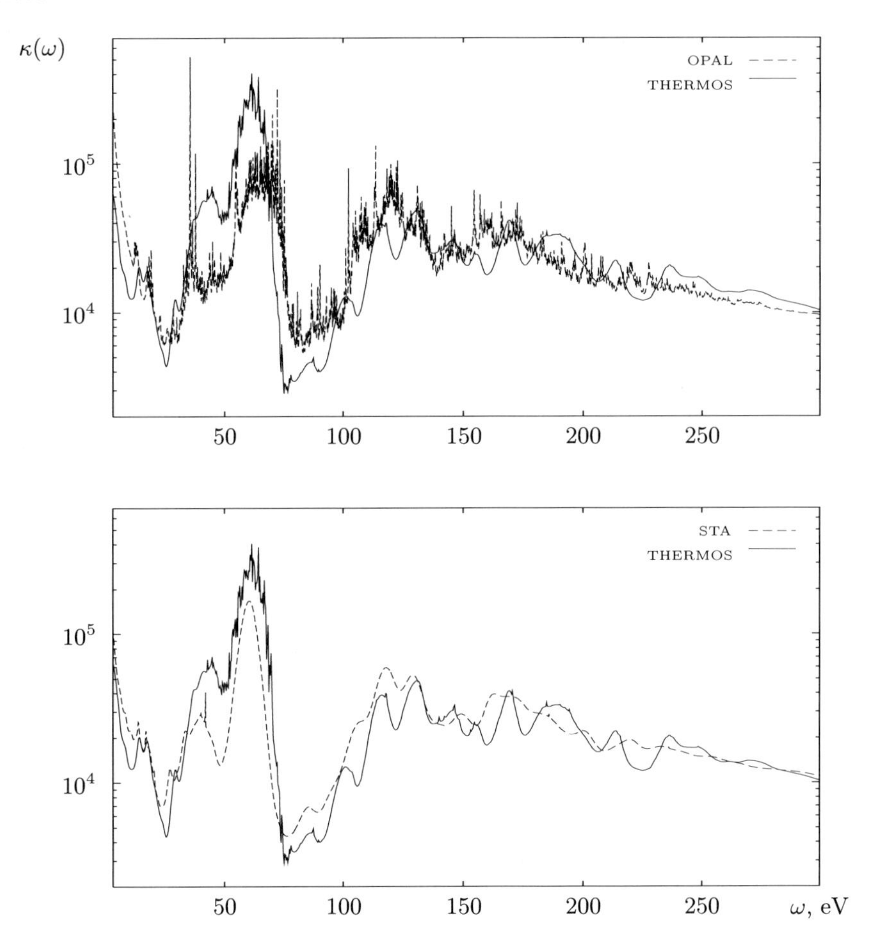

Figure 5.32: Comparison of spectral absorption coefficients in a Fe plasma at temperature T= 31.7 eV and density $\rho = 0.01$ g/cm^3, obtained with the THERMOS code (solid curve, $\kappa_{\rm R} = 1.565 \cdot 10^4$), the OPAL code (upper dashed curve, $\kappa_{\rm R} = 1.734 \cdot 10^4$) and the STA code (lower dashed curve, $\kappa_{\rm R} = 1.640 \cdot 10^4$)

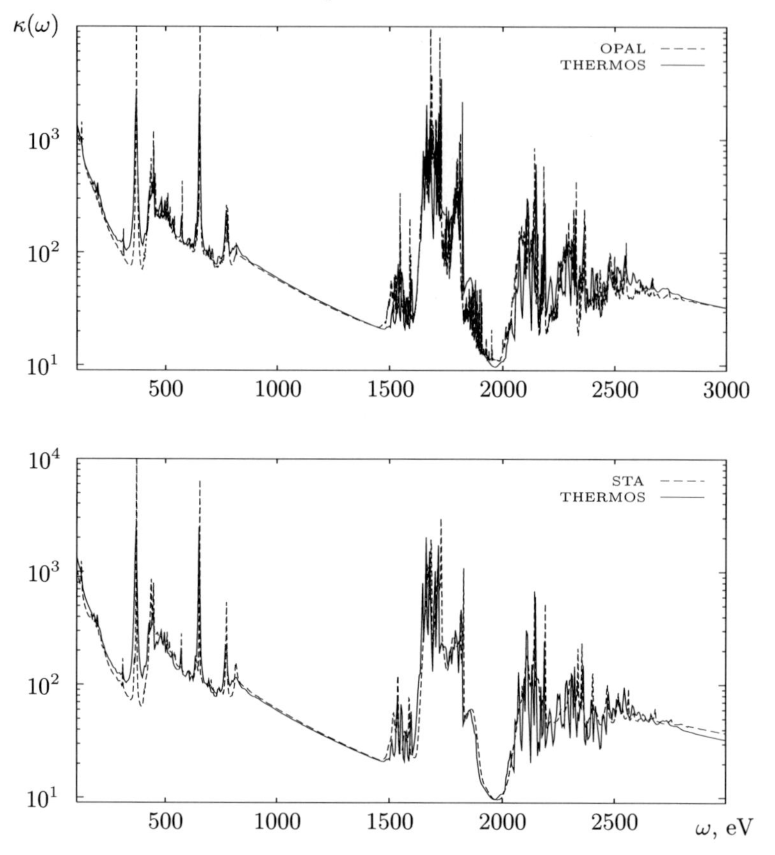

Figure 5.33: Spectral absorption coefficients in a $C_{190}H_{190}O_{20}Br$ plasma (agar with an admixture of bromine) at temperature $T=$ 270 eV and density $\rho = 0.2$ g/cm^3, computed with the THERMOS code (solid curve, $\kappa_R = 45$) and with the OPAL code (upper dashed curve, $\kappa_R = 45$) and the STA code (lower dashed curve, $\kappa_R = 46$)

As an example, let us consider a problem of major importance in the production of X-ray sources [47]. For such a source to be efficient it is necessary that it converts the maximal amount of energy injected into radiation. Such conversion requires that the materials used enjoy certain properties; in particular, in a number of problems a critical parameter is the free path of photons. To achieve a highly efficient conversion one needs a material that for specified temperature and density has minimal free path. Using databases of spectral absorption coefficients one can analyze the behavior of these coefficients as functions of the element number Z (see Figure 5.28).

As we have seen in a number of examples, the photon absorption spectrum is specific to each element. At high temperatures, for high-Z elements a basic role is played by absorption in spectral lines, which are arranged in sufficiently narrow groups with alternating windows of transparency (domains of the spectrum for which the free path is maximal). As a consequence, the transport of radiation for different elements takes place in different domains of the spectrum, though the radiative heat conductivities of elements whose numbers Z are close to one another are almost identical. By a special choice of elements with a specific Z one can close transparency windows and obtain a substance with minimal free path.

Let us consider a mixture of an arbitrary number of elements, with mass fractions m_i $(i = 1, 2, \ldots, n)$. For such a substance the Rosseland mean free path ℓ is given by the formula

$$\ell = \frac{1}{\rho\,\kappa_{\mathrm{R}}} = \int_0^\infty \frac{R(x)}{\rho \sum\limits_{i=1}^{n} m_i \kappa_i(x)}\,dx, \tag{5.211}$$

where $x = \dfrac{\hbar\omega}{kT}$,

$$R(x) = \frac{15}{4\pi^4} \frac{x^4 e^{-x}}{(1 - e^{-x})^2}$$

is the weight function, and $\kappa_i(x)$ is the spectral absorption coefficient of the i-th component of the mixture. In formula (5.211) for the mixture it is assumed that the dependence of the spectral absorption coefficients on the partial (intrinsic) densities of the elements in the mixture is weak. Therefore, one makes the assumption that approximately the partial densities of all components are equal to the matter density ρ.

We are required to find such m_i $(\sum_{i=1}^{n} m_i = 1)$ that the value of ℓ will be minimal. The conditions for minimum read

$$\frac{\partial}{\partial m_i}\left(\frac{1}{\kappa_{\mathrm{R}}}\right) = 0, \quad i = 1, 2, \ldots, n-1. \tag{5.212}$$

Since $m_n = 1 - \sum_{i=1}^{n-1} m_i$, one can recast (5.212) as

$$\int_0^\infty \frac{\kappa_i(x) - \kappa_n(x)}{\kappa^2(x)} R(x)\,dx = 0, \tag{5.213}$$

where $\kappa(x) = \sum m_i \kappa_i(x)$.

Multiplying (5.213) by m_i and summing over i, we obtain a relation for $i = n$, which together with equations (5.213) for $i = 1, 2, \ldots, n-1$ yields a system of n nonlinear equations for the determination of the mass fractions m_i:

$$\int_0^\infty \frac{\kappa_i(x) R(x)}{\left(\sum_k m_k \kappa_k(x)\right)^2}\,dx = \int_0^\infty \frac{R(x)}{\sum_k m_k \kappa_k(x)}\,dx, \quad i = 1, 2, \ldots, n. \tag{5.214}$$

We will solve this system by an iteration method, setting

$$m_i^{(s+1)} = m_i^{(s)} + \Delta m_i. \tag{5.215}$$

Note that the condition $\sum m_i = 1$ will be satisfied whenever equations (5.214) hold. Substituting (5.215) in (5.214), we obtain, in the linear approximation, a system of linear equations for the unknowns Δm_i:

$$\sum_{k=1}^{n} A_{ik} \Delta m_k = b_i, \tag{5.216}$$

where

$$A_{ik} = \int_0^\infty \frac{[2\kappa_i(x) - \kappa(x)]\kappa_k(x)}{\kappa^3(x)} R(x)dx, \quad b_i = \int_0^\infty \frac{\kappa_i(x) - \kappa(x)}{\kappa^2(x)} R(x)dx.$$

As one can see in Figure 5.28, to minimize the Rosseland mean free path we must use elements with $Z \geq 40$. We will assume that for these elements the spectral absorption coefficients $\kappa_i(x)$ are known. To solve the resulting system of ~ 40 equations for $Z = 40, 41, \ldots, 83$ (radioactive elements are excluded) several iterations according to formulas (5.215), (5.216) are needed.

Table 5.18: Mass fractions m_i of the different elements for a compound with minimal Rosseland mean free path and the corresponding Rosseland mean absorption coefficients $\kappa_R = 1/(\rho\ell)$, in cm^2/g

$T = 0.28$ keV, $\rho = 0.05$ g/cm^3			$T = 1$ keV, $\rho = 0.1$ g/cm^3		
Element	m_i	κ_R	Element	m_i	κ_R
Ce	0.097	529.4	Ce	0.116	7.1
Nd	0.074	528.0	Eu	0.072	7.4
Eu	0.103	508.5	Gd	0.072	7.2
Ho	0.081	460.4	Ho	0.063	7.5
Yb	0.080	462.0	Lu	0.108	9.2
Hf	0.062	500.8	W	0.090	11.8
Re	0.076	614.1	Au	0.167	19.1
Au	0.115	785.6	Tl	0.083	23.0
Tl	0.096	851.0	Bi	0.228	28.5
Bi	0.216	922.0			
Mixture		1512.6	Mixture		78.1

Table 5.18 lists mass fractions m_i of the different elements for the resulting compound with minimal Rosseland mean free path and the corresponding Rosseland mean absorption coefficients κ_R. The computations were carried out for two cases: $T = 0.28$ keV, $\rho = 0.05$ g/cm^3 and $T = 1$ keV, $\rho = 0.1$ g/cm^3. We see that the magnitude of the free path for a mixture can be made 2–3 smaller than that for a pure substance.

5.6.4 Radiative and electron heat conductivity

For very high densities of matter the flux of energy transported by electrons becomes comparable to or even larger than the flux of energy transported by photons. To compare the energy fluxes in the case of high temperatures it is convenient to describe the electron heat conductivity of plasma as a radiative one. To that end, let us write the heat flux $\vec{F}_e$ transported by electrons in a form analogous to the radiative heat conductivity approximation (5.34):

$$\vec{F}_e = -\kappa_e \mathrm{grad}\,\theta = -\ell_e \frac{4ac\theta^3}{3} \,\mathrm{grad}\,\theta.$$

Here κ_e is the electron heat conductivity coefficient and ℓ_e is the mean effective free path of heat-conducting electrons:

$$\ell_e = \frac{3\kappa_e}{4ac\theta^3}.$$

The total energy flux can be then written as

$$\vec{F} = -\ell_{\mathrm{eff}} \frac{4ac\theta^3}{3} \,\mathrm{grad}\,\theta,$$

where $\ell_{\mathrm{eff}} = \ell + \ell_e$ is the effective free path which includes the Rosseland mean free path ℓ.

The value of ℓ_e can be expressed through the heat conductivity coefficients, as proposed by Basko (see [21]):

$$\ell_e = \frac{(\kappa_S^2 + \kappa_H^2)^{1/2}}{\theta^3} \quad [\mathrm{cm}],$$

where κ_S is the heat conductivity coefficient of an ideal nondegenerate plasma:

$$\kappa_S = 1.17 \cdot 10^{-3} \frac{\epsilon_1 \theta^{5/2}}{Z_{\mathrm{eff}} L_1},$$

and κ_H is the heat conductivity coefficient for the case of strong degeneration:

$$\kappa_H = 3.10 \cdot 10^{-4} \frac{\epsilon_2 \theta \theta_F^{3/2}}{Z_{\mathrm{eff}} L_2}.$$

The extra (compared to the well-known expressions of Spitzer and Hubbard) factors $\epsilon_{1,2}$ introduced above arise when corrections due to the nonideal character of the plasma are accounted for [21]. The values of all the numerical factors reflect the fact that the plasma temperature θ and the Fermi temperature θ_F are measured in atomic units.

The Coulomb logarithms of electron-ion collisions can be calculated by the formulas

$$L_1 = 0.5 \ln \left[1 + 9 \frac{\langle Z^2 \rangle}{\Gamma^2} \left(\max \left\{ 1, \left[\frac{Z_{\text{eff}}}{3\Gamma(1+Z_0)} \right]^{1/2} \right\} \right)^2 \right],$$

$$L_2 = 0.5 \ln \left[\left(\frac{2\pi Z_{\text{eff}}}{3} \right)^{2/3} \left(1.5 + \frac{3}{\Gamma} \right)^{1/2} \right],$$

where Γ is the nonideality parameter:

$$\Gamma = \frac{\langle Z^2 \rangle}{r_0} \min \left(\frac{1}{\theta}; \frac{1}{\theta_{\mathrm{F}}} \right).$$

In estimations one can neglect the nonideality corrections, taking $\epsilon_1 = \epsilon_2 = 1$. As effective charge $Z_{\text{eff}} \equiv \langle Z^2 \rangle / Z_0$ we use the approximation

$$Z_{\text{eff}} = \begin{cases} 1, & \text{if } Z_0 \leq 1, \\ Z_0, & \text{if } Z_0 > 1, \end{cases}$$

where Z_0 is the average ion charge according to the Hartree-Fock-Slater model.

Remark. For photon energies $\omega \leq \omega_{\mathrm{p}}$, where ω_{p} is the plasma frequency ($\omega_{\mathrm{p}} = \sqrt{4\pi n_{\mathrm{e}}} = \sqrt{3Z_0/r_0^3}$), a large influence on the radiative transfer processes is exerted by collective effects that we left out of the picture. To describe the transport of photons in this case we need to resort to electrodynamics, describing the medium by means of its dielectric permeability ε ($\sqrt{\varepsilon} = n + ik$, where $k = \rho\kappa\, c/\omega$, κ is the absorption coefficient in cm^2/g and n is the refraction coefficient). The role of the refraction (and reflection) processes may be significant at relatively low temperatures.

5.6.5 Databases of atomic data and spectral photon absorption coefficients

The development of communication networks (Internet, etc.) yields new mechanisms for effective information exchange between network databases. This makes possible to combine the efforts of many researchers in accumulating and analyzing atomic data, obtained experimentally as well as by using various approximations and computer codes. Here is a brief list of relevant websites:

- Database of the National Institute of Standards and Technology,

  http://physics.nist.gov/cgi-bin/AtData/main_asd,

  contains experimental information on the position of spectral lines of 99 elements and the energy levels of ions of 52 elements. It is a very convenient and readily accessible source of reliable atomic data.

- Database of the International Atomic Energy Agency (Vienna, Austria),

  http://www-amdis.iaea.org/databases.htm.

  It contains information on electron-atom, ion-ion and intermolecular interactions, on interaction of particles with a surface, and also on scattering and ionization and excitation and de-excitation cross-sections. The data are presented mainly in a specialized form (in the ALADDIN format).

- The database of the National Institute for Fusion Science, Data and Planning Center (Nagoya, Japan),

  http://dbshina.nifs.ac.jp,

  is a large-volume and universal interactive network reference tool, which contains a large amount of data on ionization and excitation.

We should mention separately several computational databases, containing information on spectral and integral photon absorption coefficients for various elements and mixtures (values of opacity, Rosseland and Planck mean free paths):

- Database of the European Opacity Project (maintained by M. J. Seaton): TOPbase of *Opacity Project*, http://cdsweb.u-strasbg.fr/topbase.html.
- Database of results generated by the code LEDCOP of the Los Alamos National Laboratory: TOPS Opacities (LANL, Los–Alamos, USA), http://www.t4.lanl.gov/cgi-bin/opacity/tops.pl.
- Database of results generated by the code OPAL of the Lawrence Livermore National Laboratory: OPLIB (LLNL, Livermore, USA), http://www-phys.llnl.gov/Research/OPAL/opal.html.

Widely used in calculations of atomic data are also the codes written by R. Cowan (http://aphysics2.lanl.gov/tempweb), the code GRASP by I. P. Grant and coauthors (http://www.maths.ox.ac.uk/~ipg), and the code FAC written by Ming Feng Gu (http://kipac-tree.stanford.edu/fac).

5.7 Absorption of photons in a plasma with nonequilibrium radiation field

The investigation of hot plasmas produced in laser, beam or discharge devices requires that one takes into account its nonequilibrium character, due to nonstationary ionization and/or outgoing (incoming) radiation. The utilization in radiative gas dynamics codes of opacity coefficients that are obtained on the basis of equilibrium or quasi-equilibrium models of matter (for example, in the approximation of total escape of radiation) is not always justified, since the radiation field is formed in the dynamics of plasma and has an essential influence on the microstates of the ions in the plasma, and hence on the thermodynamic and radiation properties of the plasma. In their turn, the ion microstates determine the emissivity of the plasma and its spectral absorption coefficients. Accounting for the nonequilibrium character of the radiation field may modify considerably the composition of the plasma, its degree of ionization, the photon absorption coefficients and the emissivity at the same thermodynamic parameters of the plasma, i.e., at the same internal energy and pressure.

In practice, in addition to the local thermodynamic equilibrium approximation one often resorts to the so-called coronal equilibrium (CE) approximation, in which the ions are considered in the ground state and all collision excitations are removed simultaneously due to radiation decay because the plasma is completely transparent. A more general approximation is the collisional radiative steady state (CRSS) model, in which the radiation field is assumed to be known, for example, it is the Planckian field with the given radiation temperature.

Let us mention here also the collisional radiative equilibrium (CRE) model, in which one assumes equilibrium state for some configuration of radiating plasma [196].

5.7.1 Basic processes and relaxation times

Let us consider a plasma composed of electrons e, ions A_s^j of different charges j in some states s, and photons $\hbar\omega$. Various processes of interaction between particles and between particles and radiation take place in the plasma, for example,

1) elastic collisions between particles, i.e., collisions that do not modify their internal states, but redistribute the kinetic energies among the particles;

2) inelastic collisions:

   a) excitation due to electron impact and the inverse de-excitation process:

$$e + A_s^j \underset{\longleftarrow}{\overset{\longrightarrow}{}} e' + A_{s'}^j \,;$$

   b) ionization and the inverse three-body recombination process:

$$e + A_s^j \underset{\longleftarrow}{\overset{\longrightarrow}{}} e + e' + A_{s'}^{j+1} \,;$$

3) processes of interaction of ions and electrons with photons:

   a) absorption and emission in spectral lines

$$A_s^j + \hbar\omega \underset{\longleftarrow}{\overset{\longrightarrow}{}} A_{s'}^j \,;$$

   b) photoionization and photorecombination

$$A_s^j + \hbar\omega \underset{\longleftarrow}{\overset{\longrightarrow}{}} A_{s'}^{j+1} + e \,;$$

c) bremsstrahlung and inverse bremsstrahlung

$$e + A_s^j + \hbar\omega \rightleftarrows e' + A_s^j\,;$$

d) Auger effect and the inverse process of dielectronic capture

$$A_{s''}^j \rightleftarrows e + A_s^{j+1}\,;$$

e) Compton scattering

$$e + \hbar\omega \longrightarrow e' + \hbar\omega'.$$

Here e' is an electron with an energy different from that of e, s' is an excited state relative to s, s'' is a doubly excited state.

Needless to say, we have listed here by far not all processes that take place in a plasma. In principle, any process in which several photons, electrons and ions participate that is not forbidden by the conservation laws and selection rules may occur. In practice one singles out the most essential processes, i.e., the processes with maximal probability.

Under conditions of thermodynamic equilibrium, statistical equilibrium holds between all components of the system and all processes that take place in it. The values of the thermodynamic variables are close to their mean values, and the energy is distributed over all degrees of freedom in accordance to the distribution law.

In these circumstances the state of the matter-radiation system is characterized by a well-defined temperature, which in turn determines

(i) the distribution function of particles over energies;

(ii) the rates of the excitation, ionization and other processes;

(iii) the spectral density of the equilibrium radiation.

Therefore, under conditions of thermodynamic equilibrium, only the element composition, the temperature and the density need to be specified in order to be able, in principle, to determine all the properties of a plasma. If the system is not in an equilibrium state, then it will reach equilibrium after some time; this transition process is called relaxation. In a number of cases one speaks about incomplete equilibrium; one example is when the distribution functions of particles according to energy retain their forms, but the temperature appearing in these functions becomes a parameter. In the description of nonequilibrium processes one often assumes that the velocity distribution of ions and electrons is maxwellian, but with a different temperature. Moreover, notions like rotation temperature, oscillation temperature, excitation (distribution) temperature, population temperature of a level, ionization temperature, radiation temperature, and so on are used [125, 31]. Obviously, in the case of complete equilibrium all these temperatures coincide.

When equilibrium does not hold, knowledge of the element composition, temperature, and density of a plasma is not sufficient for determining its properties. In this case the system composed of plasma particles and radiation is nonstationary, transfer processes take place in it, and the system relaxes to equilibrium if there are no external sources acting on matter. Among the important characteristics of a nonequilibrium system are the relaxation times of the different degrees of freedom. By definition, the relaxation time τ_p of some parameter p to its equilibrium value p_0 is defined by the relation

$$\frac{dp}{dt} \simeq \frac{p_0 - p(t)}{\tau_p}.$$

As an example, let us estimate the time required to reach the equilibrium distribution for an admixture of particles of mass m_1 in a gas of particles of mass m_2 in equilibrium. Consider an arbitrary collision of particles of different masses:

$$m_1, v_1 \quad \longrightarrow \qquad \longleftarrow \quad m_2, v_2$$

In the system of coordinates attached to the particle m_2, the conservation laws of energy and momentum yield

$$\begin{cases} m_1(v_1^2 - {v'_1}^2) = m_2 {v'_2}^2, \\ m_1(v_1 + v'_1) = m_2 v'_2, \end{cases}$$

where $v_1 > 0$ and $v_2 = 0$ are the particle velocities before collision, while v'_1, v'_2 are the respective velocities after collision. This allows us to calculate the relative change in the energy of the molecule with mass m_1:

$$\frac{\Delta\varepsilon}{\varepsilon} = \frac{m_1(v_1^2 - {v'_1}^2)}{m_1 v_1^2} = \frac{4m_1 m_2}{(m_1 + m_2)^2}.$$

Suppose ν collisions take place per unit of time. Then the relaxation time τ_p can be estimated from the relation $\Delta\varepsilon\, \nu \tau_p \sim \varepsilon$, or

$$\tau_p \simeq \frac{(m_1 + m_2)^2}{4m_1 m_2} \frac{1}{\nu}.$$

We see that when $m_1 = m_2$ the relaxation time coincides with the free-flight time $\tau = 1/\nu$. If, however, $m_1 \gg m_2$ (for example, for ions and electrons $m_1/m_2 \simeq 1836\, A$, where A is the atomic weight), then the relaxation time is many times larger than the free-flight time (by a factore of about 10^4).

In real processes full equilibrium is practically nonexistent: as a rule, the state of a medium changes with time, and mass and energy fluxes occur. For this reason in practice it is customary to employ the hypothesis of local thermodynamic equilibrium (LTE) when equilibrium holds in each sufficiently small volume of matter, being characterized by a well-defined temperature. It is also assumed that changes in the system take place slower than the relaxation to equilibrium.

The condition for the applicability of the LTE approximation at high temperatures depends in an essential manner on the role of the radiation, since radiation processes are the fastest. For a large radiation density, in order for the LTE approximation to apply it is necessary that the photon free path ℓ be much smaller than the typical length over which the temperature and density change:

$$\frac{\ell}{T}|\nabla T| \ll 1, \quad \frac{\ell}{\rho}|\nabla \rho| \ll 1. \tag{5.217}$$

For low plasma densities the radiation processes usually violate the LTE, and consequently in this case the LTE approximation is applicable only when the radiation processes are negligible compared with the collision processes, i.e.,

$$w^{(\mathrm{r})} \ll w^{(\mathrm{c})}, \tag{5.218}$$

where $w^{(\mathrm{r})}$ and $w^{(\mathrm{c})}$ are the probabilities of the radiation and respectively collision processes.

In the opposite case, i.e., when

$$w^{(\mathrm{r})} \gg w^{(\mathrm{c})}, \tag{5.219}$$

the coronal approximation applies (see Subsection 6.3.4 below).

At high temperatures and low densities condition (5.217) is almost always violated in laboratory plasmas, so even if conditions (5.218) or (5.219) do not hold, one cannot, in particular, assume that the level populations are in equilibrium. In this case it is necessary, together with photon transfer processes, to take into consideration the nonequilibrium kinetics of the populations on bound levels [31].

5.7.2 Joint consideration of the processes of photon transport and level kinetics of electrons

In the steady-state approximation, if one neglects scattering processes then the transfer equation for radiation of intensity $I_\omega = I_\omega(\vec{r}, \vec{\Omega})$ and photon energy $\hbar\omega$ reads

$$(\vec{\Omega}\,\nabla) I_\omega = j(\omega, I_\omega) - \rho\kappa(\omega, I_\omega) \cdot I_\omega, \tag{5.220}$$

where $\kappa = \kappa(\omega, I_\omega)$ is the absorption coefficient, $j = j(\omega, I_\omega)$ is the emissivity and $\vec{\Omega}$ is the unit vector in the direction of propagation of the radiation.

The absorption coefficient and the emissivity of plasma are determined by the electron populations of the ground and excited states of ions or, in other words, by the number of ions with ionization degree j found in some state s; we will denote this number by $n_{js} = n_{js}(\vec{r})$ (thus the relative concentration of ions is $x_{js} = n_{js}/n_i$, where $n_i = \rho N_A/A$ is the ion density, N_A being the Avogadro number and A the atomic weight).

In the absence of macroscopic transport processes, the concentrations n_{js} must obey the balance conditions

$$\frac{dn_{js}}{dt} = \sum_{j's'} \left(n_{j's'}\, w_{j's' \to js} - n_{js}\, w_{js \to j's'} \right), \tag{5.221}$$

where $w_{js \to j's'}$ is the total probability of transition of an ion from the state js to the state $j's'$ due to the processes described in Subsection 5.7.1.

The probabilities of the collision processes are determined by the density of the free electrons and their energy distribution. The probabilities of radiation processes depend in essential manner on the photon distribution function f $\sim I_\omega/\omega^3$, more precisely, on how much of the radiation remains (is reabsorbed) in the plasma. In particular, absorption in spectral lines is proportional to f, while emission is proportional to 1+f. In particular, if a considerable part of the radiation is reabsorbed, then equations (5.220) and (5.221) must be solved simultaneously.

In the case of a single spectral line (two-level approximation), starting from these equations one can obtain the well-known Bieberman-Holstein integro-differential equation for the distribution of the electron population of an excited state [2, 31], assuming that the excited state is much less populated than the ground state. In the general case solving the equations (5.220) and (5.221) does not seem to be possible due to the huge number of ion states in the plasma and the necessity of a detailed description of the numerous elementary processes with the complex geometry of plasma formations accounted for.

To investigate the main functional dependencies of the population and radiation distributions, we will consider the simplest approximation, in which instead of the detailed computation of the absorption and emission coefficients one resorts to the effective methodology described in § 5.5, and instead of the system (5.221) one considers the corresponding system of level kinetics in the approximation of average occupation numbers of one-electron levels N_ν:

$$N_\nu = \sum_{js} n_{js} N_\nu^{js} / n_i, \tag{5.222}$$

where N_ν^{js} is the number of electrons on the level ν in the ion state js. Here a one-electron level ν is understood as the set of quantum numbers $n\ell j$ that specify the state of the electron: the principal quantum number n, the orbital quantum number ℓ, and the quantum number j of the total angular momentum of the electron.

5.7.3 Average-atom approximation

In the average-atom approximation*) it is convenient to write the spectral absorption and emission coefficients by making the dependence on the average occupation

*) As in the equilibrium case, the concept of the average atom presupposes the approximation of the average ion, i.e., of the ion with the average occupation numbers, which together with the free electrons is contained in an electrically-neutral spherical cell.

numbers N_ν and the radiation intensities I_ω explicit [164]:

$$\kappa'_\omega = \frac{N_A}{A}\left\{\sum_{\mu<\nu} N_\mu \left(1-\frac{N_\nu}{g_\nu}\right)\tilde{\sigma}^{\mathrm{bb}}_{\mu\nu}\Phi^{\mathrm{abs}}_{\mu\nu}(\omega) + \sum_{\mu} N_\mu (1-n_\varepsilon)\tilde{\sigma}^{\mathrm{bf}}_{\mu}(\omega) + \tilde{\sigma}^{\mathrm{ff}}(\omega)\right\}, \quad (5.223)$$

$$j'_\omega = \rho\frac{N_A}{A}\left\{\sum_{\mu<\nu} N_\nu \left(1-\frac{N_\mu}{g_\mu}\right)\frac{g_\mu}{g_\nu}\tilde{\sigma}^{\mathrm{bb}}_{\mu\nu}\Phi^{\mathrm{em}}_{\nu\mu}(\omega) + \sum_{\mu} n_\varepsilon \left(1-\frac{N_\mu}{g_\mu}\right) g_\mu\,\tilde{\sigma}^{\mathrm{bf}}_{\mu}(\omega) + e^{-\omega/\theta}\tilde{\sigma}^{\mathrm{ff}}(\omega)\right\}\left[\frac{\omega^3}{4\pi^3c^2} + I_\omega\right]. \quad (5.224)$$

The coefficients κ_ω and j_ω in equation (5.220) differ from κ'_ω and j'_ω in that the part of j'_ω contributed by the induced radiation and proportional to I_ω, is moved over to κ'_ω to obtain κ_ω. Note also that here we neglect scattering processes. In formulas (5.223)–(5.224), g_ν is the statistical weight, ε_ν is the energy of the level ν at given temperature T, density ρ and radiation intensity I_ω, $\varepsilon = \omega + \varepsilon_\nu$ is the energy of the electron after ionization, n_ε is the distribution function of the free electrons according to energy ε, $\tilde{\sigma}^{\mathrm{bb}}_{\mu\nu}(\omega)$, $\tilde{\sigma}^{\mathrm{bf}}_{\mu}(\omega)$, $\tilde{\sigma}^{\mathrm{ff}}(\omega)$ are the reduced line-absorption, photoionization and inverse bremsstrahlung cross-sections, respectively, and $\Phi_{\mu\nu}(\omega)$ is the line profile of absorption (with the superscript abs) or emission (with the superscript em).

Usually, for simplicity, one assumes that during emission or absorption a total redistribution over frequencies takes place inside the spectral line, and so one can consider that

$$\Phi^{\mathrm{abs}}_{\mu\nu}(\omega) = \Phi^{\mathrm{em}}_{\nu\mu}(\omega) = \Phi_{\mu\nu}(\omega). \quad (5.225)$$

Since the joint profile $\Phi_{\mu\nu}(\omega)$ is obtained by averaging a huge number of components (see (5.196), where this profile was denoted by $J_{\mu\nu}(\varepsilon)$), the approximation (5.225) is justified when the energy spread of line components is smaller than the temperature of matter.

The reduced cross-sections $\tilde{\sigma}^{\mathrm{bb}}_{\mu\nu}(\omega)$, $\tilde{\sigma}^{\mathrm{bf}}_{\mu}(\omega)$, $\tilde{\sigma}^{\mathrm{ff}}(\omega)$ differ from the usual cross-sections introduced in Subsection 5.2.1 in that the factors connected with the occupation numbers of levels and their degeneracy are now incorporated into the absorption coefficient κ' (see (5.223)). In particular,

$$\tilde{\sigma}^{\mathrm{bb}}_{\mu\nu} = 2\pi^2\alpha\, a_0^2 f_{\mu\nu}, \quad (5.226)$$

where $f_{\mu\nu}$ is the oscillator strength (see (5.180)).

In the average-atom model, for the photoionization and inverse bremsstrahlung cross-sections it is convenient to use the simple Kramers approximation

$$\tilde{\sigma}_{\mu}^{\rm bf}(\omega) = \frac{64\pi\,\alpha}{3\sqrt{6}}\,a_0^2\frac{Z_\mu\,|\varepsilon_\mu|^{3/2}}{g_\mu}\,\frac{1}{\omega^3}, \tag{5.227}$$

$$\tilde{\sigma}^{\rm ff}(\omega) = \frac{16\pi\,\alpha}{3\sqrt{3}}\,a_0^2 Z_0^2\theta\,e^{\zeta/\theta}\,\frac{1}{\omega^3}, \tag{5.228}$$

where Z_0 is the average ion charge, Z_μ is the effective charge for the level μ and ζ is the chemical potential.

In a nonequilibrium plasma the average occupation numbers N_ν are determined from the condition of balance of electrons with respect to energy levels [175, 193]:

$$\frac{dN_\nu}{dt} = (1-\frac{N_\nu}{g_\nu})S_\nu - N_\nu L_\nu. \tag{5.229}$$

Here S_ν is the total rate, (more precisely, the flux) measured in 1/sec, of the processes leading to an increase of the number electrons in the state ν, and L_ν is the total rate of the processes leading to a decrease of the number of electrons in that state:

$$S_\nu = \sum_{\mu<\nu} N_\mu(\alpha_{\mu\nu}^{\rm abs}+\alpha_{\mu\nu}^{\rm ex}) + \sum_{\mu>\nu} N_\mu(\alpha_{\mu\nu}^{\rm em}+\alpha_{\mu\nu}^{\rm dex}) + \alpha_\nu^{\rm ir} + \alpha_\nu^{\rm phr} + \alpha_\nu^{\rm dc}, \tag{5.230}$$

$$L_\nu = \sum_{\mu<\nu}\left(1-\frac{N_\mu}{g_\mu}\right)(\alpha_{\nu\mu}^{\rm em}+\alpha_{\nu\mu}^{\rm dex}) + \sum_{\mu>\nu}\left(1-\frac{N_\mu}{g_\mu}\right)(\alpha_{\nu\mu}^{\rm abs}+\alpha_{\nu\mu}^{\rm ex}) + \alpha_\nu^{\rm ii} + \alpha_\nu^{\rm phi} + \alpha_\nu^{\rm ai}. \tag{5.231}$$

We used the following notation for the rates of the processes considered (the number of the corresponding transitions occurring during one unit of time per one ion): $\alpha_{\mu\nu}^{\rm ex}$ and $\alpha_{\nu\mu}^{\rm dex}$ for the excitation and respectively the quenching of the level ν; $\alpha_\nu^{\rm ii}$ and $\alpha_\nu^{\rm ir}$ for ionization and three-body recombination; $\alpha_{\mu\nu}^{\rm abs}$, $\alpha_{\nu\mu}^{\rm em}$ for absorption and emission in lines; $\alpha_\nu^{\rm phi}$ and $\alpha_\nu^{\rm phr}$ for photoionization and photorecombination; and $\alpha_\nu^{\rm ai}$ and $\alpha_\nu^{\rm dc}$ for the Auger effect (autoionization) and dielectronic capture.

5.7.4 Rates of radiation and collision processes

To calculate the rates of elementary processes one uses the formulas obtained in the works [219, 193, 218, 124, 134]. The oscillator strengths, energy levels and other requisite quantities are calculated on the basis of the relativistic Hartree-Fock-Slater self-consistent field model, with given occupation numbers N_ν of the electron states, which satisfy the system of equations (5.229)–(5.231). The rates of the direct and inverse processes are connected by the conditions of the principle of detailed balance; at equilibrium from these conditions should follow the Fermi-Dirac distribution of level occupancies for each of the considered processes. This yields a simply connection between direct and inverse processes.

It is convenient to express the rates of the radiation processes in terms of the corresponding reduced cross-sections (5.226)–(5.228) with the radiation field accounted for [134, 164]:

$$\alpha_{\mu\nu}^{\text{abs}} = \nu_0 \int \frac{\tilde{\sigma}_{\mu\nu}^{\text{bb}} \Phi_{\mu\nu}(\omega)}{\omega} \left(\int I_\omega d\Omega \right) d\omega, \tag{5.232}$$

$$\alpha_{\nu\mu}^{\text{em}} = \nu_0 \frac{g_\mu}{g_\nu} \exp\left(\frac{\varepsilon_\nu - \varepsilon_\mu}{\theta}\right) \int \frac{\tilde{\sigma}_{\mu\nu}^{\text{bb}} \Phi_{\mu\nu}(\omega)}{\omega} \left(\frac{\omega^3}{\pi^2 c^2} + \int I_\omega d\Omega \right) e^{-\omega/\theta} d\omega, \tag{5.233}$$

$$\alpha_{\mu}^{\text{phi}} = \nu_0 \int \frac{\tilde{\sigma}_{\mu}^{\text{bf}}(\omega)}{\omega} \left(\int I_\omega d\Omega \right) d\omega, \tag{5.234}$$

$$\alpha_{\mu}^{\text{phr}} = \nu_0 g_\mu \exp\left(\frac{\zeta - \varepsilon_\mu}{\theta}\right) \int \frac{\tilde{\sigma}_{\mu}^{\text{bf}}(\omega)}{\omega} \left(\frac{\omega^3}{\pi^2 c^2} + \int I_\omega d\Omega \right) e^{-\omega/\theta} d\omega. \tag{5.235}$$

Here, in the cross-section formulas (5.226)–(5.228), one puts $a_0 = 1$; $\nu_0 = me^4/\hbar^3$ $= 4.134 \cdot 10^{16}$ 1/s is the atomic frequency, ζ is the chemical potential of the free electrons, assumed here to be in equilibrium at temperature θ; $d\Omega$ is the solid-angle element. If the intensity I_ω is expressed in TW/(cm^2·eV·steradian), then to translate it into atomic units we must multiply by $C_{\text{W}} = 4.23 \cdot 10^{-3}$. In order for $j(\omega)$ to be expressed in TW/(cm^3· eV· steradian), we must divide the expression of $j(\omega)$ by the same number C_{W}.

For low densities (see formula (1.26), and also Subsection 6.2.2)

$$e^{\zeta/\theta} \approx \frac{3}{2} \sqrt{\frac{\pi}{2}} \frac{Z_0}{r_0^3 \theta^{3/2}},$$

where $Z_0 = Z - \sum_\nu N_\nu$ is the number of free electrons per atom ($Z_0 = n_{\text{e}}/n_i$ and n_{e} is the free electron density).

The rate of the collision processes can be calculated by means of formulas analogous to those obtained in Subsection 5.4.5 for the electron widths of lines in the impact approximation (see [218]). However, because of the approximate character of the model (5.230)-(5.231), we can use the following simple formulas (see [124, 219])*):

$$\alpha_{\mu\nu}^{\text{ex}} = 1.3 \cdot 10^{16} \frac{\rho Z_0}{A\theta^{3/2}} f_{\mu\nu} \frac{E_1(x_{\mu\nu})}{x_{\mu\nu}}, \tag{5.236}$$

*) More accurate calculations relying on the distorted wave method or on the Born approximation enable one to derive simple semi-empirical formulas for rates that are specific for a given element [218].

$$\alpha_{\nu\mu}^{\rm dex} = \frac{g_\mu}{g_\nu} e^{x_{\mu\nu}} \alpha_{\mu\nu}^{\rm ex}, \tag{5.237}$$

$$\alpha_\nu^{\rm ii} = 1.3 \cdot 10^{16} \frac{\rho Z_0}{A\theta^{3/2}} \frac{E_1(x_\nu)}{x_\nu}, \tag{5.238}$$

$$\alpha_\nu^{\rm ir} = g_\nu \exp\left(\frac{\zeta - \varepsilon_\nu}{\theta}\right) \alpha_\nu^{\rm ii}. \tag{5.239}$$

In the above formulas $x_{\mu\nu} = (\varepsilon_\nu - \varepsilon_\mu)/\theta$, $x_\nu = |\varepsilon_\nu|/\theta$ and $E_1(x)$ is the integral exponential. In [193], instead of $E_1(x)$, B. F. Rozsnyai uses the function $\widetilde{E}_1(x) = E_1(x) - xE_1(x+0.3)/(x+0.3)$, which in a number of cases yields a more correct value for the rate.

To calculate the probability of autoionization per one second for an electron at the level $n\ell$ with de-excitation of another electron, $n''\ell'' \to n'\ell'$, let us use the simplest dipole approximation [237]:

$$W_{n\ell}^{\rm ai} = 4.1 \cdot 10^{16} \frac{2\pi}{g_{n\ell}} \frac{f_{n'\ell', n''\ell''}}{\omega} \frac{g_{n'\ell'}}{g_{n''\ell''}} \left(\sum_{\widetilde{\ell}} \frac{\ell + \widetilde{\ell} + 1}{2} \int_0^{r_0} \frac{R_{n\ell}(r) R_{\varepsilon\widetilde{\ell}}(r)}{r^2}\, dr\right)^2,$$

where $\omega = \varepsilon_{n''\ell''} - \varepsilon_{n'\ell'}$, $\varepsilon = \omega + \varepsilon_{n\ell} > 0$.

The quantity inside parentheses can be calculated in the hydrogen-like approximation using the connection between the matrix elements of the acceleration and of the position vector (see Subsection 5.2.1) and the expression (5.65) for the oscillator strength.

Summing over all possible de-excitation processes we obtain for the autoionization rate the expression

$$\alpha_\nu^{\rm ai} = \sum_{\mu < \mu'} N_{\mu'} \left(1 - \frac{N_\mu}{g_\mu}\right) W_{\nu\mu'\mu}^{\rm ai} \tag{5.240}$$

where $\nu = n\ell$, $\mu = n'\ell'$, $\mu' = n''\ell''$, $W_{\nu\mu'\mu}^{\rm ai} = W_{n\ell}^{\rm ai}$.

For the inverse process of dielectronic capture, similar considerations yield

$$\alpha_\nu^{\rm dc} = g_\nu \exp\left(\frac{\zeta - \varepsilon_\nu}{\theta}\right) \sum_{\mu < \mu'} N_\mu \left(1 - \frac{N'_\mu}{g'_\mu}\right) \frac{g_{\mu'}}{g_\mu} W_{\nu\mu'\mu}^{\rm ai} \tag{5.241}$$

The kinetic equations (5.230)–(5.231) include dielectronic capture processes instead of the dielectronic recombination processes, as is usually done. In the average-atom approximation the process of dielectronic recombination may be considered as a result of kinetic processes of dielectronic capture, autoionization, and radiative decay [193]. Therefore, this process depends on the relations between all the aforementioned rates at given temperature and density.

5.7.5 Radiation properties of a plasma with nonequilibrium radiation field

The simplest way to analyze the dependence of radiation characteristics of plasma (i.e., of its absorption and emission coefficients) on the levels of the nonequilibrium radiation is to make use of some quasi-stationary (steady-state) approximation. Such an approximation works well whenever the gas-dynamics processes are considerably slower than the collision and radiation processes. The type of quasi-stationarity is determined by the relations between the rates of various processes. In particular, among the widely used approximations are the coronal model (see Subsection 6.3.4), the collisional-radiative model for transparent plasma, and others [31, 44, 20, 193, 36]. A crucial step for a simplified description of the complex nonstationary processes in a plasma is to single out among all processes the most important ones that, in addition, have a sufficiently high rate, and first of all, the radiative processes.

Our goal here is to investigate the basic dependencies on the levels of the nonequilibrium radiation rather than solving a concrete problem. With this in mind, let us consider quasi-stationary solutions of the system (5.229)–(5.231) incorporating all collision and radiative processes listed above, assuming, however, that the radiation intensity I_ω is arbitrary.

In the quasi-stationary approximation ($dN_\nu/dt = 0$) we can recast the system (5.229)–(5.231) in a form that is convenient for the application of iterative methods:

$$N_\nu = \frac{g_\nu}{1 + g_\nu L_\nu / S_\nu}. \tag{5.242}$$

Here the rates S_ν and L_ν depend not only on all average occupation numbers of levels, N_ν, but also on the energies ε_ν and the oscillator strengths $f_{\mu\nu}$, which in turn are determined by the state of the average ion (i.e., its occupation numbers N_ν).

Since the calculation of energy levels and wave functions of electrons for a given state of an ion is rather laborious (this problem was solved in the approximation of the Hartree-Fock-Slater self-consistent field model), two iteration cycles were used to solve the system (5.242).

First, for given energies $\varepsilon_\nu^{(s)}$ and oscillator strengths $f_{\mu\nu}^{(s)}$, where s is the iteration number, the average occupation numbers N_ν were calculated by carrying out additional iterations according to the formula

$$N_\nu^{(p+1)} = \left. \frac{g_\nu}{1 + g_\nu L_\nu / S_\nu} \right|_{N_\nu = N_\nu^{(p)}}, \quad p = 0, 1, 2, \ldots \tag{5.243}$$

To improve the convergence, use was made of the relaxation procedure

$$\widetilde{N}_\nu = \alpha N_\nu^{(p+1)} + (1 - \alpha) N_\nu^{(p)},$$

where $\alpha = 0.3 \div 0.5$.

After the iterations (5.243) converged, the resulting occupation numbers were used to determine new energy levels $\varepsilon_\nu^{(s+1)}$ and oscillator strengths $f_{\mu\nu}^{(s+1)}$ by solving the Hartree-Fock-Slater system of equations with the given occupation numbers. Then the process was repeated for as long as the condition

$$\max_\nu \left| N_\nu \left(\varepsilon_\nu^{(s+1)}, f_{\mu\nu}^{(s+1)}\right) - N_\nu \left(\varepsilon_\nu^{(s)}, f_{\mu\nu}^{(s)}\right)\right| < 10^{-6}$$

was satisfied.

In this way we obtain average occupation numbers, electron energy levels and electron wave functions that are consistent with the given radiation field I_ω. These are then used to calculate the absorption coefficients and emissivities of the plasma. The resulting coefficients make it possible to get a more accurate value of the radiation field by solving the transfer equation (5.220). The described iteration cycle is then repeated till there will be complete consistency between the radiation field $I_\omega(\vec{r}, \vec{\Omega})$ and the level populations $N_\nu(\vec{r})$.

As an example, the method was tested to calculate the radiative properties of nonequilibrium molybdenum and tungsten plasmas at density $\rho = 0.01$ g/cm^3 and different temperatures T, which are contained in a thermostat with a given radiation temperature, i.e., the radiation field was specified by the function

$$I_\omega = B_\omega(T_\mathrm{r}) = \frac{\hbar\omega^3}{4\pi^3 c^2} \frac{1}{e^{\hbar\omega/T_\mathrm{r}} - 1}, \tag{5.244}$$

where T_r is the effective temperature of the radiation. When $T_\mathrm{r} = T$ local thermodynamic equilibrium holds, i.e., formula (5.224) must yield $j(\omega) = \rho\kappa(\omega)B_\omega(T)$.

The results of the computations for a molybdenum plasma ($T = 100$ eV, $\rho = 0.01$ g/cm^3) and a tungsten plasma ($T = 200$ eV, $\rho = 0.01$ g/cm^3) are displayed in figures 5.34 and 5.35. These show the average degree of ionization (average charge of the ion) Z_0, the Rosseland and Planck mean absorption coefficients, as well as the radiative losses (integral emissivity) of the plasma, $Q = 4\pi \int j(\omega)d\omega$, as functions of the parameter ξ, which characterizes the level of radiation in plasma relative to the equilibrium value:

$$\xi = \frac{\displaystyle\int \kappa(\omega) I_\omega/\omega^3 \, d\omega}{\displaystyle\int \kappa(\omega) B_\omega/\omega^3 \, d\omega}.$$

Note that ξ is determined not only by the intensity of the nonequilibrium radiation in plasma, but also by how much of this radiation is reabsorbed by the plasma.

As the figures show, there are two regions, depending on the parameter ξ, where the character of the behavior of the quantities considered is essentially different. In our case this takes place for $T_\mathrm{r} \simeq T/2$, which gives $\xi = \xi^* = 0.34$ for molybdenum and $\xi = \xi^* = 0.27$ for tungsten.

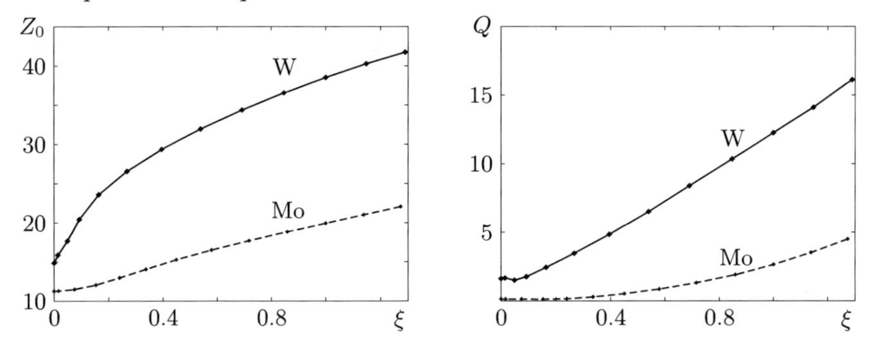

Figure 5.34: Average charge of the ion Z_0 and the radiation losses Q in 10^3 TW/cm^3 as functions of the parameter ξ for a molybdenum plasma ($T = 100$ eV, $\rho = 10^{-2}$ g/cm^3) and a tungsten plasma ($T = 200$ eV, $\rho = 10^{-2}$ g/cm^3)

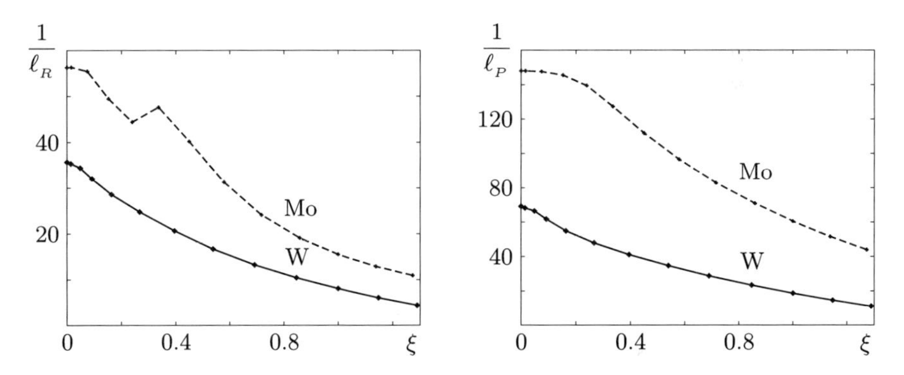

Figure 5.35: Reciprocals of the Rosseland mean free paths, $1/\ell_{\mathrm{R}}$, and of the Planck mean free path, $1/\ell_{\mathrm{P}}$, as functions of the parameter ξ. The notations are the same as in Figure 5.34

Analysis shows that for $\xi < \xi^*$ the main role is played by collision processes, whereas for $\xi > \xi^*$ everything is determined by radiation. In this connection, the character of the approximation of opacities in gas-dynamic codes in the domains where radiation plays the main role must be different from that in the domains where that role is taken over by collision processes. In particular, such approximations, which describe the data shown in figures 5.34 and 5.35, were constructed in [163].

In the numerical modeling of dynamical processes in a plasma, the radiation intensity I_ω is, of course, not defined by formula (5.244), but must be found by solving the transfer equation (5.220). The joint solution of the photon transfer equation and the equation of level kinetics of electrons, even in the average-atom approximation, is a computationally very complex problem [164]. In a number of cases one can use simpler approximations.

5.7.6 Radiative heat conductivity of matter for large gradients of temperature and density

Let us examine the case when the temperature or density change considerably over distances of the order of the free path of photons, which violates the conditions for the applicability of the diffusion approximation in the calculation of the radiation flux [147].

For simplicity, we will confine ourselves to the case of plane geometry. We will also assume that absorption in lines is essential, which allows us to neglect photon scattering. In this case it is convenient to write the radiative transfer equation in the form

$$\mu \frac{\partial I_\omega}{\partial z} + \frac{1}{\ell_\omega(z)} I_\omega = \frac{1}{\ell_\omega(z)} B_\omega(T). \tag{5.245}$$

Here z is the coordinate, $\ell_\omega = \ell(\omega) = 1/(\rho\kappa(\omega))$ the spectral free path of photons, μ the cosine of the angle between the z-axis and the direction of propagation of the radiation, $I_\omega(\mu, z)$ the spectral intensity of the radiation and $B_\omega(T)$ the spectral intensity the equilibrium radiation in matter at temperature T, normalized with respect to the ω-scale:

$$B_\omega(T) = c\hbar\omega\, f_P(\omega) = \frac{\hbar\omega^3}{4\pi^3 c^2} \frac{1}{e^x - 1}, \qquad x = \frac{\hbar\omega}{kT}$$

(Kirchhoff's radiation law is assumed to hold).

Since the volume occupied by matter is finite, equation (5.245) is supplemented with the following boundary conditions:

$$\begin{aligned} I_\omega(\mu, -\infty) &= 0 \qquad \text{if} \quad \mu > 0; \\ I_\omega(\mu, \infty) &= 0 \qquad \text{if} \quad \mu < 0. \end{aligned} \tag{5.246}$$

The solution of the boundary value problem (5.245)–(5.246) has the form

$$I_\omega(z,\mu) = \int\limits_{-\infty}^{z} \frac{1}{\mu \ell_\omega(z')} \exp\left[-\int\limits_{z'}^{-z} \frac{dz''}{\mu \ell_\omega(z'')}\right] B_\omega(T)|_{T=T(z')}\, dz' \quad \text{if } \mu > 0;$$

$$I_\omega(z,\mu) = -\int\limits_{z}^{\infty} \frac{1}{\mu \ell_\omega(z')} \exp\left[\int\limits_{z}^{z'} \frac{dz''}{\mu \ell_\omega(z'')}\right] B_\omega(T)|_{T=T(z')}\, dz' \quad \text{if } \mu < 0.$$

The flux of energy carried by photons, for some value of z, which for the sake of definiteness we take equal to zero, is given by

$$F = \int\limits_{0}^{\infty} F_\omega\, d\omega, \tag{5.247}$$

where

$$F_\omega = 2\pi \int_{-1}^{1} \mu\, I_\omega(0,\mu)\, d\mu = 2\pi \int_0^1 \mu\, d\mu \left\{ \int_{-\infty}^{0} B_\omega \exp\left[- \int_z^0 \frac{dz'}{\mu \ell_\omega(z')} \right] \frac{dz}{\mu \ell_\omega(z)} - \int_0^\infty B_\omega \exp\left[- \int_0^z \frac{dz'}{\mu \ell_\omega(z')} \right] \frac{dz}{\mu \ell_\omega(z)} \right\}. \quad (5.248)$$

Next, let us transform expression (5.248) via integration by parts with respect to the variable z:

$$F_\omega = -2\pi \int_0^1 \mu\, d\mu \left\{ \int_{-\infty}^{0} \exp\left[- \int_z^0 \frac{dz'}{\mu \ell_\omega(z')} \right] \frac{dB_\omega}{dz}\, dz + \int_0^\infty \exp\left[- \int_0^z \frac{dz'}{\mu \ell_\omega(z')} \right] \frac{dB_\omega}{dz}\, dz \right\}.$$

For the derivative $dB_\omega(T)/dz$ we have

$$\frac{dB_\omega(T)}{dz} = \frac{\hbar c}{\pi k}\, a T^2\, \frac{dT}{dz}\, R(x), \quad (5.249)$$

where a is the radiation constant and $R(x)$ is the Rosseland weight function:

$$a = \frac{\pi^2 k^4}{15 c^3 \hbar^3}, \qquad R(x) = \frac{15}{4\pi^4} \frac{x^4 e^{-x}}{(1 - e^{-x})^2}$$

(see (5.32), (5.35)). Integration with respect to μ yields the integral exponential E_3:

$$\int_0^1 \mu \exp\left(-\frac{u}{\mu} \right) d\mu = E_3(u), \quad (5.250)$$

where

$$E_n(u) = \int_1^\infty \frac{\exp(-tu)}{t^n}\, dt.$$

Now using (5.249) and (5.250), we obtain for F_ω the expression

$$F_\omega = -2 \frac{a \hbar c}{k} R(x) \int_{-\infty}^{\infty} T^2(z)\, \frac{dT(z)}{dz}\, E_3\left(\left| \int_0^z \frac{dz'}{\ell_\omega(z')} \right| \right) dz. \quad (5.251)$$

Therefore, the total flux is

$$F = \int_0^\infty F_\omega \, d\omega = -2\frac{a\hbar c}{k} \int_{-\infty}^{\infty} T^2(z) \frac{dT(z)}{dz} \left[\int_0^\infty R(x) \, E_3 \left(\left| \int_0^z \frac{dz'}{\ell_\omega(z')} \right| \right) d\omega \right] dz.$$

Changing the variable ω to $x = \hbar\omega/(kT)$ and switching the order of integration with respect to ω and z, we obtain

$$F = -2ac \int_0^\infty R(x) \left[\int_{-\infty}^{\infty} T^3(z) \frac{dT(z)}{dz} E_3 \left(\left| \int_0^z \frac{dz'}{\ell_\omega(z')} \right| \right) dz \right] dx. \tag{5.252}$$

Since the function $E_3(u)$ decays rapidly when u grows, the size of the essential part of the integration domain with respect to z in (5.252) is of the order several photon free-path lengths. From gas-dynamics calculations it is known that the heat flux is a smooth function of coordinates. Hence, in the case when the free path is small, in a domain of the order of several free-path lengths the flux may be regarded as constant. Assuming a power-like dependence of the Rosseland mean free path on temperature ($\ell \sim T^n$) and calculating the flux in the diffusion approximation, we obtain an approximate rule governing the dependence of temperature on coordinates:

$$T^{n+3}(z) \frac{dT(z)}{dz} = \text{const} = T_0^{n+3} \left. \frac{dT}{dz} \right|_{z=0}, \tag{5.253}$$

where $T_0 = T(0)$.

Using (5.253), we obtain for the total flux the expression

$$F = -\frac{\widetilde{\ell} c}{3} 4aT_0^3 \left. \frac{dT}{dz} \right|_{z=0}, \tag{5.254}$$

where

$$\widetilde{\ell} = \int_0^\infty \widetilde{\ell}_\omega \, R(x) \, dx \,, \tag{5.255}$$

$$\widetilde{\ell}_\omega = \frac{3}{2} \int_{-\infty}^{\infty} \left[\frac{T_0}{T(z)} \right]^n E_3 \left(\left| \int_0^z \frac{dz'}{\ell_\omega(z')} \right| \right) dz. \tag{5.256}$$

Comparing (5.254) with the expression of F in the diffusion approximation we see that the quantity $\widetilde{\ell}$ plays a role similar to that played by the Rosseland mean free path ℓ in the equation for the radiative heat conductivity (see (5.35)), and when the variation of $T(z)$ is small over distances of the order of ℓ_ω, then $\widetilde{\ell} = \ell$.

Since $E_3\left(\left|\int_0^z (1/\ell_\omega(z'))dz'\right|\right)$ is a rapidly varying function of the variable z even when $\ell_\omega(z)$ does not depend on z, to calculate the integral (5.256) it is convenient to recast it as

$$\widetilde{\ell}_\omega = \int\limits_{-\infty}^{0} \left(\frac{T_0}{T(z)}\right)^n \ell_\omega(z)\, d\left\{E_4\left(\int\limits_z^0 \frac{dz'}{\ell_\omega(z')}\right)\right\} - \int\limits_0^{\infty} \left(\frac{T_0}{T(z)}\right)^n \ell_\omega(z)\, d\left\{E_4\left(\int\limits_0^z \frac{dz'}{\ell_\omega(z')}\right)\right\}. \quad (5.257)$$

To integrate in (5.257) it is convenient to use the trapezoid rule with respect to the variable $y = y(z) = E_4\left(\left|\int_0^z (1/\ell_\omega(z'))dz'\right|\right)$, i.e.,

$$\int\limits_{z_1}^{z_2} \ell_\omega(z) \left(\frac{T_0}{T(z)}\right)^n d\left\{E_4\left(\left|\int\limits_0^z \frac{dz'}{\ell_\omega(z')}\right|\right)\right\} = \frac{1}{2}\left[\ell_\omega(z_1)\left(\frac{T_0}{T(z_1)}\right)^n + \ell_\omega(z_2)\left(\frac{T_0}{T(z_2)}\right)^n\right]\left[y(z_2) - y(z_1)\right].$$

Note that for this method the integration becomes exact when $dT/dz = d\rho/dz = 0$.

To interpret the result obtained above, let us recall that we are dealing with transport of photons, for which the magnitude of the free path is influenced in essential manner by absorption in spectral lines. Since the intensity of lines depends on the temperature and density of matter, large gradients of temperature and density lead to a redistribution of absorption over frequencies, which in fact translates into an additional effective broadening of such groups of lines. If in the essential domain of integration in the variable $x = \hbar\omega/(kT)$ for the integral (5.255) the lines are already overlapped, one should expect that the mean effective free path $\widetilde{\ell}$ will differ only slightly from the free path calculated without this additional broadening, i.e., from the Rosseland mean free path ℓ. Therefore, one should expect that the influence of the temperature and density gradients will be maximal in the range of values of T and ρ where absorption in lines has an essential influence on the free path, but the lines are practically not overlapped.

Chapter 6

The equation of state

The *equation of state* is a necessary complement to the thermodynamic laws, which allows the latter to be applied to real substances. By "equation of state" one usually means a set of two relations which define the *pressure* and *internal energy* of a physically homogeneous system in thermodynamic equilibrium as functions of two arbitrary parameters that specify the state of matter. An example of two such relations is

$$P = P(T, \rho), \quad E = E(T, \rho),$$

where T and ρ are the temperature and density of matter, P is the pressure and E is the specific internal energy, i.e., the energy per unit of mass.

The equations of state cannot be derived solely from the thermodynamic laws; they are determined experimentally or are calculated theoretically, based on representations about the structure of matter provided by the methods of statistical mechanics. In § 6.1 we have shown how, by using the Thomas-Fermi model for matter with given temperature and density, one can in a relatively easy manner obtain formulas for the pressure, internal energy and entropy. In the exposition that follows these relations will be verified and sharpened by means of refined models and will be compared with experiments.

Let us remark that, as it turned out, in the study of opacity of matter and the calculation of Rosseland mean free path knowledge of the characteristics of the average atom is not enough (see § 5.5), since the photon absorption process (and especially the absorption in spectral lines) takes place not in the average ion, but in some state of the ion with concrete occupation numbers. At the same time, quantum-statistical models of the average atom, which deal with one ion with average occupation numbers, allow one to obtain sufficiently accurate equations of state in a wide range of temperatures and densities.

6.1 Description of thermodynamics of matter based on quantum-statistical models

6.1.1 Formulas for the pressure, internal energy and entropy according to the Thomas-Fermi model

The electron pressure is readily found if we use the Thomas-Fermi statistics and calculate the value of the pressure on the boundary $r = r_0$ of the average atom cell, where $V(r) = 0$, $dV/dr = 0$. If we "place" a rigid wall at the boundary of the atom cell, then the change in the density of the flux of electron momenta per unit of time during their reflection in the wall gives the electron pressure P_e. Let us choose the z-axis drawn from the center of the atom cell to be perpendicular to the wall. Then for electrons with momentum $\vec{p}$ ranging in the element of phase-space volume $d\vec{p} = dp_x dp_y dp_z = p^2\, dp\, \sin\vartheta d\vartheta\, d\varphi$, the momentum-flux density changes at the rate

$$N(\vec{p})d\vec{p} \cdot p\cos\vartheta \cdot 2p\cos\vartheta,$$

where, according to formulas (1.2) and (1.6),

$$N(\vec{p})d\vec{p} = \frac{2}{(2\pi)^3} \left.\frac{d\vec{p}}{1+\exp\left(\dfrac{p^2/2 - V(r)}{\theta} + \eta\right)}\right|_{r=r_0} = \frac{2}{(2\pi)^3} \frac{p^2\, dp\, \sin\vartheta d\vartheta d\varphi}{1+\exp\left(\dfrac{p^2}{2\theta} + \eta\right)}.$$

Integration with respect to the variables φ, ϑ and p yields

$$P_\text{e} = \frac{2}{(2\pi)^3} \int\limits_0^\infty dp \int\limits_0^{\pi/2} d\vartheta \int\limits_0^{2\pi} d\varphi \frac{p^2 \sin\vartheta \cdot 2p^2\cos^2\vartheta}{1+\exp\left(\dfrac{p^2}{2\theta}+\eta\right)} =$$

$$\frac{1}{3\pi^2} \int\limits_0^\infty \frac{p^4 dp}{1+\exp\left(\dfrac{p^2}{2\theta}+\eta\right)} = \frac{(2\theta)^{5/2}}{6\pi^2} I_{3/2}(-\eta).$$

We can arrive to the same expression if we calculate the pressure as the average momentum transported by electrons per unit of time across a unit of surface:

$$P_\text{e} = \frac{2}{(2\pi)^3} \int\limits_0^\infty dp \int\limits_0^{\pi} d\vartheta \int\limits_0^{2\pi} d\varphi \frac{p^2 \sin\vartheta \cdot p^2\cos^2\vartheta}{1+\exp\left(\dfrac{p^2}{2\theta}+\eta\right)} = \frac{(2\theta)^{5/2}}{6\pi^2} I_{3/2}(-\eta). \tag{6.1}$$

The electron density on the boundary of the atom cell is given, according to formula (1.7), by

$$\rho(r_0) = \rho_{\rm e} = \frac{(2\theta)^{3/2}}{2\pi^2} I_{1/2}(-\eta). \tag{6.2}$$

Therefore, using the asymptotics of the function $I_k(x)$, we conclude that at high temperatures, when $\eta \gg 1$ and the average kinetic energy of the continuum electrons is much larger than the potential energy, we obtain

$$P_{\rm e} = \rho_{\rm e}\theta, \tag{6.3}$$

i.e., the well-known formula for the pressure of an ideal Boltzmann gas. For large negative values of $\eta = -\mu/\theta$, i.e., for high densities and low temperatures, formulas (6.1) and (6.2) yield

$$P_{\rm e} = \frac{4\sqrt{2}}{15\pi^2}\mu^{5/2}, \qquad \rho_{\rm e} = \frac{2\sqrt{2}}{3\pi^2}\mu^{3/2}, \tag{6.4}$$

and so for a degenerate Fermi gas the pressure depends on the electron density according to a power law:

$$P_{\rm e} = \frac{1}{5}(3\pi^2)^{2/3}\rho_{\rm e}^{5/3}. \tag{6.5}$$

To proceed further we need the following interpolation formula for the function $I_{3/2}(x)$, which is analogous to formula (1.37) for $I_{1/2}(x)$:

$$I_{3/2}(x) \approx \frac{3}{10} I_{1/2}(x) \left[125 + 60 I_{1/2}(x) + 18 I_{1/2}^2(x)\right]^{1/3},$$

Then from (6.1) and (6.2) we obtain a simple approximation for pressure, which holds for a Boltzmann as well as for a degenerate electron gas [98]:

$$P_{\rm e} = \rho_{\rm e}\left[\theta^3 + 3.36\rho_{\rm e}\theta^{3/2} + \frac{9\pi^4}{125}\rho_{\rm e}^2\right]^{1/3}, \tag{6.6}$$

where $\rho_e = Z_0/v$, Z_0 is the number of free electrons per atom and $v = \frac{4}{3}\pi r_0^3$ is the volume of the atom cell.

To find the total pressure we must take into account the pressure of nuclei (ion cores). At high temperatures nuclei are usually regarded as an ideal gas. In this case the total pressure is given by the formula

$$P = 2.942 \cdot 10^4 \left(P_{\rm e} + \frac{\theta}{v}\right) \text{ GPa} \tag{6.7}$$

(see Figure 6.1).

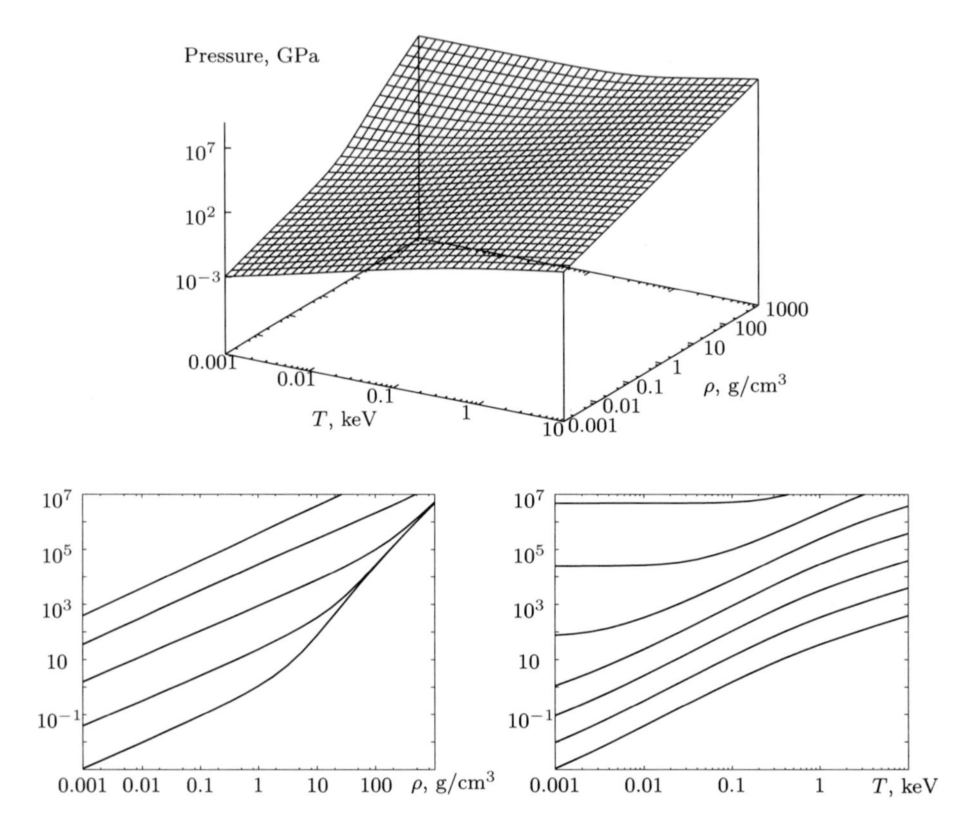

Figure 6.1: Pressure surface in the Thomas-Fermi model for gold, and the corresponding isotherms and isochores at temperatures $T = 10^i$ keV, $i = -3, -2, -1, 0, 1$ and densities $\rho = 10^k$ g/cm^3, $k = -3, -2, -1, 0, 1, 2, 3$

The expression for the internal energy of the electrons in one atom cell follows from formula (1.3):

$$E_{\mathrm{e}} = \frac{2}{(2\pi)^3} \int\int \left[\frac{p^2}{2} - \frac{Z}{r} - \frac{1}{2}\left(V(r) - \frac{Z}{r}\right)\right] n(r,p)\, d\vec{r} d\vec{p} = E_{\mathrm{k}} + E_{\mathrm{p}}, \qquad (6.8)$$

$$E_{\mathrm{k}} = \frac{2}{(2\pi)^3} \int\int \frac{p^2}{2} n(r,p)\, d\vec{r} d\vec{p} = \frac{3\sqrt{2}}{\pi^2} v\theta^{5/2} \int_0^1 x^2 I_{3/2}\left(\frac{\phi(x)}{x}\right) dx, \qquad (6.9)$$

$$E_{\mathrm{p}} = -\frac{1}{2}\int \rho(r)\left(V(r)+\frac{Z}{r}\right)d\vec{r} =$$
$$-\frac{3}{\sqrt{2}\pi^2}v\theta^{5/2}\int_0^1 x^2\left[\frac{\phi(x)}{x}+\eta+\frac{Z}{\theta r_0 x}\right]I_{1/2}\left(\frac{\phi(x)}{x}\right)dx. \quad (6.10)$$

Let us denote the integrals appearing in (6.10) by

$$\mathcal{J}_1 = \int_0^1 x\phi(x)I_{1/2}\left(\frac{\phi(x)}{x}\right)dx, \qquad \mathcal{J}_2 = \eta\int_0^1 x^2 I_{1/2}\left(\frac{\phi(x)}{x}\right)dx,$$

$$\mathcal{J}_3 = \phi(0)\int_0^1 xI_{1/2}\left(\frac{\phi(x)}{x}\right)dx.$$

To calculate these integrals we use integration by parts, attempting either to reduce them to the integral $\int_0^1 x^2 I_{3/2}(\phi(x)/x)\,dx$ that gives the kinetic energy (6.9), or to write them in a simpler form, using equation (1.17) and the boundary conditions (1.18). Since

$$\frac{d}{dx}\left[xI_{3/2}\left(\frac{\phi(x)}{x}\right)\right] = I_{3/2}\left(\frac{\phi(x)}{x}\right)+\frac{3}{2}xI_{1/2}\left(\frac{\phi(x)}{x}\right)\frac{x\phi'-\phi}{x^2},$$

we have

$$x\phi(x)I_{1/2}\left(\frac{\phi(x)}{x}\right) = -\frac{2}{3}x^2\frac{d}{dx}\left[xI_{3/2}\left(\frac{\phi(x)}{x}\right)\right]+$$
$$\frac{2}{3}x^2I_{3/2}\left(\frac{\phi(x)}{x}\right)+x^2I_{1/2}\left(\frac{\phi(x)}{x}\right)\phi'(x)$$

and consequently the integral $\mathcal{J}_1$ can be recast in the form

$$\mathcal{J}_1 = -\frac{2}{3}x^2\cdot xI_{3/2}\left(\frac{\phi(x)}{x}\right)\Big|_0^1+\frac{4}{3}\int_0^1 x^2I_{3/2}\left(\frac{\phi(x)}{x}\right)dx+$$
$$\frac{2}{3}\int_0^1 x^2I_{3/2}\left(\frac{\phi(x)}{x}\right)dx+\int_0^1 x^2I_{1/2}\left(\frac{\phi(x)}{x}\right)\phi'(x)\,dx =$$
$$-\frac{2}{3}I_{3/2}(-\eta)+2\int_0^1 x^2I_{3/2}\left(\frac{\phi(x)}{x}\right)dx+\frac{1}{a}\int_0^1 x\phi''(x)\phi'(x)\,dx.$$

The last integral above can be reduced again to $\mathcal{J}_1$. Indeed,

$$\int_0^1 x\phi''(x)\phi'(x)\,dx = \int_0^1 \frac{x}{2}\,d[\phi'(x)]^2 = \frac{x}{2}\,[\phi'(x)]^2\Big|_0^1 - \frac{1}{2}\int_0^1 \phi'(x)\,d\phi(x) =$$

$$\frac{1}{2}\eta^2 - \frac{1}{2}\phi'(x)\phi(x)\Big|_0^1 + \frac{1}{2}\int_0^1 \phi(x)\phi''(x)\,dx = \frac{1}{2}\,\phi(0)\phi'(0) + \frac{a}{2}\mathcal{J}_1.$$

Finally, we obtain the following expression for $\mathcal{J}_1$:

$$\mathcal{J}_1 = -\frac{4}{3}I_{3/2}(-\eta) + \frac{1}{a}\phi(0)\phi'(0) + 4\int_0^1 x^2 I_{3/2}\left(\frac{\phi(x)}{x}\right)\,dx. \tag{6.11}$$

After integration by parts, the integrals $\mathcal{J}_2$ and $\mathcal{J}_3$ are readily calculated to be

$$\mathcal{J}_2 = \frac{\eta}{a}\int_0^1 x\phi''(x)\,dx = \frac{\eta}{a}\left[x\phi'(x)\big|_0^1 - \int_0^1 \phi'(x)\,dx\right] = \frac{\eta}{a}\phi(0),$$

$$\mathcal{J}_3 = \frac{\phi(0)}{a}\int_0^1 \phi''(x)\,dx = -\frac{\eta}{a}\phi(0) - \frac{1}{a}\phi(0)\phi'(0).$$

Thus, the potential energy of the electrons, E_{p}, is given by

$$E_{\mathrm{p}} = \frac{2\sqrt{2}}{\pi^2}v\theta^{5/2}\left[I_{3/2}(-\eta) - 3\int_0^1 x^2 I_{3/2}\left(\frac{\phi(x)}{x}\right)dx\right]. \tag{6.12}$$

Formulas (6.1) and (6.8)–(6.12) yield the *virial theorem:*

$$2E_{\mathrm{k}} + E_{\mathrm{p}} = 3P_{\mathrm{e}}v. \tag{6.13}$$

The virial theorem holds not only in the framework of the Thomas-Fermi model, but also in the more precise quantum-mechanical approach [17]. This theorem may be used to obtain the pressure, but requires a high accuracy in calculating E_{k} and E_{p}.

When one computes the specific internal energy we have to remember that the total internal energy itself is not of interest — what counts is by how much this energy is larger than the energy corresponding to matter at null temperature and null pressure. Therefore, it makes sense to subtract from the value given by formula (6.8) the internal energy of an isolated atom, $E_0 = -0.76874512 \cdot Z^{7/3}$ [73], corresponding to $T = 0$, $\rho = 0$, and also take into account the kinetic energy of nuclei $\frac{3}{2}\,kT = \frac{3}{2}\,\theta$:

$$E = \frac{2.626 \cdot 10^3}{A} \left(E_e - E_0 + \frac{3}{2}\theta \right) \qquad \text{kJ/g} \tag{6.14}$$

(see Figure 6.2).

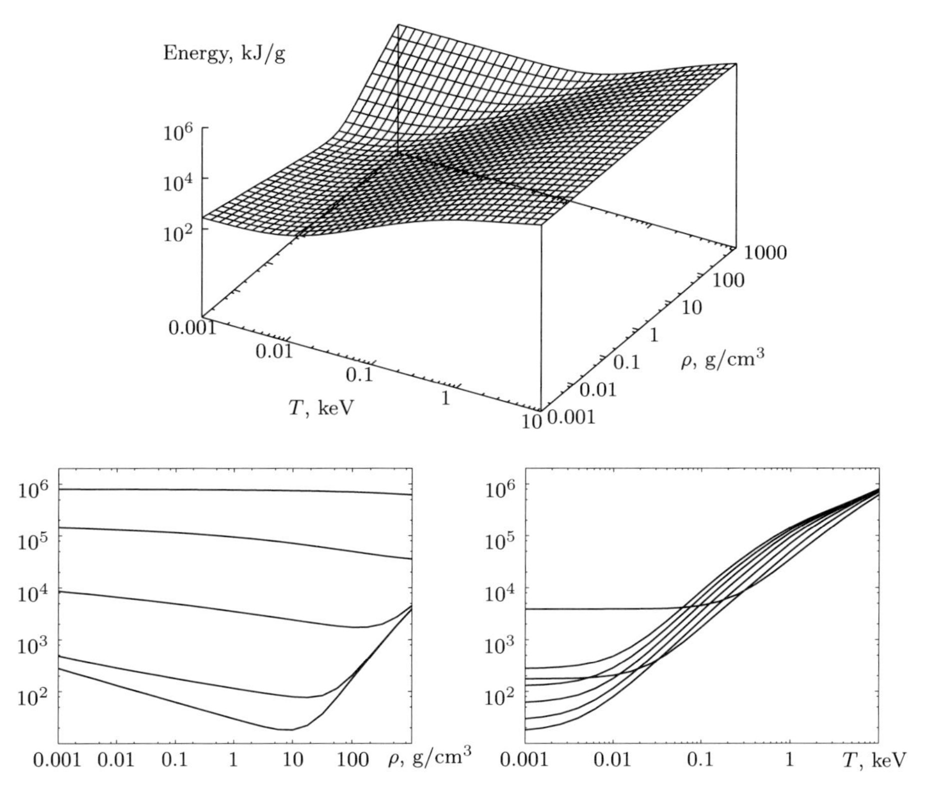

Figure 6.2: Internal energy surface according to the Thomas-Fermi model for gold, and the isotherms and isochores at temperatures $T = 10^i$ keV, $i = -3, -2, -1, 0, 1$ and densities $\rho = 10^k$ g/cm^3, $k = -3, -2, -1, 0, 1, 2, 3$

The entropy of electrons is given by the expression (see (1.4)):

$$S_e = -\frac{2}{(2\pi)^3} \int\int [n \ln n + (1-n) \ln (1-n)] \, d\vec{r} \, d\vec{p} =$$

$$- \frac{4\sqrt{2}\theta^{3/2} r_0^3}{\pi} \int_0^1 \int_0^\infty [n \ln n + (1-n) \ln (1-n)] \, y^{1/2} dy \, x^2 \, dx,$$

where

$$y = p^2/(2\theta), \quad n \equiv n(r,p) = 1 \bigg/ \left[1 + \exp\left(y - \frac{\phi(x)}{x}\right)\right].$$

The integral with respect to y can be calculated by integrating by parts twice and observing that the terms obtained by setting $y = 0$ and $y = \infty$ are equal to zero:

$$\int_0^\infty [n \ln n + (1-n)\ln(1-n)] y^{1/2} dy =$$

$$-\frac{2}{3}\int_0^\infty y^{3/2} \ln \frac{n}{1-n} \frac{\partial n}{\partial y} dy = -\frac{2}{3}\int_0^\infty y^{3/2}\left(-y + \frac{\phi(x)}{x}\right)\frac{\partial n}{\partial y} dy =$$

$$-\frac{5}{3}\int_0^\infty \frac{y^{3/2}}{1+\exp\left(y - \frac{\phi(x)}{x}\right)} dy + \frac{\phi(x)}{x}\int_0^\infty \frac{y^{1/2}}{1+\exp\left(y - \frac{\phi(x)}{x}\right)} dy =$$

$$-\frac{5}{3} I_{3/2}\left(\frac{\phi(x)}{x}\right) + \frac{\phi(x)}{x} I_{1/2}\left(\frac{\phi(x)}{x}\right).$$

Therefore,

$$S_{\rm e} = \frac{4\sqrt{2}\theta^{3/2} r_0^3}{\pi} \int_0^1 \left[\frac{5}{3} I_{3/2}\left(\frac{\phi(x)}{x}\right) - \frac{\phi(x)}{x} I_{1/2}\left(\frac{\phi(x)}{x}\right)\right] x^2 dx,$$

and using (6.11) we have

$$S_{\rm e} = \frac{4\sqrt{2}\theta^{3/2} r_0^3}{\pi}\left[\frac{4}{3} I_{3/2}(-\eta) - \frac{7}{3}\int_0^1 x^2 I_{3/2}\left(\frac{\phi(x)}{x}\right) dx\right] - Z\phi'(0). \tag{6.15}$$

If we now account for the contribution of nuclei to the entropy in the ideal-gas approximation (considering that their statistical weight is equal to 1), we finally obtain

$$S = \frac{0.9648 \cdot 10^2}{A}\left[S_{\rm e} + \frac{3}{2}\ln\left(\frac{M\theta v^{2/3}}{2\pi}\right) + \frac{5}{2}\right] \text{ kJ/(g·eV)}, \tag{6.16}$$

where $M = 1836 \cdot A$ is the ion mass (see Figure 6.3).

The graphs of pressure, internal energy and entropy for gold furnished by formulas (6.7), (6.14) and (6.16), respectively, are shown in figures 6.1–6.3. Although these graphs were obtained for gold ($Z_1 = 79$, $A_1 = 197$), for other elements (Z_2, A_2) the graphs retain their shape, but all quantities corresponding to the contribution of electrons are multiplied by scaling factors that depend on the element in question:

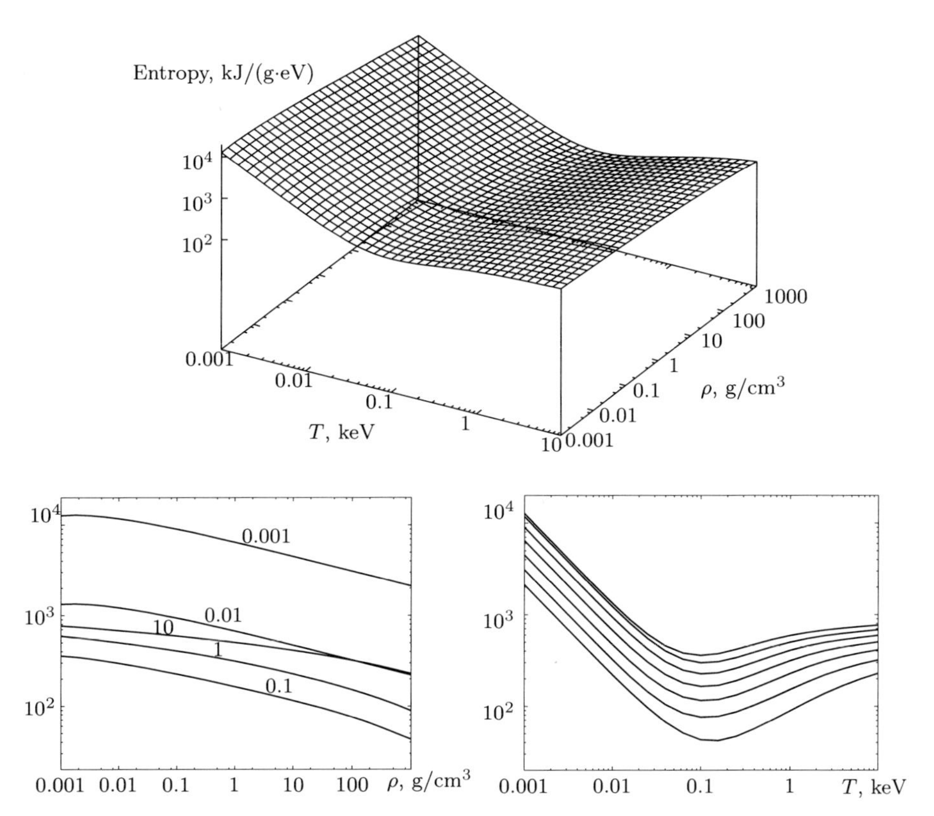

Figure 6.3: Entropy surface according to the Thomas-Fermi model for gold, and isotherms and isochores at temperatures $T = 10^i$ keV, $i = -3, -2, -1, 0, 1$ and densities $\rho = 10^k$ g/cm^3, $k = -3, -2, -1, 0, 1, 2, 3$

$$\rho \to \frac{A_2 Z_2}{A_1 Z_1} \cdot \rho, \qquad T \to \left(\frac{Z_2}{Z_1}\right)^{4/3} \cdot T,$$

$$P \to \left(\frac{Z_2}{Z_1}\right)^{10/3} \cdot P, \qquad E \to \left(\frac{Z_2}{Z_1}\right)^{7/3} \cdot E, \qquad S \to \frac{Z_2}{Z_1} \cdot S \tag{6.17}$$

(recall the self-similarity properties of the Thomas-Fermi potential discussed in § 1.1).

6.1.2 Quantum, exchange and oscillation corrections to the Thomas-Fermi model

To refine the Thomas-Fermi model one can resort to the expansion of the electron density in the Hartree-Fock approximation in powers of the Planck constant $\hbar$. Such an expansion was obtained in [103] by recasting the Hartree-Fock equation as a Poisson equation for the self-consistent Coulomb potential:

$$\Delta V(\vec{r}) = 4\pi\rho(\vec{r}) - 4\pi Z\delta(\vec{r}), \tag{6.18}$$

where Z is the charge of the nucleus. The difference between (6.18) and the analogous equation in the Thomas-Fermi model (see Subsection 1.1.2) is that here the electron density $\rho(\vec{r})$ is expressed through the density matrix $\widehat{\rho}(\widehat{H})$. If for basis functions we take plane waves, then

$$\rho(\vec{r}) = \int e^{-i\vec{p}\vec{r}/\hbar}\widehat{\rho}(\widehat{H})e^{i\vec{p}\vec{r}/\hbar}\frac{2d\vec{p}}{(2\pi)^3}, \tag{6.19}$$

where

$$\widehat{\rho}(\widehat{H}) = f\left(\frac{\widehat{H}-\mu}{\theta}\right) = \frac{1}{1+\exp\left(\dfrac{\widehat{H}-\mu}{\theta}\right)}, \tag{6.20}$$

$$\widehat{H} = \frac{1}{2}\widehat{\vec{p}}^{\,2} - V(\vec{r}) - \widehat{A}(\vec{r},\widehat{\vec{p}}),$$

$$\widehat{A}(\vec{r},\widehat{\vec{p}}) = \frac{1}{2}\int e^{-i\vec{p}\,'\vec{r}/\hbar}\,\frac{4\pi\hbar^2\widehat{\rho}(\widehat{H}')}{|\vec{p}-\vec{p}\,'|^2}\,e^{i\vec{p}\,'\vec{r}/\hbar}\,\frac{d\vec{p}\,'}{(2\pi)^3}. \tag{6.21}$$

Here $\widehat{\vec{p}} = -i\hbar\nabla$ is the momentum operator and $\widehat{A}(\vec{r},\widehat{\vec{p}})$ is the exchange-interaction operator. Although in atomic units the Planck constant $\hbar$ is equal to 1, it is partially left in formulas to facilitate the computation of the expansion in powers of $\hbar$. When we expand the expression appearing in (6.19) in powers of $\hbar$ we must keep in mind that the expansion is valid only for functions that are analytic in $\hbar$. As a matter of fact, as one can see from (6.19), the analyticity conditions are violated, and consequently the expansion leaves unaccounted for the so-called oscillation corrections (see below).

To calculate the action of the operator $\widehat{\rho}(\widehat{H})$ on the function $e^{i\vec{p}\vec{r}/\hbar}$ it suffices to replace in $\widehat{H}$ the operator $\widehat{\vec{p}}$ by $\vec{p} - i\hbar\nabla$, and instead of $e^{-i\vec{p}\vec{r}/\hbar}\,\widehat{\rho}(\widehat{H})\,e^{i\vec{p}\vec{r}/\hbar}$ in (6.19) one can write the action of the operator $f(\widehat{a}+\widehat{b})$ (see (6.20)) on the unity, where

$$\widehat{a} = \frac{(\vec{p}-i\hbar\nabla)^2}{2\theta}, \quad \widehat{b} = -\Phi(\vec{r}) = -\frac{V(\vec{r})+\mu}{\theta}$$

(here we neglected the exchange interaction).

For $f(\widehat{a}+\widehat{b})$ one has the following expansion in commutators (see [105]):

$$f(\widehat{a}+\widehat{b}) = f(a+\widehat{b}) - \frac{1}{2}f''(a+\widehat{b})[\widehat{b},\widehat{a}] + \\ \frac{1}{6}f'''(a+\widehat{b})\left\{[\widehat{b},[\widehat{b},\widehat{a}]] - [\widehat{a},[\widehat{b},\widehat{a}]\right\} + \frac{1}{8}f^{\mathrm{IV}}(a+\widehat{b})([\widehat{b},\widehat{a}])^2 + \dots, \tag{6.22}$$

which yields

$$e^{-i\vec{p}\vec{r}/\hbar} f\left(\frac{\widehat{\vec{p}}^2}{2\theta} - \Phi\right) e^{i\vec{p}\vec{r}/\hbar} = f(\varepsilon) + \frac{i\hbar}{2\theta} f''(\varepsilon)\vec{p}\nabla\Phi + \frac{\hbar^2}{4\theta} f''(\varepsilon)\,\Delta\Phi - \\ \frac{\hbar^2}{6\theta} f'''(\varepsilon)(\nabla\Phi)^2 + \frac{\hbar^2}{6\theta^2} f'''(\varepsilon)(\vec{p}\nabla)^2\Phi - \frac{\hbar^2}{8\theta^2} f^{\mathrm{IV}}(\varepsilon)(\vec{p}\nabla\Phi)^2 + \dots, \tag{6.23}$$

where we used the notation

$$\varepsilon = \frac{p^2}{2\theta} - \frac{V(\vec{r})+\mu}{\theta}.$$

First let us consider the first term of order zero in $\hbar$, discarding all the other terms. This means that, in addition to the already unaccounted for exchange interaction $\widehat{A} = \widehat{A}(\vec{r},\widehat{\vec{p}}) \sim \hbar^2$, we also neglect the fact that the operators $\widehat{\vec{p}}$ and $\Phi(\vec{r})$ do not commute. In this way we obtain the expression for the electron density in the Thomas-Fermi model:

$$\rho_{\mathrm{TF}}(\vec{r}) = \frac{2}{(2\pi)^3}\int \frac{d\vec{p}}{1+\exp\left(\dfrac{p^2}{2\theta} - \dfrac{V_{\mathrm{TF}}(\vec{r})+\mu_{\mathrm{TF}}}{\theta}\right)} = \\ \frac{\sqrt{2}\,\theta^{3/2}}{\pi^2} I_{1/2}\left(\frac{V_{\mathrm{TF}}(\vec{r})+\mu_{\mathrm{TF}}}{\theta}\right),$$

where

$$\Delta V_{\mathrm{TF}}(\vec{r}) = 4\pi\rho_{\,\mathrm{TF}}(\vec{r}) = \frac{4\sqrt{2}\,\theta^{3/2}}{\pi} I_{1/2}\left(\frac{V_{\mathrm{TF}}(\vec{r})+\mu_{\mathrm{TF}}}{\theta}\right).$$

To incorporate the corrections of second order in $\hbar$ it is necessary, in addition to (6.23), to account for the corrections connected with the exchange interaction (6.21), for which is suffices to use the linear approximation in $\widehat{A}$:

$$\delta\rho_{\text{ex}}(\vec{r}) = \int \left[f\left(\frac{p^2}{2\theta} - \Phi - \frac{\widehat{A}}{\theta}\right) - f\left(\frac{p^2}{2\theta} - \Phi\right)\right] \frac{2d\vec{p}}{(2\pi)^3} \approx$$

$$\frac{\partial}{\partial\Phi} \int \frac{\widehat{A}}{\theta} f\left(\frac{p^2}{2\theta} - \Phi\right) \frac{2\,d\vec{p}}{(2\pi)^3} =$$

$$\frac{4\pi\hbar^2}{2(2\pi)^3\theta} \frac{\partial}{\partial\Phi} \iint \frac{f\left(\frac{p^2}{2\theta} - \Phi\right) f\left(\frac{p'^2}{2\theta} - \Phi\right) d\vec{p}d\vec{p}\,'}{|\vec{p} - \vec{p}\,'|^2} =$$

$$\frac{2\theta\hbar^2}{\pi^3} \frac{\partial}{\partial\Phi} \int_{-\infty}^{\Phi} \left[I'_{1/2}(t)\right]^2 dt = \frac{2\theta\hbar^2}{\pi^3} \left[I'_{1/2}(\Phi)\right]^2 \tag{6.24}$$

(cf. the calculation of the exchange integral in Subsection 3.3.1).

Substituting the expression (6.23) in (6.19) and incorporating the correction (6.24) we obtain the expression for the electron density in the Thomas-Fermi model with corrections (TFC):

$$\rho_{\text{TFC}}(\vec{r}) = \frac{\sqrt{2}\,\theta^{3/2}}{\pi^2} \left\{ I_{1/2}(\Phi) + \frac{\hbar^2\sqrt{2}}{\pi\sqrt{\theta}} \left(I'_{1/2}(\Phi)\right)^2 + \frac{\hbar^2\Delta\Phi}{12\,\theta} I''_{1/2}(\Phi) - \frac{\hbar^2(\nabla\Phi)^2}{24\,\theta} I'''_{1/2}(\Phi) \right\}. \tag{6.25}$$

Let us set $\Phi = \Phi_0 + \hbar^2\Phi_1$, where

$$\Phi_0 = \Phi_0(\vec{r}) = \frac{V_{\text{TF}}(\vec{r}) + \mu_{\text{TF}}}{\theta},$$

$$\Phi_1 = \Phi_1(\vec{r}) = \frac{\sqrt{2}}{6\pi\sqrt{\theta}} \left[I'_{1/2}\left(\frac{V_{\text{TF}}(\vec{r}) + \mu_{\text{TF}}}{\theta}\right) + \zeta(\vec{r})\right],$$

and then substitute Φ in the equation $\Delta\Phi = 4\pi\rho/\theta$. If we retain only the terms of order $\hbar^2$, then the correction $\zeta = \zeta(\vec{r})$ is defined by the expression

$$\Delta\zeta - \frac{4\sqrt{2\theta}}{\pi} I'_{1/2}(\Phi_0) \cdot \zeta = \frac{4\sqrt{2\theta}}{\pi} Y'(\Phi_0), \tag{6.26}$$

where

$$Y(x) = I_{1/2}(x) I'_{1/2}(x) + 6 \int_{-\infty}^{x} \left[I'_{1/2}(t)\right]^2 dt. \tag{6.27}$$

Further, to calculate the integral $Y(x)$ one can use approximations for the Fermi-Dirac function [46] and formula (3.58) for the integral

$$\int_{-\infty}^{x} \left[I'_{1/2}(t)\right]^2 dt = \frac{1}{4} \int_{-\infty}^{x} \left[I_{-1/2}(t)\right]^2 dt.$$

Assuming spherical symmetry and passing to the variable $x = r/r_0$ $(0 \le x \le 1)$, we obtain the equation of the Thomas-Fermi model with corrections (see Subsection 2.1.5) for the function $\chi(x) = r\zeta(r)/r_0$:

$$\chi'' - \frac{4\sqrt{2\theta}}{\pi} r_0^2 I'_{1/2}\left(\frac{\phi(x)}{x}\right)\chi = \frac{4\sqrt{2\theta}}{\pi} r_0^2 xY'\left(\frac{\phi(x)}{x}\right), \tag{6.28}$$

$$\chi(0) = 0, \quad \chi(1) = \chi'(1), \quad \frac{\phi(x)}{x} = \Phi_0(r).$$

The corrections for the thermodynamic functions of electrons are obtained in much the same way as the thermodynamic functions of electrons in the Thomas-Fermi model. The formulas for corrections have the following form [104, 98]:

for pressure

$$\Delta P = \frac{\theta^2}{3\pi^3}\left[\chi(1)I_{1/2}(\phi(1)) + Y(\phi(1))\right]; \tag{6.29}$$

for energy

$$\Delta E = \frac{2\theta^2}{3\pi^2} r_0^2 \left[\int_0^1 x\chi(x)I_{1/2}\left(\frac{\phi(x)}{x}\right)dx + 2\int_0^1 x^2 Y\left(\frac{\phi(x)}{x}\right)dx\right] + \frac{Z\sqrt{2\theta}}{6\pi}\chi'(0) - \Delta E_0, \tag{6.30}$$

where $\Delta E_0 = -0.2690017Z^{5/3}$;

for entropy

$$\Delta S = \frac{2\theta}{3\pi^2} r_0^2 \left[\int_0^1 x\chi(x)I_{1/2}\left(\frac{\phi(x)}{x}\right)dx + 4\int_0^1 x^2 Y\left(\frac{\phi(x)}{x}\right)dx\right] + \frac{\sqrt{2}Z}{6\pi\sqrt{\theta}}\chi'(0). \tag{6.31}$$

The corrections ΔP, ΔE and ΔS enjoy the self-similarity property, like the pressure, energy and entropy themselves in the Thomas-Fermi model. In particular, if one knows their values for an element (Z_1, A_1), then for another element

(Z_2, A_2), with the appropriate scaling of the temperature and density (see (6.17)), we must carry out the transformation

$$\Delta P \to \left(\frac{Z_2}{Z_1}\right)^{8/3} \cdot \Delta P, \qquad \Delta E \to \left(\frac{Z_2}{Z_1}\right)^{5/3} \cdot \Delta E, \qquad \Delta S \to \left(\frac{Z_2}{Z_1}\right)^{1/3} \cdot \Delta S. \quad (6.32)$$

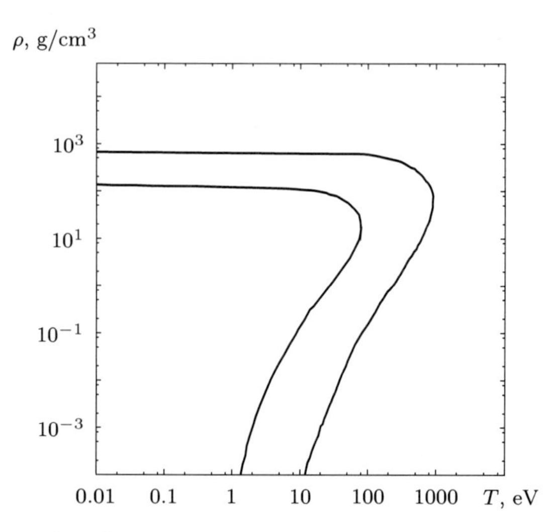

Figure 6.4: Magnitude of the corrections to the Thomas-Fermi model for aluminum. The solid curves in the (T, ρ)-plane correspond to a fixed value of the quantum and exchange corrections to pressure (see (6.29)): for the outer [resp., inner] curve the correction is about 5% [resp., 30%]

Detailed tables of thermodynamic quantities computed by means of the Thomas-Fermi model and the Thomas-Fermi model with corrections can be found in [98]. A representation about the magnitude of these corrections is provided by Figure 6.4, which shows the curves corresponding to a fixed correction to pressure for aluminum: $\Delta P/P_{\mathrm{TF}}$=0.3 and $\Delta P/P_{\mathrm{TF}}$=0.05. Analogous curves for other elements can be obtained by using the similarity properties (6.17) and (6.32). From the magnitude of these corrections one could draw conclusions about the applicability of the Thomas-Fermi model, provided that one would bring into the picture the oscillation corrections [106].

Let us illustrate on a simple example how an oscillation correction to the electron density at temperature $T = 0$ can be calculated in the one-dimensional case. We will proceed from the expression for the density of particles at temperature zero:

$$\rho(x) = 2 \sum_{\varepsilon_n \le \mu} |\psi_n(x)|^2. \quad (6.33)$$

Here the factor 2 is introduced to account for the two directions of the spin, $\psi_n(x)$ is the solution of the stationary Schrödinger equation and μ is the chemical potential (Fermi energy).

In the semiclassical approximation, the wave function $\psi_n(x)$ in the classical domain of motion $a_n < x < b_n$ has the form

$$\psi_n(x) = \frac{c_n}{\sqrt{p_n(x)}} \sin\left(\sigma_n(x) + \frac{\pi}{4}\right), \tag{6.34}$$

where

$$\sigma_n(x) = \int\limits_{a_n}^{x} p_n(x)dx, \quad p_n(x) = \sqrt{2[\varepsilon_n - U(x)]},$$

ε_n is the energy of the particle and $U(x)$ is the potential energy. The constant c_n is determined from the condition $\int_{a_n}^{b_n} \psi_n^2(x)dx = 1$, which yields $c_n^2 = 2/\tau_n^0$, with $\tau_n^0 = \tau_n(b_n) = \int_{a_n}^{b_n} [1/p_n(x)]dx$. Substituting expression (6.34) in (6.33), we obtain

$$\rho(x) = 2\sum_{\varepsilon_n \le \mu} \frac{c_n^2}{p_n(x)} \sin^2[\sigma_n(x) + \frac{\pi}{4}] = \rho_{\mathrm{TF}}(x) + \rho_{\mathrm{osc}}(x), \tag{6.35}$$

where

$$\rho_{\mathrm{TF}}(x) = \sum_{\varepsilon_n \le \mu} \frac{2}{\tau_n^0}\frac{1}{p_n(x)}, \quad \rho_{\mathrm{osc}}(x) = \sum_{\varepsilon_n \le \mu} \frac{2}{\tau_n^0}\frac{1}{p_n(x)} \sin 2\sigma_n(x).$$

The first term in (6.35), which corresponds to the Thomas-Fermi model, yields the averaged density distribution. It can be calculated with the help of the Bohr-Sommerfeld quantization rule (see Subsection 2.1.3).

The second term in (6.35), which describes shell effects, can be calculated in closed form by using the Poisson summation formula and integration by parts (see [106]):

$$\rho_{\mathrm{osc}}(x) = -\frac{1}{\tau^0 p(x)} \cos 2\sigma(x) \operatorname{ctg} \pi\frac{\tau(x)}{\tau^0}, \tag{6.36}$$

where $p(x), \sigma(x)$ and $\tau(x)$ are calculated in the same way as $p_n(x), \sigma_n(x)$ and $\tau_n(x)$, but for $\varepsilon_n = \mu$. Note that formula (6.36), like the analogous formula for the three-dimensional case, has a singularity when $\rho_{\mathrm{osc}} \to \infty$ for certain values of x. To remove these singularities tedious additional calculations are needed. Oscillation corrections have an irregular character and are specific for each element [106]. Exchange, quantum and oscillation corrections are accounted for in the equation of state in [108].

6.2 The ionization equilibrium method

6.2.1 The Gibbs distribution for the atom cell

In the case of an ideal equilibrium plasma (which is also in equilibrium with the radiation), a model widely used to calculate the composition and thermodynamic functions of the plasma is the ionization equilibrium method (the Saha-Boltzmann equations, see, e.g., [119]). The equations of this model can be derived if one starts from the Gibbs distribution and uses as a subsystem a spherical cell of volume $v = \frac{4}{3}\pi r_0^3$. In this approximation the probability of a state with energy E and number of electrons N is given by

$$W = Cg \exp\left(-\frac{E - \mu N}{\theta}\right), \tag{6.37}$$

where g is the statistical weight of the state and μ is the chemical potential.

Since we are interested only in the distribution of ions for any possible states of the free electrons, the distribution (6.37) must be averaged over all possible states of the free electrons. As an approximation we may consider that all the quantities figuring in the Gibbs distribution are taken for some average density of the free electrons. In particular, the energy E may be assumed to be equal to the sum of the ion energy E_{js} (j is the degree of ionization and s is the number of the ion state with the given degree of ionization j), the average energy of the free electrons, E_e, and the average energy of interaction of the ion under consideration with the free electrons, E_{je}, i.e.,

$$E = E_{js} + E_{je} + E_e. \tag{6.38}$$

We write the number of electrons in the cell as

$$N = Z - j + Z_0, \tag{6.39}$$

where Z is the charge of the nucleus and Z_0 is the average charge of the ion (the average number of free electrons per one ion).

Accordingly, the concentration of ions with charge j in state s, x_{js}, that is, the fraction of ion with charge j in the state s, is given by

$$x_{js} = C\, g_{js} \exp\left(-\frac{E_{js} + E_{je} + \mu j}{\theta}\right), \tag{6.40}$$

where g_{js} is the statistical weight of the corresponding state of the ion and C is a normalization constant.

Using (6.40), one obtains relations between concentrations of ions with different degrees of ionization:

$$\frac{x_{j+1,p}}{x_{js}} = \frac{g_{j+1,p}}{g_{js}} \exp\left(-\frac{\mu}{\theta}\right) \exp\left(-\frac{E_{j+1,\mathrm{e}} - E_{j,\mathrm{e}}}{\theta}\right) \exp\left(-\frac{E_{j+1,p} - E_{j,s}}{\theta}\right). \tag{6.41}$$

6.2.2 The Saha approximation

Next, let us find how the chemical potential μ and the density of the free electrons ρ_{e} are related. We shall assume that the potential $V(r)$ created by the free electrons and the ion under consideration is central.

According to the Fermi-Dirac statistics

$$\rho_{\mathrm{e}} = \rho_{\mathrm{e}}(r) = \frac{2}{(2\pi)^3} \int \frac{4\pi p^2 \, dp}{1 + \exp\left[\frac{1}{\theta}\left(\frac{p^2}{2} - V(r) - \mu\right)\right]}.$$

For a nondegenerate electron gas $\exp(-\mu/\theta) \gg 1$, and so

$$\rho_{\mathrm{e}}(r) \approx \frac{8\pi}{(2\pi)^3} \exp(\mu/\theta) \int p^2 \exp\left[-\frac{1}{\theta}\left(\frac{p^2}{2} - V(r)\right)\right] dp. \tag{6.42}$$

At sufficiently high temperatures the potential energy of the electrons can be neglected. In this approximation the electron density ρ_{e} is a constant, and we obtain

$$\exp(-\mu/\theta) = \frac{2}{\rho_{\mathrm{e}}} \left(\frac{\theta}{2\pi}\right)^{3/2}.$$

Since $\rho_{\mathrm{e}} = 3Z_0/(4\pi r_0^3)$, we have

$$\exp(-\mu/\theta) = \frac{2}{3}\sqrt{\frac{2}{\pi}} \frac{r_0^3 \theta^{3/2}}{Z_0} \tag{6.43}$$

(compare to formula (1.26)).

Using (6.41), (6.43) and neglecting the interaction between the free electrons and the ion (for small densities of the free electrons $E_{j\mathrm{e}} \approx 0$), we obtain the system of Saha-Boltzmann equations

$$Z_0 \frac{x_{j+1,p}}{x_{js}} = \frac{2}{3}\sqrt{\frac{2}{\pi}} r_0^3 \theta^{3/2} \frac{g_{j+1,p}}{g_{js}} \exp\left(-\frac{E_{j+1,p} - E_{js}}{\theta}\right). \tag{6.44}$$

These equations must be supplemented by the normalization and charge neutrality conditions

$$\sum_{js} x_{js} = 1, \qquad \sum_{js} j x_{js} = Z_0. \tag{6.45}$$

The quantities $E_{j+1,p}$ and E_{js} that appear in (6.44) are determined from experimental data, and are also computed in the Hartree-Fock (Hartree-Fock-Slater) approximation, or by means of some semi-empirical models (see, e.g., [76]). As its derivation shows, the ionization equilibrium model is applicable for sufficiently low densities of matter. Let us mention two circumstances that restrict the domain of applicability of the model. First, it may be the case that no data on energy levels of highly ionized ions are available in the literature. Second, when the density increases, the energy levels shift and may not agree with the experimental data obtained for free ions. Let us add here that, in the case of very low densities the plasma is, as a rule, in another equilibrium state — coronal equilibrium — and consequently it cannot be described by the Saha equations (see Subsection 6.2.4).

It is convenient to rewrite the system of equations (6.44)–(6.45) in terms of the ion concentrations

$$x_j = \sum_s x_{js}. \tag{6.46}$$

This is done by summing with respect to s and p in (6.44), (6.45). This yields

$$\begin{cases} Z_0 \dfrac{x_{j+1}}{x_j} = \varphi_j, \quad j = 0, 1, \dots, j_{\max}, \\ \sum\limits_j x_j = 1, \\ \sum\limits_j j x_j = Z_0, \end{cases} \tag{6.47}$$

where

$$x_{js} = \frac{x_j}{B_j} g_{js} \exp\left(-\frac{E_{js} - E_{j0}}{\theta}\right), \tag{6.48}$$

$$\varphi_j = \frac{2}{3}\sqrt{\frac{2}{\pi}}\, r_0^3 \theta^{3/2} \frac{B_{j+1}}{B_j} \exp\left(-\frac{I_j}{\theta}\right), \tag{6.49}$$

$$B_j = \sum_s g_{js} \exp\left(-\frac{E_{js} - E_{j0}}{\theta}\right), \tag{6.50}$$

$I_j = E_{j+1,0} - E_{j0}$ is the ionization potential of the ion with charge j and E_{j0} is the energy of the ground state of the ion j (the energies E_{js} are measured starting from the ground state of the atom: $j = 0$, $s = 0$).

The thermodynamic functions of matter in the ionization equilibrium model are computed according to the formulas for the ideal Boltzmann gas with the ionization loss accounted for:

$$P = (Z_0 + 1)\frac{\theta}{v}, \tag{6.51}$$

$$E = \frac{3}{2}(Z_0 + 1)\theta + \sum_{js} x_{js}(E_{js} - E_{00}). \tag{6.52}$$

6.2.3 An iteration scheme for solving the system of equations of ionization equilibrium

To solve the system (6.47) let us derive an equation for Z_0, eliminating the quantities x_j by taking the product of the first j equations of the system:

$$\frac{x_1}{x_0} \cdot \frac{x_2}{x_1} \cdots \frac{x_j}{x_{j-1}} = \prod_{k=0}^{j-1} \left(\frac{\varphi_k}{Z_0} \right) = a_j, \tag{6.53}$$

$$j = 1, 2, \dots, j_{\max}; \qquad a_0 = 1.$$

Since $x_j = x_0 a_j$ (this being a consequence of (6.53)), the normalization condition $\sum_j x_j = 1$ yields $x_0 = 1/\sum_j a_j$. Substituting this expression in the last of equations (6.47), we obtain the sought-for equation for the determination of Z_0:

$$Z_0 = \frac{\sum\limits_j j a_j}{\sum\limits_j a_j}. \tag{6.54}$$

To solve (6.54) we use Newton's iteration method

$$\overset{(s+1)}{Z_0} = \overset{(s)}{Z_0} - \left. \frac{F}{\frac{\partial F}{\partial Z_0}} \right|_{Z_0 = \overset{(s)}{Z_0}},$$

where

$$F = F(Z_0) = Z_0 - \frac{\sum\limits_j j a_j}{\sum\limits_j a_j}.$$

Observing that $\partial a_j / \partial Z_0 = -j a_j / Z_0$, we arrive at the following iteration scheme:

$$\overset{(s+1)}{Z_0} = Z_0 \left. \frac{\langle j^2 \rangle - \langle j \rangle^2 + \langle j \rangle}{\langle j^2 \rangle - \langle j \rangle^2 + Z_0} \right|_{Z_0 = \overset{(s)}{Z}_0}, \tag{6.55}$$

where

$$\langle j^k \rangle = \frac{\sum\limits_j j^k a_j}{\sum\limits_j a_j}.$$

To obtain an initial approximation we note that the quantities φ_j decrease with the growth of j, since the ionization energies I_j increase with j. Therefore, the values a_j either decrease when j grows, if $\varphi_0/Z_0 < 1$, or have a maximum for some value $j = j_0$, for which

$$\varphi_{j_0-1}/Z_0 \geq 1, \qquad \varphi_{j_0}/Z_0 \leq 1. \tag{6.56}$$

From equation (6.54) it follows that $Z_0 \approx j_0$, because a_j has a maximum for $j \approx j_0$. We conclude that as a criterion for choosing j_0 we can use, instead of (6.56), the condition

$$\varphi_{j_0-1}/j_0 \geq 1 \geq \varphi_{j_0}/j_0, \tag{6.57}$$

which in turn shows that j_0 is the smallest value of j for which $\varphi_j/j \leq 1$. Since $x_j = x_0 a_j$, to find an approximate value of Z_0 in the system (6.47) one can neglect all x_j, except for the two largest ones. If $a_{j_0+1} > a_{j_0-1}$, i.e., $\varphi_{j_0-1}\varphi_{j_0}/j_0^2 > 1$, then for the largest values of x_j we must take x_{j_0} and x_{j_0+1}. This yields the system

$$\begin{cases} Z_0 \dfrac{x_{j_0+1}}{x_{j_0}} = \varphi_{j_0}, \\ x_{j_0} + x_{j_0+1} = 1, \\ j_0 x_{j_0} + (j_0+1)x_{j_0+1} = Z_0, \end{cases}$$

whence

$$Z_0 = \frac{(j_0 - \varphi_{j_0}) + \sqrt{(j_0 + \varphi_{j_0})^2 + 4\varphi_{j_0}}}{2}. \tag{6.58}$$

In the case $a_{j_0-1} \geq a_{j_0+1}$, i.e., $\varphi_{j_0-1}\varphi_{j_0}/j_0^2 \leq 1$, when we choose an approximate value for Z_0 we must replace j_0 by $j_0 - 1$ in formula (6.58).

When one computes the statistical sums B_j one has to keep in mind that for low densities — in particular, for a free atom — such sums diverge. In practical computations one truncates these statistical sums, discarding the highly-excited states, or diminishing the role they play. As the analysis carried out in [100] has shown, the most effective approach is to take into account only those bound states for which the domain of classical motion of the electron lies inside the atom cell [149].

To estimate the radius of the electron orbit in a highly-excited state with quantum numbers n and ℓ for an ion with ionization degree j we may consider, as an approximation, that the electron moves in the field $(j+1)/r$. The domain of classical motion of such an electron is defined by the inequalities

$$r_{n\ell}^{(1)} < r < r_{n\ell}^{(2)},$$

where

$$r_{n\ell}^{(1,2)} = \frac{n^2}{j+1}\left[1 \mp \sqrt{1 - \left(\frac{\ell + 1/2}{n}\right)^2}\right].$$

To compute the statistical sums B_j we must consider only the states for which $r_{n\ell}^{(2)} < r_0$. When the density is increased further, it is necessary to account for the fact that the ionization potential decreases by the amount $\Delta I_j \simeq 3j/(2r_0)$, and also pay attention to the possible vanishing of weakly-coupled bound levels (their transition into the continuum). Note that under such circumstances the Saha-Boltzmann approximation ceases to work. When the density grows, the so-called cell models become increasingly more accurate, among them the quantum-statistical Hartree-Fock-Slater self-consistent field model discussed in Chapter IV.

6.2.4 Coronal equilibrium

For low densities, in a totally transparent plasma the equilibrium population of levels prescribed by the Gibbs distribution (6.40) may fail to be realized in the case of a real plasma. The point is that, due to the high rate of radiative processes the collisional electron excitations are rapidly eliminated by radiative decay to the ground state. For such a plasma one can assume that all ions with various ionization degrees are the most of the time in the ground state.[*)]

From (5.22) we have

$$\frac{dn_j}{dt} = \alpha_{j+1,j}^{\text{rec}} n_{j+1} - (\alpha_{j,j-1}^{\text{rec}} + \alpha_{j,j+1}^{\text{ion}}) n_j + \alpha_{j-1,j}^{\text{ion}} n_{j-1}, \tag{6.59}$$

where n_j is the concentration of ions with ionization degree j and $\alpha_{j,j'}^{\text{ion}}$, $\alpha_{j,j'}^{\text{rec}}$ are the total ionization and recombination rates of these ions. For small densities, in an optically thin plasma one can include into considerations the impact ionization (ii), photorecombination (phr) and dielectronic recombination (dr) processes only:

$$\alpha_{j,j+1}^{\text{ion}} = \sum_s \sum_{s'} \alpha_{js \to j+1,s'}^{\text{ii}}, \tag{6.60}$$

$$\alpha_{j,j-1}^{\text{rec}} = \sum_s \sum_{s'} \alpha_{js \to j-1,s'}^{\text{phr}} + \sum_s \sum_{s'} \alpha_{js \to j-1,s'}^{\text{dr}}. \tag{6.61}$$

Here the sum is taken over those states s of an ion with ionization degree j that are close to the ground state $s = 0$ ($E_{js} - E_{j0} \ll \theta$, where E_{js} is the energy of the ion state js). The excitations take place for states js^* such that $E_{js^*} - E_{j0} \gg \theta$, i.e., there exists a gap between ground and excited states, which is essential for the coronal approximation.

Putting $dn_j/dt = 0$ and grouping the direct and inverse processes for each of the ions, we obtain from (6.59) the relations

$$\frac{n_{j+1}}{n_j} = \frac{\alpha_{j,j+1}^{\text{ion}}}{\alpha_{j+1,j}^{\text{rec}}}, \quad j = 0, 1, \ldots, Z-1.$$

[*)] The Gibbs distribution and the Saha equation are valid for a low-density transparent plasma if this plasma is placed in a thermostat with the radiation temperature equal to the electron temperature.

Finally, in the coronal approximation the system of equations for the relative concentrations $x_j = n_j / \sum_j n_j$ reads

$$\frac{x_{j+1}}{x_j} = \frac{\alpha^{\mathrm{ion}}_{j,j+1}}{\alpha^{\mathrm{rec}}_{j+1,j}}, \tag{6.62}$$

$$\sum_j j x_j = Z_0, \tag{6.63}$$

$$\sum_j x_j = 1. \tag{6.64}$$

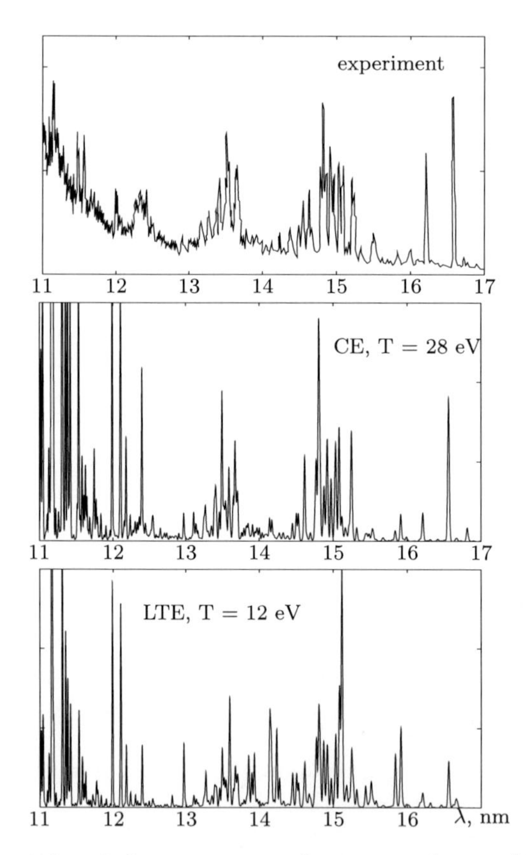

Figure 6.5: Experimental emission spectrum of a xenon plasma (the intensity is measured in arbitrary units; see [169]) and calculated values of its emissivities for density $\rho = 10^{-6}$ g/cm^3, in different approximations: LTE - Saha, CE - coronal equilibrium. The calculations were carried out with an additional broadening of 0.05 eV.

Figure 6.5 compares the spectra of a xenon plasma with density $\rho = 10^{-6}$ g/cm^3, obtained in different approximations. In the coronal model the plasma temperature $T = 28$ eV must be more than two times larger than the equilibrium temperature of the plasma, $T = 12$ eV, in order for the same degree of ionization $Z_0 = 9.6$ to be attained (see Table 6.1). The experimental data shown in the figure have an integral character; nevertheless, it is obvious that the result given by the coronal equilibrium (CE) model are considerably closer to the experimental data than those given by the Saha approximation (LTE).

The collisional radiative equilibrium (CRE) model considered in Subsection 5.7.2 describes the CE, LTE and intermediate cases, depending on the amount of reabsorbed radiation.

Table 6.1: Ionization composition of a xenon plasma for density $\rho = 10^{-6}$ g/cm^3 and two temperatures, $T = 12$ eV and $T = 28$ eV, according to the Saha model (LTE) and the coronal equilibrium approximation (CE)

$T = 12$ eV			$T = 28$ eV		
j	CE	LTE	j	CE	LTE
3	0.0013		6	0.0004	
4	0.0422		7	0.0126	
5	0.2641		8	0.0596	
6	0.2568		9	0.3675	
7	0.3909	0.0001	10	0.4337	
8	0.0444	0.0026	11	0.1181	
9	0.0002	0.4480	12	0.0078	
10		0.4892	13	0.0001	
11		0.0365	14		0.0014
12		0.0002	15		0.0464
13			16		0.3412
			17		0.4809
			18		0.1043
			19		0.0249
			20		0.0009
	$Z_0 = 6.1$	$Z_0 = 9.6$		$Z_0 = 9.6$	$Z_0 = 16.7$

6.3 Thermodynamic properties of matter in the Hartree-Fock-Slater model

Utilization of the ionization equilibrium models makes it possible to obtain reliable results for low densities and relatively low temperatures. In the case of very high temperatures and densities good results are furnished by the Thomas-Fermi model. A more precise delineation of the domain of applicability of theoretical equations of state is a rather complex task, which requires the elaboration of more refined approaches. Thus, for example, the *Thomas-Fermi model with corrections* (TFC) was constructed in an attempt to clarify what is the domain of applicability of the Thomas-Fermi model (see [104], and also Subsection 6.1.2). Later it turned out that incorporating quantum and exchange corrections in the TFC model is not sufficient — oscillation corrections must also be accounted for. This has led to the *semiclassical shell model* (SSM) [106, 108]. As it follows from the derivation of the equations of the TFC and SSM models, the domain of applicability of these models is determined by how small the incorporated corrections are.

More successful are the approaches based on the self-consistent field method, in which it is not required that the corrections be small. The first computations in the framework of the *modified Hartree-Fock-Slater model* [152], whose foundations for matter with given temperature and density were first established in [151], have demonstrated that this model can be applied to calculate *photon absorption coefficients* and *Rosseland mean free paths*, as well as to describe the *equations of state* in a wide range of temperatures and densities. Nevertheless, to make the Hartree-Fock-Slater model applicable to the computation of equations of state has required a fundamental revision of certain details of this model, including, in particular, the following:

1) In the computation of the contribution of ions to the equation of state of a dense nonideal plasma, the strong repulsive force between ions at small distances due to the fact that the ions have finite dimensions (rather than being point particles) was accounted for in [158].

2) When one approximates the exchange potential for large values of r (in the case of small densities) one has, as shown in [49], to use the asymptotics of the self-consistent potential, $V(r) \sim 1/r$, which enables one to better describe the energy spectrum of excited electron states (see Subsection 4.3.3).

3) The influence of the effective boundary of the continuum, i.e., of the value ε_0 of the electron energy above which the semi-classical approximation applies in the HFS model, turned out to be more substantial in the calculation of the equation of state than in that of the photon absorption coefficients. As shown in § 3.3, to satisfy the thermodynamic consistency condition the value of ε_0 must be chosen in accordance to the conditions of conservation of the total number of possible electron states [157].

Among the factors enumerated above, which affect the results of computations, the choice of the effective boundary of the continuum ε_0 proved to be the most essential. To make the right choice of ε_0, in the computation of tables of equations of state it is sufficient to fix the number of the root of the equation for the determination of ε_0 for all values of the temperature T and density ρ. Violation of this condition may lead to nonphysical results for some values of T and ρ. The influence of the nonuniqueness in the choice of ε_0, i.e., of the number of the fixed root, on the results of computations of the equation of state and shock adiabats is analyzed in [158]. It turned out that computational results are practically independent of ε_0 whenever the value of this parameter is sufficiently large, for example, when it lies in the region of the allowed energy bands under normal conditions, i.e., for $T = 0$, $\rho = \rho_0$, where ρ_0 is the normal density of matter.

6.3.1 Electron thermodynamic functions

Expressions for the electron thermodynamic functions—pressure P_e, energy E_e and entropy S_e—can be derived from the thermodynamic potential $\Omega = E - \theta S - \mu N$:

$$P_e = -\frac{\Omega}{v}, \tag{6.65}$$

$$E_e = \Omega - \theta\frac{\partial\Omega}{\partial\theta} - \mu\frac{\partial\Omega}{\partial\mu}, \tag{6.66}$$

$$S_e = -\frac{\partial\Omega}{\partial\theta} \tag{6.67}$$

(see [17]). Here $v = (4/3)\pi r_0^3$ is the volume of the spherical atom cell.

Formula (6.65) is sufficiently simple. However, computations relying on this formula suffer from a loss of accuracy because the expression for Ω contains large quantities with different signs. For that reason, the simplest way to derive a formula for the pressure is not from relation (6.65), but from the equivalent relation obtained by differentiating the free energy $F = E - \theta S$ with respect to volume:

$$P = -\left(\frac{\partial F}{\partial v}\right)_{T,N}.$$

The pressure will obviously be expressed through the contributions of the bound and continuum electrons, as well as through the derivative of the supplementary term (3.69) with respect to the volume v, which in view of the thermodynamic consistency condition (3.70) is equal to zero.

First of all, let us remark that for the contribution of the continuum electrons one obtains an expression that is similar to the expression (6.1) for the pressure in the Thomas-Fermi model, where the integration is carried out starting with the boundary of the continuum ε_0.

To calculate the contribution of the bound electrons we shall use the expressions for the energy and entropy in the HFS model (3.51), keeping in mind that the splitting with respect to the spin variable can be neglected at high temperatures. The states with different projections of the spin will have identical energy ε_ν and occupation numbers n_ν. To account for the spin we introduce the factor 2. We obtain

$$F = E - \theta S = 2\sum_\nu n_\nu \int \psi_\nu^*(\vec{r})\left[-\frac{1}{2}\Delta - V_a(r)\right]\psi_\nu(\vec{r})\, d\vec{r} +$$

$$\frac{1}{2}\iint \frac{\rho(\vec{r})\rho(\vec{r}\,')}{|\vec{r}-\vec{r}\,'|}\, d\vec{r}\, d\vec{r}\,' - \int \varphi(\rho)\, d\vec{r} + 2\theta\sum_\nu \Big[n_\nu \ln n_\nu + (1-n_\nu)\ln(1-n_\nu)\Big], \tag{6.68}$$

where

$$\rho(\vec{r}) = 2\sum_\nu n_\nu\, |\psi_\nu(\vec{r})|^2,$$

$$n_\nu = \frac{1}{1 + \exp\left(\dfrac{\varepsilon_\nu - \mu}{\theta}\right)}.$$

Let us change variables to make the dependence on the radius r_0 of the atom cell explicit:

$$\vec{r} = r_0\,\vec{x}, \qquad \bar{\psi}(\vec{x}\,) = r_0^{3/2}\psi(r_0\vec{x}\,),$$

$$\bar{V}(x) = r_0 V(r_0 x), \qquad \bar{\rho}(x) = r_0^3\rho(r_0 x), \qquad \bar{\varphi}(\bar{\rho}) = r_0^4\varphi(r_0^{-3}\bar{\rho}).$$

In addition, we need to separate the components of the self-consistent potential. These are the Coulomb component $\bar{V}_{\rm c}(x) = r_0 V_{\rm c}(r_0 x)$, where

$$V_{\rm c}(r) = V_{\rm a}(r) - \int \frac{\rho(\vec{r}\,')\,d\vec{r}\,'}{|\vec{r} - \vec{r}\,'|}, \tag{6.69}$$

the contribution of the nucleus $V_{\rm a}(r) = Z/r\ \big(\bar{V}_{\rm a}(x) = r_0 V_{\rm a}(r_0 x)\big)$, and the effective exchange potential

$$\frac{\partial\bar{\varphi}}{\partial\bar{\rho}} = \frac{1}{r_0}\frac{\partial\varphi}{\partial\rho}.$$

Using the Schrödinger equation (3.52) for $\psi_\nu(\vec{r}\,)$, we obtain a similar equation for $\bar{\psi}_\nu(\vec{x}\,)$:

$$-\frac{1}{2}\Delta\bar{\psi}_\nu = r_0^2\left(\varepsilon_\nu + \frac{\bar{V}}{r_0}\right)\bar{\psi}_\nu. \tag{6.70}$$

The free energy (6.68) is given by

$$F = 2\sum_\nu n_\nu \int \left(\varepsilon_\nu + \frac{\bar{V}}{r_0}\right)\left|\bar{\psi}_\nu(\vec{x})\right|^2 d\vec{x} - \frac{1}{2r_0}\int (\bar{V}_{\rm c} + \bar{V}_{\rm a})\bar{\rho}\,d\vec{x}\, -$$
$$\frac{1}{r_0}\int \bar{\varphi}\,d\vec{x} + 2\theta\sum_\nu \big[n_\nu \ln n_\nu + (1 - n_\nu)\ln(1 - n_\nu)\big]. \tag{6.71}$$

Since $\dfrac{\partial F}{\partial v} = \dfrac{1}{4\pi r_0^2}\dfrac{\partial F}{\partial r_0}$, we have $3Pv = -r_0\dfrac{\partial F}{\partial r_0}$. Differentiating expression (6.71) with respect to r_0, we obtain

$$3Pv = -r_0\frac{\partial F}{\partial r_0} = -2r_0\sum_\nu \frac{\partial n_\nu}{\partial r_0}\int\left(\varepsilon_\nu + \frac{\bar{V}}{r_0}\right)\left|\bar{\psi}_\nu(\vec{x})\right|^2 d\vec{x}\, -$$
$$2r_0\sum_\nu n_\nu\int\frac{\partial}{\partial r_0}\left(\varepsilon_\nu + \frac{\bar{V}}{r_0}\right)\left|\bar{\psi}_\nu(\vec{x})\right|^2 d\vec{x} - 2r_0\sum_\nu n_\nu\int\left(\varepsilon_\nu + \frac{\bar{V}}{r_0}\right)\frac{\partial}{\partial r_0}\left|\bar{\psi}_\nu(\vec{x})\right|^2 d\vec{x}\, -$$
$$\frac{1}{2r_0}\int(\bar{V}_{\rm c} + \bar{V}_{\rm a})\bar{\rho}\,d\vec{x} + \frac{1}{2}\int\frac{\partial}{\partial r_0}(\bar{V}_{\rm c} + \bar{V}_{\rm a})\bar{\rho}\,d\vec{x} + \frac{1}{2}\int(\bar{V}_{\rm c} + \bar{V}_{\rm a})\frac{\partial\bar{\rho}}{\partial r_0}\,d\vec{x}\, -$$
$$\frac{1}{r_0}\int\bar{\varphi}\,d\vec{x} + \int\frac{\partial\bar{\varphi}}{\partial r_0}d\vec{x} - 2r_0\sum_\nu(\mu - \varepsilon_\nu)\frac{\partial n_\nu}{\partial r_0}. \tag{6.72}$$

We will use the obvious relations

$$\sum_{\nu} \frac{\partial n_\nu}{\partial r_0} = 0, \tag{6.73}$$

and

$$2\sum_{\nu} \frac{\partial n_\nu}{\partial r_0} \int \frac{\bar{V}}{r_0} \left|\bar{\psi}_\nu(\vec{x})\right|^2 d\vec{x} + 2\sum_{\nu} n_\nu \int \frac{\bar{V}}{r_0} \frac{\partial}{\partial r_0} \left|\bar{\psi}_\nu(\vec{x})\right|^2 d\vec{x} = \int \frac{\bar{V}}{r_0} \frac{\partial \bar{\rho}}{\partial r_0} d\vec{x}. \tag{6.74}$$

Next, let us differentiate equation (6.70) with respect to r_0, multiply the result by $\bar{\psi}_\nu^*(x)$ and subtract from it the complex-conjugate equation (6.70), multiplied by $\partial\bar{\psi}_\nu(\vec{x})/\partial r_0$. Integrating the resulting relation with respect to $d\vec{x}$ and using the Green formula in the right-hand side, we obtain

$$\frac{1}{2} \int\limits_{|\vec{x}|=1} \left[\frac{\partial \bar{\psi}_\nu}{\partial r_0} \frac{\partial \bar{\psi}_\nu^*}{\partial x} - \bar{\psi}_\nu^* \frac{\partial}{\partial r_0} \frac{\partial \bar{\psi}_\nu}{\partial x} \right] ds = 2r_0 \int \left(\varepsilon_\nu + \frac{\bar{V}}{r_0}\right) \left|\bar{\psi}_\nu(\vec{x})\right|^2 d\vec{x} + r_0^2 \int \frac{\partial}{\partial r_0} \left(\varepsilon_\nu + \frac{\bar{V}}{r_0}\right) \left|\bar{\psi}_\nu(\vec{x})\right|^2 d\vec{x}. \tag{6.75}$$

The left-hand side of this expression is equal to zero if we use boundary conditions of the form (4.14), because the expression in the square brackets changes sign on the upper and lower hemispheres. From the Poisson equation for $\bar{V}_c(x)$ it follows that $4\pi x\bar{\rho} = (x\bar{V}_c - x\bar{V}_a)''$. Using this relation and integrating by parts twice we have

$$\int \frac{\partial \bar{V}_c}{\partial r_0} \bar{\rho}\, d\vec{x} = 4\pi \int \frac{\partial \bar{V}_c}{\partial r_0} \bar{\rho} x^2\, dx = \int \frac{\partial \bar{\rho}}{\partial r_0} (\bar{V}_c - \bar{V}_a)\, d\vec{x}\,. \tag{6.76}$$

Now combining relations (6.76), (6.73)–(6.74) and the fact that the right-hand side of equation (6.75) vanishes, we obtain

$$3Pv = -\int \bar{V} \frac{\partial \bar{\rho}}{\partial r_0}\, d\vec{x} + 2\sum_{\nu} n_\nu \int \left(\varepsilon_\nu + \frac{\bar{V}}{r_0}\right) \left|\bar{\psi}_\nu(\vec{x})\right|^2 d\vec{x} - \frac{1}{2r_0} \int (\bar{V}_c + \bar{V}_a) \bar{\rho}\, d\vec{x} + \frac{1}{2} \int \frac{\partial \bar{\rho}}{\partial r_0} (\bar{V}_c - \bar{V}_a)\, d\vec{x} + \frac{1}{2} \int (\bar{V}_c + \bar{V}_a) \frac{\partial \bar{\rho}}{\partial r_0}\, d\vec{x} - \frac{1}{r_0} \int \bar{\varphi}\, d\vec{x} + \int \frac{\partial \bar{\varphi}}{\partial r_0}\, d\vec{x}. \tag{6.77}$$

Since

$$\frac{\partial \bar{\varphi}}{\partial r_0} = \frac{4}{r_0} \int \bar{\varphi}\, d\vec{x} - \frac{3}{r_0} \int \bar{\rho} \frac{\partial \bar{\varphi}}{\partial \bar{\rho}}\, d\vec{x} + \int \frac{\partial \bar{\varphi}}{\partial \bar{\rho}} \frac{\partial \bar{\rho}}{\partial r_0}\, d\vec{x} \tag{6.78}$$

and $\bar{V} = \bar{V}_c + \dfrac{\partial \bar{\varphi}}{\partial \bar{\rho}}$, the terms involving the derivative $\dfrac{\partial \bar{\rho}}{\partial r_0}$ in (6.77) cancel each other. Returning to the variable r, using the following equality for the exchange term:

$$12\pi \int_0^{r_0} \left(\varphi - \rho \frac{\partial \varphi}{\partial \bar{\rho}}\right) r^2\, dr = 4\pi r_0^3 \left(\varphi - \rho \frac{\partial \varphi}{\partial \rho}\right)\Big|_{r=r_0} + 4\pi \int_0^{r_0} r^3 \frac{d}{dr}\left(\frac{\partial \varphi}{\partial \rho}\right) \rho\, dr, \quad (6.79)$$

as well as the obvious relation $V' = V_c' + \dfrac{d}{dr}\dfrac{\partial \varphi}{\partial \rho}$, we obtain

$$3Pv = 2\sum_\nu n_\nu \int (2\varepsilon_\nu + 2V)\, |\psi_\nu(\vec{r})|^2\, d\vec{r} -$$
$$\frac{1}{2}\int (V_c + V_a)\rho\, d\vec{r} + \int r(V' - V_c')\rho\, d\vec{r} + 4\pi r_0^3 \left(\varphi - \rho \frac{\partial \varphi}{\partial \rho}\right)\Big|_{r=r_0}. \quad (6.80)$$

Now let us transform (6.80) so that only the terms with $r = r_0$ remain. First of all, note that

$$2\pi \int (V_c + V_a + 2rV_c')\, \rho r^2\, dr = \frac{1}{2}\int (rV_c + rV_a + 2r^2 V_c')(rV_c)''\, dr =$$
$$\frac{1}{2}\int_0^{r_0} \frac{d}{dr}\Big[(rV_c)'\big(r(rV_c)' - rV_c + rV_a\big)\Big] = 0. \quad (6.81)$$

The radial part of the wave function for $\varepsilon = \varepsilon_\nu$ satisfies the equation

$$-\frac{1}{2}R''_{\varepsilon\ell} + \left(\frac{\ell(\ell+1)}{2r^2} - V(r)\right) R_{\varepsilon\ell} = \varepsilon R_{\varepsilon\ell}, \quad (6.82)$$

Let us differentiate this equation with respect to r, multiply the result by $rR_{\varepsilon\ell}$ and subtract from it equation (6.82) twice, first multiplied by $rR'_{\varepsilon\ell}$, and then multiplied by $2R$. This yields

$$(2\varepsilon + 2V + rV')R^2_{\varepsilon\ell} = -\frac{1}{2}(rR_{\varepsilon\ell}R'''_{\varepsilon\ell} - rR'_{\varepsilon\ell}R''_{\varepsilon\ell} + 2R_{\varepsilon\ell}R''_{\varepsilon\ell}) =$$
$$-\frac{1}{2}\frac{d}{dr}\Big[R_{\varepsilon\ell}R'_{\varepsilon\ell} + rR_{\varepsilon\ell}R''_{\varepsilon\ell} - r(R'_{\varepsilon\ell})^2\Big]. \quad (6.83)$$

Using (6.83), instead of (6.80) we obtain

$$P = \left\{\varphi - \rho\frac{\partial \varphi}{\partial \rho} - \frac{1}{4\pi r_0^3}\sum_\nu n_\nu \Big[R_{\varepsilon\ell}R'_{\varepsilon\ell} + rR_{\varepsilon\ell}R''_{\varepsilon\ell} - r(R'_{\varepsilon\ell})^2\Big]\right\}\Bigg|_{r=r_0}, \quad (6.84)$$

where $\varepsilon = \varepsilon_\nu$.

In [224, 95] several other formulas for the electron pressure were obtained, which differ between them and also from (6.84). However, all these formulas are equivalent when periodic boundary conditions of the type (4.14) are satisfied. Recalling that $V_{\text{ex}} = \rho \dfrac{\partial \varphi}{\partial \rho}$, the final formulas for the electron components of the pressure, energy and entropy can be recast in the following forms:

$$P_{\text{e}} = \frac{2\sqrt{2}}{3\pi^2}\theta^{5/2} \int\limits_{y_0(r_0)}^{\infty} \frac{y^{3/2}dy}{1+\exp\left(y - \dfrac{V(r_0)+\mu}{\theta}\right)} + \int\limits_0^{\rho(r_0)} V_{\text{ex}}(\rho')d\rho' - \rho(r_0)V_{\text{ex}}[\rho(r_0)] - \frac{3}{4\pi r_0^3 k_0^3} \sum_{n\ell m\ell'} \int\limits_0^{k_0} k^2 dk A^2_{n\ell m\ell'}(k) n_\varepsilon \left[R_{\varepsilon\ell'}R'_{\varepsilon\ell'} + rR_{\varepsilon\ell'}R''_{\varepsilon\ell'} - r(R'_{\varepsilon\ell'})^2\right]\Big|_{r=r_0}. \quad (6.85)$$

Here

$$y_0 = y_0(r) = \max\left\{0; \frac{V(r)+\varepsilon_0}{\theta}\right\}, \quad n_\varepsilon = \left[1+\exp\left(\frac{\varepsilon-\mu}{\theta}\right)\right]^{-1}, \quad \varepsilon = \varepsilon_{n\ell m}(k),$$

and the sum with respect to n, ℓ, m, ℓ' and the integration with respect to k are carried out for states with energy $\varepsilon < \varepsilon_0$. In the next two formulas for the internal energy and the entropy the summation and integration are carried out in the same manner:

$$E_{\text{e}} = \frac{6}{k_0^3} \sum_{n\ell m} \int\limits_0^{k_0} k^2 dk\, n_\varepsilon \int\limits_0^{r_0} [\varepsilon + V(r)] \sum_{\ell'} A^2_{n\ell m\ell'}(k)\, R^2_{\varepsilon\ell'}(r)\, dr + \frac{4\sqrt{2}}{\pi}\theta^{5/2} \int\limits_0^{r_0} r^2\, dr \int\limits_{y_0}^{\infty} \frac{y^{3/2}dy}{1+\exp\left(y - \dfrac{V(r)+\mu}{\theta}\right)} - 2\pi \int\limits_0^{r_0} \left[\frac{Z}{r} + V(r) - V_{\text{ex}}(r)\right] \rho(r) r^2 dr - 4\pi \int\limits_0^{r_0} \left[\int\limits_0^{\rho(r)} V_{\text{ex}}(\rho')\, d\rho' - \theta \frac{\partial}{\partial\theta} \int\limits_0^{\rho(r)} V_{\text{ex}}(\rho')\, d\rho'\right] r^2 dr - E_0, \quad (6.86)$$

where E_0 is the energy of the free atom (so that for $T = 0$, $\rho = 0$ we have $E_e = 0$);

$$S_e = \frac{6}{k_0^3} \sum_{n\ell m} \int_0^{k_0} [n_\varepsilon \ln n_\varepsilon + (1 - n_\varepsilon) \ln(1 - n_\varepsilon)]\, k^2\, dk - \frac{Z_0 \mu}{\theta} -$$

$$\frac{4\pi}{\theta} \int_0^{r_0} V(r)\rho(r) r^2\, dr + 4\pi \int_0^{r_0} \frac{\partial}{\partial \theta} \int_0^{\rho(r)} V_{ex}(\rho') d\rho' r^2\, dr +$$

$$\frac{20\sqrt{2}}{\pi} \theta^{3/2} \int_0^{r_0} r^2 dr \int_{y_0}^{\infty} \frac{y^{3/2}\, dy}{1 + \exp\left(y - \dfrac{V(r) + \mu}{\theta}\right)} +$$

$$\frac{8\sqrt{2}}{3\pi} \theta^{3/2} \ln\left[n_{\varepsilon_0}(1 - n_{\varepsilon_0})\right] \int_0^{r_0} y_0^{3/2}(r) r^2\, dr. \quad (6.87)$$

6.3.2 Accounting for the thermal motion of ions in the charged hard-sphere approximation

In a strongly-interacting nonideal plasma one cannot separate the thermodynamic functions into electron and ion components [65]. However, at high temperatures, where the main contribution to the equation of state is made by electrons, the contribution of ions can be regarded as a supplementary term. At the foundation of such a separation lies the assumption about the average atom, whose characteristics — in particular, the intra-atomic potential — are defined by averaging the quantity one is interested in over all possible states of the ion in question as well as over the position of the other ions. As a result of this approximation, the electron structure of the average atom is calculated in the charge-neutral atom cell (see Chapter IV). As it turns out, in this manner the main part of the electron-ion interaction is effectively accounted for (for a more accurate discussion see [90]).

The influence of the motion and interaction of ions can be accounted for by using the characteristics of the average atom and applying perturbation theory in the adiabatic approximation [186, 187]. To account for the interaction between ions it is necessary to go beyond the framework of the atom-cell approximation. One of the ways to do this is to regard the plasma as an idealized gas of interacting particles. In particular, as such an idealized model one can consider a gas of positively *charged hard spheres* (CHS) that move on the background of a uniformly distributed negative charge of electrons [167].

The parameters of the CHS model, in particular, the dimensions and charges of ions, depend on the results of computations inside atom cells. Therefore, in the general case, to obtain the equations of a thermodynamically consistent model it is necessary to impose the minimum condition for the total thermodynamic potential of the system of ions and electrons, which complicates the problem considerably. In the domain where the main contribution to the thermodynamic potential is due to the electrons, the minimum condition for the thermodynamic potential is almost identical to the minimum condition for the thermodynamic potential of the electrons.

To obtain the total thermodynamic functions let us represent them as a sum of contributions due to electrons (indicated by the subscript "e") and to ions (indicated by the subscript "i"). Then, if the electron and ion components are expressed in atomic units, the total pressure (measured in GPa) and the total energy (measured in kJ/g) are given by:

$$P = 2.942 \cdot 10^4 (P_{\mathrm{e}} + P_{\mathrm{i}}), \tag{6.88}$$

and respectively

$$E = \frac{2.626 \cdot 10^3}{A}(E_{\mathrm{e}} + E_{\mathrm{i}}). \tag{6.89}$$

The simplest approximation that accounts for the motion of ions is the ideal-gas approximation (see formulas (6.4), (6.11) and (6.13)). The ideal-gas approximations can be used for sufficiently high temperatures and gas densities. For dense and condensed media, however, this model is rather crude. Approximations that account for the interaction between ions are required. Thus, for example, in the *one-component plasma* (OCP) model it is assumed that ions with identical charge move in a homogeneous medium that carries a charge of opposite sign [81, 116]. The ion charge Z_0, or the average number of free electrons in the atom cell, can be approximated by the formula $Z_0 = (4\pi/3) r_0^3 \rho(r_0)$, where $\rho(r_0)$ is the electron density on the boundary of the cell.

For a one-component plasma methods of molecular dynamics have been used to calculate the distribution function of ions and their interaction energy and, based on these results, the corresponding interpolation formulas were derived. In the OCP approximation [81, 116]

$$P_{\mathrm{i}} = \left[1 + \frac{\Delta E_{\mathrm{i}}}{3\theta}\right] \frac{\theta}{v}, \qquad E_{\mathrm{i}} = \frac{3}{2}\theta + \Delta E_{\mathrm{i}}, \tag{6.90}$$

where ΔE_{i} is the correction to the internal energy in the OCP model (see [81]) and v is the volume of the atom cell.

In computations of the equation of state it is convenient to modify somewhat the interpolation formulas given in [81] for the interaction energy ΔE_i given by the OCP model in order that they continue to work for arbitrary temperatures and densities. The point is that in the OCP model, at values of the nonideality parameter Γ close to 158, where $\Gamma = Z_0^2/(\theta r_0)$, a phase transition occurs in the solid state. An accurate description of this phase transition requires a detailed analysis of the thermodynamic functions near the transition point. Since the physical accuracy of the OCP model is not beyond dispute, in practical computations, to simplify the calculation of the ion contribution to the equation of state, one requires that at large values of the nonideality parameter Γ the ion energies $E_{\rm i}$ will not exceed the asymptotic value 3θ:

$$\Delta E_{\rm i} = \begin{cases} \Delta E_{\rm OCP}, & \text{if } \Delta E_{\rm OCP} < \dfrac{3}{2}\theta, \\ \dfrac{3}{2}\theta, & \text{if } \Delta E_{\rm OCP} \geq \dfrac{3}{2}\theta, \end{cases} \tag{6.91}$$

where for $\Gamma > 0.1$

$$\Delta E_{\rm OCP} = \theta\Gamma^{3/2} \sum_{i=1}^{4} \frac{a_i}{(b_i + \Gamma)^{i/2}} - \theta\Gamma a_1,$$

$$\begin{aligned} a_1 &= -0.895929, & b_1 &= 4.666486, \\ a_2 &= 0.11340656, & b_2 &= 13.675411, \\ a_3 &= -0.90872827, & b_3 &= 1.8905603, \\ a_4 &= -0.11614773, & b_4 &= 1.0277554. \end{aligned}$$

The proposed modification practically does not affect the final results of computations of the equation of state, yet it allows one to avoid a detailed analysis of the phase transition in the indicated range of temperatures and densities.

In the OCP model the ions are regarded as point particles and their interaction potentials is assumed to be Z_0^2/r (see [116]). The point-charge approximation fails at high densities of matter, when the dimensions of the ion cores are comparable with the interatomic distances [77]. For example, such a situation arises for a shock compression of factor $\sigma \simeq 3 \div 4$, where $\sigma = \rho/\rho_0$, ρ_0 being the normal density of matter.

To account approximately for the influence of the strong repulsion between ions at arbitrary temperatures and high densities one can use the model of charged hard spheres (CHS) [167, 158, 144], or the simpler approximation in which to the OCP model one adds the correction furnished by the model of hard spheres:

$$P_{\rm i} = \left[1 + \frac{\eta(4+2\eta)}{(1-\eta)^2} + \frac{\Delta E_{\rm i}}{3\theta}\right]\frac{\theta}{v}, \qquad E_{\rm i} = \frac{3}{2}\theta + \Delta E_{\rm i}, \tag{6.92}$$

where $\eta = (r^*/r_0)^3$ and r^* is the effective radius of the ion core.

Formulas (6.90), (6.92) are not thermodynamically consistent, since the packing parameter η and the average ion charge Z_0 are functions of temperature and density. Thermodynamically consistent expressions for energy and pressure can be obtained by starting from the expression for the free energy in the CHS approximation [167]. However, the increase in the complexity of formulas is not matched by an increase in their level of physical accuracy.

6.3.3 Effective radius of the average ion

For given values of the temperature T and density ρ, the value of r^* can be estimated based on the following considerations. We shall assume that all the electrons of the atom cell can be divided into so-called free electrons, which are uniformly distributed over the cell with a density $\rho(r_0)$ that is equal to the density of electrons on the boundary of the atom cell, and the remaining, bound electrons, with density $\rho(r) - \rho(r_0)$. Outside the domain $0 < r < r^*$ the function $\rho(r) - \rho(r_0)$ is required to decay rapidly.

The interaction between ion cores becomes essential at distances r^* for which the density of the bound electrons, $\rho(r) - \rho(r_0)$, is sufficiently large. Obviously, some part of the bound electrons will be found beyond the limits of the core, i.e., for $r > r^*$. If the number of such electrons is taken to be equal to the number of the free electrons that penetrate inside the sphere $r < r^*$, then we arrive at the expression

$$4\pi \int_{r^*}^{r_0} (\rho(r) - \rho(r_0))r^2 \, dr = 4\pi \int_0^{r^*} \rho(r_0)r^2 \, dr. \tag{6.93}$$

Formula (6.93) yields good results for sufficiently high temperatures, when more than one electrons are ionized. In particular, for metals the radii of the ion cores computed by means of formula (6.93) agree well with experimental results under nearly-normal conditions. For dielectrics and inert gases, when under normal conditions there is no ionization, formula (6.93) gives $r^* = r_0$. Therefore, in their case we will require that, in addition to (6.93), the quantity r^* be bounded, say, by the radius of the last shell that hosts electrons. One can put approximately $r^* < 2n^2/Z_{n\ell}$, where n, ℓ are the quantum numbers of that shell and $Z_{n\ell}$ is the corresponding effective charge.

Since $4\pi \int_0^{r_0} \rho(r)r^2 \, dr = Z$, condition (6.93) can be recast as

$$4\pi \int_0^{r^*} \rho(r)r^2 \, dr = Z - Z_0. \tag{6.94}$$

The effective radius of the ion, as given by formula (6.94), depends on the matter temperature and density. When T and ρ grow, r^* decreases rapidly. In the case of extremely high temperatures and densities, when full ionization takes place, formula (6.94) yields $r^* \cong 0$. Let us mention that in some cases, for a weakly ionized plasma the ratio of r^* to the radius of the atom cell r_0 may grow with T and ρ.

Figure 6.6 shows the graph of the dependence of the packing parameter $\eta = (r^*/r_0)^3$ on temperature on the shock compression curve for aluminum. As

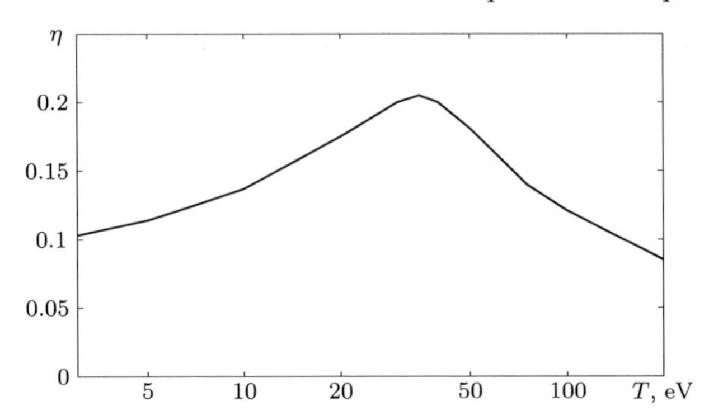

Figure 6.6: Dependence of the packing parameter η on temperature T on the shock adiabat of aluminum

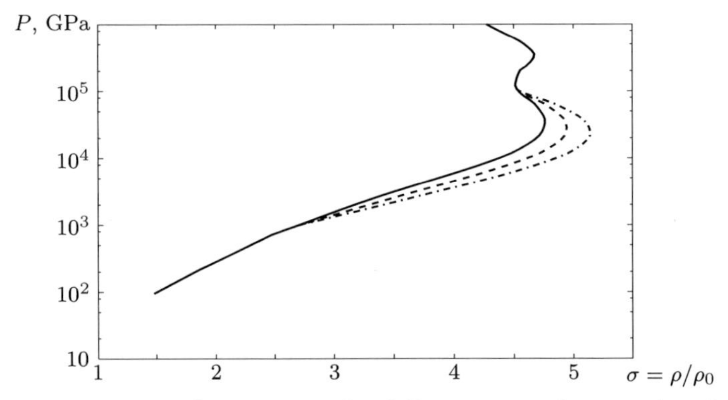

Figure 6.7: Shock adiabat of aluminum, for different way of accounting for the ion contribution to the equation of state (solid curve — charged hard spheres, dashed curve — one-component plasma, dashed-dotted curve — ideal gas)

results of the computation of the shock adiabat for aluminum show, the maximum value $\eta \approx 0.2$ is attained at temperature $T \sim 35$ eV and density $\rho \sim 12$ g/cm^3. At such a temperature one may expect that the contribution of ions will have a large influence on the shock adiabat. And indeed this is plainly seen in Figure 6.7, which displays the shock adiabat of aluminum, with the contribution of ions accounted for according to the ideal gas, OCP and CHS models (note the differences in the behavior of the curves for $\sigma \sim 4.5$, $P \sim 10^4$ GPa).

6.3.4 On methods for deriving wide-range equations of state

Since the domain of applicability of the equations of state constructed on the basis of a specific model is limited, in practical computations of tables in a wide range of temperatures and densities one resorts to various methods of approximation and interpolation of the data provided by different models. Such methods can be divided into three main groups:

1) Semi-empirical models of equations of state, in which the functional form of the equation and the constant coefficients are chosen from the requirement that the computational results should match known experimental data and asymptotics [117, 4, 41, 42].

2) Phenomenological models of the equations of state, which contain parameters determined from experimental data that are not directly connected with the equations of state (for example, the ionization equilibrium model, which involves the energy levels of excited ion states) [183, 136, 76, 21, 184].

3) Computations based on statistical and quantum-statistical models of matter that do not rely on experimental data [122, 98, 152, 202, 188].

One of the first and most successful method proved to be the interpolation of data of the Thomas-Fermi model with experiments (see, e.g., [58]). Currently the most widely used is the method in which in order to construct wide-range equations of state one interpolates between the results obtained by means of the TF (TFC) model for high densities, the ionization equilibrium model for small densities and semi-empirical equations of state for nearly normal conditions (see, e.g., [58, 136, 214, 97], as well as the well-known library of equations of state SESAME [245, 246]).

The equations of state constructed in this way are verified by comparison with experiment. Note that for dense hot plasmas the available experimental data are relatively poor and they do not provide a completely enough information on the reliability of the equations of state. The evaluation of the accuracy of the derived and applied in practice equations of state in a wide range of temperatures and densities is based not only on experimental verification, but also on calculations using theoretical models with different levels of accuracy.

6.4 Computational results

6.4.1 General description

The Hartee-Fock-Slater model has been used to compute equations of state of various substances in a wide range of temperatures and densities. In these computations, in order to determine the effective boundary of the continuum ε_0 for an arbitrary temperature and arbitrary density, it suffices to fix the number of the root of equation (4.20), labelling the roots according to the growth of the electron energy (see Figure 3.10). It is clear, on physical grounds, that the energy value ε_0 in the normal state ($T = 0$, $\rho = \rho_0$) should lie in the region of the upper level of electron energies (or in the conduction band for metals). This condition yields the number N of the root, close to the number of the electron shells (counted with quantum numbers n and ℓ) in the atom (in particular, for aluminum $N = 4$, for iron $N = 6$, and for gold $N = 18$).

Violation of the thermodynamic consistency condition for the chosen ε_0 may lead to physically incorrect results, such as those obtained, for example, in [107, 240]. In [107] the authors used the value $\varepsilon_0 = -\sqrt{\rho(r_0)}$, where $\rho(r_0)$ is the electron density on the boundary of the atom cell. For bound states [resp., states with energy $\varepsilon > \varepsilon_0$] they used semi-classical wave functions (2.12) [resp., the semi-classical approximation (4.5)].

In calculations based on the method of [107], when the plasma density ρ is increased, bound energy levels $\varepsilon = \varepsilon_{n\ell}$ successively cross the effective boundary of the continuum ε_0, because as ρ increases the energies of the levels also increase, while the effective boundary $\varepsilon_0 = -\sqrt{\rho(r_0)}$ decreases. This results in jumps in the thermodynamic functions of electrons, which are interpreted as phase transitions (see Figure 6.8). As one can see from equation (4.20), after $\varepsilon_{n\ell}$ becomes larger than ε_0, to account correctly for the electrons found in the states with quantum numbers $n\ell$ it is necessary to modify the position of the effective boundary ε_0 (see Subsection 3.3.4). This results in the smoothing of the thermodynamic functions and absence of phase transitions at extremely high compressions, though the character of the dependence on density may differ from the smooth curve furnished by the Thomas-Fermi model.

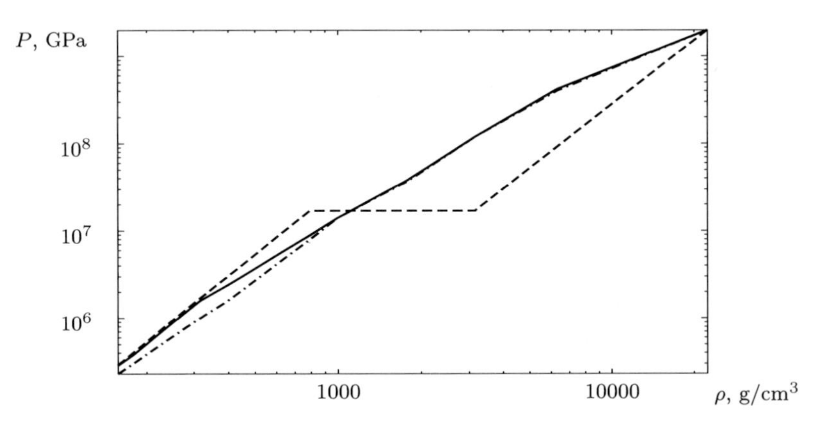

Figure 6.8: Cold compression curves for iron: the pressure P as a function of density ρ at temperature $T = 0$ (solid curve — HFS model, dashed-dotted curve, which passes into the solid curve when $\rho > 10^3$ g/cm^3 — TF model, dashed curve with a plateau — according to [107])

Figure 6.8 compares, in the pressure region $P \cong 10^8$ GPa, the cold compression curves obtained by means of the HFS and TF models, as well as according to the model of [107]. By selecting the parameter ε_0 in accordance to condition (4.20) the cold compression curve becomes smoother in the range of extremely high pressures, regardless of whether the broadening of discrete levels into bands is taken into account or not.

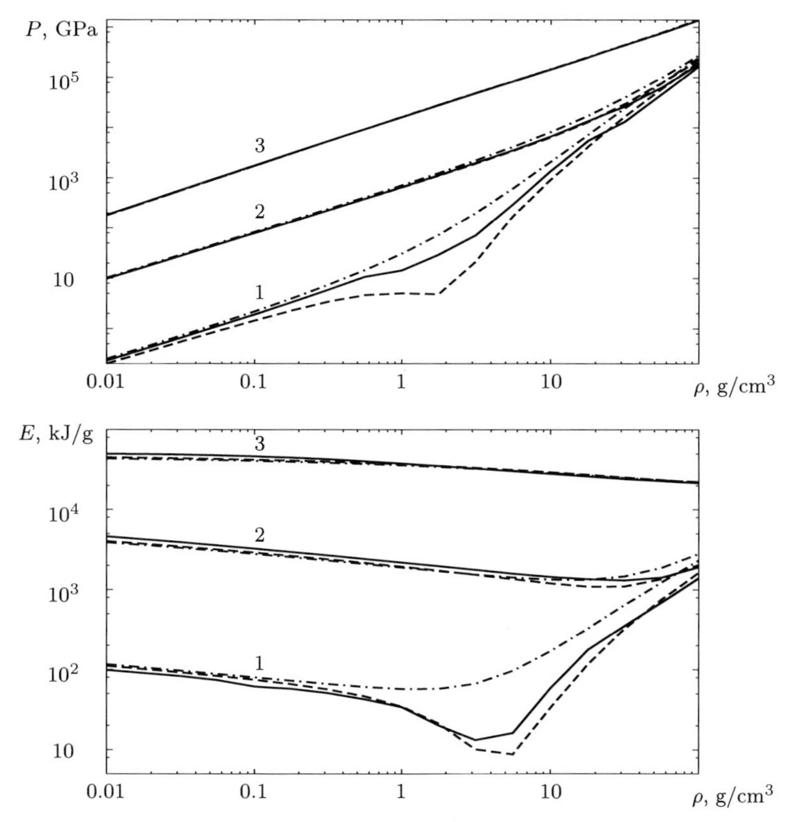

Figure 6.9: The electron pressure P and energy E in aluminum according to different models. Shown are the isotherms at several temperatures: 1 — T=3.98 eV, 2 — T=39.8 eV, 3 — T=398 eV, as functions of the density ρ in g/cm^3 (solid curve — HFS, dashed curve — TFC, dashed-dotted curve — TF)

For a detailed comparison of the computational results obtained by different models in a wide range of temperatures and densities we take aluminum as an example [144]. Figure 6.9 shows for comparison three isotherms of the electron components of pressure and energy for the temperatures 3.98, 39.8 and 398 eV, obtained by means of three different models: HFS, TF and TFC. As one can see in this figure, the largest differences between models (some of the results are several times larger that the other!) occur for low temperatures, near the normal density; there the results of the HFS computations occupy an intermediate position between the results of the TF and TFC models. Note that the corrections in the TFC model to the initial TF data turned out to be here of the order of the ground quantity, which indicates that the TFC model is not applicable in the considered range of densities ρ. With the growth of the temperature the differences become less significant.

We should point out to a peculiarity in the behavior of the isotherms computed by means of the HFS model for $T = 3.98$ eV. For densities $\rho \sim 30$ g/cm^3 these curves intersect the corresponding curves of the TFC model before turning into the asymptote. This feature is connected with the restructuring of the energy spectrum of electrons and the related change in the occupancy of bands $3s$ and $3d$ (see Figure 4.4).

6.4.2 Cold compression curves

Accounting for the band structure of the electron energy spectrum in the Hartree-Fock-Slater model makes it possible to apply this model not only for the calculation of properties of rarefied and dense gas plasmas, but also of those of solid materials under strong compression. The results of experimental studies of equations of state of solid materials are mainly concentrated in the region of normal conditions ($P \leq 100$ GPa). For extremely high pressures ($P > 10$ TPa) the TF model, or the more accurate TFC model, become applicable. Since equations of state are needed in a wide range of compression factors, methods for the computation of equations of state using available experimental data in combination with asymptote extrapolation according to the statistical TF and TFC models became widely used in practice. However, such methods do not always yield reliable results, because in the intermediate domain, where the Thomas-Fermi model is not applicable and no experimental data are available, the irregular shell effects neglected in the statistical models may have a large contribution.

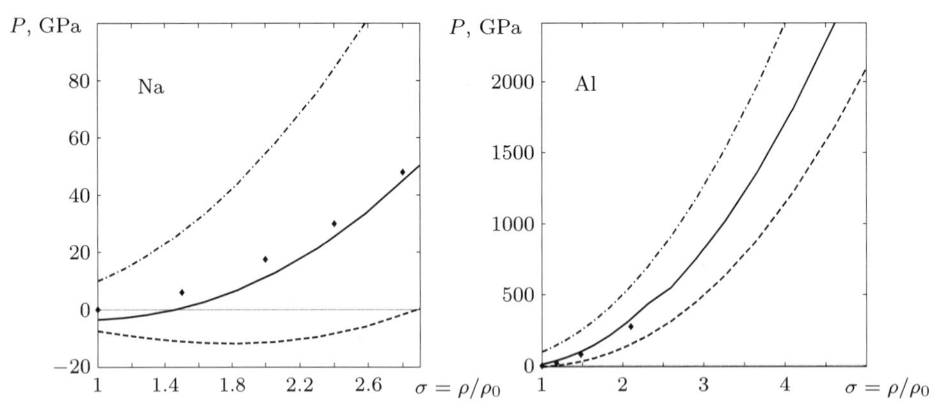

Figure 6.10: Cold compression curves of sodium and aluminum (pressure as a function of compression $\sigma = \rho/\rho_0$; solid curve — HFS, dashed curve — TFC, dashed-dotted curve — TF, rhombuses — experiment [5, 15])

First let us examine some results of computations in different models for alkaline metals. It is known that for such metals the Fermi surface is nearly spherical, and so apparently here the Wigner-Zeits spherical-cells approximation should not lead to significant errors, because in this approximation the Fermi surface is also a sphere. At the same time, for alkaline metals the TF and TFC models are indeed not applicable due to the strong influence of shell effects.

The comparison done in Figure 6.10 of the cold compression curve of sodium obtained by means of the HFS model with the experimental data of [15] shows that the HFS model describes the situation relatively well. The computed theoretical value of the normal density of sodium, $\rho_0 = 1.36$ g/cm^3 (i.e., the density for which the pressure $P = 0$), is closer to the experimental value of $\rho_0 = 0.97$ g/cm^3 than the value $\rho_0 = 2.8$ g/cm^3 provided by the TFC model. For lithium the normal density according to the HFS model, $\rho_0 = 0.54$ g/cm^3, differs from the experimental value $\rho_0 = 0.53$ g/cm^3 by only 2%.

Figure 6.11 compares computational results for cold compression of helium and neon obtained by means of the HFS model with those obtained by means of the TFC model and with experimental data generated by the method of molecular beams [121]. Note the good agreement between HFS calculations and experimental data for sufficiently large compressions. However, while for helium the results agree with the data provided by the TFC model, for neon the TFC results are worse than the HFS ones.

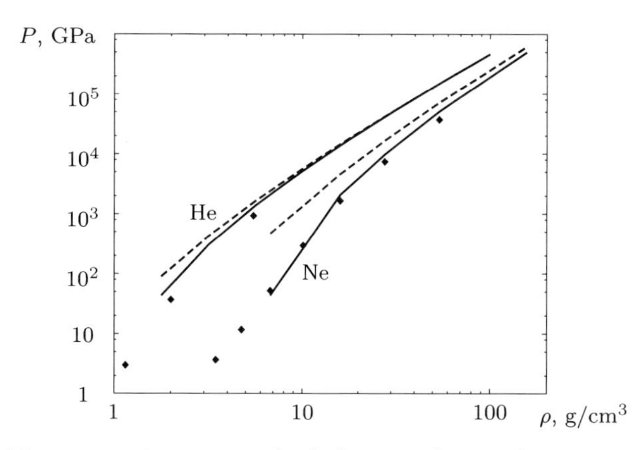

Figure 6.11: Cold compression curves for helium and neon (pressure as a function of density ρ; solid curve — HFS, dashed curve — TFC, rhombuses — experiment [121])

In the case of the cold compression curves of aluminum, iron and gold (see, for example, Figure 6.10 for aluminum), there is a notable deviation from experimental data [5, 238] for weak compressions, namely, of about 10% for aluminum and gold, and 20% for iron.

6.4.3 Shock adiabats

Detailed tables of equations of state, obtained by means of the Hartree-Fock-Slater model, allow one to build compression curves for substances. These curves are alternatively known as *shock adiabats* (the spelling *adiabatics* is used as well) or *Hugoniot adiabats*. The last term is connected with the fact that, in accordance with conservation laws, under compression of a substance the following Hugoniot relation holds across the shock-wave front [234]:

$$2(E - E_0) = (P + P_0)\left(\frac{1}{\rho_0} - \frac{1}{\rho}\right), \tag{6.95}$$

where E_0, P_0, ρ_0 are the internal energy, the pressure and the density of the undisturbed medium and E, P, ρ are the corresponding quantities behind the shock wave, for the compressed medium. The mass velocity U and the wave velocity D of the shock wave are connected with P and E by the relations

$$U = \sqrt{\frac{2(E - E_0)(P - P_0)}{P + P_0}}, \quad D = \frac{P - P_0}{\rho_0 U}. \tag{6.96}$$

Since, in accordance to the equation of state, $P = P(\rho, T)$ and $E = E(\rho, T)$, it follows that once the value of the density ρ is given, one can find the temperature T and the pressure P on the shock adiabat (6.95).

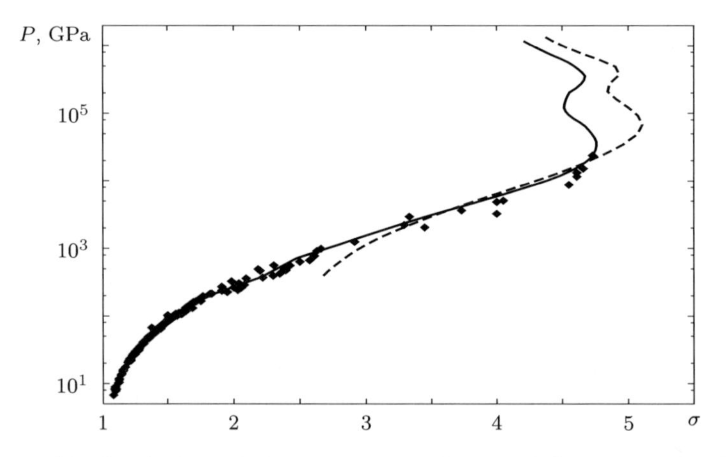

Figure 6.12: Shock adiabat of aluminum according to different models (pressure in GPa as a function of compression $\sigma = \rho/\rho_0$): solid curve — HFS, dashed curve — QZI, rhombuses — experimental data

Shock adiabats of aluminum and iron, calculated by means of the Hartree-Fock-Slater model, are shown in figures 6.12 and 6.13. Also shown are computational results provided by the TFC model for iron as well as results obtained

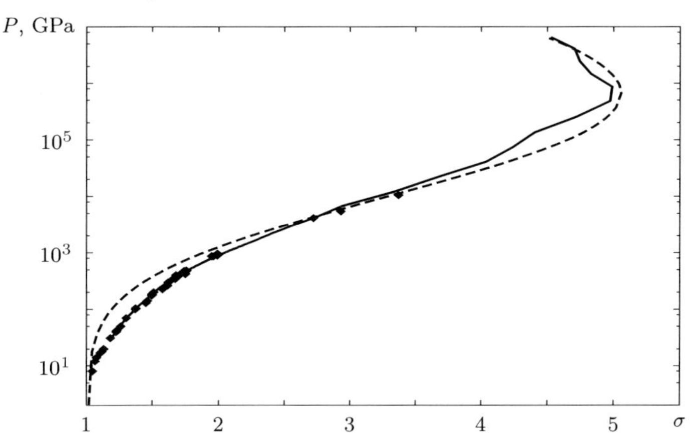

Figure 6.13: Shock adiabat of iron (solid curve — HFS, dashed curve — TFC [101]). Also shown are experimental data of [6, 7, 8, 9]

by means of the interpolation of the TFC and Saha equations of state [97] (the so-called quasi-zone interpolation, QZI) for aluminum. For comparison, the figures also display experimental data. The abscissae [resp. ordinates] represent compression $\sigma = \rho/\rho_0$ ($\rho_0 = 2.71$ g/cm^3 for aluminum and $\rho_0 = 7.85$ g/cm^3 for iron) [resp., pressure in GPa].

One sees that shell effects play a rather essential role after the substance is compressed almost to the limit density and begins to heat intensely. The contribution of the temperature ionization of each shell manifests itself as a slowly varying oscillation about the TFC-computed curve.

An analysis of the data given in Figure 6.12 shows that in the domain of compressions $\sigma \simeq 1 \div 1.8$ (pressure $P \leq 200$ GPa), the position of the shock adiabat of aluminum based on available experimental data and semi-empirical equations of state is established with sufficient reliability. In the compression range $\sigma \simeq 1.8-2.8$, the experimental results partly contradict each other, so that a critical analysis of data obtained in various experiments in this domain of compressions becomes necessary. For higher compressions ($\sigma > 3$) part of the experimental data are affected by hard to control and rather high errors, and therefore we may consider that the computational results obtained by means of the HFS model, as well as those obtained by the QZI method, do not contradict these data. Overall, from the results displayed above it is seen that the HFS shock adiabat for compressions $\sigma \geq 1.8$ agrees well with the experimental results [200]. The QZI adiabat has, in the domain of compressions $\sigma < 3$, the tendency to deviate considerably from the experimental data and the computational results cited.

For high compression factors ($\sigma \simeq 5$), the first oscillation of the QZI shock adiabat is shifted relative to the first oscillation according to the HFS model towards the domain of higher compression factors. The second oscillations on the adiabats, which correspond to the ionization of the K-shell at temperatures $T \simeq 1$ keV, lie considerably closer to one another. The shift of the oscillation on the QZI adiabat relative to the HFS adiabat is apparently due to the fact that in the QZI method, in computations according to the Saha model a certain approximation is used to lower the ionization potentials at high temperatures, whereas in the HFS model such a decrease in the ionization potentials is obtained without any supplementary assumptions. The reliability of the HFS shock adiabat in this domain of compressions is confirmed by the good agreement between the results of computations of the equation of state for a hot rarefied plasma with the results provided by the ionization equilibrium model with nonideality accounted for in the Debye-Huckel approximation of the grand canonical ensemble (Saha-GDH [76], see below).

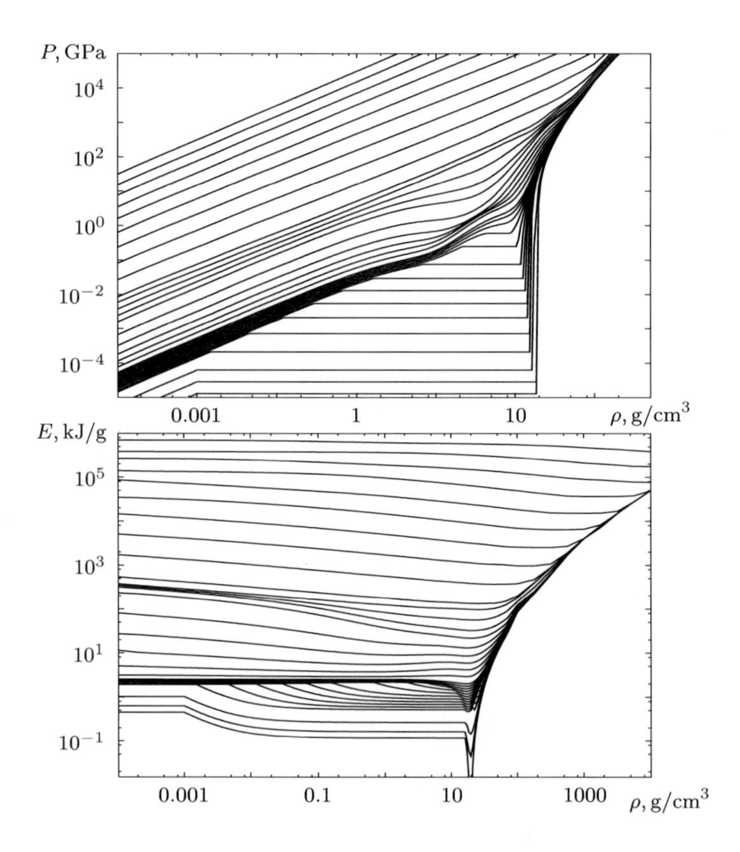

Figure 6.14: Isotherms of pressure P and internal energy E for gold as functions of density ρ

To apply the equations of state obtained in the Hartree-Fock-Slater model in practice, they must be made more accurate for nearly-normal conditions. The simplest way to achieve this is to rely on semi-empirical equations of state, which are compatible with experimental data.

Using computational results obtained with the Hartree-Fock-Slater model in conjunction with semi-empirical models — in particular, the model discussed in [43], one can construct equations of state that are valid in a wide range of temperatures and densities. Such equations of state are indispensable for the numerical modelling of gas dynamic processes. Figure 6.14 above shows pressure and internal energy isotherms for gold obtained by means of equations of state constructed in the manner described above. These results were kindly provided to us by V. V. Val′ko. The corresponding shock adiabat is shown in Figure 6.15.

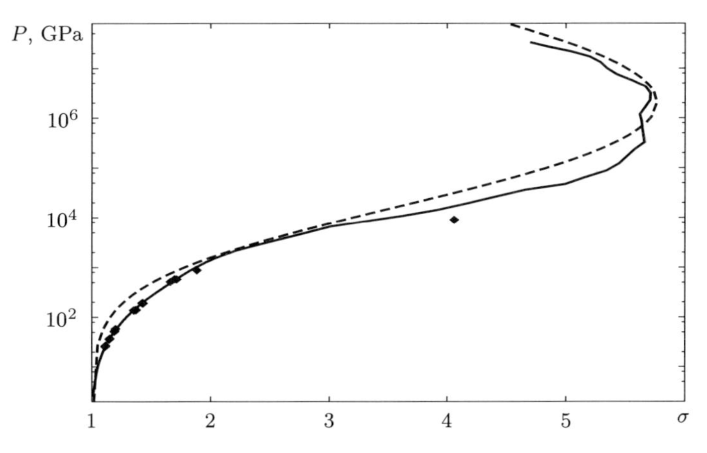

Figure 6.15: Shock adiabat of gold (solid curve — the HFS model combined with a semi-empirical equation of state; dashed curve — the TFC model [101]). Also shown are experimental results reported in [96, 244, 9, 8, 5, 131]

6.4.4 Comparison with the Saha model

For an ideal or weakly nonideal plasma the equation of state and other characteristics of matter can be calculated independently within the framework of the ionization equilibrium model, using experimental values of the energy levels of ions. The very fact that the ionization equilibrium model uses experimental information makes it possible to achieve sufficiently reliable results in many cases. Therefore, it is very interesting to compare the existing theoretical models with the result of computations according to the Saha model in the range where its applicability is beyond doubt.

Table 6.2: Comparison of some characteristics of aluminum for $T = 24.28$ eV, $\rho = 1.49 \cdot 10^{-4}$ g/cm^3, obtained with the TF, HFS and Saha models

Model	Average occupation numbers of bound electron states				Number of free electrons	Chemical potential
	$n=1$	$n=2$	$n=3$	$n=4$		
TF	2.000	4.870	0.040	0.010	6.048	9.177
HFS	2.000	4.157	0.016	0.005	6.804	9.270
Saha	2.000	4.317	0.015	0.005	6.674	9.276

Table 6.2 compares the average occupation numbers obtained by the TF, HFS and Saha models for aluminum at temperature $T \simeq 25$ eV and density $\rho \simeq 1.5 \cdot 10^{-4}$ g/cm^3. In the TF and HFS models these numbers were obtained, for states with principal quantum number n and orbital number ℓ, by the formula

$$N_{n\ell} = \frac{2(2\ell+1)}{1+\exp\left(\frac{\varepsilon_{n\ell}-\mu}{\theta}\right)},$$

where μ is the chemical potential and $\varepsilon_{n\ell}$ are the energy eigenvalues calculated by means of the TF or HFS potential, respectively (the results for the TFC potential practically coincide with the TF results). In the Saha model the occupation numbers were obtained by averaging the populations on a level with given values of n and ℓ over all ion states. The probabilities of possible ion states, which are necessary in the averaging, were determined from the full system of Saha-Boltzmann equations (see (6.47)). Note that the average occupation numbers of electron states constitute one of the most important characteristics of matter; indeed, they determine the degree of ionization, and hence the equation of state.

As one can see in Table 6.2, the computational results in the HFS model agree well with those in the Saha model, whereas the TF model yields, for example, for the shell with principal quantum number $n = 3$, an average number of electrons that is larger by a factor of 2.5. Let us remark that for a sufficiently high temperature the results provided by the Thomas-Fermi model agree well with those provided by the Hartree-Fock-Slater model. As the temperature decreases, in the TF model the inaccuracy in the determination of the occupation numbers — and consequently of other characteristics of matter — grows rapidly.

To characterize the possible magnitude of the deviations we take as an example the results of the calculation of the internal energy of lithium for pressure $P = 0.1$ GPa. Figure 6.16 below shows the results of computations according to the TF, TFC, Saha and HFS models. In the considered range of temperatures and densities, for $T \geq 1$ eV the results for the Saha model may be practically regarded as experimental results, because the energy levels of the excited states of lithium are well known, and under these conditions the plasma may be considered as ideal. As Figure 6.16 shows, the computational results using the HFS model agree well with those using the Saha model. At the same time, one observes a considerable deviation from the results generated by the TF and TFC models, which in the considered range give practically identical results. For example, for temperature $T = 5$ eV the energy according to the HFS model is 2.5 smaller than the energy according to the TFC model, while for $T = 15$ eV the opposite holds, namely, the energy according to the HFS model 1.3 time larger than that according to the TFC model.

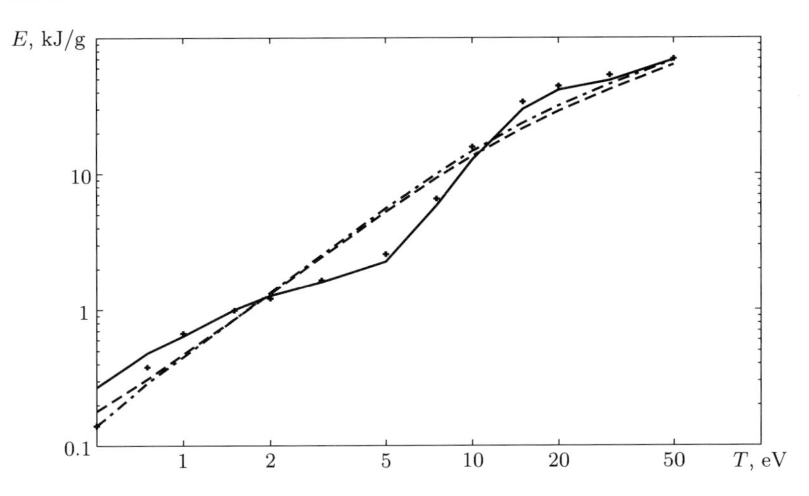

Figure 6.16: Specific internal energy of lithium at pressure $P = 0.1$ GPa, according to different models (solid curve — HFS, dashed curve — TFC, dashed-dotted curve —TF, dots — Saha [76])

The differences between the thermodynamic functions for the indicated values of the temperature and pressure are caused by the major contribution of the shell effects, which is considerably larger than the quantum and exchange corrections of the TFC model. The results of HFS computations agree with those of the TFC computations only starting at temperatures $T \geq 50$ eV, i.e., only when the lithium becomes fully ionized. The influence of the shell effects on the plasma thermodynamic functions remains essential for heavier elements as well. Figure 6.17 below compares the pressures and internal energies of aluminum for densities $\rho = 0.01$ g/cm^3 and temperatures $T = 3.98 \div 700$ eV obtained with the HFS and TFC models with the results obtained with the Saha-GDH model (more precisely, according to the Saha model where the effects of nonideality are accounted for in the approximation of the grand canonical ensemble and using the Debye-Huckel potential [76]).

The good agreement between the data provided by the HFS model and the computational results relying on the Saha model should come as no surprise, since the highest ionization potentials of ions are quite accurately described by the HFS model, and the accuracy of the average-atom approximation increases with

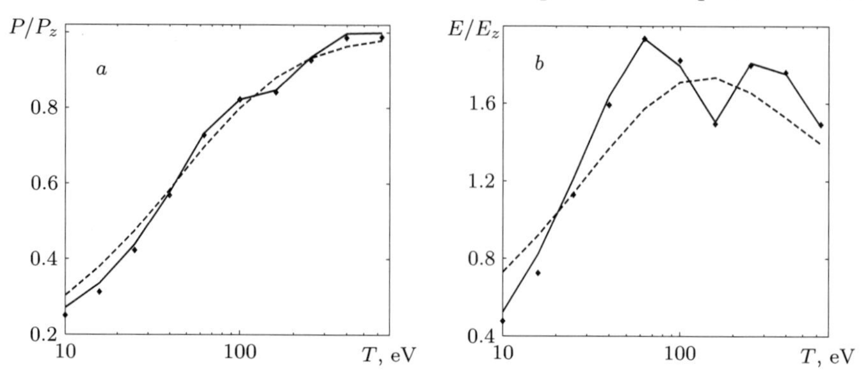

Figure 6.17: Ratio of the pressure (a) and internal energy (b) in aluminum with density ρ=0.01 g/cm^3 to the corresponding quantities for the fully ionized ideal gas, according to different models: solid curve — HFS, dashed curve — TFC, dots — Saha; here $P_z = (Z+1)\,\theta/v$, $E_z = \frac{3}{2}(Z+1)\,\theta$

number of different ion states that are actually realized. It is precisely the last two circumstances that are characteristic for highly ionized rarefied plasma.

An analysis of the results obtained shows that in the region of liquid-metallic, dense and rarefied plasma with temperature $T > 5$ eV , the HFS model yields sufficiently reliable results, which is confirmed by a comparison with the experimental data in the shock compression region, as well as with computations according to various semi-empirical models and the Saha model. The difference between the HFS data and the TFC data at low temperatures ($T \simeq 5 \div 10$ eV) reach $\simeq 100\%$, while for high temperatures ($T \simeq 50 \div 100$ eV) the difference is of $\simeq 15\%$. As for the solid-state characteristics of a substance, especially the thermodynamic description of phase transitions, here it is necessary to more accurately account for the band structure of the electron energy spectrum and the exchange and correlation effects than it is currently done in the HFS model.

6.5 Approximation of thermophysical-data tables

Let us address several questions concerning the utilization of equations of state and photon free paths in gas-dynamic computations. The requisite thermophysical data are usually maintained in the form of tables, which can be readily modified and made more accurate whenever necessary. How detailed are the meshes used to build the tables depends on how accurately one wishes to render specific features of the behavior of physical quantities. Usually the most economical in this respect are the nonuniform meshes, in particular, for temperature and density. The tabular data obtained are then used on more detailed meshes, determined by the numerical methods for solving the problems of radiative gas dynamics. This makes it necessary to multiply interpolate data given in tabular form.

The approximation of two-dimensional thermophysical-data tables must satisfy certain smoothness requirements. In particular, the function $F(x, y)$ interpolating the initial data must be continuous together with its first derivatives with respect to the variables x, y (where, say, $x = \log_{10}T$, $y = \log_{10}\rho$). Furthermore, quite often the initial physical quantities satisfy conditions such as sign constancy or absence of oscillations; they may enjoy properties such as monotonicity or convexity with respect to one of the variables in some ranges of x and y. Interpolation splines do not, in the general case, preserve such properties, and so in order to construct approximations that reproduce correctly the geometric properties of the initial dependencies one is advised to use *locally approximating splines* [233, 118]. The application of these splines in gas dynamic codes has also certain advantages owing to their relative simplicity and the local character of the coefficients of the spline, which are determined only by the tabular values of the given function that are the closest to the point under consideration.

6.5.1 Construction of an approximating spline that preserves geometric properties of the initial function

In [231] the authors used local splines of minimal defect to construct an approximation $F(x, y)$ of a function $f(x, y)$ that is given in tabular manner on a uniform mesh in a rectangular domain. The construction of the approximating spline $F(x, y)$ is rather natural and is transparently carried out by representing the mixed derivative F_{xy} of the sought-for spline as a bilinear form in each cell of the two-dimensional grid in such a way that the first derivatives in the center of the cell will coincide with the corresponding difference derivatives. By integrating the mixed derivative F_{xy} it is relatively easy to obtain the partial derivatives F_x, F_y and the approximating spline $F(x, y)$ in the rectangular domain considered.

The representation of the function $F(x, y)$ through normalized one-variable B-splines reads

$$F(x, y) = \sum_{k=i}^{i+p} \sum_{\ell=j}^{j+q} b_{k\ell}^{pq} B_k^{(p)}(x) B_\ell^{(q)}(y). \tag{6.97}$$

Here the indices i, j and the orders p, q of the normalized B-splines are determined by the position of the point (x, y) in the considered rectangular domain where the table $\{f_{k\ell}\}$ is given (inside the table $p = q = 2$; near the borders one of the numbers p or q is equal to 1; near corners $p = q = 1$). Normalized B-splines are nonnegative functions with maximum value equal to 1 (see [118]).

The spline $B_i^{(1)}(x)$ is given on a mesh x_i and is a piecewise linear continuous function on the interval (x_i, x_{i+2}), satisfying $B_i^{(1)}(x) = 0$ outside this interval:

$$B_i^{(1)}(x) = \begin{cases} \bar{\alpha}_i(x), & \text{if } x_i \le x < x_{i+1}; \\ 1 - \bar{\alpha}_{i+1}(x), & \text{if } x_{i+1} \le x \le x_{i+2}, \end{cases}$$

where $\bar{\alpha}_i(x) = (x - x_i)/h_x$, h_x is the x-step of the mesh.

The spline $B_i^{(2)}(x)$ is given on the mesh $x_i^* = (x_i + x_{i+1})/2$, is polynomial of degree two on each of the subintervals (x_k^*, x_{k+1}^*), $k = i, i+1, i+2$ of the interval (x_i^*, x_{i+3}^*), has there a continuous first derivative and vanishes outside this interval:

$$B_i^{(2)}(x) = \begin{cases} \frac{1}{2}\alpha_i^2(x), & \text{if } x_i^* \le x < x_{i+1}^*; \\ \frac{1}{2} + \alpha_{i+1}(x)(1 - \alpha_{i+1}(x)), & \text{if } x_{i+1}^* \le x < x_{i+2}^*; \\ \frac{1}{2}(1 - \alpha_{i+2}(x))^2, & \text{if } x_{i+2}^* \le x \le x_{i+3}^*, \end{cases}$$

where $\alpha_i(x) = (x - x_i^*)/h_x$.

The splines $B_i^{(1)}(y)$ and $B_i^{(2)}(y)$ are defined in the much the same way. For a uniform mesh the spline coefficients $b_{k\ell}^{pq} = f_{k\ell}$, where $f_{k\ell} = f(x_k, y_\ell)$ are the tabular data. The approximating function (6.97) is a biquadratic spline in the "inner" part of the table, a bilinear function near the "corners" of the table, and has a mixed order near its borders. The results of the investigation of the spline $F(x, y)$ demonstrate that it provides a smooth approximation of the initial tabular-given function $f(x, y)$ and reproduces well its geometric structure, i.e., it preserves those geometric properties (sign constancy, monotonicity, convexity, existence of a "plateau", etc.) that the initial functional dependence $f(x, y)$ possesses, in the entire rectangular domain considered, or in some part of it. Moreover, the approximating spline (6.97) smoothes out singularities of "break" type (if such singularities are present in the initial functional dependence) on the 1-2 node, without introducing additional structural changes like oscillations, loss of monotonicity, and so on.

In the case of nonuniform meshes the spline takes on a more complicated form. Namely, if we denote

$$\bar{\alpha}_i(x) = (x - x_i)/h_{x,i}, \quad \alpha_i(x) = (x - x_i^*)/h_{x,i}^*,$$

and

$$h_{x,i} = x_{i+1} - x_i, \quad h_{x,i}^* = (h_{x,i} + h_{x,i+1})/2,$$

then

$$B_i^{(1)}(x) = \begin{cases} \bar{\alpha}_i(x), & \text{if } x_i \le x < x_{i+1}, \\ 1 - \bar{\alpha}_{i+1}(x), & \text{if } x_{i+1} \le x \le x_{i+2}, \\ 0, & \text{if } x < x_i \text{ or } x > x_{i+2}, \end{cases}$$

$$B_i^{(2)}(x) = \begin{cases} \dfrac{h^*_{x,i}}{h^*_{x,i} + h^*_{x,i+1}} \alpha_i^2(x), & \text{if } x_i^* \le x < x_{i+1}^*, \\ \dfrac{h^*_{x,i}}{h^*_{x,i} + h^*_{x,i+1}} + 2 \dfrac{h^*_{x,i+1}}{h^*_{x,i} + h^*_{x,i+1}} \alpha_{i+1}(x) - \\ \left(\dfrac{h^*_{x,i+1}}{h^*_{x,i} + h^*_{x,i+1}} + \dfrac{h^*_{x,i+1}}{h^*_{x,i+1} + h^*_{x,i+2}} \right) \alpha_{i+1}^2(x), & \text{if } x_{i+1}^* \le x < x_{i+2}^*, \\ \dfrac{h^*_{x,i+2}}{h^*_{x,i+1} + h^*_{x,i+2}} (1 - \alpha_{i+2}(x))^2, & \text{if } x_{i+2}^* \le x \le x_{i+3}^*, \\ 0, & \text{if } x < x_i^* \text{ or } x > x_{i+3}^*. \end{cases}$$

For the coefficients $b_{k\ell}^{pq}$ one takes a linear combination, with the appropriate weights, of the values that are the closest to the (k, ℓ)-node in the mesh of tabular data:

$$b_{k\ell}^{22} = (1 - \gamma_k)\ (1 - \delta_\ell)\ f_{k\ell} + \gamma_k\ (1 - \delta_\ell)\ f_{k+m,\ell} + \\ (1 - \gamma_k)\ \delta_\ell\ f_{k,\ell+n} + \gamma_k\ \delta_\ell\ f_{k+m,\ell+n}. \quad (6.98)$$

The weights γ_k, δ_ℓ and the index shifts m, n are given by the formulas

$$\gamma_k = \frac{|h_{x,k} - h_{x,k-1}|}{4 \max\{h_{x,k}; h_{x,k-1}\}}, \qquad \delta_\ell = \frac{|h_{y,\ell} - h_{y,\ell-1}|}{4 \max\{h_{y,\ell}; h_{y,\ell-1}\}},$$

$$m = \operatorname{sign}(h_{x,k} - h_{x,k-1}), \qquad n = \operatorname{sign}(h_{y,\ell} - h_{y,\ell-1}),$$

where $h_{x,k} = x_{k+1} - x_k$, $h_{y,\ell} = y_{\ell+1} - y_\ell$.

The coefficients $b_{k\ell}^{12}$, $b_{k\ell}^{21}$ and $b_{k\ell}^{11}$ required for the regions adjacent to the boundaries are found from the requirement that $F(x, y)$ be smooth. The representation (6.97)–(6.98) on nonuniform meshes is in fact a certain symmetrization based on a piecewise bilinear interpolation of the data $\{f_{k\ell}\}$ in the two-dimensional case [232].

The results of the numerical investigation of the spline (6.97)–(6.98) on uniform and nonuniform meshes have shown that this approximation:

- ensures that $F(x, y)$ has the requisite smoothness, i.e., it is continuous together with its first derivatives;
- goes over into a one-dimensional approximation if in the table $\{f_{ij}\}$ there is no dependence on one of the variables in some subdomain of the domain D under consideration;
- guarantees a precise rendering of a plateau if one is present in the initial functional dependence $f(x, y)$;

- coincides with the tabular data in the corners of the rectangular domain D;
- makes it possible to reproduce geometric features (sign constancy, monotonicity, convexity) present in the initial functional dependence $f(x, y)$;
- ensures an exact approximation for bilinear functions.

The qualities listed above are in a certain sense achieved at the cost of some loss of accuracy in the case of smooth functions. At the same time it is obvious that, as a rule, the error of the applicable models exceed the interpolation error. One could have reached a higher accuracy by using approximations that are exact for polynomials of higher degrees, for instance, are exact for quadratic functions. However, already in the one-dimensional case one can provide examples of functions, meshes and approximations which are exact for quadratic functions, but which do not reproduce geometric properties of the initial function (for instance, its sign constancy).

6.5.2 Numerical results

First of all, let us compare approximation (6.97)-(6.98) with the well known methods of approximation by cubic splines [222] and interpolation in the mean by second-order splines [233] for functions of one variable $f(x)$ (here in the case of the spline (6.97) we assume that there is no dependence in y). The computations of the coefficients of interpolating cubic splines and interpolating-in-the-mean second-order splines were carried out by solving the corresponding systems of linear algebraic equations that differ only through their right-hand sides. To close the system of equations for the interpolating cubic spline, additional boundary conditions were imposed using the exact values of the second derivative $f''(x)$ at endpoints of the segment $[a, b]$, i.e., $f''(a)$ and $f''(b)$. For the interpolating-in-the-mean spline the values $f'(a)$ and $f'(b)$ of the first derivative were given. The interpolating cubic [resp., the interpolating-in-the mean] spline was constructed from the values $f_i = f(x_i)$ of the function $f(x)$ in the nodes of the mesh $\{x_i\}$ [resp., from the exact values of the integral $S_i = \int_{x_i}^{x_{i+1}} f(x)dx$]. For the considered local spline (6.97) only the values $f(x_i)$ were used. These approximations were constructed on different nonuniform meshes, the same for all three splines.

The results of computations for smooth functions demonstrate a satisfactory accuracy of the approximation for both the function and its derivative. The smooth test function used was $f(x) = \sin x$. A nonuniform mesh with slowly varying step: $h_{\max} \approx 0.3$, $h_{\min} \approx 0.2$, with the rates of decrease and increase of the step equal to 0.9 and 1.1, respectively, was used. In particular, in the uniform metric the error of the local approximation lies within 0.5% for the function and 1% for the derivative. As expected, the cubic spline gives the smallest approximation error, both for the function and its derivative. The error of the interpolating-in-the-mean spline is one order lower than the error of the local spline (6.97). For the derivative,

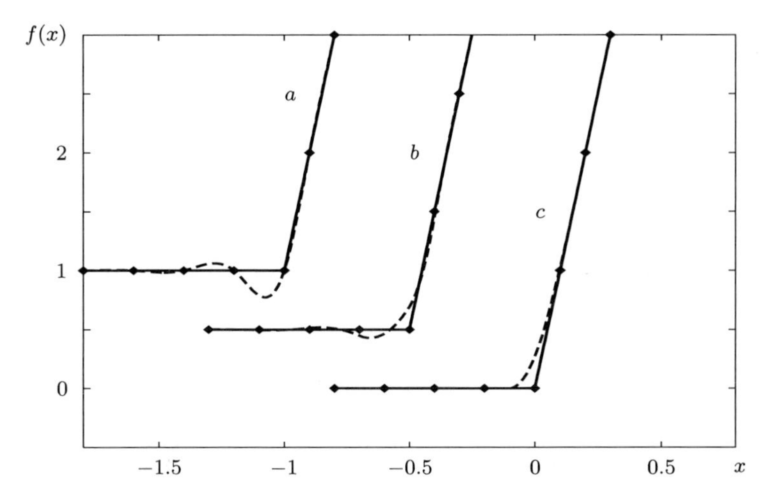

Figure 6.18: Results of the approximation of a "break" (dashed curves): (a) — interpolating cubic spline, (b) — interpolating-in-the-mean spline, (c) — local spline (6.97). The solid broken line represents the initial functional dependence $f(x)$, while the rhombuses show the tabular data

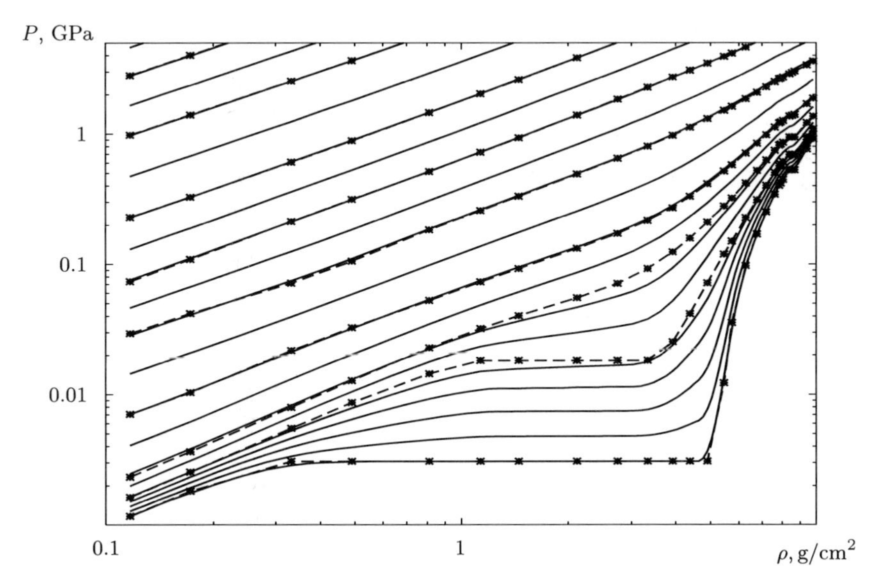

Figure 6.19: Pressure isotherms of iron at temperatures from 1 eV to 10 eV. Solid curves — approximation, asterisks (through which the dashed curves are drawn) — tabular data)

the error of the interpolating-in-the-mean spline and that of the local spline are practically identical. Thus, in the case of smooth functions on meshes with relatively detailed rendering of the character of the behavior of the initial functional dependence $f(x)$ the three different types of approximation yield a completely satisfactory accuracy and are practically indistinguishable on the graphs.

For functions with a "break-type" singularity (an abrupt change in the derivative, typical for equations of state) the results of the three approximations discussed above are displayed in Figure 6.18. A piecewise uniform mesh was used, with step $h = 0.2$ to the left of the break point and $h = 0.1$ to its right. As expected, the local spline (6.97)–(6.98), in contrast to interpolating splines, reproduces the behavior of the function $f(x)$ near the break point, preserving the sign constancy, monotonicity and convexity.

The enumerated advantages of the spline (6.97) are quite important, especially when the character of the dependence of the tabulated functions employed may change abruptly. Figure 6.19 shows a part of the pressure isotherms of the equation of state of iron in the regions of two phase transitions: vapor-liquid and liquid-solid. Since the initial quantities change by several orders, the interpolation was carried out in logarithmic variables.

As one can see from Figure 6.19, despite the fact that the mesh selected is rather crude near the liquid-vapor phase transition, the accuracy provided by the approximation using the local spline (6.97) is completely satisfactory for practical problems.

Part III

APPENDIX

Methods for solving the Schrödinger and Dirac equations

A.1 Quantum mechanical problems that can be solved analytically

Many problems of quantum mechanics lead to differential equations of the form

$$u'' + \frac{\tilde{\tau}(x)}{\sigma(x)} u' + \frac{\tilde{\sigma}(x)}{\sigma^2(x)} u = 0, \qquad a < x < b, \tag{A.1}$$

where $\sigma(x)$, $\tilde{\sigma}(x)$ are polynomials of degree at most 2 and $\tilde{\tau}(x)$ is a polynomial of degree at most 1. Equation (A.1) can be obtained from an equation that is well known in the analytic theory of differential equations and the theory of special functions, namely, the Riemann equation with three singular points, via a change of variables that sends one of the singular points to infinity [128]. Equations of the form (A.1) allow one to describe the harmonic oscillator, the motion of a particle in a central field, the hydrogen atom, or a hydrogen-like ion in the relativistic and nonrelativistic settings, as well as many other problems of atomic, molecular and nuclear physics. Thus, for instance, the Schrödinger equation for the harmonic oscillator,

$$-\frac{1}{2}\psi'' + \frac{x^2}{2}\psi = \varepsilon\psi, \qquad -\infty < x < \infty, \tag{A.2}$$

is a particular case of equation (A.1) with $\sigma(x) = 1$, $\tilde{\tau}(x) = 0$, $\tilde{\sigma}(x) = 2\varepsilon - x^2$.

When the Schrödinger equation for a central potential is solved by the method of separation of variables one obtains the following equation for the angular part of the wave function:

$$(1 - x^2)u'' - 2xu' + \left(\mu - \frac{m^2}{1 - x^2}\right) u = 0 \tag{A.3}$$

where

$$x = \cos\vartheta, \quad m = 0, \pm 1, \pm 2 \ldots, \quad \mu = \text{const.}$$

In this case $\sigma(x) = 1 - x^2$, $\tilde{\tau}(x) = -2x$, $\tilde{\sigma}(x) = \mu(1 - x^2) - m^2$.

To equation (A.1) one can reduce also many other model problems connected, in particular, with the study of scattering and interaction of neutrons with heavy nuclei, or the analysis of the rotation-vibration spectra of molecules, and so on (for example, the solutions of the Schrödinger equation with the Morse, Kratzer, Woods-Saxon, Pöschl-Teller potentials [63]). All these problems can be solved from a unified standpoint on the basis of a simple method [154].

A.1.1 Equations of hypergeometric type

Equation (A.1) can be simplified using the change of variable $u = \varphi(x)y$ with a special choice of the function $\varphi(x)$. Substituting $u = \varphi(x)y$ in (A.1), we obtain

$$y'' + \left(\frac{\widetilde{\tau}}{\sigma} + 2\frac{\varphi'}{\varphi}\right) y' + \left(\frac{\widetilde{\sigma}}{\sigma^2} + \frac{\widetilde{\tau}}{\sigma}\frac{\varphi'}{\varphi} + \frac{\varphi''}{\varphi}\right) y = 0. \tag{A.4}$$

To ensure that equation (A.4) will be not more complicated than equation (A.1), it is natural to require from the very beginning that the coefficient of y' have the form $\tau(x)/\sigma(x)$, where $\tau(x)$ is a polynomial of degree at most 1. This yields the following equation for the function $\varphi(x)$:

$$\frac{\varphi'}{\varphi} = \frac{\pi(x)}{\sigma(x)}, \tag{A.5}$$

where

$$\pi(x) = \frac{1}{2}[\tau(x) - \widetilde{\tau}(x)] \tag{A.6}$$

is a polynomial of degree at most 1. Then equation (A.4) takes on the form

$$y'' + \frac{\tau(x)}{\sigma(x)} y' + \frac{\overline{\sigma}(x)}{\sigma^2(x)} y = 0, \tag{A.7}$$

where

$$\overline{\sigma}(x) = \widetilde{\sigma}(x) + \pi^2(x) + \pi(x)[\widetilde{\tau}(x) - \sigma'(x)] + \pi'(x)\sigma(x).$$

Since $\tau(x)$ and $\overline{\sigma}(x)$ are polynomials of degree at most one and two, respectively, equation (A.7) has the same form as (A.1). To simplify further, let us pick the polynomial $\pi(x)$ so that

$$\overline{\sigma}(x) = \lambda\sigma(x), \tag{A.8}$$

where λ is some constant. In this way we arrive at the *equation of hypergeometric type*

$$\sigma(x)y'' + \tau(x)y' + \lambda y = 0, \tag{A.9}$$

particular solutions of which can be found in explicit form [154]. The name of the equation comes from the fact that among these particular solutions one finds the Gauss hypergeometric functions and the confluent hypergeometric functions. Condition (A.8) gives an equation for the polynomial $\pi = \pi(x)$:

$$\pi^2 + (\widetilde{\tau} - \sigma')\pi + \widetilde{\sigma} - k\sigma = 0,$$

where

$$k = \lambda - \pi'. \tag{A.10}$$

If we assume that the constant k is known, then solving the quadratic equation yields

$$\pi(x) = \frac{\sigma' - \widetilde{\tau}}{2} \pm \sqrt{\left(\frac{\sigma' - \widetilde{\tau}}{2}\right)^2 - \widetilde{\sigma} + k\sigma}. \tag{A.11}$$

In order to be able to extract the root and make $\pi(x)$ a polynomial of degree at most 1, the discriminant of the polynomial of second degree under the radical sign in (A.11) must be equal to zero. This condition yields an equation for the constant k that typically is quadratic*). Once k is determined, we find $\pi(x)$ using formula (A.11), and then $\tau(x)$ and λ using formulas (A.6) and (A.10). Then from equation (A.5) we find the function $\varphi(x)$, which determines the asymptotic behavior of the solutions of equation (A.1). Obviously, the reduction of equation (A.1) to equation (A.9) can be achieved in several ways, in accordance with the choice of different values of k and the choice of different signs in formula (A.11) for $\pi(x)$. By imposing additional requirements, in particular, that the solution be square integrable, the choice of the function $\varphi(x)$ becomes unique (see below).

Let us find the polynomial solutions of equation (A.9) in explicit form. To this end, we will first show that the derivative $v_1(x) = y'(x)$ satisfies an equation of the same type. Indeed, differentiating (A.9) we have

$$\sigma(x)v_1'' + \tau_1(x)v_1' + \mu_1 v_1 = 0, \tag{A.12}$$

where $\tau_1(x) = \tau(x) + \sigma'(x)$ is a polynomial of degree at most 1 and $\mu_1 = \lambda + \tau'(x)$ is a constant.

Similarly, differentiating (A.9) n times, we easily obtain for the function $v_n(x) = y^{(n)}(x)$ the equation of hypergeometric type

$$\sigma(x)v_n'' + \tau_n(x)v_n' + \mu_n v_n = 0, \tag{A.13}$$

where

$$\tau_n(x) = \tau(x) + n\sigma'(x),$$

$$\mu_n = \lambda + n\tau' + \frac{n(n-1)}{2}\sigma''.$$

This property enables us to construct a family of particular solutions of equation (A.9), corresponding to certain values of λ. Indeed, for $\mu_n = 0$ equation (A.13) has the obvious particular solution $v_n(x) = \text{const}$. Since $v_n(x) = y^{(n)}(x)$, this means that for

$$\lambda = \lambda_n = -n\tau' - \frac{n(n-1)}{2}\sigma'' \tag{A.14}$$

---

*) Except for the one case when for $\sigma(x) = \text{const}$ the polynomial under the radical sign in (A.11) is of degree 1. In this case the solutions of equation (A.1) can be expressed in terms of Bessel functions [154].

there exists a particular solution $y = y_n(x)$ of the equation of hypergeometric type which is a polynomial of degree n.

To write $y_n(x)$ explicitly, let us multiply equations (A.9) and (A.13) by functions $\rho(x)$ and $\rho_n(x)$ such that the resulting equations can be written in self-adjoint form

$$(\sigma \rho y')' + \lambda \rho y = 0, \tag{A.15}$$

$$(\sigma \rho_n v_n')' + \mu_n \rho_n v_n = 0. \tag{A.16}$$

The functions $\rho(x)$ and $\rho_n(x)$ satisfy the differential equations

$$(\sigma \rho)' = \tau \rho, \tag{A.17}$$

$$(\sigma \rho_n)' = \tau_n \rho_n. \tag{A.18}$$

Using the explicit expression for $\tau_n(x)$, we readily establish a connection between $\rho_n(x)$ and $\rho_0(x) = \rho(x)$. Namely,

$$\frac{\rho_n'}{\rho_n} = \frac{\tau_n - \sigma'}{\sigma} = \frac{\tau - \sigma'}{\sigma} + n\frac{\sigma'}{\sigma},$$

whence

$$\frac{\rho_n'}{\rho_n} = \frac{\rho'}{\rho} + n\frac{\sigma'}{\sigma}$$

and consequently

$$\rho_n(x) = \sigma^n(x)\rho(x). \tag{A.19}$$

Since $\sigma(x)\rho_n(x) = \rho_{n+1}(x)$ and $v_n(x) = y^{(n)}(x)$, equation (A.16) may be recast as the recursion relation

$$\rho_n v_n = -\frac{1}{\mu_n}(\rho_{n+1} v_{n+1})'.$$

This yields successively

$$\rho y = \rho_0 v_0 = -\frac{1}{\mu_0}(\rho_1 v_1)' = \left(-\frac{1}{\mu_0}\right)\left(-\frac{1}{\mu_1}\right)(\rho_2 v_2)' = \ldots = \frac{1}{A_n}(\rho_n v_n)^{(n)},$$

where

$$A_n = (-1)^n \prod_{k=0}^{n-1} \mu_k, \quad A_0 = 1.$$

Thus, for the solutions of equation (A.9) in the form of a polynomial of degree n, i.e., for $y = y_n(x)$ corresponding to $\lambda = \lambda_n$, with the property that $v_n = y^{(n)}(x) = \text{const}$, we obtain the following formula (known as the *Rodrigues formula* [154]):

$$y_n(x) = \frac{B_n}{\rho(x)} \frac{d^n}{dx^n}\left[\sigma^n(x)\rho(x)\right], \tag{A.20}$$

where B_n is a normalization constant, $n = 0, 1, 2, \ldots$.

The polynomial solutions (A.20) of equation (A.9) that have the orthogonality property are called *classical orthogonal polynomials.* Such polynomials are widely used in quantum mechanics. By inverting formula (A.20) for arbitrary values of λ one can obtain a unified integral representation for the solutions of equation (A.1) and their derivatives, in particular, for the Bessel functions and the Gauss hypergeometric functions [154].

A.1.2 Bound state wave functions and classical orthogonal polynomials

Consider a particle moving in some time-independent force field. If the external forces confine the particle to a bounded domain of space, so that it cannot escape to infinity, then one speaks about *bound states* of the particle. To find the wave functions $\psi(\vec{r})$ that describe such states and the corresponding energy levels E, one solves the stationary Schrödinger equation

$$-\frac{\hbar^2}{2m}\Delta\psi + U(\vec{r})\psi = E\psi, \tag{A.21}$$

where $\hbar$ is the Planck constant, m the mass of the particle, and $U(\vec{r})$ the potential energy. The wave function $\psi(\vec{r})$ must satisfy the normalization condition

$$\int |\psi(\vec{r})|^2 \, d\vec{r} = 1. \tag{A.22}$$

For many problems of quantum mechanics that can be solved by the method of separation of variables the Schrödinger equation (A.21) can be reduced to the form of a *generalized equation of hypergeometric type* (A.1). The energy E appears in the coefficients of equation (A.1) as a parameter. Here one can assume that $\sigma(x) > 0$ for $x \in (a, b)$, and that in the endpoints of the interval (a, b), if they are not infinity, the polynomial $\sigma(x)$ is equal to zero. Among the possible ways of reducing equation (A.1) to the form (A.9) there is one for which the condition $\int_a^b |y(x)|^2 \rho(x)dx < \infty$ (the weight function $\rho(x) > 0$), and hence the normalization condition for the wave function, is satisfied. One has the following

THEOREM. *Let $y = y(x)$ be a solution of the equation of hypergeometric type* (A.9). *Assume that the function $\rho(x)$ satisfies the equation $[\sigma(x)\rho(x)]' = \tau(x)\rho(x)$, is bounded and has non-negative values on an interval (a, b), and in the endpoints of the interval satisfies the conditions*

$$\sigma(x)\rho(x)x^k\big|_{x=a,b} = 0, \quad k = 0, 1, \ldots.$$

Then equation (A.9) *has nontrivial solutions $y(x)$ with the property that $y(x)$ is* *bounded and the function* $y(x)\sqrt{\rho(x)}$ *is square integrable on* (a,b) *only for*

$$\lambda = \lambda_n = -n\tau' - \frac{n(n-1)}{2}\sigma'', \quad n = 0, 1, \ldots .$$

These solutions have the form

$$y(x, \lambda_n) = y_n(x) = \frac{B_n}{\rho(x)} \frac{d^n}{dx^n}[\sigma^n(x)\rho(x)] \quad (\textit{with } B_n \textit{ a constant}),$$

i.e., they are classical orthogonal polynomials of degree n*, which are orthogonal with weight* $\rho(x)$ *on the interval* (a,b)*.*

To satisfy the conditions imposed on the function $\rho(x)$ for the classical orthogonal polynomials, the polynomial $\tau(x)$ must have a negative derivative and a zero in the interval (a,b) if $\sigma(x) > 0$ for $x \in (a,b)$. Indeed, from the Rodrigues formula (A.20) for the polynomials $y_n(x)$ it follows that $y_1(x) = B_1\tau(x)$, and hence that $\tau(x)$, like any orthogonal polynomial, has a zero in the orthogonality interval (a,b). Multiplying equation (A.15) by $y(x)$ and integrating the left-hand side by parts, we obtain the following relation for the function $y_1(x)$:

$$\int_a^b \sigma(x)\rho(x)[y_1'(x)]^2\,dx = \lambda_1 \int_a^b \rho(x)y_1^2(x)\,dx.$$

Since, by (A.14), for $n = 1$ we have $\lambda_1 = -\tau'$, it is clear that $\tau' < 0$ if $\sigma(x) > 0$ for $x \in (a,b)$. It follows that for the wave function $\psi(\vec{r})$, which satisfies the Schrödinger equation (A.21) and the normalization condition (A.22), the boundedness requirements are met [154].

Let us apply the methods described above to the solution of a number of typical problems of quantum mechanics.

EXAMPLE 1. To determine the spectrum of the harmonic oscillator, we have to find, for the Schrödinger equation (A.2), the values of the energy ε for which the wave function $\psi(x)$ satisfies the normalization condition (A.22).

In the present case $\sigma(x) = 1$, $\widetilde{\tau}(x) = 0$, $\widetilde{\sigma}(x) = 2\varepsilon - x^2$. Using (A.11), we obtain the following expression for the polynomial $\pi(x)$:

$$\pi(x) = \pm\sqrt{x^2 - 2\varepsilon + k}.$$

The constant k is found from the condition that the expression under the radical sign has a double root, i.e., $k = 2\varepsilon$. From the two possible forms $\pi(x) = \pm x$, we need to choose the one for which the function $\tau(x) = \widetilde{\tau}(x) + 2\pi(x)$ has a negative derivative. Therefore, $\tau(x) = -2x$, which correspond to (see formulas (A.5)–(A.11))

$$\pi(x) = -x, \quad \varphi(x) = \exp(-x^2/2), \quad \rho(x) = \exp(-x^2), \quad \lambda = 2\varepsilon - 1.$$

The energy eigenvalues are found from equation (A.14):

$$\lambda + n\tau' + \frac{n(n-1)}{2}\sigma'' = 0,$$

which yields

$$\varepsilon = \varepsilon_n = n + \frac{1}{2}, \quad n = 0, 1, \ldots . \tag{A.23}$$

The eigenfunctions $y_n(x)$ are given by (A.20):

$$y_n(x) = B_n \exp(x^2) \frac{d^n}{dx^n}[\exp(-x^2)]. \tag{A.24}$$

If $B_n = (-1)^n$, then the functions $y_n(x)$ coincide with the Hermite polynomials $H_n(x)$. Hence, the wave function $\psi(x)$ are given by

$$\psi_n(x) = C_n \exp(-x^2/2) H_n(x), \quad C_n = \frac{1}{\sqrt{\sqrt{\pi}2^n n!}}. \tag{A.25}$$

In the present case the requirement of square integrability for the wave function $\psi(x)$ is the same as for the function $y_n(x)\sqrt{\rho(x)}$.

A.1.3 Solution of the Schrödinger equation in a central field

A fundamental problem in quantum mechanics of atoms is that of the motion of an electron in a central field of attraction. This is connected with the fact that the description of the motion of electrons in an atom which uses the central-field approximation is the basis of computations of various properties of atomic structures [29, 83, 49]. Such a description allows one to understand in a satisfactory manner specific aspects of the behavior of atoms and find their energy states, without solving the very difficult quantum-mechanical many-body problem.

In order to find the wave function $\psi(\vec{r})$ of a particle moving in a central field with potential energy $U(r)$ we need to solve the Schrödinger equation (A.21)

$$\Delta\psi + \frac{2m}{\hbar^2}[E \quad U(r)]\psi = 0. \tag{A.26}$$

We will seek particular solutions of (A.26) by separating variables in spherical coordinates, setting

$$\psi(\vec{r}) = F(r)W(\vartheta, \varphi).$$

For the functions $F(r)$ and $W(\vartheta, \varphi)$ we obtain the equations

$$\left[\frac{1}{\sin\vartheta}\frac{\partial}{\partial\vartheta}\left(\sin\vartheta\frac{\partial}{\partial\vartheta}\right) + \frac{1}{\sin^2\vartheta}\frac{\partial^2}{\partial\varphi^2}\right]W + \mu W = 0 \tag{A.27}$$

and

$$\frac{1}{r^2}\frac{d}{dr}\left(r^2\frac{dF}{dr}\right) + \left[\frac{2m}{\hbar^2}(E - U(r)) - \frac{\mu}{r^2}\right]F(r) = 0, \tag{A.28}$$

where μ is a constant.

As we can convince ourselves by separating the variables ϑ, φ and using the Theorem in A.1.2, equation (A.27) has solutions that satisfy the boundedness and single-valuedness conditions for $0 \leq \vartheta \leq \pi$, $\mu = \ell(\ell+1)$. Moreover, in this case $W(\vartheta, \varphi) = (-1)^m Y_{\ell m}(\vartheta, \varphi)$, where $Y_{\ell m}(\vartheta, \varphi)$ are spherical harmonics of order ℓ, with $\ell = 0, 1, \ldots$ and $m = 0, \pm 1, \ldots, \pm \ell$:

$$Y_{\ell m}(\vartheta, \varphi) = \Theta_{\ell m}(x) \Phi_m(\varphi) \quad (x = \cos\vartheta),$$

$$\Phi_m(\varphi) = \frac{1}{\sqrt{2\pi}} e^{im\varphi} \quad (m = 0, \pm 1, \ldots, \pm \ell),$$

$$\Theta_{\ell m}(x) = C_{\ell m}(1-x^2)^{m/2} P_{\ell-m}^{(m,m)}(x) =$$

$$\frac{(-1)^{\ell-m}}{2^\ell \ell!} \sqrt{\frac{2\ell+1}{2} \frac{(\ell+m)!}{(\ell-m)!}} (1-x^2)^{-m/2} \frac{d^{\ell-m}}{dx^{\ell-m}} (1-x^2)^\ell \quad (m \geq 0).$$

Here $P_{\ell-m}^{(m,m)}(x)$ are the Jacobi polynomials $P_n^{(\alpha,\beta)}(x)$, which for $m = 0$ coincide with the Legendre polynomials $P_\ell(x)$, and for $m > 0$ coincide with the associated Legendre functions of the first kind. For $m < 0$ we have, by definition, $\Theta_{\ell,-m}(x) = (-1)^m \Theta_{\ell m}(x)$. The normalization constant $C_{\ell m}$ for spherical harmonics is chosen in the simplest way, as done, for example, in the well-known monograph of Bethe and Salpeter [28]:

$$C_{\ell m} = \frac{1}{2^m \ell!} \sqrt{\frac{2\ell+1}{2} (\ell-m)!(\ell+m)!}\ .$$

By introducing the factor $(-1)^m$ in the angular part $W(\vartheta, \varphi)$ of the wave function, as is done in most quantum mechanics books, many relations for the angular momentum operator can be written in a simpler form. Thus, for example, for the operators $\widehat{L}_x \pm i\widehat{L}_y$ we have (see § A.2 below):

$$(\widehat{L}_x \pm i\widehat{L}_y)\psi_{\ell m} = \sqrt{(\ell \mp m)(\ell \pm m + 1)}\, \psi_{\ell, m\pm 1}.$$

We should mention also that in a number of quantum mechanics books, in order to simplify notation the solution of the Schrödinger equation in a central field is written from the very beginning in the form $\psi(\vec{r}) = r^{-1} R_{n\ell}(r) Y_{\ell m}(\vartheta, \varphi)$, with the understanding that $Y_{\ell m}(\vartheta, \varphi)$ already contains the factor $(-1)^m$. However, when considering the Dirac equation it is anyway necessary to introduce supplementary phase factors in the angular parts of each of the four components of the relativistic wave function (see § A.2). Therefore, it is logically justified to keep for $Y_{\ell m}(\vartheta, \varphi)$ the classical notation adopted in mathematical physics.

Next, let us examine the radial part of the wave function. The substitution $R(r) = rF(r)$ reduces (A.28) to the equation

$$R'' + \left[\frac{2m}{\hbar^2}(E - U(r)) - \frac{\ell(\ell+1)}{r^2}\right] R = 0. \tag{A.29}$$

For bound states the wave function must satisfy the normalization condition $\int |\psi(\vec{r})|^2 r^2 dr\, d\Omega = 1$. Since

$$\int Y_{\ell m}(\vartheta, \varphi) Y^*_{\ell' m'}(\vartheta, \varphi) d\Omega = \delta_{\ell\ell'}\delta_{mm'},$$

the normalization condition for the function $R(r)$ becomes

$$\int_0^\infty R^2(r)dr = 1. \tag{A.30}$$

Here it is assumed that the function $F(r) = R(r)/r$ is bounded for $r \to 0$.

A.1.4 Radial part of the wave function in a Coulomb field

The only atom for which the Schrödinger equation can be solved exactly is the simplest possible — the hydrogen atom. This, however, does not diminish, but rather increases the value of the exact solution of the problem for the hydrogen atom, since analytic, closed-form solutions can be used as a point of departure in approximate calculations for more complex quantum-mechanical systems.

For the quantum-mechanical description of the hydrogen atom we need to consider the relative motion of an electron (of mass m and charge $-e$) and a nucleus (of mass M and charge e). We will solve a more general problem, assuming that the charge of the nucleus is Ze.

The problem we are dealing with, namely, the motion of an electron and a nucleus, can be readily reduced to the problem of a single body, a particle with the reduced mass

$$\mu = \frac{mM}{m+M},$$

moving in the Coulomb field $U(r) = -Ze^2/r$, i.e., to the solution of the equation

$$\Delta\psi + \frac{2\mu}{\hbar^2}\left(E + \frac{Ze^2}{r}\right)\psi = 0.$$

Since $m/M \approx 1/(1836\,A) \ll 1$ (where A denotes the atomic weight), we will assume from now on that $\mu = m$.

After passing to spherical coordinates, the radial function $R(r)$ will obviously satisfy equation (A.29). In that equation it is convenient to pass to dimensionless variables, using the atomic system of units, in which for the units of charge, length and energy one takes the charge e of the electron and the quantities

$$a_0 = \frac{\hbar^2}{\mu e^2} = 0.529 \cdot 10^{-8}\,\text{cm}, \quad E_0 = \frac{e^2}{a_0} = 27.2\,\text{eV}.$$

Then equation (A.29) becomes

$$R'' + \left[2\left(E + \frac{Z}{r}\right) - \frac{\ell(\ell+1)}{r^2}\right] R = 0. \tag{A.31}$$

The requirement that the wave function $\psi(\vec{r})$ be bounded and square integrable can be reduced to the requirement that the function $R(r)/r$ be bounded for $r \to 0$ and that the normalization condition (A.30) be satisfied.

Equation (A.31) is a particular case of the generalized equation of hypergeometric type (A.1), in which we now have

$$\sigma(r) = r, \quad \widetilde{\tau}(r) = 0, \quad \widetilde{\sigma}(r) = 2Er^2 + 2Zr - \ell(\ell+1).$$

Let us reduce (A.31) to an equation of hypergeometric type (A.9). In our case the polynomial $\pi(r)$ is given by (see (A.5)-(A.11)):

$$\pi(r) = \frac{1}{2} \pm \sqrt{\frac{1}{4} - 2Er^2 - 2Zr + \ell(\ell+1) + kr}.$$

The constant k is chosen from the condition that the expression under the radical sign have multiple roots. This yields the following possible values for k:

$$k = 2Z \pm (2\ell+1)\sqrt{-2E}$$

(from the form of the potential it follows that bound states are possible only for $E < 0$). Among all admissible forms of the polynomial $\pi(r)$ we choose one for which the function $\tau(r)$ has a zero in the interval $(0, \infty)$ and a negative derivative. These conditions are satisfied by $\tau(r) = 2(\ell + 1 - \sqrt{-2E}r)$, which yields

$$k = 2Z - (2\ell+1)\sqrt{-2E}, \quad \pi(r) = \ell + 1 - \sqrt{-2E}\, r,$$

$$\varphi(r) = r^{\ell+1} \exp(-\sqrt{-2E}\, r), \quad \rho(r) = r^{2\ell+1} \exp(-2\sqrt{-2E}\, r),$$

$$\lambda = 2\left[Z - (2\ell+1)\sqrt{-2E}\right].$$

The energy eigenvalues E are determined from the condition (A.14), i.e.,

$$\lambda + n\tau' + \frac{n(n-1)}{2}\sigma'' = 0 \quad (n = 0, 1, \ldots),$$

where n is the degree of the polynomial $y(r)$ and, accordingly, the number of zeroes of the radial wave function. Using the expressions for λ, $\tau(r)$ and $\sigma(r)$ obtained above, we get

$$E = -\frac{Z^2}{2(n+\ell+1)^2}. \tag{A.32}$$

The value of E is completely determined by the number $n + \ell + 1$, which is known as the *principal quantum number*.

The number of zeroes of the radial function $R(r)$ is usually denoted by n_r, while the principal quantum number is denoted by n ($n = n_r + \ell + 1$). To use these notations, we need to replace n by $n_r = n - \ell - 1$ in the preceding formulas. Then, by (A.20), the functions $y(r) = y_{n\ell}(r)$ (where now n is the principal quantum number) take on the form

$$y_{n\ell}(r) = \frac{B_{n\ell}}{r^{2\ell+1}\exp\left(-\frac{2Z}{n}r\right)} \frac{d^{n-\ell-1}}{dr^{n-\ell-1}}\left[r^{n+\ell}\exp\left(-\frac{2Z}{n}r\right)\right]$$

and up to a factor they coincide with the Laguerre polynomials $L^{2\ell+1}_{n-\ell-1}(x)$, where $x = 2Zr/n$. For the radial function $R(r) = R_{n\ell}(r)$ we obtain the expression

$$R_{n\ell}(r) = C_{n\ell}e^{-x/2}x^{\ell+1}L^{2\ell+1}_{n-\ell-1}(x). \tag{A.33}$$

It is readily verified that the functions $R_{n\ell}(r)$ satisfy the original requirement that $\int\limits_0^\infty R^2_{n\ell}(r)dr < \infty$. The constant $C_{n\ell}$ is found from the normalization condition (A.22):

$$\frac{n}{2Z}C^2_{n\ell}\int\limits_0^\infty e^{-x}x^{2\ell+2}\left[L^{2\ell+1}_{n-\ell-1}(x)\right]^2 dx = 1. \tag{A.34}$$

The integral in (A.34) can be calculated by means of the recursion relation for Laguerre polynomials,

$$xL^\alpha_n(x) = -(n+1)L^\alpha_{n+1}(x) + (2n+\alpha+1)L^\alpha_n(x) - (n+\alpha)L^\alpha_{n-1}(x).$$

Using the orthogonality relation for the Laguerre polynomials, this yields

$$\int\limits_0^\infty e^{-x}x^{2\ell+2}\left[L^{2\ell+1}_{n-\ell-1}(x)\right]^2 dx = 2n\int\limits_0^\infty e^{-x}x^{2\ell+1}[L^{2\ell+1}_{n-\ell-1}(x)]^2 dx = 2nd^2_{n\ell},$$

where $d^2_{n\ell}$ denotes the square of the norm of the polynomial $L^{2\ell+1}_{n-\ell-1}(x)$. Therefore,

$$C_{n\ell} = \sqrt{\frac{Z(n-\ell-1)!}{n^2(n+\ell)!}}\ . \tag{A.35}$$

The simplest radial function corresponds to the case $n_r = 0$, i.e., $\ell = n-1$:

$$R_{n\ell}(r) = \frac{1}{\ell+1}\sqrt{\frac{Z}{(2\ell+1)!}}\, e^{-x/2}x^{\ell+1}.$$

The most complicated form of the radial function occurs for $\ell = 0$, when the function has the maximum possible number of zeroes for the given principal

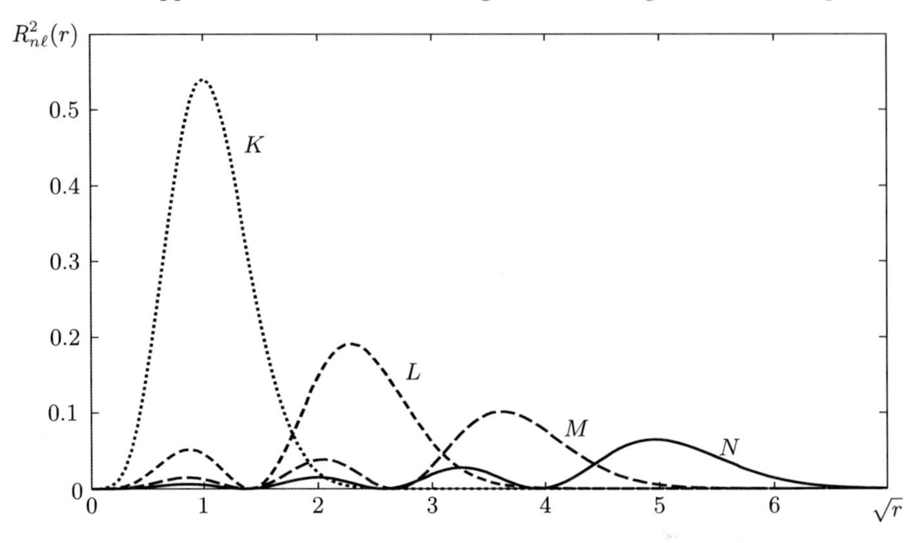

Figure A.1: Distribution of the electron charge in the hydrogen atom for the K, L, M, and N shells for $\ell = 0$, as function of the variable $x = \sqrt{r}$. The position of the main maximum corresponds to the average distance between electron and nucleus

quantum number n. However, in this case the dependence of the wave function on the angles ϑ, φ is the simplest, namely, for $\ell = 0$ the wave function is spherically-symmetric, since $Y_{00}(\vartheta, \varphi) = 1/\sqrt{4\pi}$. Figure A.1 shows the distribution of the electron charge $R^2_{n\ell}(r)$ for some states of the hydrogen atom with $\ell = 0$, for shells with principal quantum numbers $n = 1, 2, 3, 4$, i.e., for the K, L, M, N shells.

EXAMPLE 2. Once the radial function $R_{n\ell}(r)$ is known, we can calculate various characteristics of a hydrogen-like atom, in particular, the average distance between nucleus and electron in the states with quantum numbers n, ℓ:

$$\langle r_{n\ell} \rangle = \int_0^\infty r R^2_{n\ell}(r) dr = C^2_{n\ell} \left(\frac{n}{2Z} \right)^2 \int_0^\infty e^{-x} x^{2\ell+1} \left[x L^{2\ell+1}_{n-\ell-1}(x) \right]^2 dx.$$

To calculate the integral it is sufficient to use the recursion relation for the Laguerre polynomials and their orthogonality:

$$\langle r_{n\ell} \rangle = \frac{1}{2Z} \left[3n^2 - \ell(\ell+1) \right]. \tag{A.36}$$

Knowledge of wave functions allows us to obtain a visual representation of the hydrogen atom with the electron in different excited states. Figure A.2 shows the distribution of the radial electron density $R^2_{n\ell}(r) |Y_{\ell m}(\vartheta, \varphi)|^2$ for the states of the hydrogen atom with $n = 7$, $m = 0$ for $\ell = 0, 1, \ldots, 6$. Each picture displays the values of the function $R_{n\ell}^2(r)P_\ell^2(\cos\vartheta)$ in a plane passing through the nucleus. The curves where the charge density vanishes are concentric circles with the nucleus as center and lines passing through the nucleus. In three-dimensional space these curves would be spheres and cones, respectively.

The fact that the successive maxima of the radial electron density monotonically increase or decrease along lines passing through the nucleus in the ϑ-direction and on circles of a given radius r can be explained by investigating the behavior of the functions $R_{n\ell}^2(r)$ and $P_\ell^2(\cos\vartheta)$. To do this, let us consider the differential equation

$$[k(x)y']' + q(x)y = 0,$$

which has the particular solutions $R_{n\ell}(r) = C_{n\ell}e^{-x/2}x^{\ell+1}L_{n-\ell-1}^{2\ell+1}(x)$ $(x = 2r/n)$ and $P_\ell(x)$ $(x = \cos\vartheta)$ for special choices of $k(x)$ and $q(x)$. To examine the qualitative behavior of a solution $y(x)$ on an interval where $k(x) > 0$, $q(x) > 0$, it is convenient to introduce the function

$$v(x) = y^2(x) + a(x)[k(x)y'(x)]^2,$$

where the factor $a(x)$ will be chosen in a special way. We have

$$v'(x) = a'(x)[k(x)y']^2 + 2yy'[1 - a(x)k(x)q(x)].$$

Choosing $a(x)$ so that the condition $1 - a(x)k(x)q(x) = 0$ is satisfied, we obtain

$$v(x) = y^2(x) + \frac{k(x)}{q(x)}[y'(x)]^2, \quad v'(x) = \left[\frac{1}{k(x)q(x)}\right]'[k(x)y'(x)]^2.$$

This shows that the intervals of monotone increase or decrease of the function $v(x)$ will coincide with the analogous intervals for the function $1/(k(x)q(x))$. Note that the values of the functions $v(x)$ and $y^2(x)$ coincide in the points of maximum of $y(x)$ as well as in the points where $k(x) = 0$. This allows us to find the intervals in which the successive maxima of the functions $P_\ell^2(x)$ and $R_{n\ell}^2(r)$ increase or decrease. For example, in the case of the Legendre polynomials $P_\ell(x)$,

$$k(x) = 1 - x^2, \quad q(x) = \ell(\ell+1), \quad v'(x) = \frac{2x}{\ell(\ell+1)}[y'(x)]^2.$$

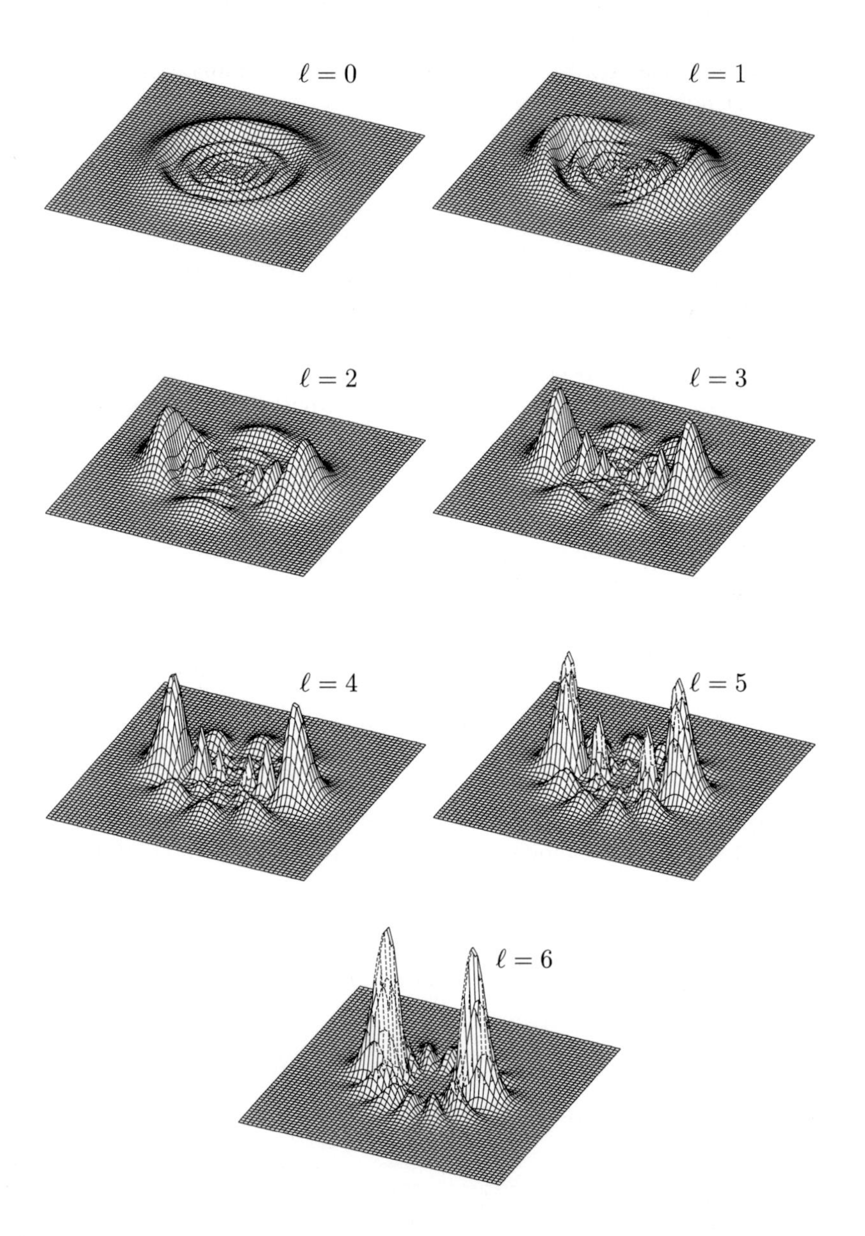

Figure A.2: Space distribution of the electron charge in the hydrogen atom for the states $n = 7$, $\ell = 0, 1, \ldots, n-1$ for $m = 0$. The plots show the graphs of the function $z = f(x, y)$, where $x = r\cos\vartheta$, $y = r\sin\vartheta$, $f(x,y) = R^2_{n\ell}(r)P^2_\ell(\cos\vartheta)$

Therefore, if $0 \leq \vartheta < \pi/2$ the magnitudes of the successive maxima of the function $P_\ell^2(\cos\vartheta)$ decrease when ϑ increases, whereas for $\pi/2 < \vartheta \leq \pi$ they decrease. The behavior of the functions $R_{n\ell}^2(r)$ is analyzed in the same manner, and one concludes that for $r > \ell(\ell+1)/Z$ the successive maxima increase (see figures A.1 and A.2).

EXAMPLE 3. Let us find the electrostatic potential generated by a hydrogen-like atom.

Suppose an electron moving in the Coulomb field of a nucleus with charge Z is in the steady state with wave function $\psi_{n\ell m}(\vec{r})$. If we consider that the nucleus does not move and is located in the point $r = 0$, then the potential of the nucleus is Z/r. Hence, for the average potential $V(r)$ produced in the point r by the electron and the nucleus we obtain

$$V(r) = \frac{Z}{r} - \int \frac{|\psi_{n\ell m}(\vec{r}')|^2}{|\vec{r} - \vec{r}'|} (r')^2 \, dr' d\Omega'.$$

To calculate the integral we use the generating function for the Legendre polynomials and the addition theorem for spherical harmonics:

$$\frac{1}{|\vec{r} - \vec{r}'|} = \sum_{s=0}^{\infty} \frac{r_<^s}{r_>^{s+1}} P_s(\cos\omega) = \sum_{s=0}^{\infty} \frac{r_<^s}{r_>^{s+1}} \left[\frac{4\pi}{2s+1} \sum_{m'=-s}^{s} Y_{sm'}^*(\vartheta', \varphi') Y_{sm'}(\vartheta, \varphi) \right],$$

where ω is the angle between the vector $\vec{r}$ and $\vec{r}'$, $r_< = \min(r, r')$, $r_> = \max(r, r')$. Since

$$\psi_{n\ell m}(\vec{r}') = \frac{1}{r'} R_{n\ell}(r')(-1)^m Y_{\ell m}(\vartheta', \varphi'),$$

we have

$$\int \frac{|\psi_{n\ell m}(\vec{r}')|^2}{|\vec{r} - \vec{r}'|} (r')^2 \, dr' d\Omega' = \sum_{s=0}^{\infty} \frac{4\pi}{2s+1} \sum_{m'} Y_{sm'}(\vartheta, \varphi) \int \frac{r_<^s}{r_>^{s+1}} R_{n\ell}^2(r') \, dr' \times$$
$$\int Y_{\ell m}(\vartheta', \varphi') Y_{\ell m}^*(\vartheta', \varphi') Y_{sm'}^*(\vartheta', \varphi') \, d\Omega'. \quad \text{(A.37)}$$

If we use the explicit form of the spherical harmonics:

$$Y_{\ell m}(\vartheta, \varphi) = \frac{1}{\sqrt{2\pi}} e^{im\varphi} \Theta_{\ell m}(\cos\vartheta),$$

then after integration with respect to φ' in the sum over m' only the term with $m' = 0$ remains, and so

$$V(r) = \frac{Z}{r} - \sum_{s=0}^{\infty} \frac{4\pi}{2s+1} Y_{s0}(\vartheta, \varphi) \int_0^{\infty} \frac{r_<^s}{r_>^{s+1}} R_{n\ell}^2(r') \, dr' \times$$
$$\int Y_{\ell m}(\vartheta', \varphi') Y_{\ell m}^*(\vartheta', \varphi') Y_{s0}^*(\vartheta', \varphi') \, d\Omega'.$$

The integration with respect to r' gives

$$\int_0^\infty \frac{r_<^s}{r_>^{s+1}} R_{n\ell}^2(r')\,dr' = \frac{1}{r^{s+1}} \int_0^r (r')^s R_{n\ell}^2(r')\,dr' + r^s \int_r^\infty \frac{R_{n\ell}^2(r')}{(r')^{s+1}}\,dr'.$$

The integral of a product of three spherical harmonics reduces to the integral of a product of three functions $\Theta_{\ell m}(\cos\vartheta)$:

$$\int Y_{\ell m}(\vartheta',\varphi')Y_{\ell m}^*(\vartheta',\varphi')Y_{s0}^*(\vartheta',\varphi')\,d\Omega' = \frac{1}{\sqrt{2\pi}} \int_{-1}^1 \Theta_{\ell m}^2(x)\Theta_{s0}(x)\,dx.$$

The last integral is different from zero only for $s = 0, 2, \ldots, 2\ell$, i.e., the sum over s in (A.37) contains only a finite number of terms. This follows from the connection between the functions $\Theta_{\ell m}(x)$ and orthogonal polynomials:

$$\Theta_{\ell m}(x) = C_{\ell m}(1-x^2)^{m/2} P_{\ell-m}^{(m,m)}(x).$$

Here $P_n^{(\alpha,\beta)}(x)$ are the Jacobi polynomials, which for $\alpha = \beta = 0$ coincide with the Legendre polynomials $P_n(x)$, and $C_{\ell m}$ are constants. Consequently,

$$\Theta_{\ell m}^2(x)\Theta_{s0}(x) = C_{\ell m}^2(1-x^2)^m \left[P_{\ell-m}^{(m,m)}(x)\right]^2 P_s(x).$$

It is known that an orthogonal polynomial (in particular, a Legendre polynomial) is orthogonal to any polynomial of smaller degree. This means that for $s > 2\ell$ the integral in question is equal to zero. Further, for s odd the function under the integral sign is odd because

$$P_s(-x) = (-1)^s P_s(x).$$

Therefore, the integral is zero also for s odd.

In the case when the electron is in the background state ($n = 1$, $\ell = 0$), all the integrals are readily calculated. We obtain

$$V(r) = \frac{Z-1}{r} + \left(Z + \frac{1}{r}\right)\exp(-2Zr). \tag{A.38}$$

For small r, $V(r) \approx Z/r$, as expected, while for $r \to \infty$, $V(r) \approx (Z-1)/r$ (the potential of the nucleus screened by the electron).

Let us remark that the integral of a product of three spherical harmonics $Y_{\ell_1 m_1}(\vartheta,\varphi)Y_{\ell_2 m_2}(\vartheta,\varphi)Y_{\ell m}(\vartheta,\varphi)$ can be expressed through the Clebsch-Gordan coefficients $C_{\ell_1 m_1 \ell_2 m_2}^{\ell m} = \langle \ell_1 m_1 \ell_2 m_2 | \ell m\rangle$, or the more symmetrized $3jm$-symbols

$$\begin{pmatrix} \ell_1 & \ell_2 & \ell \\ m_1 & m_2 & -m \end{pmatrix} = \frac{(-1)^{\ell_1-\ell_2+m}}{\sqrt{2\ell+1}} C_{\ell_1 m_1 \ell_2 m_2}^{\ell m}, \tag{A.39}$$

for which tables and formulas of various kind are available [220]. The Clebsch-Gordan coefficients and the related Racah coefficients are remarkably simply connected with the classical orthogonal polynomials of a discrete variable on a linear or quadratic mesh [146]. This connection allows one, using the basic properties of orthogonal polynomials, to study the properties of the vector-coupling coefficients and derive needed formulas.

A.2 Solution of the Dirac equation for the Coulomb potential

For high-Z atoms the electron energies become comparable with the rest energy mc^2. In such situations, to calculate energy levels and wave functions one needs to use the Dirac equation. For an electron in a central field with potential energy $U(r)$ the Dirac equation reads [28, 52]

$$\widehat{H}\Psi = \left[c(\widehat{\vec{\alpha}}\widehat{\vec{p}}) + mc^2\widehat{\beta} + U(r)\right]\Psi = E\Psi,$$

where Ψ is the 4-component wave function

$$\Psi = \begin{pmatrix}\Psi_1\\ \Psi_2\\ \Psi_3\\ \Psi_4\end{pmatrix}, \qquad (\widehat{\vec{\alpha}}\widehat{\vec{p}}) = \sum_{k=1}^{3}\widehat{\alpha}_k\widehat{p}_k \qquad \left(\widehat{p}_k = \frac{\hbar}{i}\frac{\partial}{\partial x_k}\right),$$

$$\widehat{\alpha}_k = \begin{pmatrix}O & \sigma_k\\ \sigma_k & O\end{pmatrix}, \quad \widehat{\beta} = \begin{pmatrix}I & O\\ O & -I\end{pmatrix}, \quad I = \begin{pmatrix}1 & 0\\ 0 & 1\end{pmatrix}, \quad O = \begin{pmatrix}0 & 0\\ 0 & 0\end{pmatrix}.$$

Usually for σ_k one takes the Pauli matrices

$$\sigma_1 = \begin{pmatrix}0 & 1\\ 1 & 0\end{pmatrix}, \quad \sigma_2 = \begin{pmatrix}0 & -i\\ i & 0\end{pmatrix}, \quad \sigma_3 = \begin{pmatrix}1 & 0\\ 0 & -1\end{pmatrix}.$$

Let us write the Dirac equations for each of the components of the wave function:

$$\begin{aligned}
-\frac{i}{\hbar c}\left(E - U(r) - E_0\right)\Psi_1 + \frac{\partial\Psi_3}{\partial z} + \left(\frac{\partial}{\partial x} - i\frac{\partial}{\partial y}\right)\Psi_4 &= 0,\\
-\frac{i}{\hbar c}\left(E - U(r) - E_0\right)\Psi_2 - \frac{\partial\Psi_4}{\partial z} + \left(\frac{\partial}{\partial x} + i\frac{\partial}{\partial y}\right)\Psi_3 &= 0,\\
-\frac{i}{\hbar c}\left(E - U(r) + E_0\right)\Psi_3 + \frac{\partial\Psi_1}{\partial z} + \left(\frac{\partial}{\partial x} - i\frac{\partial}{\partial y}\right)\Psi_2 &= 0,\\
-\frac{i}{\hbar c}\left(E - U(r) + E_0\right)\Psi_4 - \frac{\partial\Psi_2}{\partial z} + \left(\frac{\partial}{\partial x} + i\frac{\partial}{\partial y}\right)\Psi_1 &= 0
\end{aligned}$$

($E_0 = mc^2$ is the rest energy of the electron).

It is convenient to use the relativistic system of units, in which the electron mass m, the Planck constant $\hbar$ and the speed of light c are all equal to 1. Then the Dirac equations for an electron in the Coulomb field $U(r) = -Ze^2/r$ become

$$
\begin{aligned}
&-i\left(E+\frac{\mu}{r}-1\right)\Psi_1+\frac{\partial\Psi_3}{\partial z}+\left(\frac{\partial}{\partial x}-i\frac{\partial}{\partial y}\right)\Psi_4=0,\\
&-i\left(E+\frac{\mu}{r}-1\right)\Psi_2-\frac{\partial\Psi_4}{\partial z}+\left(\frac{\partial}{\partial x}+i\frac{\partial}{\partial y}\right)\Psi_3=0,\\
&-i\left(E+\frac{\mu}{r}+1\right)\Psi_3+\frac{\partial\Psi_1}{\partial z}+\left(\frac{\partial}{\partial x}-i\frac{\partial}{\partial y}\right)\Psi_2=0,\\
&-i\left(E+\frac{\mu}{r}+1\right)\Psi_4-\frac{\partial\Psi_2}{\partial z}+\left(\frac{\partial}{\partial x}+i\frac{\partial}{\partial y}\right)\Psi_1=0.
\end{aligned}
\qquad (A.40)
$$

Here $\mu = \alpha Z$, $\alpha = e^2/(\hbar c)$ is the fine structure constant, $1/\alpha = 137.036$, and the energy E includes the rest energy mc^2 $(0 < E < 1)$.

A.2.1 The system of equations for the radial parts of the wave functions

In spherical coordinates (r, ϑ, φ) the variables in the system of equations (A.40) can be separated if we use the fact that in a central field the angular momentum of a particle is conserved. Then one can construct eigenfunctions corresponding to specified eigenvalues of the operators $\widehat{H}, \widehat{J}^2$ and $\widehat{J}_z$ (here $\widehat{J}^2$ is the square of the angular momentum operator and $\widehat{J}_z$ is the operator of the projection of the angular momentum on the z-axis).

Since the spin interacts with the current generated by the orbital motion of the electron, in the general case the electron cannot have a determined value of the spin or orbital momentum. However, its wave function with the requisite symmetry properties can be constructed from a mixture of states with orbital momentum $\ell = j - 1/2$ and $\ell = j + 1/2$, adding to them the spin in the appropriate manner.

Accordingly, the wave function of a state in which the total momentum of the particle is equal to j and its projection on the z-axis is equal to m can be written in terms of the Clebsch-Gordan coefficients as a sum of products of the space and spin parts of the wave function:

$$
\begin{pmatrix}\Psi_1\\ \Psi_2\end{pmatrix} = \sum_{m_\ell, m_s} C^{jm}_{\ell m_\ell \frac{1}{2} m_s} f_{N\ell j}(r) W_{\ell m_\ell}(\vartheta,\varphi)\chi_{s m_s}(\sigma), \qquad (A.41)
$$

$$
\begin{pmatrix}\Psi_3\\ \Psi_4\end{pmatrix} = \sum_{m_{\ell'}, m_s} C^{jm}_{\ell' m_{\ell'} \frac{1}{2} m_s} (-1)^{(\ell-\ell'+1)/2} g_{N\ell' j}(r) W_{\ell' m_{\ell'}}(\vartheta,\varphi)\chi_{s m_s}(\sigma). \qquad (A.42)
$$

Here the spin wave functions $\chi_{sm_s}(\sigma)$ are the basic two-component spinors, corresponding to $s = 1/2$, $m_s = \pm 1/2$ (σ is the spin variable):

$$\chi_{1/2,1/2} = \begin{pmatrix} 1 \\ 0 \end{pmatrix}, \quad \chi_{1/2,-1/2} = \begin{pmatrix} 0 \\ 1 \end{pmatrix};$$

$W_{\ell m_\ell}(\vartheta, \varphi) = i^\ell (-1)^{m_\ell} Y_{\ell m_\ell}(\vartheta, \varphi)$ are the angular parts of the wave functions; $f_{N\ell j}(r)$ and $(-1)^{(\ell-\ell'+1)/2} g_{N\ell' j}(r)$ are the radial parts of the wave functions (the index N is connected with the number of zeroes of the function $f_{N\ell j}(r)$); and, finally, $C^{jm}_{\ell m_\ell \frac{1}{2} m_s} = \langle \ell\, m_\ell\, \frac{1}{2}\, m_s \,|\, j\, m \rangle$ are the Clebsch-Gordan coefficients ($m = m_\ell + m_s$, see Table A.1). The angular parts of the wave functions satisfy the relation

$$(\widehat{L}_x \pm i\widehat{L}_y) W_{\ell m_\ell}(\vartheta, \varphi) = [(\ell \mp m_\ell)(\ell \pm m_\ell + 1)]^{1/2}\, W_{\ell, m_\ell \pm 1}(\vartheta, \varphi).$$

In order to satisfy the last relation, we introduce the factor $(-1)^{m_\ell}$ in the angular component of the wave function $W_{\ell m_\ell}(\vartheta, \varphi)$, assuming that the spherical harmonics $Y_{\ell m_\ell}(\vartheta, \varphi)$ are normalized in the simplest way [28, 154]. The factor i^ℓ is introduced to ensure that the functions $f(r) = f_{N\ell j}(r)$ and $g(r) = g_{N\ell' j}(r)$ will satisfy differential equations with real coefficients. As we will see below, introducing in (A.42) the factor $(-1)^{(\ell-\ell'+1)/2}$ in the radial part of the wave function allows one to derive a system of equations for the radial parts of the wave functions $f(r)$ and $g(r)$ in a form that works for $\ell = j - 1/2$ as well as for $\ell = j + 1/2$.

Table A.1: Values of the Clebsch-Gordan coefficients $C^{jm}_{\ell m_\ell \frac{1}{2} m_s}$

m_s / j	$\frac{1}{2}$	$-\frac{1}{2}$
$\ell + \frac{1}{2}$	$\sqrt{\frac{j+m}{2j}}$	$\sqrt{\frac{j-m}{2j}}$
$\ell - \frac{1}{2}$	$-\sqrt{\frac{j-m+1}{2j+2}}$	$\sqrt{\frac{j+m+1}{2j+2}}$

Note that for given values of j, m and ℓ in equation (A.42) we must put $\ell' = 2j - \ell$. This is connected with the fact that after we substitute the expressions (A.41) for Ψ_1 and Ψ_2 in the last two of the Dirac equations (A.40) we obtain for Ψ_3 and Ψ_4 the representation (A.42) for $\ell' = 2j - \ell$ (see formulas (A.45) and (A.46) below). For this reason the radial part of the wave function $g_{N\ell' j}(r)$ will henceforth be denoted by $g_{N\ell j}(r)$.

Recall that in equations (A.41) and (A.42) j is the quantum number characterizing the total angular momentum of the electron ($j = 1/2, 3/2, \ldots$). The quantum number m takes half-integer values ranging from $-j$ to j. Clearly, for $\ell = j \mp 1/2$ we have $\ell' = j \pm 1/2$.

Upon substituting in (A.41) and (A.42) the expressions for the spin and angular parts of the wave functions $\chi_{sm_s}(\sigma)$, $W_{\ell m_\ell}(\vartheta,\varphi)$, $W_{\ell' m_{\ell'}}(\vartheta,\varphi)$, as well as the values of the Clebsch-Gordan coefficients, we obtain for $\ell = j - 1/2$ the formula

$$\begin{pmatrix}\Psi_1\\ \Psi_2\\ \Psi_3\\ \Psi_4\end{pmatrix} = (-1)^{m+1/2}\, i^\ell \begin{pmatrix} -\sqrt{\dfrac{j+m}{2j}}\; f_{N\ell j}(r)\, Y_{\ell,m-1/2}(\vartheta,\varphi) \\ \sqrt{\dfrac{j-m}{2j}}\; f_{N\ell j}(r)\, Y_{\ell,m+1/2}(\vartheta,\varphi) \\ \sqrt{\dfrac{j-m+1}{2j+2}}\; i g_{N\ell j}(r)\, Y_{\ell+1,m-1/2}(\vartheta,\varphi) \\ \sqrt{\dfrac{j+m+1}{2j+2}}\; i g_{N\ell j}(r)\, Y_{\ell+1,m+1/2}(\vartheta,\varphi) \end{pmatrix}. \tag{A.43}$$

Similarly, setting $\ell = j + 1/2$ in (A.41) and (A.42), we get

$$\begin{pmatrix}\Psi_1\\ \Psi_2\\ \Psi_3\\ \Psi_4\end{pmatrix} = (-1)^{m+1/2}\, i^\ell \begin{pmatrix} \sqrt{\dfrac{j-m+1}{2j+2}}\; f_{N\ell j}(r)\, Y_{\ell,m-1/2}(\vartheta,\varphi) \\ \sqrt{\dfrac{j+m+1}{2j+2}}\; f_{N\ell j}(r)\, Y_{\ell,m+1/2}(\vartheta,\varphi) \\ \sqrt{\dfrac{j+m}{2j}}\; i g_{N\ell j}(r)\, Y_{\ell-1,m-1/2}(\vartheta,\varphi) \\ -\sqrt{\dfrac{j-m}{2j}}\; i g_{N\ell j}(r)\, Y_{\ell-1,m+1/2}(\vartheta,\varphi) \end{pmatrix}. \tag{A.44}$$

Now let us substitute the expressions (A.43) and (A.44) in the system of equations (A.40) and use the following formulas, which are given in [28]:

$$\begin{aligned} \frac{\partial}{\partial z}\left[f(r) Y_{\ell m}(\vartheta,\varphi)\right] = {} & \\ & \sqrt{\frac{(\ell+m+1)(\ell-m+1)}{(2\ell+3)(2\ell+1)}}\, Y_{\ell+1,m}(\vartheta,\varphi)\left[\frac{df}{dr} - \ell\frac{f(r)}{r}\right] + \\ & \qquad \sqrt{\frac{(\ell+m)(\ell-m)}{(2\ell+1)(2\ell-1)}}\, Y_{\ell-1,m}(\vartheta,\varphi)\left[\frac{df}{dr} + (\ell+1)\frac{f(r)}{r}\right], \end{aligned} \tag{A.45}$$

$$\left(\frac{\partial}{\partial x} \pm i\frac{\partial}{\partial y}\right)[f(r)Y_{\ell m}(\vartheta,\varphi)] =$$
$$\pm\sqrt{\frac{(\ell\pm m+2)(\ell\pm m+1)}{(2\ell+3)(2\ell+1)}}\,Y_{\ell+1,m\pm 1}(\vartheta,\varphi)\left[\frac{df}{dr} - \ell\frac{f(r)}{r}\right] \mp$$
$$\sqrt{\frac{(\ell\mp m)(\ell\mp m-1)}{(2\ell+1)(2\ell-1)}}\,Y_{\ell-1,m\pm 1}(\vartheta,\varphi)\left[\frac{df}{dr} + (\ell+1)\frac{f(r)}{r}\right]. \quad \text{(A.46)}$$

This yields the following system of equations for $f(r) = f_{N\ell j}(r)$ and $g(r) = g_{N\ell j}(r)$:

$$\begin{cases} f' + \dfrac{1+\kappa}{r}f - \left(E + 1 + \dfrac{\mu}{r}\right)g = 0, \\ \\ g' + \dfrac{1-\kappa}{r}g + \left(E - 1 + \dfrac{\mu}{r}\right)f = 0, \end{cases} \quad \text{(A.47)}$$

where $\mu = \alpha Z$ and the constant κ depends on ℓ:

$$\kappa = \begin{cases} -(\ell+1) & \text{for} \quad \ell = j - 1/2, \\ \ell & \text{for} \quad \ell = j + 1/2. \end{cases}$$

Clearly, $|\kappa| = j+1/2$. Let us mention here that in the nonrelativistic approximation $|f(r)| \gg |g(r)|$ (this will be shown later, in Subsection A.2.5).

The functions $f(r)$ and $g(r)$ for bound states must satisfy the normalization condition

$$\int_0^\infty r^2\left[f^2(r) + g^2(r)\right]dr = 1, \quad \text{(A.48)}$$

which in turn follows from the normalization condition for the 4-component wave function Ψ.

A.2.2 Reduction of the system of equations for the radial functions to an equation of hypergeometric type

Let us write the system of equations (A.47) in matrix form, setting

$$u = \begin{pmatrix} u_1 \\ u_2 \end{pmatrix} = \begin{pmatrix} f(r) \\ g(r) \end{pmatrix}.$$

Then

$$u' = Au, \quad \text{(A.49)}$$

where

$$A = \begin{pmatrix} a_{11} & a_{12} \\ a_{21} & a_{22} \end{pmatrix} = \begin{pmatrix} -\dfrac{1+\kappa}{r} & 1+E+\dfrac{\mu}{r} \\ 1-E-\dfrac{\mu}{r} & -\dfrac{1-\kappa}{r} \end{pmatrix}.$$

To find the function $u_1(r)$, let us eliminate the function $u_2(r)$ from the system (A.49). We obtain for $u_1(r)$ the second-order differential equation

$$u_1'' - \left(a_{11} + a_{22} + \frac{a_{12}'}{a_{12}}\right) u_1' + \left(a_{11}a_{22} - a_{12}a_{21} - a_{11}' - \frac{a_{12}'}{a_{12}} a_{11}\right) u_1 = 0. \quad \text{(A.50)}$$

Similarly, eliminating $u_1(r)$, we obtain the following equation for $u_2(r)$:

$$u_2'' - \left(a_{11} + a_{22} + \frac{a_{21}'}{a_{21}}\right) u_2' + \left(a_{11}a_{22} - a_{12}a_{21} - a_{22}' + \frac{a_{21}'}{a_{21}} a_{22}\right) u_2 = 0. \quad \text{(A.51)}$$

The coefficients of the matrix A have the form

$$a_{ik} = b_{ik} + \frac{c_{ik}}{r},$$

where b_{ik} and c_{ik} are constants. Equations (A.50) and (A.51) are not generalized equations of hypergeometric type (see (A.1) in § A.1). This is because

$$\frac{a_{12}'}{a_{12}} = -\frac{c_{12}}{c_{12}r + b_{12}r^2},$$

and so the coefficients of $u_1'(r)$ and $u_1(r)$ in equation (A.50) have the form

$$a_{11} + a_{22} + \frac{a_{12}'}{a_{12}} = \frac{p_1(r)}{r} - \frac{c_{12}}{c_{12}r + b_{12}r^2} \quad (b_{12} = 1 + E \neq 0, \quad c_{12} = \mu \neq 0),$$

$$a_{11}a_{22} - a_{12}a_{21} - a_{11}' + \frac{a_{12}'}{a_{12}} a_{11} = \frac{p_2(r)}{r^2} - \frac{c_{12}}{c_{12}r + b_{12}r^2} \, \frac{c_{11} + b_{11}r}{r}$$

(here $p_1(r)$ and $p_2(r)$ are polynomials of degree at most one and two, respectively).

Equation (A.50) would be a generalized equation of hypergeometric type (A.1), with $\sigma(r) = r$, if one of the coefficients b_{12} or c_{12} would be equal to zero. Here it is convenient to argue as follows. A linear transformation

$$\begin{pmatrix} v_1 \\ v_2 \end{pmatrix} = T \begin{pmatrix} u_1 \\ u_2 \end{pmatrix}$$

with a nonsingular matrix T that does not depend on r takes our system (A.49) into a system of the same form for the functions $v_1(r)$ and $v_2(r)$, namely

$$v' = \widetilde{A}v,$$

where

$$\widetilde{A} = TAT^{-1} = \begin{pmatrix} \widetilde{a}_{11} & \widetilde{a}_{12} \\ \widetilde{a}_{21} & \widetilde{a}_{22} \end{pmatrix}.$$

Obviously, the coefficients $\widetilde{a}_{ik}$ are linear combinations of a_{ik}, and consequently

$$\widetilde{a}_{ik} = \widetilde{b}_{ik} + \frac{\widetilde{c}_{ik}}{r},$$

where again $\widetilde{b}_{ik}$ and $\widetilde{c}_{ik}$ are constants. It follows that the equations for $v_1(r)$ and $v_2(r)$ will be similar to (A.50) and (A.51):

$$v_1'' - \left(\widetilde{a}_{11} + \widetilde{a}_{22} + \frac{\widetilde{a}_{12}'}{\widetilde{a}_{12}}\right) v_1' + \left(\widetilde{a}_{11}\widetilde{a}_{22} - \widetilde{a}_{12}\widetilde{a}_{21} - \widetilde{a}_{11}' + \frac{\widetilde{a}_{12}'}{\widetilde{a}_{12}}\widetilde{a}_{11}\right) v_1 = 0, \quad \text{(A.52)}$$

$$v_2'' - \left(\widetilde{a}_{11} + \widetilde{a}_{22} + \frac{\widetilde{a}_{21}'}{\widetilde{a}_{21}}\right) v_2' + \left(\widetilde{a}_{11}\widetilde{a}_{22} - \widetilde{a}_{12}\widetilde{a}_{21} - \widetilde{a}_{22}' + \frac{\widetilde{a}_{21}'}{\widetilde{a}_{21}}\widetilde{a}_{22}\right) v_2 = 0. \quad \text{(A.53)}$$

To make (A.52) a generalized equation of hypergeometric type it suffices to put $\widetilde{b}_{12} = 0$, or $\widetilde{c}_{12} = 0$. Similarly, for (A.53) it suffices to put $\widetilde{b}_{21} = 0$, or $\widetilde{c}_{21} = 0$. These conditions give certain constraints on the choice of the matrix T. Let

$$T = \begin{pmatrix} \alpha & \beta \\ \gamma & \delta \end{pmatrix}.$$

Then

$$T^{-1} = \frac{1}{\Delta}\begin{pmatrix} \delta & -\beta \\ -\gamma & \alpha \end{pmatrix}, \quad \Delta = \alpha\delta - \beta\gamma,$$

and so

$$\widetilde{A} = \begin{pmatrix} a_{11}\alpha\delta - a_{12}\alpha\gamma + a_{21}\beta\delta - a_{22}\beta\gamma & a_{12}\alpha^2 - a_{21}\beta^2 + (a_{22} - a_{11})\alpha\beta \\ a_{21}\delta^2 - a_{12}\gamma^2 + (a_{11} - a_{22})\gamma\delta & -a_{11}\beta\gamma + a_{12}\alpha\gamma - a_{21}\beta\delta + a_{22}\alpha\delta \end{pmatrix}.$$

There are several possible choices for $\alpha, \beta, \gamma, \delta$. As a matter of fact, in books on quantum mechanics only one of the choices corresponding to the conditions $\widetilde{b}_{12} = 0$, $\widetilde{b}_{21} = 0$ is used [123]. We will consider the case when the constants $\alpha, \beta, \gamma, \delta$ are chosen from the conditions $\widetilde{c}_{12} = 0$, $\widetilde{c}_{21} = 0$, which give the equations

$$2\kappa\alpha\beta + \mu(\alpha^2 + \beta^2) = 0,$$

$$2\kappa\gamma\delta + \mu(\gamma^2 + \delta^2) = 0.$$

We will show that this requirement is preferable over the requirement $\widetilde{b}_{12} = 0$, $\widetilde{b}_{21} = 0$. The conditions $\widetilde{c}_{12} = 0$, $\widetilde{c}_{21} = 0$ will be satisfied if we take the matrix T to be of the form

$$T = \begin{pmatrix} \mu & \nu - \kappa \\ \nu - \kappa & \mu \end{pmatrix},$$

where $\mu = \alpha Z$, $\nu = \sqrt{\kappa^2 - \mu^2}$. In this way we arrive at the following system of equations for the functions $v_1(r)$ and $v_2(r)$:

$$v_1' = \left(-\frac{\nu+1}{r} + \frac{E\mu}{\nu}\right) v_1 + \left(1 + \frac{E\kappa}{\nu}\right) v_2, \tag{A.54}$$

$$v_2' = \left(1 - \frac{E\kappa}{\nu}\right) v_1 + \left(\frac{\nu-1}{r} - \frac{E\mu}{\nu}\right) v_2. \tag{A.55}$$

If $1 + E\kappa/\nu \neq 0$, then from the system (A.54), (A.55) one can eliminate $v_2(r)$ and obtain a generalized equation of hypergeometric type for the function $v_1(r)$:

$$v_1'' + \frac{2}{r} v_1' + \frac{(E^2-1)r^2 + 2E\mu r - \nu(\nu+1)}{r^2} v_1 = 0. \tag{A.56}$$

In the special case $E = -\nu/\kappa$ (i.e., $1 + E\kappa/\nu = 0$), the general solution of equation (A.54) has the form

$$v_1(r) = C_1 r^{-\nu-1} e^{E\mu r/\nu}.$$

This function $v_1(r)$ satisfies the integrability condition (A.48) only when $C_1 = 0$. Then (A.55) yields

$$v_2(r) = C_2 r^{\nu-1} e^{-E\mu r/\nu}. \tag{A.57}$$

Clearly, the functions $v_1(r) \equiv 0$ and $v_2(r)$ with $C_2 \neq 0$ satisfy the conditions of our problem. Note that such a solution is possible only for $\kappa < 0$ (i.e., for $\ell = j - 1/2$), because $\nu > 0$ and $E > 0$.

A.2.3 Equations of hypergeometric type for the bound states and their solution

Consider equation (A.56). First, let us examine the behavior of the function $v_1(r)$ for small r. When $r \to 0$,

$$\left|(E^2-1)r^2 + 2E\mu r\right| \ll \nu(\nu+1),$$

and so in the vicinity of the point $r = 0$ the behavior of $v_1(r)$ will be approximately described by the Euler equation

$$r^2 v_1'' + 2r v_1' - \nu(\nu+1) v_1 = 0,$$

whose solution has the form

$$v_1(r) = C_1 r^{\nu} + C_2 r^{-\nu-1}.$$

From the conditions imposed on $v_1(r)$ it follows that $C_2 = 0$. Hence, for $r \sim 0$ we have $v_1(r) \approx C_1 r^{\nu}$.

Equation (A.56) is an equation of type (A.1), with $\sigma(r)=r$, $\widetilde{\tau}(r)=2$, $\widetilde{\sigma}(r)=(E^2-1)r^2+2E\mu r-\nu(\nu+1)$. We can reduce (A.56) to an equation of hypergeometric type (A.9) by means of the substitution $v_1 = \varphi(r)y$, where $\varphi(r)$ satisfies the equation $\varphi'/\varphi = \pi(r)/\sigma(r)$, with $\pi(r)$ a polynomial of degree at most one.

Among the four possible forms for the polynomial $\pi(r)$ we choose the one for which the function $\tau(r) = \widetilde{\tau}(r) + 2\pi(r)$ has a negative derivative and a zero in the interval $(0, \infty)$. When $\pi(r) = -ar + \nu$, where $a = \sqrt{1-E^2}$, $\nu = \sqrt{\kappa^2 - \mu^2}$, the function $\tau(r) = 2(\nu + 1 - ar)$ satisfies these requirements, and we get

$$\varphi(r) = r^{\nu} e^{-ar}, \quad \rho(r) = r^{2\nu+1} e^{-2ar}, \quad \lambda = 2\left[\mu E - (\nu+1)a\right].$$

The energy values $E = E_N$ are found from the equation

$$\lambda + N\tau' + \frac{1}{2}N(N-1)\sigma'' = 0, \quad N = 0, 1, \ldots,$$

whence

$$\mu E - (N + \nu + 1)a = 0. \tag{A.58}$$

The eigenfunctions are given by the Rodrigues formula:

$$y_N(r) = \frac{C_N}{\rho(r)} \frac{d^N}{dr^N}\left[\sigma^N(r)\rho(r)\right] = C_N\, r^{-2\nu-1} e^{2ar} \frac{d^N}{dr^N}\left(r^{N+2\nu+1} e^{-2ar}\right) \tag{A.59}$$

(we denote the degree of the polynomial $y(r)$ by N, because we want to preserve the notation n for the principal quantum number, which will be needed below). The functions $y_N(r)$ coincide, up to a constant factor, to the Laguerre polynomials $L_N^{2\nu+1}(x)$, where $x = 2ar$.

The energy eigenvalues $E = -\nu/\kappa$ found above satisfy the equation (A.58) for $N = -1$. It is therefore natural to replace N by $N - 1$ in formulas (A.58) and (A.59) and define the energy eigenvalues by means of the equality

$$\mu E - (N + \nu)a = 0, \quad N = 0, 1, \ldots. \tag{A.60}$$

For $N = 1, 2, \ldots$ the function $v_1(r)$ has the form

$$v_1(r) = A_N x^{\nu} e^{-x/2} L_{N-1}^{2\nu+1}(x), \quad x = 2ar \tag{A.61}$$

($v_1(r) = 0$ if $N = 0$). It is readily verified that the functions $rv_1(r)$ are square integrable, as required.

Equation (A.54) with $E = E_N$ $(N = 1, 2, \ldots)$ yields

$$v_2(r) = \frac{1}{1 + (E\kappa)/\nu}\left[v_1'(r) + \left(\frac{\nu+1}{r} - \frac{E\mu}{\nu}\right)v_1(r)\right].$$

Substituting here the expression for $v_1(r)$, we get

$$v_2(r) = x^{\nu-1}e^{-x/2}y(x),$$

where $y(x)$ is a polynomial of degree N. To find $y(x)$, we first write the equation for the function $v_2(r)$ obtained by eliminating $v_1(r)$ from the system (A.54), (A.55):

$$v_2'' + \frac{2}{r}v_2' + \frac{(E^2-1)r^2 + 2E\mu r + \nu(1-\nu)}{r^2}v_2 = 0. \tag{A.62}$$

This translates into the following differential equation for the polynomial $y(x)$:

$$xy'' + (2\nu - x)y' + Ny = 0. \tag{A.63}$$

Equation (A.63) is of hypergeometric type. Its unique polynomial solution is the Laguerre polynomial $y(x) = B_N L_N^{2\nu-1}(x)$, and so

$$v_2(r) = B_N\, x^{\nu-1}e^{-x/2}L_N^{2\nu-1}(x). \tag{A.64}$$

It is readily verified that the solution (A.57) obtained earlier for $E = -\nu/\kappa$ is a particular case of this formula for $N = 0$. As expected, expression (A.64) for the function $v_1(r)$ can be obtained from the expression (A.61) for $v_1(r)$ if we replace ν by $\nu - 1$. This is connected with the fact that equation (A.56) becomes equation (A.62) when ν is replaced by $\nu - 1$.

To find how the constants A_N and B_N in (A.61) and respectively (A.64) are related, let us compare the behavior of the left- and right-hand sides of (A.54) when $r \to 0$, using the formula

$$L_N^\alpha(0) = \frac{\Gamma(N+\alpha+1)}{N!\,\Gamma(\alpha+1)}.$$

We have

$$2a\nu A_N\, L_{N-1}^{2\nu+1}(0) = -2a(\nu+1)A_N L_{N-1}^{2\nu+1}(0) + \left(1 + \frac{E\kappa}{\nu}\right)B_N L_N^{2\nu-1}(0),$$

whence

$$A_N = \frac{\nu + E\kappa}{aN(N+2\nu)}B_N, \quad N = 1, 2, \ldots.$$

Since, by (A.60),

$$N(N+2\nu) = (N+\nu)^2 - \nu^2 = \frac{E^2\mu^2}{a^2} - \nu^2 = \frac{E^2\kappa^2 - \nu^2}{a^2},$$

we conclude that

$$A_N = \frac{a}{E\kappa - \nu}B_N.$$

A.2.4 Energy levels and radial functions

Once the functions $v_1(r)$ and $v_2(r)$ are known, we can find $f(r)$ and $g(r)$:

$$\begin{pmatrix} f \\ g \end{pmatrix} = T^{-1} \begin{pmatrix} v_1 \\ v_2 \end{pmatrix}, \quad T^{-1} = \frac{1}{2\nu(\kappa - \nu)} \begin{pmatrix} \mu & \kappa - \nu \\ \kappa - \nu & \mu \end{pmatrix}.$$

Therefore,

$$\begin{pmatrix} f(r) \\ g(r) \end{pmatrix} = \frac{B_N}{2\nu(\kappa-\nu)} x^{\nu-1} e^{-x/2} \begin{pmatrix} f_1 & f_2 \\ g_1 & g_2 \end{pmatrix} \begin{pmatrix} x L_{N-1}^{2\nu+1}(x) \\ L_N^{2\nu-1}(x) \end{pmatrix},$$

where

$$f_1 = \frac{a\mu}{E\kappa - \nu}, \quad f_2 = \kappa - \nu, \quad g_1 = \frac{a(\kappa-\nu)}{E\kappa - \nu}, \quad g_2 = \mu;$$

$$E = E_{Nj} = \frac{1}{\sqrt{1 + (\mu/(N+\nu))^2}}, \quad \nu = \sqrt{(j+1/2)^2 - \mu^2},$$

$$x = 2ar, \quad a = \frac{\mu/(N+\nu)}{\sqrt{1 + (\mu/(N+\nu))^2}}, \quad \mu = \alpha Z.$$

In the case $N = 0$ $(\ell = j - 1/2)$ the formulas for $f(r)$ and $g(r)$ remain valid if we formally set the terms containing $L_{-1}^{2\nu+1}(x)$ equal to zero. When $\ell = j + 1/2$, we have $N = 1, 2, \ldots$. The number N is determined by the number of zeroes of the functions $f(r)$ and $g(r)$ (as we shall see below, the number of zeroes of $f(r)$ is N for $j = \ell + 1/2$ and $N - 1$ for $j = \ell - 1/2$).

Let us calculate the normalization coefficient B_N, using the normalization condition (A.48). We have

$$\int_0^\infty r^2[f^2(r) + g^2(r)]dr = \frac{B_N^2}{4\nu^2(\kappa-\nu)^2(2a)^3} \int_0^\infty e^{-x} x^{2\nu} \times$$
$$\times \left\{ \left[f_1 x L_{N-1}^{2\nu+1}(x) + f_2 L_N^{2\nu-1}(x)\right]^2 + \left[g_1 x L_{N-1}^{2\nu+1}(x) + g_2 L_N^{2\nu-1}(x)\right]^2 \right\} dx = 1.$$

We need to calculate integrals of two type:

$$J_1 = \int_0^\infty e^{-x} x^{\alpha+1} \left[L_n^\alpha(x)\right]^2 dx$$

and

$$J_2 = \int_0^\infty e^{-x} x^\alpha L_{n-1}^\alpha(x) L_n^{\alpha-2}(x) dx.$$

The integral J_1 can be expressed in terms of the squared norm of the Laguerre polynomials

$$d_n^2 = \int_0^\infty e^{-x} x^\alpha \left[L_n^\alpha(x)\right]^2 dx = \frac{1}{n!}\Gamma(n+\alpha+1).$$

This is done by using the recursion relation

$$xL_n^\alpha(x) = -(n+1)L_{n+1}^\alpha(x) + (2n+\alpha+1)L_n^\alpha(x) - (n+\alpha)L_{n-1}^\alpha(x)$$

and the orthogonality property

$$\int_0^\infty e^{-x} x^\alpha L_n^\alpha(x) L_m^\alpha(x)\, dx = 0 \quad (m \neq n).$$

This yields

$$J_1 = (2n+\alpha+1)\int_0^\infty e^{-x} x^\alpha \left[L_n^\alpha(x)\right]^2 dx = \frac{1}{n!}(2n+\alpha+1)\,\Gamma(n+\alpha+1).$$

To calculate J_2 it suffices to expand the polynomial $L_n^{\alpha-2}(x)$ with respect to the polynomials $L_k^\alpha(x)$:

$$L_n^{\alpha-2}(x) = C_1 L_n^\alpha(x) + C_2 L_{n-1}^\alpha(x) + \dots .$$

The coefficients C_1 and C_2 are easily found by comparing the coefficients of x^n and x^{n-1} in both sides of this equality. This yields $C_1 = 1$, $C_2 = -2$, and so

$$J_2 = -2\int_0^\infty e^{-x} x^\alpha \left[L_{n-1}^\alpha(x)\right]^2 dx = -2\,\frac{\Gamma(n+\alpha)}{(n-1)!}.$$

Therefore,

$$B_N = 2a^2\sqrt{\frac{(\kappa-\nu)(E\kappa-\nu)N!}{\mu\Gamma(N+2\nu)}}.$$

We finally obtain

$$\begin{aligned} f(r) &= \frac{B_N}{2\nu(\kappa-\nu)} x^{\nu-1} e^{-x/2}\left[f_1 x L_{N-1}^{2\nu+1}(x) + f_2 L_N^{2\nu-1}(x)\right], \\ g(r) &= \frac{B_N}{2\nu(\kappa-\nu)} x^{\nu-1} e^{-x/2}\left[g_1 x L_{N-1}^{2\nu+1}(x) + g_2 L_N^{2\nu-1}(x)\right], \end{aligned} \tag{A.65}$$

where

$$f_1 = \frac{a\mu}{E\kappa-\nu}, \quad f_2 = \kappa-\nu, \quad g_1 = \frac{a(\kappa-\nu)}{E\kappa-\nu}, \quad g_2 = \mu$$

(for $N = 0$ we must set $xL_{-1}^{2\nu+1}(x) = 0$, in accordance with (A.61)). Let us mention that for the solutions of the Dirac equation with the Coulomb potential one usually resorts to a more complicated representation in terms of confluent hypergeometric functions (see, e.g., [28]).

The wave functions (A.65) correspond to the energy values

$$E = E_{Nj} = \frac{1}{\sqrt{1 + (\mu/(N+\nu))^2}}, \quad \mu = \alpha Z, \quad \nu = \sqrt{(j+1/2)^2 - \mu^2}, \tag{A.66}$$

where for $j = \ell + 1/2$ [resp., $j = \ell - 1/2$] N takes the values $N = 0, 1, 2, \ldots$[resp., $N = 1, 2, \ldots$].

A.2.5 Connection with the nonrelativistic theory

To get the nonrelativistic limit, let us pass from the Dirac equation to the Schrödinger equation. This is ordinarily done using atomic units ($e = 1$, $m = 1$, $\hbar = 1$; then $c = 1/\alpha = 137.036$). The expression for the energy levels (A.66), originally written in mc^2 units, becomes

$$E_{Nj} = \frac{1}{\sqrt{1 + \left(\frac{\alpha Z}{N+\nu}\right)^2}} \frac{mc^2}{me^4/\hbar^2} = \frac{1}{\alpha^2\sqrt{1 + \left(\frac{\alpha Z}{N+\nu}\right)^2}} \tag{A.67}$$

in atomic units. If the quantity αZ is small, then

$$\nu = \sqrt{(j+1/2)^2 - (\alpha Z)^2} \approx j + 1/2.$$

Eliminating the rest energy $E_0 = mc^2$ from E_{Nj}, we obtain

$$\varepsilon_{n\ell j} = E_{Nj} - E_0 \approx -\frac{Z^2}{2n^2},$$

where

$$n = N + j + \frac{1}{2} = \begin{cases} N + \ell + 1, & \text{if} \quad j = \ell + 1/2 \quad (N = 0, 1, \ldots), \\ N + \ell, & \text{if} \quad j = \ell - 1/2 \quad (N = 1, 2, \ldots). \end{cases}$$

This formula obviously coincides with formula (A.23) of § A.1 for the Schrödinger equation, which gives the energy levels of an electron in a Coulomb field; n plays the role of the principal quantum number.

Now let us consider the nonrelativistic limit for the radial parts of the wave functions, $f(r)$ and $g(r)$. To this end we pass to atomic units in formula (A.65), setting $N = n - j - 1/2$. Since the system of atomic [resp., relativistic] units uses $\hbar^2/(me^2)$ [resp., $\hbar/(mc)$] for the unit of length, it follows, according to (A.48), that in order to switch to atomic units we must multiply the right-hand side of (A.65) by

$$\left(\frac{\hbar/mc}{\hbar^2/me^2}\right)^{-3/2} = \left(\frac{e^2}{\hbar c}\right)^{-3/2} = \alpha^{-3/2}.$$

After that, the formulas for the radial functions $F(r) = rf(r)$ and $G(r) = rg(r)$ in atomic units are obtained by multiplying (A.65) by $r\alpha^{-3/2} = x/(2a\alpha^{1/2})$:

$$\begin{aligned} F(r) \equiv F_{n\ell j}(r) &= C_{n\ell j}x^{\nu}e^{-x/2}\left[f_1 x L^{2\nu+1}_{n-j-3/2}(x) + f_2 L^{2\nu-1}_{n-j-1/2}(x)\right], \\ G(r) \equiv G_{n\ell j}(r) &= C_{n\ell j}x^{\nu}e^{-x/2}\left[g_1 x L^{2\nu+1}_{n-j-3/2}(x) + g_2 L^{2\nu-1}_{n-j-1/2}(x)\right]. \end{aligned} \tag{A.68}$$

Here

$$x = 2ar/\alpha, \quad a = \frac{\dfrac{\alpha Z}{n-j-1/2+\nu}}{\sqrt{1+\left(\dfrac{\alpha Z}{n-j-1/2+\nu}\right)^2}}, \quad \nu = \sqrt{(j+1/2)^2 - (\alpha Z)^2},$$

$$f_1 = \frac{a(\alpha Z)^2}{\kappa a(n-j-1/2+\nu) - \alpha Z\nu}, \quad f_2 = \kappa - \nu, \quad g_1 = \frac{\kappa-\nu}{\alpha Z}f_1, \quad g_2 = \alpha Z,$$

$$C_{n\ell j} = \frac{a}{2\nu(\kappa-\nu)Z\alpha^{3/2}}\sqrt{\frac{(\kappa-\nu)(n-j-1/2)!\,[a\kappa\,(n-j-1/2+\nu) - \alpha Z\nu]}{\Gamma(n-j-1/2+2\nu)}},$$

$$\kappa = \begin{cases} -(\ell+1), & \text{if} \quad \ell = j-1/2, \\ \ell, & \text{if} \quad \ell = j+1/2, \end{cases} \qquad (\ell = 0, 1, \dots, n-1).$$

The number of zeroes of the large component of the relativistic radial function $F_{n\ell j}(r)$ is equal to $n_r = n - \ell - 1$, like in the nonrelativistic case, although for $\ell = j + 1/2$ the degree of the polynomial

$$f_1 x L^{2\nu+1}_{n-j-3/2}(x) + f_2 L^{2\nu-1}_{n-j-1/2}(x)$$

is $n - j - 1/2 = n_r + 1$. In this case one of the zeroes of this polynomial is negative, which does not change the number of zeroes of the radial function $F_{n\ell j}(r)$ for $r > 0$.

Let us estimate the order of magnitude of the coefficients f_1, f_2 and g_1, g_2 for small values of αZ in formula (A.68). Here

$$a = \sqrt{1-E^2} \approx \frac{\alpha Z}{n}, \quad \nu \approx j + \frac{1}{2}.$$

(1) Let $\ell = j - 1/2$. Then $\kappa = -(\ell+1)$, $\kappa - \nu = \kappa - \sqrt{\kappa^2 - (\alpha Z)^2} \approx 2\kappa$, $E\kappa - \nu \approx 2\kappa$, whence

$$\begin{pmatrix} f_1 & f_2 \\ g_1 & g_2 \end{pmatrix} \approx \begin{pmatrix} -\dfrac{(\alpha Z)^2}{2(\ell+1)n} & -2(\ell+1) \\ \dfrac{\alpha Z}{n} & \alpha Z \end{pmatrix}.$$

(2) Let $\ell = j + 1/2$. Then

$$\kappa = \ell, \quad \kappa - \nu \approx \frac{(\alpha Z)^2}{2\ell}, \quad E\kappa - \nu = (E-1)\kappa + \kappa - \nu \approx \frac{(\alpha Z)^2}{2\ell n^2}(n^2 - \ell^2),$$

whence

$$\begin{pmatrix} f_1 & f_2 \\ g_1 & g_2 \end{pmatrix} \approx \begin{pmatrix} \dfrac{2\ell n}{n^2 - \ell^2} & \dfrac{(\alpha Z)^2}{2\ell} \\ \dfrac{\alpha Z n}{n^2 - \ell^2} & \alpha Z \end{pmatrix}.$$

These considerations show that in both cases $|G(r)| \ll |F(r)|$ for $\alpha Z \ll 1$. In case (1), i.e., for $\ell = j - 1/2$ we have

$$F(r) \approx \sqrt{\frac{Z(n-\ell-1)!}{n^2(n+\ell)!}}\, x^{\ell+1} e^{-x/2} L_{n-\ell-1}^{2\ell+1}(x),$$

where $x = 2ar/\alpha \approx 2Zr/n$, which coincides with the corresponding solution of the Schrödinger equation (A.33) of § A.1. Clearly, in this case the radial function $F(r)$ has $n_r = N = n - \ell - 1$ zeroes. Similarly, in case (2), when $\ell = j + 1/2$ we again obtain expression (A.33) from § A.1, and $F(r)$ has $n_r = N - 1 = n - \ell - 1$ zeroes.

The representation of the functions $F(r)$ and $G(r)$ in the form (A.68) is convenient for passing to the nonrelativistic limit, since only one of the coefficients f_1, f_2, g_1, g_2 is much larger than all the others when $\alpha Z \to 0$. By contrast, in the traditional representation for $F(r)$ and $G(r)$ there are several coefficients of the same order of smallness, and so to show that the nonrelativistic limit coincides with the solution of the Schrödinger equation one needs to resort, in addition, to the recursion relations for hypergeometric functions.

The passage to the limit considered above allows us to write a formula for $\varepsilon_{n\ell j}$ in a computationally convenient form. Subtracting the rest energy of the electron from (A.66) and passing to atomic units, we have

$$\varepsilon_{n\ell j} = \frac{1}{\alpha^2}\left\{\left[1 + \frac{(\alpha Z)^2}{\left(n - (j+1/2) + \sqrt{(j+1/2)^2 - (\alpha Z)^2}\right)^2}\right]^{-1/2} - 1\right\},$$

or

$$\varepsilon_{n\ell j} = -\frac{Z^2}{\left(1 + \dfrac{(\alpha Z)^2}{\widetilde{n}^2} + \sqrt{1 + \dfrac{(\alpha Z)^2}{\widetilde{n}^2}}\right)\widetilde{n}^2}, \tag{A.69}$$

where

$$\widetilde{n} = N + \nu = n - \frac{(\alpha Z)^2}{j + 1/2 + \sqrt{(j+1/2)^2 - (\alpha Z)^2}},$$

n is the principal quantum number, $j = \frac{1}{2}, \frac{3}{2}, \ldots, n - \frac{1}{2}$.

As formula (A.69) shows, for a given j the energy eigenvalues of the hydrogen-like ion are independent of ℓ. Therefore, in addition to the degeneracy with respect to the projection of the momentum j on the z-axis (the energy levels are independent of m), there is also a two-fold degeneracy with respect to ℓ for $j < n - 1/2$.

Table A.2 lists, for the K, L, M, N shells, the quantum numbers n, j and ℓ, which characterize the possible electron states in a hydrogen-like gold ion, and the corresponding energy level. For the arrangement of the quantum numbers in the indicated order the energy levels of the electron, given by formula (A.69), increase with n and j.

Table A.2: Energy levels of the hydrogen-like gold atom ($Z = 79$) in atomic units. For comparison, the nonrelativistic values of the energy $\varepsilon_n = -Z^2/(2n^2)$ and their deviation Δ (in percents) from the relativistic values are also shown.

n	j	ℓ	ε_n	$\varepsilon_{n\ell j}$	Δ	n	j	ℓ	ε_n	$\varepsilon_{n\ell j}$	Δ
1	1/2	0	−3121	−3435	9.1	4	1/2	0	−195	−211	7.4
						4	1/2	1		−211	7.4
2	1/2	0	−780	−879	11.3	4	3/2	1		−200	2.6
2	1/2	1		−879	11.3	4	3/2	2		−200	2.6
2	3/2	1		−797	2.1	4	5/2	2		−197	1.2
						4	5/2	3		−197	1.2
3	1/2	0	−347	−381	9.1	4	7/2	3		−196	0.5
3	1/2	1		−381	9.1						
3	3/2	1		−357	2.8	5	1/2	0	−125	−133	6.2
3	3/2	2		−357	2.8	5	1/2	1		−133	6.2
3	5/2	2		−350	0.9	5	3/2	1		−128	2.4

APPROXIMATION METHODS

It is well known that only aver small number of quantum-mechanical problems, which all concern simple systems, can be solved by analytic methods. Generally, for real physical systems one cannot manage without a suitable approximation method. In a number of cases the approximate calculation of energy levels and wave functions can be successfully carried out by means of a variational method and on the basis of the semiclassical approximation. Approximation methods enable us to determine the relative role played by various factors in the given physical system, to choose a suitable initial approximation, and to construct efficient numerical methods for the further, more precise investigation of the problem under consideration.

A.3 The variational method and the method of the trial potential

A.3.1 Main features of the variational method

The variational method for computing the ground-state energy ε_0 of the system under study amounts to using the inequality

$$\varepsilon_0 \leq \int \psi^*(\xi)\widehat{H}\psi(\xi)d\xi, \tag{A.70}$$

where $\widehat{H}$ is the total Hamiltonian operator of the system, ξ is the collection of values of all independent coordinates (ξ is a point in configuration space), $d\xi$ is the volume element in configuration space, and $\psi(\xi)$ is an arbitrary function subject to the normalization condition

$$\int \psi\psi^* d\xi = 1.$$

Indeed, let $\widehat{H}\varphi_n = \varepsilon_n\varphi_n$, $\int \varphi_m^*(\xi)\varphi_n(\xi)\, d\xi = \delta_{mn}$. Then

$$\psi = \sum_{n=0}^{\infty} a_n\varphi_n; \quad \sum_n |a_n|^2 = 1;$$

$$\int \psi^*\widehat{H}\psi\, d\xi = \sum_{n=0}^{\infty} |a_n|^2\varepsilon_n \geq \varepsilon_0 \sum_{n=0}^{\infty} |a_n|^2 = \varepsilon_0.$$

Thus, the calculation of the ground-state energy reduces to the calculation of the minimum of the integral (A.70) for variations of the normalized wave function $\psi(\xi)$, i.e.,

$$\varepsilon_0 = \min \int \psi^* \widehat{H}\, \psi d\xi \quad \text{under the condition} \quad \int \psi^* \psi\, d\xi = 1. \tag{A.71}$$

In practice, the calculation of ε_0 reduces to choosing a trial function that contains a number of parameters $\alpha, \beta, \ldots$. These parameters are found by solving the system of equations

$$\frac{\partial J}{\partial \alpha} = 0, \quad \frac{\partial J}{\partial \beta} = 0, \; \ldots\,, \tag{A.72}$$

where

$$J(\alpha, \beta, \ldots) = \int \psi^*(\xi, \alpha, \beta, \ldots) \widehat{H} \psi(\xi, \alpha, \beta, \ldots)\, d\xi.$$

For a successful choice of the form of the trial function one obtains a value of ε_0 that is close to the true value even for a small number of parameters. The wave function φ_0 will approximately coincide with the function $\psi(\xi, \alpha, \beta, \ldots)$.

The calculation of the energy ε_1 of the first excited state reduces to solving the variational problem

$$\varepsilon_1 = \min \int \psi_1^* \widehat{H} \psi_1\, d\xi$$

under the additional orthonormalization conditions

$$\int \psi_1^* \psi_1\, d\xi = 1, \quad \int \psi_1^* \psi_0\, d\xi = 0.$$

In the calculation of the next excited states the number of additional conditions increases, and the problem becomes more complex. In some cases the requisite orthogonality conditions are satisfied for an appropriate choice of trial functions thanks to symmetry properties. For example, in the investigation of the motion of a particle in a central field the orthogonality of the states with different values of ℓ is a consequence of the orthogonality of the spherical harmonics $Y_{\ell m}(\theta, \varphi)$.

The variational method described above is also known as the *Ritz method.*

EXAMPLE. Let us find the ground-state energy of a two-electron system in the field of a nucleus with charge Z, in the nonrelativistic approximation. In this case the Schrödinger equation reads

$$\widehat{H}\psi = E\psi,$$

where the Hamiltonian is

$$\widehat{H} = -\frac{1}{2}\Delta_1 - \frac{1}{2}\Delta_2 - \frac{Z}{r_1} - \frac{Z}{r_2} + \frac{1}{r_{12}}.$$

Since the Hamiltonian does not contain spin variables, the complete wave function is a product of a function depending on the space variables and a function depending on the spin variables, such that the complete wave function is anti-symmetric.

When one calculates the ground-state energy, the anti-symmetric spin part of the wave function is not required. As a symmetric trial function, depending on the space variables, one takes the product of two elementary hydrogen-like functions with effective charge $Z' < Z$:

$$\psi(\vec{r}_1, \vec{r}_2) = \frac{C}{4\pi} e^{-Z'(r_1+r_2)}.$$

The constant C is found from the normalization condition:

$$\int |\psi(\vec{r}_1, \vec{r}_2)|^2 \, d\vec{r}_1 \, d\vec{r}_2 = C^2 \left[\int_0^\infty e^{-2Z'r} \, r^2 dr \right]^2 = \frac{4C^2}{(2Z')^6} = 1,$$

i.e., $C = 4(Z')^3$.

The expression for the energy takes on the form

$$E(Z') = \int \psi^*(\vec{r}_1, \vec{r}_2) \widehat{H} \psi(\vec{r}_1, \vec{r}_2) \, d\vec{r}_1 d\vec{r}_2 = \frac{Z'^2}{2} + \frac{Z'^2}{2} - ZZ' -$$

$$ZZ' + \int \frac{\psi^*(\vec{r}_1, \vec{r}_2) \psi(\vec{r}_1, \vec{r}_2)}{r_{12}} \, d\vec{r}_1 \, d\vec{r}_2 = Z'^2 - 2ZZ' + J,$$

where

$$J = C^2 \int_0^\infty dr_2 \left[r_2^2 e^{-2Z'r_2} \int_0^\infty \frac{r_1^2 e^{-2Z'r_1}}{r_>} \, dr_1 \right] = \frac{5}{8} Z',$$

because

$$\int_0^\infty \frac{r_1^2 e^{-2Z'r_1}}{r_>} \, dr_1 = \frac{1}{4Z'^3} \left[\frac{1}{r_2} - \left(\frac{1}{r_2} + Z' \right) e^{-2Z'r_2} \right];$$

here $r_> = \max(r_1, r_2)$ (see Example 3 in § A.1).

Therefore,

$$E(Z') = Z'^2 - 2ZZ' + \frac{5}{8} Z'.$$

The minimum condition for $E(Z')$,

$$\frac{dE}{dZ'} = 2Z' - 2Z + \frac{5}{8} = 0,$$

yields

$$Z' = Z - \frac{5}{16}, \quad E = -\left(Z - \frac{5}{16} \right)^2.$$

Here an experimentally measurable quantity is not the total energy of the atom, but its ionization potential I. For a helium-like ion the ionization potential is $I = E_0 - E$, where $E_0 = -Z^2/2$. In particular, for helium ($Z = 2$)

$$I = -\frac{(2)^2}{2} + \left(2 - \frac{5}{16}\right)^2 = 0.848 \text{ a.e.} = 186114 \text{ cm}^{-1}.$$

Further refinements of the variational method provided for the ionization potential the value (see [30]):

$$I = 198317.974 \pm 0.022 \text{ cm}^{-1},$$

while the experimental value is

$$I_{\text{exp}} = 198310.82 \pm 0.15 \text{ cm}^{-1}.$$

If the motion of the nucleus, relativistic corrections and the interaction of the electron with its own field (Lamb shift) are accounted for, one obtains the corrected value of the ionization potential

$$I_{\text{theor}} = 198310.665 \text{ cm}^{-1}.$$

A.3.2 Calculation of hydrogen-like wave functions

The variational method can be applied also for the calculation of energy and wave functions of multi-electron atoms. In the simplest approximation one assumes that each electron in the atom is acted upon by a unique potential generated by the other electrons and the nucleus. In connection with this assumption let us consider the following problem.

Suppose we need to solve the Schrödinger equation

$$-\frac{1}{2}R''_{n\ell} + \left[-V(r) + \frac{\ell(\ell+1)}{2r^2}\right] R_{n\ell} = \varepsilon R_{n\ell} \quad (0 < r < \infty) \tag{A.73}$$

with a given potential $V(r)$. As trial functions for an electron with quantum numbers n, ℓ we will use the radial functions $\widetilde{R}_{n\ell}(r)$ corresponding to the trial potential

$$\widetilde{V}(r) = \frac{Z_{n\ell}}{r} - A_{n\ell}, \tag{A.74}$$

where $Z_{n\ell}$ is the effective charge of the nucleus and $A_{n\ell}$ is the external screening constant. The function $\widetilde{R}_{n\ell}(r)$ satisfies the equation

$$-\frac{1}{2}\widetilde{R}''_{n\ell} + \left[-\widetilde{V}(r) + \frac{\ell(\ell+1)}{2r^2}\right] \widetilde{R}_{n\ell} = \widetilde{\varepsilon}\,\widetilde{R}_{n\ell}, \tag{A.75}$$

where

$$\widetilde{\varepsilon} = \widetilde{\varepsilon}_{n\ell} = -\frac{Z^2_{n\ell}}{2n^2} + A_{n\ell}, \quad \widetilde{R}_{n\ell}(r) = C_{n\ell}\, x^{\ell+1}\, e^{-x/2} L^{2\ell+1}_{n-\ell-1}(x),$$

$$x = \frac{2Z_{n\ell}\, r}{n}, \quad C_{n\ell}^2 = \frac{Z_{n\ell}}{n^2} \frac{(n-\ell-1)!}{(n+\ell)!}, \quad \int_0^\infty \widetilde{R}_{n\ell}^2(r) dr = 1.$$

The representation of the atomic potential in the form (A.74) is in agreement with the Thomas-Fermi model considered in Chapter I and the uniform free-electron density model that follows from the latter for high temperatures. Indeed, a bound electron with quantum numbers n and ℓ is essentially located in the spherical layer $(r, r + \Delta r)$, in the vicinity of the main maximum of the function $R_{n\ell}(r)$. As a rule, the volume of such a spherical layer constitutes only a small fraction of the volume of the atom cell. The outer electrons with respect to this layer generate in the region of the principal maximum a constant potential, determined by the external screening constant $A_{n\ell}$. The inner electrons decrease the attraction of the nucleus ($Z_{n\ell} < Z$). The parameters $Z_{n\ell}$ and the related Slater screening parameters are widely used in atomic calculations [83, 50, 135].

When the variational method is used, the constant $Z_{n\ell}$, which determines the wave function $\widetilde{R}_{n\ell}(r)$, can be found directly from the condition

$$\varepsilon_{n\ell} = \min_{Z_{n\ell}} \int_0^\infty \widetilde{R}_{n\ell}(r) \widehat{H} \widetilde{R}_{n\ell}(r)\, dr.$$

Here the additional mutual orthogonality conditions can be disregarded, because

(i) if the values of ℓ are different, the wave functions will be orthogonal thanks to the orthogonality of the spherical harmonics;

(ii) if the values of ℓ are identical, while the quantum numbers n are different, then the principal maxima of the functions $\widetilde{R}_{n\ell}(r)$ and $\widetilde{R}_{n'\ell}(r)$ are located quite far from one another, which yields the approximate equality $\int \widetilde{R}_{n\ell}(r)\, \widetilde{R}_{n'\ell}(r)\, dr \approx 0$ (see Figure 2.5).

Thus, we have

$$\varepsilon_{n\ell} = \min \int_0^\infty \widetilde{R}_{n\ell} \left(-\frac{1}{2} \widetilde{R}_{n\ell}'' - V(r) \widetilde{R}_{n\ell} + \frac{\ell(\ell+1)}{2r^2} \widetilde{R}_{n\ell} \right) dr =$$

$$= \min \int_0^\infty [\widetilde{\varepsilon}_{n\ell} + \widetilde{V}(r) - V(r)] \widetilde{R}_{n\ell}^2\, dr = \min \left[\frac{Z_{n\ell}^2}{2n^2} - \int_0^\infty V(r) \widetilde{R}_{n\ell}^2(r, Z_{n\ell})\, dr \right].$$

Since

$$\widetilde{R}_{n\ell}^2(r, Z_{n\ell}) \frac{dr}{r} = \frac{Z_{n\ell}}{n^2} e^{-x} Q_{n\ell}(x)\, dx,$$

where

$$Q_{n\ell}(x) = \frac{(n-\ell-1)!}{(n+\ell)!} x^{\ell+1} \left(L_{n-\ell-1}^{2\ell+1}(x) \right)^2, \quad x = \frac{2Z_{n\ell}\, r}{n},$$

we conclude that

$$\varepsilon_{n\ell} = \min \left[\frac{Z_{n\ell}^2}{2n^2} - \frac{Z_{n\ell}}{n^2} \int_0^\infty e^{-x} Q_{n\ell}(x) rV(r) \Big|_{r=nx/(2Z_{n\ell})} dx \right]. \tag{A.76}$$

The minimum condition $d\varepsilon_{n\ell}/dZ_{n\ell} = 0$ yields

$$Z_{n\ell} = \int_0^\infty e^{-x} Q_{n\ell}(x) \left[-r^2 \frac{dV}{dr} \right] \Bigg|_{r=nx/(2Z_{n\ell})} dx,$$

or

$$Z_{n\ell} = \int_0^\infty e^{-x} Q_{n\ell}(x) \left[rV(r) - r\frac{d}{dr}(rV(r)) \right] \Bigg|_{r=nx/(2Z_{n\ell})} dx. \tag{A.77}$$

To find $Z_{n\ell}$, we use the simplest iteration scheme that results from (A.77):

$$\overset{(s+1)}{Z}_{n\ell} = \overset{(s)}{\alpha}_{n\ell} + \overset{(s)}{\beta}_{n\ell}, \tag{A.78}$$

where

$$\overset{(s)}{\alpha}_{n\ell} = \int_0^\infty e^{-x} Q_{n\ell}(x) rV(r) \Bigg|_{r=nx/(2\overset{(s)}{Z}_{n\ell})} dx,$$

$$\overset{(s)}{\beta}_{n\ell} = \int_0^\infty e^{-x} Q_{n\ell}(x) \left[-r\frac{d}{dr}(rV(r)) \right] \Bigg|_{r=nx/(2\overset{(s)}{Z}_{n\ell})} dx,$$

s being the iteration number. As the initial approximation one can use the preceding values of $Z_{n\ell}$:

$$\overset{(0)}{Z}_{10} = Z, \quad \overset{(0)}{Z}_{20} = Z_{10}, \quad \overset{(0)}{Z}_{21} = Z_{20}, \ \ldots ,$$

since usually $Z > Z_{10} > Z_{20} > Z_{21}$ and so on.

To calculate the quantities $\overset{(s)}{\alpha}_{n\ell}$ and $\overset{(s)}{\beta}_{n\ell}$ we use the Gauss-type quadrature formulas [154]

$$\int_a^b f(x)\rho(x)dx = \sum_{j=1}^N a_j f(x_j). \tag{A.79}$$

In our case it is natural to take $\rho(x) = e^{-x}$. Then x_j are the zeroes of the Laguerre polynomial $L_N^0(x)$ and

$$f(x) = Q_{n\ell}(x) rV(r) \ \text{ or } \ f(x) = -Q_{n\ell}(x) r\frac{d}{dr}\left[rV(r)\right],$$

where $r = nx/(2Z_{n\ell})$. The integral (A.77) can be calculated in closed form whenever $f(x)$ is a polynomial of degree at most $2N-1$. We need to choose N in such a manner that the functions $rV(r)$ and $r\frac{d}{dr}[rV(r)]$ will be well enough approximated by polynomials in the essential part of the domain of integration. As a rule, for $n < 10$ one can take $N = 12$. Note that the numbers $Q_{n\ell}(x_j)$ are independent of the potential $V(r)$, which is convenient in calculations.

Once the values $Z_{n\ell}$ are found, we can use (A.76) to find the energy values $\varepsilon_{n\ell}$ and the external screening constant $A_{n\ell}$:

$$\varepsilon_{n\ell} = \frac{Z_{n\ell}}{n^2}\left(\frac{Z_{n\ell}}{2} - \alpha_{n\ell}\right), \quad A_{n\ell} = \frac{Z_{n\ell}}{n^2}\left(Z_{n\ell} - \alpha_{n\ell}\right). \tag{A.80}$$

A.3.3 Method of the trial potential for the Schrödinger and Dirac equations

A method that is close to the variational method is the so-called method of the trial potential [217], which is based on the fact that the function $R_{n\ell}(r)$ is notably different from zero only in a small interval of r. For that reason, the behavior of the potential $V(r)$ outside this interval has apparently little influence on the character of the function $R_{n\ell}(r)$.

In the method of the trial potential the effective charge $Z_{n\ell}$ is found from the condition of minimum of the functional

$$\Phi(Z_{n\ell}) = \int \left[rV(r) - r\widetilde{V}(r, Z_{n\ell}, A_{n\ell})\right]^2 \widetilde{R}^2_{n\ell}(r, Z_{n\ell})\, dr, \tag{A.81}$$

where the screening constant $A_{n\ell}$ is determined from the additional condition that the energy eigenvalue $\widetilde{\varepsilon}$ should be close to ε. As such a condition it is natural to ask that the first correction to the energy $\widetilde{\varepsilon}$ according to perturbation theory vanishes:

$$\int [V(r) - \widetilde{V}(r, Z_{n\ell}, A_{n\ell})]\widetilde{R}^2_{n\ell}(r, Z_{n\ell})\, dr = 0. \tag{A.82}$$

This yields

$$A_{n\ell} = \frac{Z^2_{n\ell}}{n^2} - \int V(r)\widetilde{R}^2_{n\ell}(r, Z_{n\ell})\, dr.$$

Figures A.3, A.4 and tables A.3, A.4 show some results obtained by solving the Schrödinger equation with the Thomas-Fermi potential for gold with density $\rho = 0.1$ g/cm^3 and temperatures $T = 1$ keV and $T = 0.01$ keV. As one can see, the results obtained by the method of the trial potential are somewhat superior to those obtained by the variational method. Even in the case of low temperatures, when the wave functions are far from the hydrogen-like wave functions (see Figure A.4), the method of the trial potential reproduces well the position of the principal maximum of the wave function.

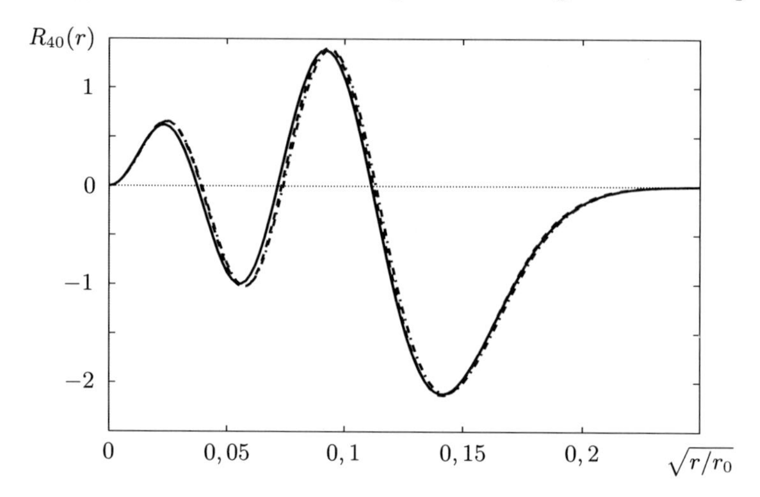

Figure A.3: Dependence of the radial function $R_{40}(r)$ on the variable $x = \sqrt{r/r_0}$ for gold ($Z = 79$) in the TF potential with temperature $T = 1$ keV and density $\rho = 0.1$ g/cm^3 (the solid, dashed and dotted-dashed curves represent the numerical solution, the result of the method of the trial potential, and the result of the variational method, respectively). Here the variational method and the method of the trial potential give practically identical results

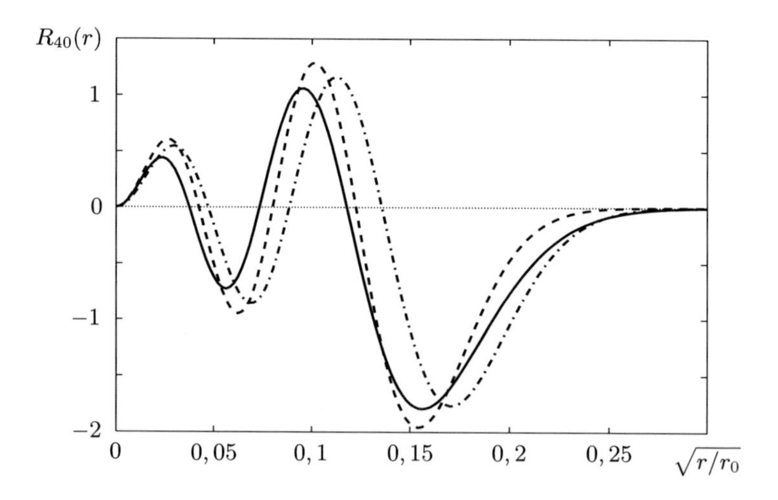

Figure A.4: Dependence of the radial function R_{40} on $x = \sqrt{r/r_0}$ for gold ($Z = 79$) with temperature $T = 0.01$ keV and density $\rho = 0.1$ g/cm^3 (the solid, dashed and dotted-dashed curves represent the numerical solution, the result of the method of the trial potential, and the result of the variational method, respectively)

Due to its computational simplicity, the method of the trial potential is efficient in finding the parameters $Z_{n\ell j}$ and $A_{n\ell j}$ in the relativistic case as well. To implement it, it suffices to replace $\widetilde{R}^2_{n\ell}(r)$ by $\widetilde{F}^2_{n\ell j}(r) + \widetilde{G}^2_{n\ell j}(r)$ in formulas (A.81) and (A.82) (see formulas (A.68) in § A.2). Let us mention that the computer time required by this method is roughly 50–100 smaller than the time needed for the numerical integration of the corresponding equations.

The results of computations by the method of the trial potential for the Schrödinger and Dirac equations are discussed in more detail in § 2.2.

Table A.3: Energy eigenvalues $\varepsilon_{n\ell}$ (in atomic units), computed by the method of the trial potential and the variational method, and also obtained by integrating numerically the Schrödinger equation in the TF potential for gold ($Z = 79$) at temperature $T = 1$ keV and density $\rho = 0.1$ g/cm^3. Since the quantities $\varepsilon_{n\ell} < 0$, we give their absolute values

n	ℓ	$\|\varepsilon_{n\ell}\|$ trial	var.	Schröd.
1	0	2848.956	2848.733	2849.178
2	0	624.644	624.667	624.434
2	1	608.827	608.869	609.049
3	0	258.848	258.950	258.662
3	1	252.887	252.906	252.937
3	2	244.108	244.120	244.180
4	0	139.554	139.587	139.432
4	1	136.949	136.951	136.992
4	2	133.562	133.565	133.623
4	3	131.390	131.392	131.395
5	0	85.948	85.956	85.855
5	1	84.586	84.586	84.609
5	2	82.917	82.919	82.953
5	3	81.868	81.871	81.873
5	4	81.568	81.569	81.569
6	0	57.328	57.333	57.252
6	1	56.521	56.521	56.534
6	2	55.578	55.579	55.597
6	3	54.984	54.986	54.988
6	4	54.802	54.802	54.802
6	5	54.760	54.757	54.757

n	ℓ	$\|\varepsilon_{n\ell}\|$ trial	var.	Schröd.
7	0	40.349	40.353	40.209
7	1	39.791	39.792	39.795
7	2	39.166	39.167	39.177
7	3	38.794	38.795	38.799
7	4	38.683	38.678	38.677
7	5	38.648	38.645	38.645
7	6	38.631	38.630	38.628
8	0	29.502	29.507	29.247
8	1	29.004	29.004	28.947
8	2	28.543	28.544	28.560
8	3	28.307	28.308	28.309
8	4	28.225	28.225	28.225
8	5	28.204	28.206	28.201
8	6	28.190	28.188	28.187
8	7	28.179	28.176	28.176
9	0	22.118	22.120	21.788
9	1	21.617	21.617	21.578
9	2	21.285	21.286	21.308
9	3	21.133	21.133	21.132
9	4	21.071	21.071	21.071
9	5	21.057	21.057	21.052
9	6	21.045	21.042	21.041
9	7	21.034	21.030	21.031
9	8	21.024	21.019	21.020

Table A.4: Energy eigenvalues $\varepsilon_{n\ell}$ in the TF potential for gold with temperature $T = 0.01$ keV and density $\rho = 0.1$ g/cm^3. For $n > 5$ the variational method considered here does not work

n ℓ	$\lvert\varepsilon_{n\ell}\rvert$ trial	var.	Schröd.
1 0	2655.040	2655.019	2654.842
2 0	441.965	442.042	441.559
2 1	423.276	423.089	423.464
3 0	105.494	106.452	104.885
3 1	96.456	96.763	96.562
3 2	80.303	80.549	81.159
4 0	25.053	27.309	24.677
4 1	20.933	22.284	21.159
4 2	13.988	14.272	14.748
4 3	3.136	5.393	6.235
5 0	5.323	7.891	5.270
5 1	3.819	5.421	4.095
5 2	2.028	2.101	2.199

n ℓ	$\lvert\varepsilon_{n\ell}\rvert$ trial	var.	Schröd.
5 3	0.149	—	0.570
6 0	1.671	—	1.408
6 1	1.592	—	1.088
6 2	0.333	—	0.610
6 3	-0.010	—	0.164
7 0	1.012	—	0.485
7 1	0.630	—	0.364
7 2	0.014	—	0.178
8 0	0.559	—	0.140
8 1	0.232	—	0.088
8 2	0.016	—	0.010
9 0	—	—	0.005

A.4 The semiclassical approximation

The semiclassical approximation has proven to be quite effective in the analysis of many problems of quantum mechanics. The basic problem that lies at the foundation of the semiclassical approximation is to obtain a uniform asymptotics for $\lambda \to \infty$ for the solutions of a differential equation of the form

$$[k(x)y']' + \lambda r(x)y = 0.$$

The first investigations of this problem were carried out by Liouville and Green in the first half of the 19th century (see Chapter 6 in [165]). The method was further developed by Wentzel, Kramers, Brillouin and Jeffreys (the WKB methods) and also by many other authors.

A.4.1 Semiclassical approximation in the one-dimensional case

Let us examine the behavior of the solutions of the equation

$$[k(x)y']' + \lambda r(x)y = 0 \tag{A.83}$$

as $\lambda \to \infty$. As the simple example of equation (A.83) with $k(x) =$const and $r(x) =$const shows, this behavior depends on the sign of the functions $k(x)$ and $r(x)$. For this reason, in what follows we will consider equation (A.83) on intervals where $k(x)$ and $r(x)$ are of constant sign. First let us examine the case when on some interval (a, b) the functions $k(x)$ and $r(x)$ have the same sign, say $k(x) > 0$, $r(x) > 0$. We assume that $k(x)$ and $r(x)$ have continuous first and second derivatives.

A.4.1.1. To simplify equation (A.83), we use the change of variables

$$y(x) = \varphi(x)u(s), \quad s = s(x). \tag{A.84}$$

We have

$$y'_x = \varphi'_x u + \varphi u'_s s'_x,$$

$$(ky')' = (k\varphi' u + k\varphi u' s')' = (k\varphi')' u + k\varphi' u' s' + (ks')'\varphi u' + ks'(\varphi' u' + \varphi u'' s').$$

Substitution in (A.83) yields

$$u'' + f(s)u' + [\lambda g(s) - q(s)]u = 0,$$

where

$$f(s) = \frac{2k(x)s'(x)\varphi'(x) + [k(x)s'(x)]'\varphi(x)}{k(x)\varphi(x)[s'(x)]^2},$$

$$g(s) = \frac{r(x)}{k(x)[s'(x)]^2}, \quad q(s) = -\frac{[k(x)\varphi'(x)]'}{k(x)\varphi(x)[s'(x)]^2}.$$

To investigate the behavior of $y(x)$ when $\lambda \to \infty$ it is convenient to choose the functions $s(x)$ and $\varphi(x)$ from the conditions $f(s) = 0$ and $g(s) = 1$, which gives

$$[s'(x)]^2 = \frac{r(x)}{k(x)},$$

$$\frac{\varphi'}{\varphi} = -\frac{(ks')'}{2ks'} = -\frac{1}{4}[\ln(ks')^2]' = -\frac{1}{4}(\ln kr)' = -\frac{1}{4}\left(\frac{k'}{k} + \frac{r'}{r}\right). \tag{A.85}$$

Thus, the equation for $u(s)$ takes on the form

$$u'' + [\lambda - q(s)]u = 0, \tag{A.86}$$

where

$$q(s) = -\frac{[k(x)\varphi'(x)]'}{r(x)\varphi(x)}\Big|_{x=x(s)} = -\frac{k'\varphi' + k\varphi''}{r\varphi} = -\frac{k'}{r}\frac{\varphi'}{\varphi} - \frac{k}{r}\left[\left(\frac{\varphi'}{\varphi}\right)' + \left(\frac{\varphi'}{\varphi}\right)^2\right] =$$

$$\frac{k}{4r}\left[\left(\frac{k'}{k} + \frac{r'}{r}\right)' + \left(\frac{3}{4}\frac{k'}{k} - \frac{1}{4}\frac{r'}{r}\right)\left(\frac{k'}{k} + \frac{r'}{r}\right)\right]. \tag{A.87}$$

In the case when the functions $k(x)$ and $r(x)$ have different signs on (a, b), equation (A.83), with the aid of the substitution $s'^2 = -r(x)/k(x)$, is transformed into an equation similar to (A.86):

$$u''(s) - [\lambda + q(s)]u(s) = 0, \tag{A.88}$$

where $q(s)$ is given by the same formula as in the first case. Since the investigation of the behavior of solutions when $\lambda \to +\infty$ for equation (A.88) is basically carried out with the same methods as for equation (A.86), we will confine ourselves to the first case, when equation (A.83) reduces to the form (A.86) by means of the substitution (A.84).

From (A.85) it follows that

$$s(x) = \int_{x_0}^{x} \sqrt{\frac{r(t)}{k(t)}}dt \quad (a \le x_0 \le b), \quad \varphi(x) = [k(x)r(x)]^{-1/4}.$$

Let $s(a) = c \quad (c \le 0), \quad s(b) = d \quad (d \ge 0)$. The function $s(x)$ is continuous and monotonically increasing on (a, b). Hence, there exists the inverse function $x = x(s)$, which is also continuous and monotonically increasing on (c, d).

A.4.1.2. It is natural to expect that in the limit $\lambda \to +\infty$ the solutions of (A.86) will coincide with solutions of the simpler equation

$$u'' + \lambda u = 0,$$

i.e.,

$$u(s) \approx A \cos \mu s + B \sin \mu s,$$

where $\mu = \sqrt{\lambda}$, A and B are some constants.

To prove this assertion we use a method proposed by V. A. Steklov [211]. Namely, let us solve the equation

$$u'' + \mu^2 u = q(s)u$$

regarding the right-hand side as a known function. In this way we obtain

$$u(s) = \overline{u}(s) + R_\mu(s), \tag{A.89}$$

where

$$\bar{u}(s) = A \cos \mu s + B \sin \mu s,$$

$$R_\mu(s) = \frac{1}{\mu} \int_0^s \sin \mu(s - s')\, q(s')u(s')ds'.$$

We claim that for $c \le s \le d$ the term $R_\mu(s)$ can be neglected compared with $\max |\overline{u}(s)|$ when $\mu \to \infty$, i.e.,

$$\lim_{\mu \to \infty} \frac{R_\mu(s)}{\overline{M}(\mu)} = 0, \tag{A.90}$$

where

$$\overline{M}(\mu) = \max_{c \le s \le d} |\overline{u}(s)| \le |A| + |B|.$$

From the expression for $R_\mu(s)$ it follows that

$$|R_\mu(s)| \le \frac{1}{\mu} L M(\mu), \tag{A.91}$$

where

$$L = \int\limits_c^d |q(s')|\, ds', \quad M(\mu) = \max_{c \le s \le d} |u(s)|.$$

Let us estimate $M(\mu)$ for $\mu \to +\infty$. From relations (A.89) and (A.91) it follows that, for $c \le s \le d$,

$$|u(s)| \le \overline{M}(\mu) + \frac{1}{\mu} L M(\mu),$$

whence

$$M(\mu) \le \overline{M}(\mu) + \frac{1}{\mu} L M(\mu).$$

Therefore, if $\mu > L$, then

$$M(\mu) \le \frac{\overline{M}(\mu)}{1 - L/\mu}$$

which in view of (A.91) yields

$$\frac{R_\mu(s)}{\overline{M}(\mu)} \le \frac{L}{\mu - L}.$$

This proves (A.90).

The uniform estimate obtained above remains valid also in the case when the interval (c, d) is infinite, provided that the integral $\int_c^d |q(s)| ds$ converges.

Returning to the original variables, we conclude that when $k(x) > 0$, $r(x) > 0$ on (a, b) the solutions of equation (A.83) admit for $\lambda \to +\infty$ the representation

$$y(x) = \frac{1}{\sqrt{k(x) p(x)}} [A \cos \xi(x) + B \sin \xi(x)], \tag{A.92}$$

where

$$p(x) = \sqrt{\lambda \frac{r(x)}{k(x)}}, \quad \xi(x) = \int\limits_{x_0}^{x} p(x') dx'.$$

The replacement of the solution of equation (A.83) by the approximate solution (A.92) lies at the foundation of the semiclassical method for solving equation (A.83).

In the case when $k(x) > 0$, $r(x) < 0$, we similarly obtain

$$y(x) = \frac{1}{\sqrt{k(x)p(x)}}[Ae^{\xi(x)} + Be^{-\xi(x)}], \tag{A.93}$$

where

$$p(x) = \sqrt{\lambda\left|\frac{r(x)}{k(x)}\right|}, \quad \xi(x) = \int_{x_0}^{x} p(x')\,dx'.$$

When the exact solution is replaced by the approximate one, what really matters is that the inequality $L \ll \mu$ (where $L = \int_c^d |q(s)|ds$) holds for equation (A.86). Therefore, the approximate solutions (A.92) and (A.93) can be used not only when λ is large, but also when $\lambda \sim 1$, provided that $L \ll 1$. As seen from formula (A.87), this is the case when the derivatives of the functions $k(x)$ and $r(x)$ are small, i.e., when the coefficients of equation (A.83) vary in a slow, smooth manner. Note that the right-hand side of formula (A.87) contains the logarithmic derivatives of $k(x)$ and $r(x)$, as well as derivatives of these logarithmic derivatives. When the coefficients of equation (A.83) vary in a smooth manner, the derivatives of the logarithmic derivatives of $k(x)$ and $r(x)$ are usually neglected, and one considers that the most essential term in the right-hand side of (A.87) is the second one, which must satisfy the condition

$$\left|\frac{k}{4r}\left(\frac{3}{4}\frac{k'}{k} - \frac{1}{4}\frac{r'}{r}\right)\left(\frac{k'}{k} + \frac{r'}{r}\right)\right| \ll 1. \tag{A.94}$$

This condition is somewhat more crude than the condition $L \ll 1$. However, if we consider the Schrödinger equation, written in the form

$$\frac{d^2\psi}{dx^2} + p^2(x)\psi = 0,$$

where

$$p(x) = \sqrt{2[\varepsilon - U(x)]},$$

then for this equation the condition (A.94) will coincide with the condition of applicability of the semiclassical approximation ordinarily used in quantum mechanics,

$$\left|\frac{p'}{p^2}\right| \ll 1, \tag{A.95}$$

provided that in (A.94) we put $k(x) = 1$ and $r(x) = p^2(x)$.

A.4.1.3. In practice it is of interest to obtain an approximate solution of equation (A.83) for $\lambda \to +\infty$ whose validity extends all the way to the endpoints of the interval (a, b) in case the functions $r(x)$ and $k(x)$ are singular or vanish in a and *b*. For example, let us consider the problem of the approximate representation of the solution of equation (A.83) for $a \le x < b$ when

$$k(x) = (x-a)^{\alpha} k_0(x), \quad r(x) = (x-a)^{\beta} r_0(x),$$

where the functions $k_0(x) > 0$ and $r_0(x) > 0$ have continuous second-order derivatives for $a \le x < b$. In order for $s(a)$ to be finite, we will assume that

$$\frac{1}{2}(\beta - \alpha) > -1. \tag{A.96}$$

Let us rewrite the expressions for $s(x)$ and $q(x)$ for $x_0 = a$:

$$s(x) = \int_a^x \sqrt{\frac{r_0(t)}{k_0(t)}} (t-a)^{(\beta-\alpha)/2} dt, \tag{A.97}$$

$$\begin{aligned} q(s) = (x-a)^{\alpha-\beta-2} \frac{k_0(x)}{4r_0(x)} \times \\ \times \Bigg\{ \frac{(\alpha+\beta)(3\alpha-\beta-4)}{4} + \frac{x-a}{2}\left[(3\alpha+\beta)\frac{k_0'}{k_0} + (\alpha-\beta)\frac{r_0'}{r_0}\right] + \\ (x-a)^2 \left[\left(\frac{k_0'}{k_0} + \frac{r_0'}{r_0}\right)' + \left(\frac{3}{4}\frac{k_0'}{k_0} - \frac{1}{4}\frac{r_0'}{r_0}\right)\left(\frac{k_0'}{k_0} + \frac{r_0'}{r_0}\right)\right]\Bigg\}. \end{aligned}$$

If $x \approx a$, then

$$s(x) \approx \sqrt{\frac{r_0(a)}{k_0(a)}} \frac{(x-a)^{(\beta-\alpha+2)/2}}{[(\beta-\alpha+2)/2]},$$

and consequently the expression for $q(s)$ can be recast as

$$q(s) = \frac{\nu^2 - 1/4}{s^2} + s^{\gamma-2} f(s),$$

where

$$\gamma = \frac{2}{\beta-\alpha+2} > 0, \quad \nu = \frac{|\alpha-1|}{\beta-\alpha+2} > 0$$

and the function $f(x)$ is continuous for $0 \le s < s(b)$. Here $q(s)$ is singular for $s \to 0$. To apply Steklov's method it is convenient to isolate the principal singularity of the function $q(s)$, i.e., rewrite (A.86) in the form

$$u'' + \left(\mu^2 - \frac{\nu^2 - 1/4}{s^2}\right) u = s^{\gamma-2} f(s) u \quad (\mu = \sqrt{\lambda}), \tag{A.98}$$

and then solve this equations, regarding its right-hand side as a known function. Since the equation

$$u'' + \left(\mu^2 - \frac{\nu^2 - 1/4}{s^2}\right) u = 0$$

has the solution

$$u = Av_\nu(\mu s) + Bv_{-\nu}(\mu s),$$

where $v_\nu(x) = \sqrt{x}J_\nu(x)$, A and B are constants, $J_\nu(x)$ is the Bessel function of the 1st kind, we obtain the solution of equation (A.98) in the form

$$u(s) = Av_\nu(\mu s) + Bv_{-\nu}(\mu s) + R_\mu(s), \tag{A.99}$$

where

$$R_\mu(s) = \int_{s_0}^{s} K_\mu(s, s')(s')^{\gamma-2} f(s')u(s')ds',$$

$$K_\mu(s, s') = \frac{\pi}{2\mu \sin \pi\nu}[v_\nu(\mu s)v_{-\nu}(\mu s') - v_\nu(\mu s')v_{-\nu}(\mu s)].$$

One can show that the term $R_\mu(s)$ in (A.99) can be neglected when $\mu \to +\infty$ [154].

Returning to the original variables, we obtain the solution of equation (A.83) in the semiclassical approximation:

$$y(x) = \sqrt{\frac{\xi(x)}{k(x)p(x)}}\Big\{AJ_\nu[\xi(x)] + BJ_{-\nu}[\xi(x)]\Big\}, \tag{A.100}$$

$$p(x) = \sqrt{\lambda\frac{r(x)}{k(x)}}, \quad \xi(x) = \int_a^x p(x')dx'.$$

For integer values of ν in (A.100) we must replace $J_{-\nu}(\xi)$ by $Y_\nu(\xi)$. Note that for $\xi(x) \gg 1$, replacing the Bessel function in (A.100) by the first term of the asymptotic expansion leads to a formula equivalent to (A.92).

If $k_0(x) > 0$ and $r_0(x) < 0$, representation (A.100) must be replaced by

$$y(x) = \sqrt{\frac{\xi(x)}{k(x)p(x)}}\Big\{CI_\nu[\xi(x)] + DK_\nu[\xi(x)]\Big\}, \tag{A.101}$$

$$p(x) = \sqrt{\lambda\left|\frac{r(x)}{k(x)}\right|}, \quad \xi(x) = \int_a^x p(x')dx',$$

where $I_\nu(z)$ and $K_\nu(z)$ are modified Bessel functions.

Analogous formulas can be obtained for the domain $a < x \le b$ when the functions $k(x)$ and $r(x)$ have the form

$$k(x) = (b - x)^\alpha k_0(x), \quad r(x) = (b - x)^\beta r_0(x).$$

A.4.2 Application of the WKB method to an equation with singularity. Semiclassical approximation for a central field

In the study of the motion of a particle in a central field it is of interest to derive the semiclassical approximation for the equation

$$y'' + r(x)y = 0, \tag{A.102}$$

where the function $x^2 r(x)$ is continuous together with its first and second derivatives for $0 \leq x \leq b$. The approximation (A.100) obtained earlier is not applicable for equation (A.102), because $\alpha = 0$, $\beta = -2$ and condition (A.96) is not satisfied. However, the change of variables $x = e^z$, $y = e^{z/2}v(z)$ reduces (A.102) to the form [137]

$$v''(z) + r_1(z)v = 0, \tag{A.103}$$

where

$$r_1(z) = -\frac{1}{4} + x^2 r(x)|_{x=e^z}.$$

The function $r_1(z)$ differs only slightly from the constant $-\frac{1}{4} + \lim_{x\to 0} x^2 r(x)$ when $z \to -\infty$ (which corresponds to $x \to 0$). Furthermore, one can show that $\lim_{z\to-\infty} r_1^{(k)}(z) = 0$ $(k = 1, 2, \ldots)$. For example, if $k = 1$ we have

$$\frac{d}{dz} r_1(z) = \frac{d}{dx}[x^2 r(x)]\frac{dx}{dz} = x\frac{d}{dx}[x^2 r(x)] \to 0, \quad x \to 0.$$

It follows that for negative values of z with $|z|$ sufficiently large the function $r_1(z)$ and its derivatives will vary slowly and the semiclassical approximation will apply to equation (A.103). If the condition of applicability of the semiclassical approximation for equation (A.103) is satisfied for all requisite values of z, then upon returning to the old variables we obtain an approximate solution of equation (A.102) in the form considered earlier, but with $r(x)$ replaced by the function

$$r(x) - \frac{1}{4x^2}.$$

Thus, for instance, when one solves in spherical coordinates the Schrödinger equation

$$-\frac{1}{2}R'' + \left[U(r) + \frac{\ell(\ell+1)}{2r^2}\right] R = \varepsilon R$$

for the radial part $R(r)$ of the wave function, where $U(r)$ is the potential energy, ε is the total energy of the particle and $\ell = 0, 1, 2, \ldots$ is the orbital quantum number, the semiclassical approximation yields the expression

$$R(r) = \begin{cases} \sqrt{\dfrac{\xi}{|p|}}[CI_{1/3}(\xi) + DK_{1/3}(\xi)], & \text{if } r \leq \tilde{r}, \\[2ex] \sqrt{\dfrac{\xi}{p}}[AJ_{1/3}(\xi) + BJ_{-1/3}(\xi)], & \text{if } r \geq \tilde{r}. \end{cases} \tag{A.104}$$

Here

$$p = p(r) = \sqrt{2\left[\varepsilon - U(r) - \frac{\left(\ell+\frac{1}{2}\right)^2}{2r^2}\right]}, \tag{A.105}$$

$$\xi = \xi(r) = \left|\int_{\tilde{r}}^{r} p(r')\,dr'\right|$$

and $\tilde{r}$ is the root of the equation $p(r) = 0$ that is the closest to $r = 0$ (we assume that this root is simple). The approximation (A.104) remains valid also for $\ell = 0$, provided that $x^2 r(x)$ has a limit as $x \to 0$.

Since the function $R(r)$ is required to be bounded for $r \to 0$, i.e., $\xi \to \infty$, we must put $C = 0$. From the junction condition in $r = \tilde{r}$ for the functions $R(r)$ and $R'(r)$ the constants A and B can be expressed through the constant D. Expanding the expression under the radical sign in the formula for $p(r)$ in powers of $r - \tilde{r}$ one can readily verify that the functions $p(r)/\sqrt{|r-\tilde{r}|}$, $\xi(r)/|r-\tilde{r}|^{3/2}$ and their first derivatives are continuous in the point $r = \tilde{r}$. Hence, from the junction condition for $R(r)$ and $R'(r)$ at the point $r = \tilde{r}$ one obtains analogous junction conditions for the function $\Phi(r) = (\xi/2)^{1/3}\sqrt{|p|/\xi}\,R(r)$ and its derivatives. We have

$$\Phi(r) = \begin{cases} \dfrac{A\,(\xi/2)^{2/3}}{\Gamma\,(4/3)} + \dfrac{B}{\Gamma\,(2/3)} + O[(r-\tilde{r})^3], & \text{if } r \geq \tilde{r}, \\[2ex] \dfrac{D\pi}{2\sin(\pi/3)}\left[\dfrac{1}{\Gamma\,(2/3)} - \dfrac{(\xi/2)^{2/3}}{\Gamma\,(4/3)}\right] + O[(r-\tilde{r})^3], & \text{if } r \leq \tilde{r}. \end{cases}$$

From the condition that the function $\Phi(r)$ and its derivative be continuous in $r = \tilde{r}$ and the fact that $\xi \sim |r-\tilde{r}|^{3/2}$ it follows that

$$A = B = \frac{\pi}{\sqrt{3}}\,D.$$

A.4.3 The Bohr-Sommerfeld quantization rule

Let us use the semiclassical approximation to compute the energy levels of an electron that moves in a one-dimensional potential well, a typical graph of which is shown in Figure A.5.

We will assume that for any energy $\varepsilon > U_{\min}$ there are only two turning points, x_1 and x_2, determined by the condition $U(x_1) = U(x_2) = \varepsilon$. As we have shown, in the semiclassical approximation, in the domain $x_1 \leq x < x_2$ the wave function of the particle has the form

$$\psi_1(x) = A\sqrt{\frac{\xi_1}{p}}[J_{1/3}(\xi_1) + J_{-1/3}(\xi_1)],$$

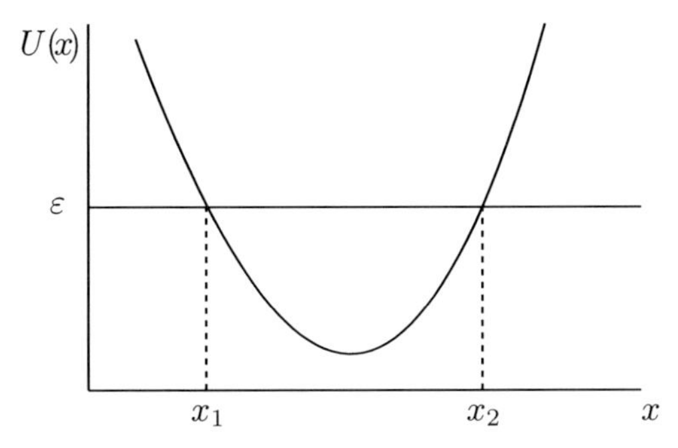

Figure A.5: Potential well

$$\xi_1 = \xi_1(x) = \int_{x_1}^{x} p(x')dx', \quad p = p(x) = \sqrt{2[\varepsilon - U(x)]}.$$

If for the fixed limit in the integral we take the turning point x_2 instead of x_1, then for $x_1 < x \leq x_2$ we get

$$\psi_2(x) = B\sqrt{\frac{\xi_2}{p}}\,[J_{1/3}(\xi_2) + J_{-1/3}(\xi_2)], \quad \xi_2 = \xi_2(x) = \int_{x}^{x_2} p(x')dx'.$$

In any point x lying between x_1 and x_2 one can impose the junction conditions

$$\psi_1(x) = \psi_2(x), \quad \psi_1'(x) = \psi_2'(x). \tag{A.106}$$

Let us choose the point x so that it is far from the turning points and the conditions $\xi_1 \gg 1$, $\xi_2 \gg 1$ are satisfied. Then, using the asymptotics of the Bessel function,

$$J_{\pm 1/3}(\xi) \approx \sqrt{\frac{2}{\pi\xi}}\cos\left(\xi \mp \frac{\pi}{6} - \frac{\pi}{4}\right),$$

we derive from (A.106) the relations

$$\begin{aligned} A\cos\left(\xi_1 - \frac{\pi}{4}\right) &= B\cos\left(\xi_2 - \frac{\pi}{4}\right), \\ A\sin\left(\xi_1 - \frac{\pi}{4}\right) &= -B\sin\left(\xi_2 - \frac{\pi}{4}\right), \end{aligned} \tag{A.107}$$

where in the differentiation we used the condition (A.95) of applicability of the semiclassical approximation.

Since $\xi_2 = \int_{x_1}^{x_2} p(x)dx - \xi_1$, upon dividing the second relation in (A.107) by the first one obtains

$$\tan\left(\xi_1 - \frac{\pi}{4}\right) = \tan\left(\xi_1 + \frac{\pi}{4} - \int_{x_1}^{x_2} p(x)dx\right),$$

which yields the *Bohr-Sommerfeld quantization rule*

$$\int_{x_1(\varepsilon)}^{x_2(\varepsilon)} p(x)dx = \pi\left(n+\frac{1}{2}\right) \tag{A.108}$$

for the energy levels of the discrete spectrum $\varepsilon = \varepsilon_n$, $n = 0, 1, 2, \ldots$. Since the zeroes of the wave function lie inside the interval $(x_1,\ x_2)$ (as clearly follows from the form of the semiclassical solution outside this interval), the n in (A.108) is the number of zeroes of the wave function $\psi(x)$. Setting $\xi_2 = \pi(n+1/2) - \xi_1$ in any of the relations (A.107), we conclude that $B = (-1)^n A$.

According to (A.105) and (A.108), for a particle moving in the central field $U(r)$ the Bohr-Sommerfeld condition reads

$$\int_{r_1(\varepsilon)}^{r_2(\varepsilon)} p(r)dr = \pi\left(n_r+\frac{1}{2}\right), \tag{A.109}$$

where n_r is the number of zeroes of the radial function.

EXAMPLE 1. Let us find the energy levels of a hydrogen-like atom, with $U(r) = -Z/r$, in the semiclassical approximation.

To this end we calculate the integral in the left-hand side of (A.109), using integration by parts:

$$\int_{r_1}^{r_2} p(r)dr = rp(r)\Big|_{r_1}^{r_2} - \int_{r_1}^{r_2} \frac{r\left(-\dfrac{Z}{r^2}+\dfrac{(\ell+1/2)^2}{r^3}\right)}{\sqrt{2\left[\varepsilon+\dfrac{Z}{r}-\dfrac{(\ell+1/2)^2}{2r^2}\right]}}dr =$$

$$Z\int_{r_1}^{r_2}\frac{dr}{\sqrt{2\varepsilon r^2+2Zr-(\ell+1/2)^2}} + \int_{x_1}^{x_2}\frac{(\ell+1/2)^2dx}{\sqrt{2\varepsilon+2Zx-(\ell+1/2)^2x^2]}}\Bigg|_{x=1/r} =$$

$$\frac{Z}{\sqrt{-2\varepsilon}}\int_{-a}^{a}\frac{dx}{\sqrt{a^2-x^2}} - (\ell+1/2)\int_{-b}^{b}\frac{dx}{\sqrt{b^2-x^2}} = \pi\left(\frac{Z}{\sqrt{-2\varepsilon}}-\ell-1/2\right).$$

A comparison with (A.109) yields

$$\varepsilon = \varepsilon_{n\ell} = -\frac{Z^2}{2n^2},$$

where $n = n_r + \ell + 1$ is the principal quantum number. Thus, the energy levels of a hydrogen-like ion provided by the semiclassical approximation coincide with the exact ones. A similar assertion holds for the linear harmonic oscillator.

A.4.4 Using the semiclassical approximation to normalize the continuum wave functions

The semiclassical continuum wave functions provide a rather good approximation, because the higher the energy ε, the smoother the behavior of the function $p(r)$. However, a known difficulty arises, connected with the normalization of the numerical solutions of the Schrödinger equation [45, 49]. In the semiclassical approximation, the continuum wave functions $R_{\varepsilon\ell}(r)$, normalized by the condition (see § 2.3)

$$\int_0^\infty R_{\varepsilon\ell}(r)R_{\varepsilon'\ell}(r)dr = \delta(\varepsilon - \varepsilon'), \tag{A.110}$$

have the form

$$\tilde{R}_{\varepsilon\ell}(r) = \begin{cases} \dfrac{1}{\pi}\sqrt{\dfrac{\xi}{|p|}}K_{1/3}(\xi), & \text{if } r \le r_{\varepsilon\ell}, \\ \dfrac{1}{\sqrt{3}}\sqrt{\dfrac{\xi}{p}}[J_{-1/3}(\xi) + J_{1/3}(\xi)], & \text{if } r \ge r_{\varepsilon\ell}, \end{cases} \tag{A.111}$$

where $r_{\varepsilon\ell}$ is a turning point. The first zero $r = r^*$ of the function $\tilde{R}_{\varepsilon\ell}(r)$ corresponds to the first zero $\xi = \xi^*$ of the equation

$$J_{-1/3}(\xi^*) + J_{1/3}(\xi^*) = 0,$$

where $\xi^* = 1.929379$. For the derivative of the semiclassical function (A.111) we have

$$\frac{d}{dr}\tilde{R}_{\varepsilon\ell}(r)\Big|_{r=r^*} \approx \sqrt{\frac{\xi^* p(r^*)}{3}}\,[J'_{-1/3}(\xi^*) + J'_{1/3}(\xi^*)] = -0.801952\sqrt{p(r^*)}. \tag{A.112}$$

This condition can be used in the numerical integration of the Schrödinger equation to normalize the continuum wave functions: the value of $R'_{\varepsilon\ell}$ in the first zero of $R_{\varepsilon\ell}(r)$ is set equal to $\tilde{R}'_{\varepsilon\ell}(r)$, in accordance with formula (A.112).

EXAMPLE 2. Consider the normalization of the continuum wave function for the Coulomb potential $U(r) = -Z/r$. For $\varepsilon = 0$ the Schrödinger equation has an exact solution, which can be expressed in terms of Bessel functions. In particular, for $\ell = 1$ we have

$$R_{\varepsilon\ell}(r) = \sqrt{2r}J_3(\sqrt{8Z}\,r).$$

In this case computations based on formula (A.112) give the normalization coefficient 0.99998 instead of 1.

NUMERICAL METHODS

A.5 The phase method for calculating energy eigenvalues and wave functions

A.5.1 Equation for the phase and the connection with the semiclassical approximation

To find the energy levels and wave functions of a electron moving in a given central field with potential $V(r)$, we consider the problem of solving the Schrödinger equation

$$-\frac{1}{2}R'' + \left[-V(r) + \frac{\ell(\ell+1)}{2r^2}\right] R = \varepsilon R, \quad 0 < r < \infty, \tag{A.113}$$

with the boundary conditions $R(0) = R(\infty) = 0$ and the normalization condition $\int_0^\infty R^2(r)dr = 1$. Here the function $R(r)$ must have a prescribed number of zeroes $n_r = n - \ell - 1$ in the interval $(0, \infty)$, n being the principal quantum number.

The problem posed above is usually solved as follows. For some energy value ε one integrates equation (A.113) in the direction of increasing values of r, starting at $r = 0$ with initial conditions corresponding to the behavior of the the wave function $R(r)$ for $r \sim 0$, i.e., $R(r) \sim r^{\ell+1}$. Denote the resulting solution by $R_0(r)$. For the sake of definiteness, we will assume, as is customary, that $R_0(r) > 0$ for small r. In much the same way, integrating equation (A.113) in the direction decreasing values of r, one can find a solution $R_\infty(r)$, which for $r \to \infty$ behaves like $\exp(-\sqrt{-2\varepsilon}\, r)$, if $V(r) \to 0$ as $r \to \infty$.

The energy eigenvalues $\varepsilon = \varepsilon_{n\ell}$ are usually found from the conditions that the functions $R_0(r)$ and $R_\infty(r)$ and their first derivatives be continuous in some intermediate point $r = r^*$. Generally, the direct use of these conditions in the calculation of energy eigenvalues requires a sufficiently large number of iterations, since the functions $R(r)$ and $R'(r)$ depend on ε in a rather complex manner. To simplify the dependence on ε we resort to the semiclassical approximation.

In the semiclassical approximation the solution of equation (A.113) in the domain of classical motion $r_1 < r < r_2$ has the form

$$R(r) = \frac{C}{\sqrt{p(r)}} \sin\left(\int_{r_1}^{r} p(r)dr + \frac{\pi}{4}\right), \tag{A.114}$$

where

$$p(r) = p_{\varepsilon\ell}(r) = \sqrt{2\left[\varepsilon + V(r) - \frac{(\ell+1/2)^2}{2r^2}\right]}, \tag{A.115}$$

$$p(r_1) = p(r_2) = 0.$$

Starting from representation (A.114), one obtains the Bohr-Sommerfeld condition for the determination of the energy eigenvalues in the semiclassical approximation (see the derivation of formulas (A.109) in §A.4).

Let us find a representation of the function $R(r)$ with the property that in the semiclassical approximation it will go over, in a natural manner, into representation (A.114). To this end we will seek the function $R(r)$ in the form

$$R(r) = g(r)\sin\varphi(r). \tag{A.116}$$

For the derivative $R'(r)$ we use the expression

$$R'(r) = a(r)g(r)\cos\varphi(r), \tag{A.117}$$

because in the semiclassical approximation

$$R'(r) = C\sqrt{p(r)}\cos\left(\int_{r_1}^{r} p(r)dr + \frac{\pi}{4}\right).$$

Let us require that in the semiclassical approximation the function $a(r)$ will go over into $p(r)$. Then obviously the phase $\varphi(r)$ will go over into

$$\int_{r_1}^{r} p(r')dr' + \frac{\pi}{4}$$

which readily leads to conditions analogous to the Bohr-Sommerfeld condition.

Assuming the function $a(r)$ is given, let us write the equations that the functions $g(r)$ and $\varphi(r)$ must satisfy. Differentiating expression (A.116) for $R(r)$ yields

$$g'\sin\varphi + g\varphi'\cos\varphi = ag\cos\varphi. \tag{A.118}$$

Similarly, differentiating expression (A.117) and using the Schrödinger equation (A.113), we have

$$ag'\cos\varphi - ag\varphi'\sin\varphi + a'g\cos\varphi = -fg\sin\varphi,$$

where

$$f(r) = 2[\varepsilon + V(r)] - \frac{\ell(\ell+1)}{r^2}.$$

Therefore,

$$\varphi' = a - \sin\varphi \left[\left(a - \frac{f}{a} \right) \sin\varphi - \frac{a'}{a} \cos\varphi \right], \tag{A.119}$$

$$\frac{g'}{g} = \cos\varphi \left[\left(a - \frac{f}{a} \right) \sin\varphi - \frac{a'}{a} \cos\varphi \right]. \tag{A.120}$$

Equation (A.120) shows that the function $g(r)$ has no zeroes, i.e., it indeed has the character of an amplitude factor in (A.116).

A.5.2 Construction of an iteration scheme for the calculation of eigenvalues

In accordance with the semiclassical approximation, it is natural to put $a(r) = p(r)$ in the domain of classical motion. Outside that domain we take for $a(r)$ a bounded function that does not vanish, so that $d\varphi/dr > 0$ for all r, in agreement with the meaning of the phase φ. For definiteness, we put $a(\infty) = \sqrt{-2\varepsilon}$. Then, by (A.116) and (A.117),

$$\tan\varphi = a(r)\,\frac{R(r)}{R'(r)}. \tag{A.121}$$

Taking into account the way $R(r)$ behaves when $r \to 0$ and $r \to \infty$, we find that

$$\tan\varphi(0) = \lim_{r\to 0} \frac{a(r)}{\ell+1} r = 0, \tag{A.122}$$

$$\tan\varphi(\infty) = -\lim_{r\to\infty} \frac{a(r)}{\sqrt{-2\varepsilon}} = -1. \tag{A.123}$$

Let $\varphi_0(r)$ and $\varphi_\infty(r)$ be two solutions of equation (A.119), with the first satisfying condition (A.122) and the second satisfying condition (A.123), i.e., $\varphi_0(0) = 0$, $\varphi_\infty(\infty) = -\pi/4$. For the energy eigenvalue $\varepsilon = \varepsilon_{n\ell}$, similarly to the Bohr-Sommerfeld condition obtained above, we have the condition $\tan\varphi_0(r) = \tan\varphi_\infty(r)$, or $\varphi_0(r) - \varphi_\infty(r) = \pi k$, where k is an integer. Obviously, $k = n_r + 1 = n - \ell$, i.e., the relation

$$\varphi_0(r^*) - \varphi_\infty(r^*) = \pi(n-\ell) \tag{A.124}$$

must hold in any intermediate point $r = r^*$ $(r_1 < r^* < r_2)$.

Note that the value of the phase $\varphi(r)$ in the nodes of the function $R(r)$ is equal to πm, with m an integer, while its value in the nodal points of the derivative $R'(r)$ is $\pi(m + 1/2)$, for any choice of the function $a(r)$. In any intermediate point $r = r^*$ the value of the phase $\varphi(r)$ is determined by the value of $a(r^*)$.*)

---

*) Strictly speaking, to derive equality (A.124) it is not necessary to give precisely the conditions (A.122)–(A.123) for $r = 0$ and $r = \infty$ and require that the functions $a(r)$

In the domain of classical motion, away from the turning points, we follow the semiclassical approximation and put $a(r) = p(r)$. Then for the sewing point one can take a point $r = r^*$ $(r_1 < r^* < r_2)$ for which $p'(r^*) = 0$, i.e., (see (A.115))

$$-r^3 \frac{dV}{dr}\bigg|_{r=r^*} = \left(\ell + \frac{1}{2}\right)^2.$$

Near $r = r^*$ the condition $|p'| \ll p^2$ for the applicability of the semiclassical approximation is obviously satisfied.

Let us remark that the semiclassical approximation (A.114) can be obtained directly from equation (A.119), setting $a(r) = p(r) \approx \sqrt{f(r)}$:

$$\varphi' = \frac{d\varphi}{dr} \approx p + \frac{p'}{2p} \sin 2\varphi,$$

whence

$$\varphi \approx \int p(r)dr + \int \frac{p'}{2p^2}\, p(r) \sin 2\varphi(r)\, dr.$$

Here the second term is small not only because $|p'/p^2| \ll 1$, but also due to the presence of the factor $\sin 2\varphi(r)$ in the integrand. From this formula it also follows that in the domain of classical motion $\varphi'(r) \approx p(r) > 0$, which was assumed in the derivation of condition (A.124).

In the semiclassical approximation equation (A.124) coincides with the Bohr-Sommerfeld quantization rule. To solve this equation numerically (by Newton's method, for instance), it would be convenient to reduce it to an equation of the form $F(\varepsilon) = 0$ such that the function $F(\varepsilon)$ is almost linear in ε. It is clear, however, that in practice such a reduction is not an easy task. For this reason, the procedure described has been implemented for the typical case of the Coulomb potential with external screening $V(r) = Z/r - A$ (see Example 1 in §A.4, and also §A.3). If, in this case, instead of equation (A.124) we take the equivalent equation

$$\frac{\pi^2}{[\varphi_0(r^*) - \varphi_\infty(r^*) + \pi\ell]^2} - \frac{1}{n^2} = 0, \tag{A.125}$$

---

and $g(r)$ have a specific behavior outside the domain of classical motion. Indeed, representation (A.116)–(A.117) is actually used only inside the domain of classical motion, where the function $R(r)$ oscillates and where all the nodal points of the functions $R(r)$ and $R'(r)$ lie. For example, one can use the following boundary conditions for the phase functions $\varphi_0(r)$ and $\varphi_\infty(r)$:

$$\varphi_0(\tilde{r}_1) = \frac{\pi}{4}, \quad \varphi_\infty(\tilde{r}_2) = -\frac{\pi}{4},$$

where the points $\tilde{r}_1$ and $\tilde{r}_2$ are solutions of the equation $f(r) = 0$, i.e., $R''(r) = 0$, and determine the domain of oscillations of the function $R(r)$. Since $p^2(r) \approx f(r)$, this domain of oscillations, $\tilde{r}_1 < r < \tilde{r}_2$, is almost identical to the domain of classical motion, $r_1 < r < r_2$.

we conclude that, in the semiclassical approximation, it has the form

$$\frac{2(A-\varepsilon)}{Z^2} - \frac{1}{n^2} = 0,$$

i.e., the requisite linearity in ε holds. One can expect that in the general case, too, the right-hand side of equation (A.125) will be almost linear in ε. Thus, we arrive at the following iteration scheme for determining the eigenvalue $\varepsilon = \varepsilon_{n\ell}$:

$$\overset{(s+1)}{\varepsilon} = \overset{(s)}{\varepsilon} + \frac{1}{2}\frac{\varphi_0(r^*) - \varphi_\infty(r^*) + \pi\ell}{\dfrac{\partial\varphi_0(r^*)}{\partial\varepsilon} - \dfrac{\partial\varphi_\infty(r^*)}{\partial\varepsilon}}\left\{1 - \left(\frac{\varphi_0(r^*) - \varphi_\infty(r^*) + \pi\ell}{\pi n}\right)^2\right\}\Bigg|_{\varepsilon=\overset{(s)}{\varepsilon}} \tag{A.126}$$

(s is the iteration number).

The values of the functions $\varphi_0(r)$ and $\varphi_\infty(r)$, and also of $\partial\varphi_0(r)/\partial\varepsilon$ and $\partial\varphi_\infty(r)/\partial\varepsilon$, in the point $r = r^*$ can be found with the help of the values of $R_0(r)$, $R_0'(r)$ and $R_\infty(r)$, $R_\infty'(r)$, which in turn are obtained via the numerical integration of equation (A.113). Starting from (A.121), let us construct continuous functions $\varphi_0(r)$ and $\varphi_\infty(r)$, using the principal branch of the function arctan and setting $a(r) = p(r)$:

$$\varphi_0(r) = \begin{cases} \operatorname{arctg}\left(p(r)\dfrac{R_0(r)}{R_0'(r)}\right) + \pi k_1, & \text{if } \dfrac{R_0(r)}{R_0'(r)} > 0, \\[2ex] \operatorname{arctg}\left(p(r)\dfrac{R_0(r)}{R_0'(r)}\right) + \pi(k_1+1), & \text{if } \dfrac{R_0(r)}{R_0'(r)} < 0, \end{cases}$$

$$\varphi_\infty(r) = \begin{cases} \operatorname{arctg}\left(p(r)\dfrac{R_\infty(r)}{R_\infty'(r)}\right) - \pi(k_2+1), & \text{if } \dfrac{R_\infty(r)}{R_\infty'(r)} > 0, \\[2ex] \operatorname{arctg}\left(p(r)\dfrac{R_\infty(r)}{R_\infty'(r)}\right) - \pi k_2, & \text{if } \dfrac{R_\infty(r)}{R_\infty'(r)} < 0. \end{cases}$$

Here k_1 [resp., k_2] is the number of zeroes of $R_0(r)$ [resp., $R_\infty(r)$] in the interval $(0, r^*)$ [resp., (r^*, ∞)].

To calculate $\partial\varphi/\partial\varepsilon$ we resort again to formula (A.121), setting $a(r) = p(r)$:

$$\frac{1}{\cos^2\varphi}\frac{\partial\varphi}{\partial\varepsilon} = \frac{1}{p(r)}\frac{R(r)}{R'(r)} + p(r)\left[\frac{1}{R'(r)}\frac{\partial R(r)}{\partial\varepsilon} - \frac{R(r)}{(R'(r))^2}\frac{\partial R'(r)}{\partial\varepsilon}\right],$$

whence

$$\frac{\partial\varphi}{\partial\varepsilon} = \frac{\dfrac{1}{p(r)}R'(r)R(r) + p(r)\left[R'(r)\dfrac{\partial R(r)}{\partial\varepsilon} - R(r)\dfrac{\partial R'(r)}{\partial\varepsilon}\right]}{(R'(r))^2 + p^2(r)R^2(r)}. \tag{A.127}$$

Using the Schrödinger equation (A.113) one can show that

$$\frac{d}{dr}\left[R'(r)\frac{\partial R(r)}{\partial \varepsilon} - R(r)\frac{\partial R'(r)}{\partial \varepsilon}\right] = 2R^2(r). \tag{A.128}$$

To this end it suffices to differentiate equation (A.113) with respect to ε and then multiply the resulting equation by $R(r)$ and subtract from it equation (A.113), multiplied by $\partial R(r)/\partial\varepsilon$. Now integrating (A.128) we obtain

$$R_0'(r)\frac{\partial R_0(r)}{\partial \varepsilon} - R_0(r)\frac{\partial R_0'(r)}{\partial \varepsilon} = 2\int_0^r R_0^2(r)\,dr,$$

$$R_\infty'(r)\frac{\partial R_\infty(r)}{\partial \varepsilon} - R_\infty(r)\frac{\partial R_\infty'(r)}{\partial \varepsilon} = -2\int_r^\infty R_\infty^2(r)\,dr.$$

Therefore, in view of (A.127),

$$\begin{cases} \dfrac{\partial\varphi_0(r^*)}{\partial\varepsilon} = \left.\dfrac{\dfrac{1}{p(r)}R_0'(r)R_0(r) + 2p(r)\displaystyle\int_0^r R_0^2(r)\,dr}{(R_0'(r))^2 + p^2(r)R_0^2(r)}\right|_{r=r^*}, \\ \\ \dfrac{\partial\varphi_\infty(r^*)}{\partial\varepsilon} = \left.\dfrac{\dfrac{1}{p(r)}R_\infty'(r)R_\infty(r) - 2p(r)\displaystyle\int_r^\infty R_\infty^2(r)\,dr}{(R_\infty'(r))^2 + p^2(r)R_\infty^2(r)}\right|_{r=r^*}. \end{cases} \tag{A.129}$$

We see that when the iteration scheme (A.126) is used, it suffices to know the values of the radial function $R_{n\ell}(r)$ and of its derivative $R_{n\ell}'(r)$. Hence, in the actual calculations of energy levels and wave functions it is natural to resort directly to a difference scheme for the Schrödinger equation (A.113), without using the equation for the phase $\varphi(r)$.

Remark 1. To understand the role played by the function $a(r)$, let us consider the following model eigenvalue problem:

$$y'' + k^2 y = 0, \quad 0 < x < 1.$$

We need to select the values of k for which the boundary conditions $y(0) = y(1) = 0$ are satisfied; clearly, $k = n\pi$, where $n = 1, 2, \ldots$.

Let us pass to polar coordinates:

$$\begin{cases} y = g(x)\sin\varphi(x), \\ y' = g(x)\cos\varphi(x). \end{cases}$$

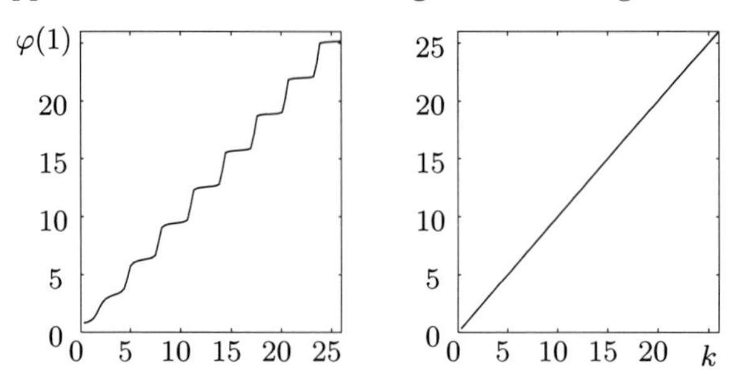

Figure A.6: Dependence of the value $\varphi(1)$ on k for $a = 1$ (left) and $a = k$ (right)

For the solution $y = \sin kx$, which satisfies the boundary condition $y(0) = 0$, we have

$$g \sin \varphi = \sin kx, \quad g \cos \varphi = k \cos kx,$$

whence

$$\tan \varphi = \frac{1}{k} \tan kx.$$

Imposing the boundary condition $y(1) = 0$ at $x = 1$, we obtain

$$\tan \varphi(1) = \frac{\tan k}{k}, \quad \left.\frac{\partial \varphi}{\partial k}\right|_{x=1} = \frac{k - \sin k \cos k}{k^2 \cos^2 k + \sin^2 k}.$$

In particular,

$$\left.\frac{\partial \varphi}{\partial k}\right|_{x=1} = \begin{cases} k, & \text{if} \quad k = \dfrac{\pi}{2} + n\pi, \\ 1/k, & \text{if} \quad k = n\pi. \end{cases}$$

The equation that determines the eigenvalues k reads

$$\varphi(1) = \pi n.$$

Newton's method proves ineffective in solving this equation, because the value of $\varphi(x)$ for $x = 1$ depends on k in an highly nonlinear manner (see Figure A.6 (left), which shows the values of $\varphi(x)|_{x=1}$ as a function of k). As one can see from this figure, convergence may fail due to the stepwise character of the function. However, if we put $y' = a(x)g(x)\cos\varphi(x)$ and substitute $a(x) = k$ in this expression, then the phase is smoothed out and for $x = 1$ we obtain $\tan\varphi = \tan k$, i.e., $\varphi = k$ (see Figure A.6 (right)). Now Newton's method will converge after one iteration.

Remark 2. When the energy levels are calculated in the semiclassical approximation by means of Newton's method directly from the Bohr-Sommerfeld quantization rule

$$\int_{r_1(\varepsilon)}^{r_2(\varepsilon)} p_{\varepsilon\ell}(r)dr = \pi\left(n_r + \frac{1}{2}\right),$$

it is natural to write the latter in a form similar to (A.125):

$$\frac{\pi^2}{[\varphi(\varepsilon)+\pi\ell]^2}-\frac{1}{n^2}=0, \quad \varphi(\varepsilon)=\int\limits_{r_1(\varepsilon)}^{r_2(\varepsilon)} p_{\varepsilon\ell}(r)dr+\frac{\pi}{2}.$$

Then, by (A.126), we obtain the iteration scheme

$$\overset{(s+1)}{\varepsilon}=\overset{(s)}{\varepsilon}+\left.\frac{[\varphi(\varepsilon)+\pi\ell]\left\{1-[(\varphi(\varepsilon)+\pi\ell)/(\pi n)]^2\right\}}{2\dfrac{\partial\varphi(\varepsilon)}{\partial\varepsilon}}\right|_{\varepsilon=\overset{(s)}{\varepsilon}}, \tag{A.130}$$

where

$$\frac{\partial\varphi(\varepsilon)}{\partial\varepsilon}=\int\limits_{r_1(\varepsilon)}^{r_2(\varepsilon)}\frac{dr}{p_{\varepsilon\ell}(r)}.$$

A.5.3 Difference schemes for calculating radial functions

To find the values of the solution $R(r)$ of equation (A.113) one usually employs difference schemes of sufficiently high order of accuracy, with a constant or variable step in r. First let us consider a very simple difference scheme of 4-th order of accuracy for the equation $v''=g(x)$, on a uniform grid $x_{i+1}=x_i+h \quad (i=1,2,\ldots,N)$. Integrating this equation we obtain

$$v(x)=\int\limits_{x_i}^{x}(x-t)g(t)\,dt+v_i+C_i(x-x_i), \quad \text{where } v_i=v(x_i).$$

This yields

$$v_{i+1}-2v_i+v_{i-1}=\int\limits_{x_{i-1}}^{x_i}(t-x_{i-1})g(t)\,dt+\int\limits_{x_i}^{x_{i+1}}(x_{i+1}-t)g(t)\,dt. \tag{A.131}$$

To calculate the integrals we can approximate the function $g(x)$ on the interval $x\in(x_{i-1},x_{i+1})$ in various ways. Let us use the quadratic interpolation

$$g(x)=g_i+\frac{g_{i+1}-g_{i-1}}{2h}(x-x_i)+\frac{g_{i+1}-2g_i+g_{i-1}}{2h^2}(x-x_i)^2.$$

Substituting this expression in (A.131), we obtain

$$\frac{v_{i+1}-2v_i+v_{i-1}}{h^2}=g_i+\frac{1}{12}(g_{i+1}-2g_i+g_{i-1}). \tag{A.132}$$

Let us find the order of accuracy of this scheme. By Taylor's formula

$$v_{i\pm1} = v_i \pm v_i' h + v_i'' \frac{h^2}{2} \pm v_i''' \frac{h^3}{6} + v_i^{(\text{iv})} \frac{h^4}{24} + O(h^5),$$

and so

$$\frac{v_{i+1} - 2v_i + v_{i-1}}{h^2} = v_i'' + v_i^{(\text{iv})} \frac{h^2}{12} + O(h^3) = g_i + g_i'' \frac{h^2}{12} + O(h^3).$$

Since, by Taylor formula for $g(x)$,

$$g_{i+1} - 2g_i + g_{i-1} = g_i'' h^2 + O(h^3),$$

we conclude that

$$\frac{v_{i+1} - 2v_i + v_{i-1}}{h^2} = g_i + \frac{1}{12}(g_{i+1} - 2g_i + g_{i-1}) + O(h^3).$$

A comparison with (A.132) shows that, for given values of v_i and v_{i-1}, we indeed obtained a 4-th order scheme for v_{i+1}.

Let us apply the results obtained above to the numerical integration of an equation of the form $v'' = F(x)v$ (for example, of the one-dimensional Schrödinger equation). Setting $g(x) = F(x)v$ in (A.132), we have

$$\frac{v_{i+1} - 2v_i + v_{i-1}}{h^2} = F_i v_i + \frac{1}{12}(F_{i+1} v_{i+1} - 2F_i v_i + F_{i-1} v_{i-1}),$$

which leads to Numerov's scheme [83]:

$$\left(1 - \frac{h^2}{12} F_{i+1}\right) v_{i+1} - 2\left(1 + \frac{5}{12} h^2 F_i\right) v_i + \left(1 - \frac{h^2}{12} F_{i-1}\right) v_{i-1} = 0. \quad \text{(A.133)}$$

To apply Numerov's scheme for the numerical integration of equation (A.113) on a grid uniform in r we must set

$$v_i = R(r_i), \quad F_i = -2[\varepsilon + V(r_i)] + \frac{\ell(\ell+1)}{r_i^2}, \quad h = r_{i+1} - r_i, \quad i = 2, 3, \ldots, N-1.$$

This yields

$$R_{i+1} = 2 \frac{1 + \frac{5}{12} h^2 F_i}{1 - \frac{1}{12} h^2 F_{i+1}} R_i - \frac{1 - \frac{1}{12} h^2 F_{i-1}}{1 - \frac{1}{12} h^2 F_{i+1}} R_{i-1}. \quad \text{(A.134)}$$

If the values of $R(r)$ for $r = r_1$ and $r = r_2 = r_1 + h$ are given, the scheme (A.134) allows us to calculate the values of $R_0(r)$ at given points of the grid. To calculate $R_\infty(r)$, we can apply the scheme (A.133) for $r < r_N$ in the form

$$R_{i-1} = 2 \frac{1 + \frac{5}{12} h^2 F_i}{1 - \frac{1}{12} h^2 F_{i-1}} R_i - \frac{1 - \frac{1}{12} h^2 F_{i+1}}{1 - \frac{1}{12} h^2 F_{i-1}} R_{i+1}. \quad \text{(A.135)}$$

A.5.4 The radial functions near zero and for large values of r

To be able to use formulas (A.134) and (A.135) we must know the values of the function $R(r)$ in two initial points r_1, r_2 (and, correspondingly, in the points r_N and r_{N-1}). First let us consider the problem when the integration of equation (A.113) is carried out starting from the point $r = 0$ in the direction of increasing values of r.

To determine the values of $R(r)$ in the neighborhood of zero (i.e., for $r \approx 0$), we recast equation (A.113) as

$$R'' - \frac{\ell(\ell+1)}{r^2} R = f(r)R, \quad \text{where } f(r) = -2[\varepsilon + V(r)]). \tag{A.136}$$

In the problems at hand $\lim_{r\to 0} rV(r) = Z$. If we regard the right-hand side of (A.136) as a known function, then using Duhamel's principle, which enables us to express the solution of the nonhomogeneous equation in terms of the solution of the homogeneous equation, we obtain

$$R(r) = Cr^{\ell+1} + \frac{r^{\ell+1}}{2\ell+1} \int_0^r \left[1 - \left(\frac{r'}{r}\right)^{2\ell+1}\right] f(r')(r')^{-\ell} R(r')\, dr'.$$

Setting

$$R(r) = Cr^{\ell+1} y(r), \quad \text{where } \lim_{r\to 0} y(r) = 1,$$

we have

$$y(r) = 1 + \frac{1}{2\ell+1} \int_0^r \left[1 - \left(\frac{r'}{r}\right)^{2\ell+1}\right] r' f(r') y(r') dr'.$$

Since the function $rf(r)y(r)$ has a limit when $r \to 0$, to calculate the integral for small values of r one can replace $rf(r)y(r)$ by its linear approximation,

$$rf(r)y(r) = a + br,$$

where $a = -2Z$. After integration we obtain

$$y(r) = 1 + r\left(-\frac{Z}{\ell+1} + \frac{br}{2(2\ell+3)}\right).$$

Setting here $br = 2Z + rf(r)y(r)$ and $C = C_0$, we see that for small r

$$R_0(r) = C_0 r^{\ell+1} \frac{1 - \dfrac{(\ell+2)Zr}{(\ell+1)(2\ell+3)}}{1 - \dfrac{r^2 f(r)}{2(2\ell+3)}} \quad (C_0 > 0). \tag{A.137}$$

Now using (A.137) one can find the values $R_0(r_1)$ and $R_0(r_2)$, and then employ formula (A.134) to calculate $R_0(r)$.

For $r = r_N$, in accordance with the choice of the sign of the function $R(r)$ for small r, we put $R_\infty(r_N) = (-1)^{n-\ell-1}C_\infty$ (where $C_\infty > 0$). The value $R_\infty(r_{N-1})$ is calculated by using the semiclassical approximation, which works well if the point r_N is sufficiently far from the second turning point $r = \tilde{r}_2$, i.e., $\int_{\tilde{r}_2}^{r_N} |p(r)|dr \gg 1$ (here $\tilde{r}_1$ and $\tilde{r}_2$ are the turning points, which should not be confused with the grid points r_1 and r_2). On the other hand, note that for too large values of r_N the computation of $R_\infty(r^*)$ is plagued by an accumulation of rounding errors. For this reason, in computations it is convenient to choose r_N from the approximate equality $\int\limits_{\tilde{r}_2}^{r_N} |p(r)|dr \approx 10$.

Thus, in the semiclassical approximation

$$R_\infty(r_{N-1}) = \sqrt{\left|\frac{p(r_N)}{p(r_{N-1})}\right|} \exp\left(\int\limits_{r_{N-1}}^{r_N} |p(r)|dr\right) R_\infty(r_N). \tag{A.138}$$

When the iteration scheme (A.126) is used, in addition to the values $R_0(r^*)$ and $R_\infty(r^*)$, which can be found with the aid of (A.134) and (A.135), we need to calculate $R_0'(r^*)$ and $R_\infty'(r^*)$. The value of the derivative $R'(r)$ can be recovered from the values of the function $R(r)$ by means of Taylor's formula, recalling that $R''(r) = F(r)R(r)$:

$$R(r \pm h) = R(r) \pm hR'(r) + \frac{h^2}{2}F(r)R(r) \pm \frac{h^3}{3!}[F(r)R(r)]' + \frac{h^4}{4!}[F(r)R(r)]'' + O(h^5).$$

Solving the expression for $R(r+h) - R(r-h)$ with respect to $R'(r)$, we obtain the requisite formula with accuracy $O(h^4)$:

$$R'(r) = \frac{(R(r+h) - R(r-h))/(2h) - h^2F'(r)R(r)/6}{1 + h^2F(r)/6}.$$

After the energy eigenvalues are calculated via the scheme (A.126), the radial function $R(r)$, which must obey the normalization condition $\int\limits_0^\infty R^2(r)dr = 1$ and be continuous in the point $r = r^*$, can be found in terms of the functions $R_0(r)$ and $R_\infty(r)$ calculated following the recipe

$$R(r) = D\frac{R_0(r)}{R_0(r^*)} \quad \text{for } 0 < r \le r^*, \quad R(r) = D\frac{R_\infty(r)}{R_\infty(r^*)} \quad \text{for } r \ge r^*,$$

where

$$D = \left[\frac{1}{R_0^2(r^*)}\int\limits_0^{r^*} R_0^2(r)dr + \frac{1}{R_\infty^2(r^*)}\int\limits_{r^*}^{\infty} R_\infty^2(r)dr\right]^{-1/2}.$$

A.5.5 Computational results

To calculate the function $R(r)$ by means of the difference scheme (A.133) with a constant step $\Delta r = h$ we need a sufficiently large number of points r_i. This is connected with the fact that in the region of small values of r the function $U_\ell(r) = -V(r) + \ell(\ell+1)/(2r^2)$ varies rapidly. It turns out that more reliable are the difference schemes with variable step Δr_i used in numerical integration, for example, the Runge-Kutta scheme with automatic choice of step for a given order of accuracy.

In practice, instead of r it is more convenient to take $x = \sqrt{r}$ or $x = \ln r$ as independent variable. Then one can use, as above, the Numerov scheme with Δx =const. For example, in the case $x = \sqrt{r}$, the change of variables $x = \sqrt{r}$, $R(r) = \sqrt{x}W(x)$ transforms (A.113) into an equation that does not contain the first derivative:

$$\frac{d^2W}{dx^2} = F(x)W(x), \quad F(x) = -8r\left[\varepsilon + V(r)\right] + \left.\frac{4\ell(\ell+1)+\frac{3}{4}}{r}\right|_{r=x^2}.$$

Similarly, the change of variables $x = \ln r$, $R(r) = e^{x/2}W(x)$ transforms (A.113) into the equation

$$\frac{d^2W}{dx^2} = F(x)W(x), \quad F(x) = \left(\ell + \frac{1}{2}\right)^2 - 2r^2\left[\varepsilon + V(r)\right]\Big|_{r=e^x}.$$

The integration of these equations by Numerov's method is carried out in the intervals $(x_1, x^*+\Delta x)$ and $(x^*-\Delta x, x_N)$ using the initial values $W(x_1)$ and $W(x_2)$, obtained by means of formulas (A.137), and the values $W(x_N)$ and $W(x_{N-1})$, obtained by means of the semiclassical approximation (A.138).

Computations have shown that the iteration process (A.126) is computationally efficient [155]. To illustrate the accuracy of the computations and the rate of convergence of the iterations tables A.5 – A.6 list the successive values of the energy, $\overset{(s)}{\varepsilon}_{n\ell}$, for the Thomas-Fermi potential (with $Z = 79$, $T = 1$ keV, $\rho = 0.1$ g/cm^3), obtained by Numerov's method. Results are shown for different numbers N of points of a grid that is uniform with respect to $\sqrt{r/r_0}$, $r_0 = 17.4$. The exact value of $\varepsilon_{n\ell}$ was obtained by means of a Runge-Kutta method with accuracy control (error smaller than 10^{-6}). To simplify the tables the results are displayed for $n \leq 4$ and $n = 9$. For $n > 4$, as computations have shown, the convergence rate does not get worse.

The initial value $\overset{(0)}{\varepsilon}_{n\ell} = \widetilde{\varepsilon}_{n\ell}$ was calculated in the semiclassical approximation by means of formula (A.130). The influence of the initial approximation on the rate of convergence of the iterations is illustrated by Table A.6, in which in the first column one takes $\overset{(0)}{\varepsilon}_{n\ell} = 1.5\,\widetilde{\varepsilon}_{n\ell}$, in the second one takes $\overset{(0)}{\varepsilon}_{n\ell} = 0.5\,\widetilde{\varepsilon}_{n\ell}$, and in the third, for comparison, one takes $\overset{(0)}{\varepsilon}_{n\ell} = \widetilde{\varepsilon}_{n\ell}$ (the number of grid points is $N = 500$).

Table A.5: Iterations $\overset{(s)}{\varepsilon}_{n\ell}$ for the Thomas-Fermi potential ($Z = 79$, $T = 1$ keV ρ=0.1 g/cm^3)

n	ℓ	$N = 125$	$N = 250$	$N = 500$	$\varepsilon_{n\ell}$ (Runge-K.)
1	0	−2661.98638940	−2902.33409669	−2852.65239995	−2849.17788058
		−2852.68543528	−2848.54090217	−2849.04011851	
		−2858.53546993	−2848.92042311	−2849.04118595	
		−2858.95317352	−2848.92011080	−2849.04118589	
		−2858.98357033	−2848.92011107	−2849.04118589	
		−2858.98578532	−2848.92011107		
		−2858.98594674			
		−2858.98595851			
		−2858.98595936			
		−2858.98595942			
		−2858.98595943			
2	0	−595.37770290	−628.38311020	−626.10921430	−624.43437793
		−626.58135038	−624.41279501	−624.42145950	
		−625.54395006	−624.42421953	−624.42095247	
		−625.54288033	−624.42418770	−624.42095251	
		−625.54288041	−624.42418779	−624.42095251	
		−625.54288041	−624.42418779		
2	1	−599.25361751	−604.09464834	−607.58726534	−609.04915129
		−609.22795813	−609.07079676	−609.05232485	
		−609.15574781	−609.06222871	−609.05155792	
		−609.15607315	−609.06223107	−609.05155793	
		−609.15607167	−609.06223107	−609.05155793	
		−609.15607168			
3	0	−253.66806911	−258.23657499	−259.25176745	−258.66158054
		−259.18033446	−258.66602217	−258.65927606	
		−259.02899288	−258.66463146	−258.65876493	
		−259.03004624	−258.66463557	−258.65876498	
		−259.03003834	−258.66463556	−258.65876498	
		−259.03003840	−258.66463556		
		−259.03003839			
3	1	−256.83614063	−254.16789552	−252.72990557	−252.93686050
		−253.02176538	−252.94593219	−252.93777058	
		−253.00603191	−252.94287109	−252.93767461	
		−253.00611075	−252.94287199	−252.93767461	
		−253.00611035	−252.94287199		
		−253.00611036			
3	2	−242.92187909	−244.07358263	−244.48978619	−244.18008049
		−244.19337568	−244.18113102	−244.18084667	
		−244.18867280	−244.18110016	−244.18066156	
		−244.18867866	−244.18110016	−244.18066156	
		−244.18867865			
4	0	−137.81652997	−139.71655719	−139.66788434	−139.43180899
		−139.60745024	−139.43457064	−139.43126146	
		−139.61361153	−139.43518570	−139.43079378	
		−139.61357070	−139.43518385	−139.43079379	
		−139.61357098	−139.43518386	−139.43079379	
		−139.61357097			
9	0	−21.63165056	−21.79779582	−21.80696817	−21.78817276
		−21.81917858	−21.78936719	−21.78817983	
		−21.81724104	−21.78936920	−21.78819009	
		−21.81725231	−21.78936920	−21.78819009	
		−21.81725225			
		−21.81725225			

Table A.6: Same as in Table A.5, but for different values of $\overset{(0)}{\varepsilon}_{n\ell}$ (the number of grid points is $N = 500$)

n	ℓ	$\overset{(0)}{\varepsilon}_{n\ell} = 1.5\tilde{\varepsilon}_{n\ell}$	$\overset{(0)}{\varepsilon}_{n\ell} = 0.5\tilde{\varepsilon}_{n\ell}$	$\overset{(0)}{\varepsilon}_{n\ell} = \tilde{\varepsilon}_{n\ell}$
1	0	−4278.97859992	−1426.32619997	−2852.65239995
		−2901.57021032	−2306.53497848	−2849.04011851
		−2848.87180002	−2811.28303029	−2849.04118595
		−2849.04119440	−2848.94523986	−2849.04118589
		−2849.04118589	−2849.04119117	−2849.04118589
		−2849.04118589	−2849.04118589	
			−2849.04118589	
2	0	−939.16382146	−313.05460715	−626.10921430
		−716.82873046	−583.03029883	−624.42145950
		−630.81321159	−624.21186547	−624.42095247
		−624.43131044	−624.42098171	−624.42095251
		−624.42095156	−624.42095251	−624.42095251
		−624.42095251	−624.42095251	
		−624.42095251		
2	1	−911.38089801	−303.79363267	−607.58726534
		−707.84877667	−542.48283307	−609.05232485
		−615.49440987	−609.39429220	−609.05155792
		−609.06707413	−609.05159433	−609.05155793
		−609.05155778	−609.05155793	−609.05155793
		−609.05155793	−609.05155793	
		−609.05155793		
3	0	−388.87765117	−129.62588372	−259.25176745
		−279.42283108	−304.63958151	−258.65927606
		−260.04599856	−267.38720905	−258.65876493
		−258.66189055	−258.83742839	−258.65876498
		−258.65876469	−258.65879800	−258.65876498
		−258.65876498	−258.65876498	
		−258.65876498	−258.65876498	
3	1	−379.09485836	−126.36495279	−252.72990557
		−242.77148845	−261.07316148	−252.93777058
		−253.14347454	−253.08076261	−252.93767461
		−252.93776185	−252.93771602	−252.93767461
		−252.93767461	−252.93767461	
		−252.93767461	−252.93767461	
3	2	−366.73467928	−122.24489309	−244.48978619
		−329.73651407	−247.25717717	−244.18084667
		−281.47443353	−244.20017403	−244.18066156
		−248.77557031	−244.18066221	−244.18066156
		−244.22545811	−244.18066156	
		−244.18066524	−244.18066156	
		−244.18066156		
		−244.18066156		
4	0	−209.50182650	−69.83394217	−139.66788434
		−120.09361223	−145.24226617	−139.43126146
		−142.05794293	−139.65009864	−139.43079378
		−139.48393238	−139.43119357	−139.43079379
		−139.43081592	−139.43079378	−139.43079379
		−139.43079379	−139.43079379	
		−139.43079379	−139.43079379	
9	0	−32.71045225	−10.90348408	−21.80696817
		−21.72346630	−21.37967389	−21.78817983
		−21.78806957	−21.78248161	−21.78819009
		−21.78819009	−21.78818932	−21.78819009
		−21.78819009	−21.78819009	
			−21.78819009	

The relatively large value of N is connected with the fact that the same grid was used for all values of n ($n = 1, 2, \ldots, 9$). Fixing the number of grid points to be the same N for all wave functions is convenient in the computation of the self-consistent atomic potential, electron density, and so on (in doing so, the function $R_{10}(r)$, for example, is different from zero only in the first 20 points of the grid).

As the third column of Table A.5 shows, already the first iteration practically allows us to obtain the five true digits for the energy eigenvalue. A manifestly bad initial approximation (see Table A.6) increases only slightly the number of iterations. It is clear that the phase method can be applied to the solution of many other eigenvalue problems.

A.5.6 The phase method for the Dirac equation

Dirac's system of equations reads

$$\begin{cases} \dfrac{d}{dr}F(r) + \dfrac{\kappa}{r}F(r) = \alpha\left[\varepsilon + V(r) + \dfrac{2}{\alpha^2}\right]G(r), \\[2ex] \dfrac{d}{dr}G(r) - \dfrac{\kappa}{r}G(r) = -\alpha[\varepsilon + V(r)]F(r), \end{cases} \tag{A.139}$$

where $\varepsilon = \varepsilon_{n\ell j}$ is the energy eigenvalue, $F(r) = F_{n\ell j}(r)$ and $G(r) = G_{n\ell j}(r)$ are the large and respectively the small radial wave functions times the radius r, ℓ is the orbital quantum number, $j = \ell \pm 1/2$ is the total momentum quantum number ($j > 0$) and $\kappa = -2(j - \ell)(j + 1/2)$.

The system (A.139) is easily integrated numerically. To construct an iteration scheme for finding energy levels we follow the earlier considerations for the Schrödinger equation.

Using the functions

$$\eta(r) = \alpha\left[\varepsilon + V(r) + \frac{2}{\alpha^2}\right], \quad \chi(r) = \alpha[\varepsilon + V(r)],$$

introduced in [80, 236], and the change of variables

$$F(r) = \sqrt{\eta(r)}\,P(r), \quad G(r) = \sqrt{\chi(r)}\,Q(r)$$

we obtain for $P(r)$ the equation (see (2.37))

$$\frac{d^2P}{dr^2} + k^2(r)P = 0, \tag{A.140}$$

where

$$k^2(r) = \eta\chi - r^{\kappa}\sqrt{\eta}\left(\frac{1}{r^{\kappa}\sqrt{\eta}}\right)'' = \eta\chi - \frac{3}{4}\left(\frac{\eta'}{\eta}\right)^2 + \frac{1}{2}\frac{\eta''}{\eta} - \frac{\kappa}{r}\frac{\eta'}{\eta} - \frac{\kappa(\kappa+1)}{r^2}.$$

To calculate the energy eigenvalues one can use an iteration scheme entirely similar to that for the Schrödinger equation:

$$\overset{(s+1)}{\varepsilon} = \overset{(s)}{\varepsilon} + \frac{1}{2}\frac{\varphi_0(r^*) - \varphi_\infty(r^*) + \pi\ell}{\dfrac{\partial\varphi_0(r^*)}{\partial\varepsilon} - \dfrac{\partial\varphi_\infty(r^*)}{\partial\varepsilon}}\left\{1 - \left(\frac{\varphi_0(r^*) - \varphi_\infty(r^*) + \pi\ell}{\pi n}\right)^2\right\}\Bigg|_{\varepsilon = \overset{(s)}{\varepsilon}}, \tag{A.141}$$

where

$$\varphi_0(r^*) = \begin{cases} \arctan\left[\overline{k}(r^*)\dfrac{P_0(r^*)}{P_0'(r^*)}\right] + \pi k_1, & \text{if } \dfrac{P_0(r^*)}{P_0'(r^*)} > 0; \\ \arctan\left[\overline{k}(r^*)\dfrac{P_0(r^*)}{P_0'(r^*)}\right] + \pi(k_1 + 1), & \text{if } \dfrac{P_0(r^*)}{P_0'(r^*)} < 0; \end{cases}$$

$$\varphi_\infty(r^*) = \begin{cases} \arctan\left[\overline{k}(r^*)\dfrac{P_\infty(r^*)}{P_\infty'(r^*)}\right] - \pi(k_2 + 1), & \text{if } \dfrac{P_\infty(r^*)}{P_\infty'(r^*)} > 0; \\ \arctan\left[\overline{k}(r^*)\dfrac{P_\infty(r^*)}{P_\infty'(r^*)}\right] - \pi k_2, & \text{if } \dfrac{P_\infty(r^*)}{P_\infty'(r^*)} < 0. \end{cases}$$

Here $\overline{k}^2(r) = k^2(r) - \dfrac{1}{4r^2}$, k_1 is the number of zeroes of $P_0(r)$ from $r = 0$ to $r = r^*$, and k_2 is the number of zeroes of $P_\infty(r)$ from $r = r^*$ to $r = r_0$. The position of the point r^* is determined by the condition that it must lie in the region of oscillation of the function $P(r)$, and that $\frac{d}{dr}\overline{k}(r)|_{r=r^*} \approx 0$. The values of the derivatives $\partial\varphi_0(r^*)/\partial\varepsilon$ and $\partial\varphi_\infty(r^*)/\partial\varepsilon$ are calculated by means of the formulas

$$\begin{cases} \dfrac{\partial\varphi_0(r^*)}{\partial\varepsilon} = \dfrac{\left(\dfrac{\partial}{\partial\varepsilon}\overline{k}(r)\right)P_0'(r)P_0(r) + \overline{k}(r)\displaystyle\int_0^r\left(\dfrac{\partial}{\partial\varepsilon}k^2(r)\right)P_0^2(r)\,dr}{(P_0'(r))^2 + \overline{k}^2(r)P_0^2(r)}\Bigg|_{r=r^*}, \\ \dfrac{\partial\varphi_\infty(r^*)}{\partial\varepsilon} = \dfrac{\left(\dfrac{\partial}{\partial\varepsilon}\overline{k}(r)\right)P_\infty'(r)P_\infty(r) - \overline{k}(r)\displaystyle\int_r^\infty\left(\dfrac{\partial}{\partial\varepsilon}k^2(r)\right)P_\infty^2(r)dr}{(P_\infty'(r))^2 + \overline{k}^2(r)P_\infty^2(r)}\Bigg|_{r=r^*}. \end{cases}$$

For small r the value of $P_0(r)$ can be determined by isolating the principal singularity for $r \sim 0$. Writing (A.140) in the form

$$\frac{d^2P_0}{dr^2} - \frac{\nu(\nu+1)}{r^2}P_0 = \mu(r)P_0,$$

$$\nu = -\frac{1}{2} + \sqrt{\kappa^2 - \alpha^2 Z^2}$$

and proceeding much like in the case of the Schrödinger equation (i.e., setting $P_0(r) = C_0 r^{\nu+1} y(r), y(0) = 1$), we obtain

$$P_0(r) = C_0 r^{\nu+1} \frac{1 + (\nu + 2)ar/[(2\nu + 2)(2\nu + 3)]}{1 - \mu(r)r^2/(4\nu + 6)},$$

where C_0 is the normalization constant and $a = \lim_{r \to 0} r\mu(r)$. In order to find the asymptotic behavior of $P_\infty(r)$ when $r \to \infty$ it is convenient to resort to the semiclassical approximation.

Let us emphasize that it is advisable to integrate numerically the system (A.139), and then to use $P(r) = F(r)/\sqrt{\eta(r)}$ in the iteration scheme (A.141).

Bibliography

[1] Abdallah, J. Jr., Clark R. E., X-*ray transmission calculations for an aluminium plasma*, J. Appl. Phys., **69** (1991), no. 1, 23–26.

[2] Abramov, V. A., Kogan, V. I., Lisitsa, V. S., *Radiative transfer in lines*, in: *Problemy Teorii Plazmy*, Vol. 12, M. A. Leontovich and B. B. Kadomtsev, editors, Energoatomizdat, Moscow, 1982. (Russian)

[3] Adams, M. L., Scott, H. A., Lee, R. W., et al, *Application of magnetically-broadened hydrogen line profiles to computational modeling of a plasma experiment*, J. Quantit. Spectr. Radiat. Transfer **71** (2001), nos. 2-6, 117-128.

[4] Al′tshuler, L. V., *Use of shock waves in high-pressure physics*, Uspekhi Fiz. Nauk **85** (1965), 197–258 (Russian); English transl. in Soviet Physics Uspekhi **8** (1965), no. 1, 52–91.

[5] Al′tshuler, L. V., Bakanova, A. A., Dudoladov, I. P., et al, *Shock adiabats of metals. New data, statistical analysis, and general regularities*, Prikl. Mekh. Tekhn. Fiz. (1981), no. 2, 3–34. (Russian); English transl. in J. Appl. Mech. Techn. Phys. **22** (1981), 145.

[6] Al′tshuler, L. V., Bakanova, A. A., Trunin, R. F., *Shock adiabats and null isotherms of seven metals at high pressures*, Zh. Eksp. Teoret. Fiz. **42** (1962), 91–104. (Russian)

[7] Al′tshuler, L. V., Kormer, S. B., Bakanova, A. A., et al, *Equation of state for aluminium, copper and lead in the high pressures domain*, Zh. Eksp. Teoret. Fiz. **38** (1960), no. 3, 790–798. (Russian)

[8] Al′tshuler, L. V., Krupnikov, K. K., Brazhnik, M. I., *Dynamic compressibility of metals under presssures ranging from four thousand to four million atmospheres*, Zh. Eksp. Teoret. Fiz. **34** (1958), 886–893. (Russian)

[9] Al′tshuler, L. V., Krupnikov, K. K., Ledenev, V. N., et al, *Dynamic compressibility and equations of state of iron at high pressures*, Zh. Eksp. Teoret. Fiz. **34** (1958), 874–885. (Russian)

[10] Andrews, G. E., Askey, R., Roy, R., *Special Functions*, Encyclopedia of Mathematics and its Applications, 71. Cambridge University Press, Cambridge, 1999.

[11] Audebert, P., Shepherd, R., Fournier, K. B., et al, *Time-resolved plasma spectroscopy of thin foils heated by a relativistic-intensity short-pulse laser*, Phys. Rev. E **66** (2002), no. 6, 066412.

[12] Avrorin, E. N., Vodolaga, V. K., Simonenko, V. A., Fortov, V. E., *Strong Shock Waves and Extreme States of Matter*, IVTAN, Moscow, 1990. (Russian)

[13] Avrorin, E. N., Vodolaga, B. K., Simonenko, V. A., Fortov, V. E., et al, *Report on the International Conference on High Pressure Science and Technology*, Padeborn University, Germany, 1989.

[14] Avrorin, E. N., Vodolaga, V. K., Voloshin, N. P., et al, *Experimental study of the effect of electron shell on shock adiabats of condensed materials*, Zh. Eksp. Teoret. Fiz. **93** (1987), no. 2(8), 613–626 (Russian); English transl. Soviet Physics – JETP **66** (1987), no. 2, 347–354.

[15] Bakanova, A. A., Dudoladov, I. P., Trunin, R. F., *Shock-wave compression of alkaline metals*, Fiz. Tverdogo Tela **7** (1965), no. 7, 1615–1622. (Russian)

[16] Bakhvalov, N. S., *Numerical Methods: Analysis, Algebra, Ordinary Differential Equations*, Translated from the Russian, Mir Publishers, Moscow, 1977.

[17] Balescu, R., *Equilibrium and Nonequilibrium Statistical Mechanics*, John Wiley, New York, 1975.

[18] Baranger, M., *Spectral line broadening in plasmas in atomic and molecular processes*, in: *Atomic and molecular processes*, D. R. Bates, ed., Academic Press, New York, 1962.

[19] Bar-Shalom, A., Oreg, J, Goldstein, W. H., et al, *Super-transition-arrays: a model for the spectral analysis of hot, dense plasma*, Phys. Rev. A **40** (1989), no. 6, 3183–3193.

[20] Bar-Shalom, A., Oreg, J., Klapisch, M., *Collisional radiative model for heavy atoms in hot non-local-thermodynamical-equilibrium plasmas*, Phys. Rev. E **56** (1997), no. 1, R70–R73.

[21] Basko, M. M., *Equation of state for metals in the mean ion approximation*, Teplofizika Vysokikh Temperatur **23** (1985), no. 3, 483–491 (Russian); English transl. High Temp. **23** (1985), no. 3, 388-396.

[22] Bauche, J., Bauche-Arnoult, C., Klapisch, M., *Transition arrays in the spectra of ionized atoms. Review*, in: *Advances in Atomic and Molecular Physics*, Vol 23, p. 131, D. Bates and B. Bederson, eds. Academic Press, 1987.

[23] Bauche-Arnoult, C., Bauche, J., Klapisch, M., *Variance of the distributions of energy levels and of the transitions arrays in atomic spectra*, I, Phys. Rev. A **20** (1979), no. 6, 2424–2439.

[24] Bauche-Arnoult, C., Bauche, J., Klapisch, M., *Variance of the distributions of energy levels and of the transitions arrays in atomic spectra*, II. *Configurations with more than two open subshells*, Phys. Rev. A **25** (1982), no. 5, 2641–2646.

[25] Bauche-Arnoult, C., Bauche, J., Klapisch, M., *Variance of the distributions of energy levels and of the transitions arrays in atomic spectra*, III. *Case of spin-orbit-split arrays*, Phys. Rev. A **31** (1985), no. 4, 2248–2259.

[26] Bauche-Arnoult, C., Bauche, J., Klapisch, M., *Simulation of atomic transition arrays for opacity calculations*, Phys. Rev. A **44** (1991), no. 9, 5707–5714.

[27] Benattar, R., Zakharov, S. V., Nikiforov, A. F., et al, *Influence of magneto-hydrodynamic Rayleigh-Taylor instability on radiation of imploded heavy ion plasmas*, Phys. Plasmas **6** (1999), no. 1, 175–187.

[28] Bethe, H. A., Salpeter, E. E., *Quantum Mechanics of One- and Two-Electron Atoms*, Springer-Verlag, Berlin; Academic Press, New York, 1957.

[29] Bethe, H. A., *Intermediate Quantum Mechanics*, W. A. Benjamin, Inc., New York–Amsterdam, 1964.

[30] Bethe, H. A., Jackiw, R. W., *Intermediate Quantum Mechanics*, 3rd edition, Benjamin/Cummings Publishing Co., Menlo Park, CA, 1986.

[31] Biberman, L. M., Vorob′ev, V. S., Yakubov, I. T., *Kinetics of Nonequilibrium Low-Temperature Plasma*, Translated from the Russian, Kluwer Academic/Plenum Publishers, Dordrect, 1987.

[32] Blenski, T., Cichocki, B., *Density-functional approach to the absorption bands in a dense, partially ionized plasma*, Phys. Rev. A **41** (1990), no. 12, 6973–6981.

[33] Blenski, T., Ligou, J., *Average atom calculations of radiation opacities for gold at high density and temperature*, Journal de Physique, Colloque C7 **49** (1988), suppl. aux no. 12, 259–265.

[34] Blenski, T., Ligou, J., *Radiation opacities for high Z elements*, in: *Laser Interaction with Matter*, G. Velarde, E. Minguez and J. M. Perlado, eds., World Scientific, Singapore, 1989.

[35] Born, M. *Atomic Physics*, Translation from the German original, 8th edition, Blackie, London, 1969.

[36] Bowen, C., NLTE *emissivities via an ionisation temperature*, J. Quantit. Spectr. Radiat. Transfer **71** (2001), nos. 2–6, 201–214.

[37] Braun, M. A., Gurchumelya, A. D., Safronova, U. I., *Relativistic Theory of the Atom*, Nauka, Moscow, 1984. (Russian)

[38] Broyles, A. A., *Stark fields from ions in a plasma*, Phys. Rev. **100** (1955), no. 4, 1181–1187.

[39] Bruneau, J., Decoster, A., Desenne, D., et al, *Time-resolved study of hot dense germanium by L-shell absorption spectroscopy*, Phys. Rev. A **44** (1991), no. 2, R832–R835.

[40] Bureeva, L. A., Lisitsa, V. S., *The Excited Atom*, IzdAT, Moscow, 1997. (Russian)

[41] Bushman, A. V., Gryaznov, V. K., Kanel, G. I., et al, *Thermodynamic properties of materials at high pressures and temperatures*, Preprint, Institute of Chemical Physics, Russian Academy of Sciences, Chernogolovka, 1983. (Russian)

[42] Bushman, A. V., Fortov, V. E., *Models of equation of state for matter*, Uspekhi Fiz. Nauk **140** (1983), 177–232. (Russian); English transl. Soviet Physics Uspekhi **26** (1983), 465–520.

[43] Bushman, A. V., Lomonosov, I. V., Fortov, V. E., *Equations of State of Metals at High Energy Densities*, Institute of Chemical Physics, Russian Academy of Sciences, Chernogolovka, 198 pages, 1992. (Russian)

[44] Busquet, M., *Radiation-dependent ionization model for laser-created plasmas*, Phys. Fluids B **5** (1993), no. 11, 4191–4206.

[45] Carson, T. R., Mayers, D. F., Stibbs, D. W. N. *The calculation of stellar radiative opacity*, Month. Notes Royal Astron. Soc., **140**, (1968), no. 4, 483–536.

[46] Cody, W. J., Thacher, H. C., Jr., *Rational Tschebyshev approximation for Fermi–Dirac integrals of orders* $-1/2$, $1/2$ *and* $3/2$, Math. of Computation **21** (1967), no. 97, 30–40.

[47] Colombant, D., Klapisch, M., Bar-Shalom, A., *Increase in Rosseland mean opacity for inertial fusion hohlraum walls*, Phys. Rev. E **57** (1998), no. 3, 3411–3416.

[48] Condon, E. U., Shortley, G. H., *The Theory of Atomic Spectra*, Cambridge University Press, Cambridge, 1953.

[49] Cowan, R. D., *The Theory of Atomic Structure and Spectra*, University of California Press, Berkeley, 1981.

[50] Cox A. N., *Stellar absorption coefficients and opacities*, in: *Stars and Stellar Systems*, Vol. 8, p. 195, L. H. Allen and D. B. McLaughlin, eds., University of Chicago Press, Chicago, 1965.

[51] Crowley, B. J. B., *Average-atom quantum-statistical cell model for hot plasma in local thermodynamic equilibrium over a wide range of densities*, Phys. Rev. A **41** (1990), no. 4, 2179–2191.

[52] Davydov, A. S., *Quantum Mechanics*, Translated from the Russian, 2nd edition, Pergamon Press, Oxford, New York, 1985.

[53] Demkov, Yu. N., Monozon, B. S., Ostrovskii, V. N., *Energy levels of the hydrogen atom in crossing electric and magnetic fields*, Zh. Eksp. Teoret. Fiz., **57** (1969), p. 1431. (Russian); English transl. Engl. transl. 1969 Soviet Physics – JETP **30** (1969), 775–776.

[54] Dirac, P. A. M., *Note on exchange phenomena in the Thomas atom*, Proc. Cambridge. Phil. Soc. **26** (1930), 376–385.

[55] Dragalov, V. V., Nikiforov, A. F, Novikov, V. G., Uvarov, V. B., *Statistical method for calculating the absorption of photons in dense high-temperature plasmas*, Fizika Plazmy **16** (1990), no. 1, 77–85 (Russian); English transl. Soviet J. Plasma Phys **16** (1990), no. 1, 44–49.

[56] Dragalov, V. V., Novikov, V. G., *Distribution of spectral lines in plasma with respect to fluctuations of the occupation numbers*, Teplofizika Vysokikh Temperatur **24** (1987), no. 6, 1057–1061 (Russian); English transl. High Temp. **25** (1987), no. 6, 762–766.

[57] Dragalov, V. V., Novikov, V. G., *Approximate ion-configuration correction in calculating photoionization cross sections for a dense hot plasma*, Teplofizika Vysokikh Temperatur **27** (1989), no. 2, 214–219 (Russian); English transl. High Temp. **27** (1989), no. 2, 161–165.

[58] Eliseev, G. M., Klinishov, G. E., *The equation of state of solid substances and its spline approximation*, Preprint, Keldysh Institute of Applied Mathematics, Russian Academy of Sciences, Moscow, 1982, 24 p. (Russian)

[59] Erdélyi, A., Magnus, W., Oberhettinger, F., Tricomi, F. G., *Higher Transcendental Functions*. Vols. 1–3, Based, in part, on notes left by Harry Bateman, McGraw-Hill, New York–Toronto–London, 1953, 1955.

[60] Fermi E., *A statistical method for the determination of some atomic properties and the application of this method to the theory of the periodic system of elements*, Z. für Phys., **48** (1928), 73–79.

[61] Feynman, R., Metropolis, N., Teller, E., *Equations of state of elements based on the generalized Fermi-Thomas theory*, Phys. Rev. **75** (1949), no. 10, 1561–1573.

[62] Fischer, C. F., *The Hartree-Fock Method for Atoms: A Numerical Approach*, John Wiley, New York, 1977.

[63] Flügge, S., *Practical Quantum Mechanics*, Vols. I, II, Translated from the 1947 German edition, Reprint of the 1994 English edition, Classics in Mathematics, Springer-Verlag, Berlin, 1999.

[64] Fock, V. A., *Fundamentals of Quantum Mechanics*, Translated from the Russian, Mir Publishers, Moscow, 1978.

[65] Fortov, V. E., Iakubov, I. T., *The Physics of Nonideal Plasma*, translation from the Russian 1994 original, World Scientific, Singapore, 1999.

[66] Foster, J. M., Hoarty, D. J., Smith, C. C., et al, *L-shell absorption spectrum of an open-M-shell germanium plasma: Comparison of experimental data with a detailed configuration-accounting calculation*, Phys. Rev. Lett. **67** (1991), no. 23, 3255–3258.

[67] Frank-Kamenetskii, D. A., *Physical Processes in Stellar Interiors*, Fizmatgiz, Moscow, 1959. (Russian); English translation, Israel Program for Scientific Translations, Jerusalem, 1962 (available from the Office of Technical Services, U.S. Dept. of Commerce, Washington)

[68] Glauber, R. J., *Some notes on multiple-boson processes*, Phys. Rev. **84** (1951), no. 3, 395–400.

[69] Glenzer, S. H., Fournier, K. B., Wilson B. G., et. al, *Ionization balance in inertial confinement fusion hohlraum plasmas*, J. Quantit. Spectr. Radiat. Transfer **71** (2001), nos. 2–7, 355–363.

[70] Godunov, S. K., Ryabenkii, V. S., *The Theory of Difference Schemes – An Introduction to the Underlying Theory*, Translated from the Russian, North Holland, Ansterdam, 1987.

[71] Goldberg, A., Rozsnyai, B. F., Thompson, P., *Intermediate-coupling calculation of atomic spectra from hot plasma*, Phys. Rev. A **34** (1986), no. 1, 421–427.

[72] Goldin, V. Ya, *A quasi-diffusion method of solving the kinetic equation*, Zh. Vychisl. Mat. i Mat. Fiz. (1964), no. 4, 1078–1087 (Russian); English transl. U.S.S.R. Comput. Math. Math. Phys. **4** (1964), no.6, 136–149.

[73] Gombas, P., *Die Statistische Theorie des Atoms und ihre Anwendungen*, Springer, Berlin, 1949.

[74] Grant, I. P., *Relativistic calculation of atomic structures*, Advances in Physics **19** (1970), 747–811.

[75] Griem, H. R. *Spectral Line Broadening by Plasmas*, Academic Press, New York, 1974.

[76] Gryaznov, V. K., Fortov, V. E., Iosilevskii, I. L., et al, *Thermophysical Properties of Working Media in a Gas-Phase Nuclear Reactor*, Atomizdat, Moscow, 1980. (Russian)

[77] Gryaznov, V. K., Iosilevskii, I. L., Fortov, V. E., *Thermodynamics of a highly compressed plasma in the megabar range*, Pis′ma Zh. Tekh. Fiz. **8** (1982), no. 22, 1378–1380 (Russian); English transl. Sov. Tekh. Phys. Lett. (1982), no. 8, 592–593.

[78] Gunnarsson, O., Johnson, M., Lundqvist, B. I., *Descriptions of exchange and correlation effects in inhomogeneous electron systems*, Phys. Rev. B **20** (1979), no. 8, 3136–3164.

[79] Gupta, U., Rajagopal, A. K., *Inhomogeneous electron gas at nonzero temperatures: Exchange effects*, Phys. Rev. A **21** (1980), no. 6, 2064–2068.

[80] Hagelstein, P. L., Jung, R. K., *Relativistic distorted-wave calculations of electron collision cross-sections and rate coefficients for* Ne-*like ions*, Atomic Data Nucl. Data Tables. **37** (1987), p. 121.

[81] Hansen, J. P., *Statistical mechanics of dense ionized matter*. I. *Equilibrium properties of the classical one-component plasma*, Phys. Rev. A **8** (1973), no. 6, 3096–3109.

[82] Hartree, D. R., *The wave mechanics of an atom with a non-Coulomb central field.* II. *Some results and discussion*, Proc. Cambridge Philos. Soc. **24** (1927), 111–132.

[83] Hartree, D. R., *The Calculation of Atomic Structures*, John Wiley, New York, 1957.

[84] Heading, D. J., Wark, J. S., Lee, R. W., et al, *Comparison of the semiclassical and modified semiempirical method of spectral calculation*, Phys. Rev. E **56** (1997), no. 1, 936–946.

[85] Heitler, W, Ma, S. T., *Quantum theory of radiation damping for discrete states*, Proceed. Royal Irish Academy **52** (1949), p. 109.

[86] Heitler, W., *The Quantum Theory of Radiation*, 3rd edition, Dover, New York, 1984.

[87] Hohenberg, P., Kohn, W., *Inhomogeneous electron gas*, Phys. Rev. **136** (1964), no. 3B, 864–870.

[88] Huebner, W. F., Merts, A. L., Magee, N. H., Argo, M. F., *Astrophysical Opacity Library*, Los Alamos National Laboratory Manual LA-6760-M, New Mexico, 1977.

[89] Huebner, W. F., *Atomic and radiative processes in the solar interior*, in: *Physics of the Sun*, Vol. 1, P. A. Sturrock et al, eds., Reidel, Dordrecht, 1986.

[90] Ichimaru, S., *Basic Principles of Plasma Physics: A Statistical Approach*, 2nd print, Benjamin/Cummings, Reading, MA, 1980.

[91] Iglesias, C. A., Hooper, C. F., DeWitt, H. E., *Some approximate microfield distributions for multiply ionized plasmas: a critique*, Phys. Rev. A. **28** (1983), no. 1, 361–369.

[92] Iglesias, C. A., Rogers, F. J., *Updated* OPAL *opacities* Astrophys. J., **464** (1996), p. 943.

[93] Iglesias, C. A., Rogers, F. J., Wilson, B. G., *Spin-orbit interaction effects on the Rosseland mean opacity*, Astrophys. J. **397** (1992), p. 717.

[94] Ilyin, V. P., Poznyak, I. G., *Linear Algebra*, translated from the Russian, Mir, Moscow, 1986.

[95] Johnson, W. R., *Pressure in the average-atom model*, to appear; see http://www.nd.edu/ johnson/Publications/press.pdf

[96] Jones, A. H., Isbell, W. H., Maiden, C. J., *Measurements of the very-high-pressure properties of materials using a light-gas gun*, J. Appl. Phys. **37** (1966), 3493-3499.

[97] Kalitkin, N. N., *Quasizone equation of state*, Mat. Modelirovnie **1** (1989), no. 2, 64–108. (Russian)

[98] Kalitkin, N. N., Kuz′mina, L. V., *Tables of thermodynamic functions of matter at high energy concentrations*, Preprint, Keldysh Institute of Applied Mathematics, Russian Academy of Sciences, Moscow, 1975, 73 p. (Russian)

[99] Kalitkin, N. N., Kuz′mina, L. V., *Quantum-statistical equation of state*, Fizika Plazmy **2** (1976), no. 5, 858–868. (Russian)

[100] Kalitkin, N. N., Kuz′mina, L. V., *Comparison of atom models in the plasma domain*, Preprint, Keldysh Institute of Applied Mathematics, Russian Academy of Sciences, Moscow, 1988, 13 p. (Russian)

[101] Kalitkin, N. N., Kuz′mina, L. V., *Quantum-statistical shock adiabats of porous substances*, Matemat. Modelirovnie **10** (1998), no. 7, 111-123. (Russian)

[102] Karazija, R., *Introduction to the Theory of* X-*Ray and Electronic Spectra of Free Atoms*, translated from the Russian by W. Robert Welsh, Plenum Press, New York, 1996.

[103] Kirzhnits, D. A., *Quantum corections to the Thomas-Fermi equation*, Zh. Eksp. Teoret. Fiz. **32** (1957), no. 1, 115–123 (Russian); English transl. Sov. Phys. JETP **5** (1957), no. 1, 64–71.

[104] Kirzhnits, D. A., *On the limits of applicability of the semiclassical equation of state*, Zh. Eksp. Teoret. Fiz. **35** (1959), no. 6, 1545–1557 (Russian); English transl. Sov. Phys. – JETP **8** (1959), no. 6, 1081–1089.

[105] Kirzhnits, D. A., *Field Theoretical Methods in Many-Body Systems*, Translated from the Russian, Pergamon Press, Oxford, New York, 1967.

[106] Kirzhnits, D. A., Shpatakovskaya, G. V., *Atomic structure oscillation effects*, Zh. Eksp. Teoret. Fiz. **62** (1972), no. 6, 2082–2096 (Russian); English transl. Sov. Phys. – JETP **35** (1972), no. 6, 1088–1094.

[107] Kirzhnits, D. A., Shpatakovskaya, G. V., *Oscillations of the elastic parameters of compressed matter*, Zh. Eksp. Teoret. Fiz. **66** (1974), no. 5, 1828–1843 (Russian); English transl. Sov. Phys. – JETP **39** (1974), 1121.

[108] Kirzhnits, D. A., Shpatakovskaya, G. V., *Wide-range equation of state of matter based on an improved statistical model*, Preprint, P. N. Lebedev Physics Institute of the Russian Academy of Sciences, Moscow, 1998, 46 p. (Russian)

[109] Kittel, C., *Elementary Statistical Physics*, John Wiley, New York, 1958.

[110] Kittel, C., *Introduction to Solid State Physics*, 7th edition, John Wiley, New York, 1996.

[111] Kittel, C., Kroemer, H., *Thermal Physics*, W. H. Freeman, San Francisco, 1980.

[112] Kivel, B., Bloom, S., Margenau, H., *Electron impact broadening of spectral lines*, Phys. Rev. **98** (1955), no. 2, 495–514.

[113] Kohn, W., Sham, L. J., *Selfconsistent equations including exchange and correlation effects*, Phys. Rev. A **140** (1965), no. 4, 1133–1141.

[114] Kogan, V. I., Lisitsa, V. S., Sholin, G. V., *Broadening of spectral lines in plasma*, in: Voprosy Fiziki Plazmy, no. 13, B. V. Kadomtsev, ed., Energoatomizdat, Moscow, 1984. (Russian)

[115] Kompaneyets, A. S., *A Course of Theoretical Physics*, Vol. 2 *Statistical Laws*, Translated from the Russian, Mir Publishers, Moscow, 1978.

[116] Kopyshev, V. P., *On the thermodynamics of nuclei of a monoatomic substance*, Preprint, Keldysh Institute of Applied Mathematics, Russian Academy of Sciences, Moscow, 1984, 18 p. (Russian)

[117] Kormer, S. V., Funtikov, A. I., Urlin, V. D., Kolesnikova, A. N., *Dynamic compression of porous metals and the equation of state with variable specific heat at high tempreatures*, Zh. Eksp. Teoret. Fiz. **42** (1962), no. 3, 686–702 (Russian); English transl. Sov. Phys. JETP **15** (1962), no. 3, 477–488.

[118] Korneichuk, N. P., *Splines in Approximation Theory*, Nauka, Moscow, 1984. (Russian)

[119] Landau, L. D., Lifshits, E. M., *Statistical Physics*, Translated from the Russian, 3rd edition, Butterworth-Heinemann, Oxford, 1980.

[120] Landau, L. D., Lifshits, E. M., *Quantum Mechanics: Non-relativistic Theory*, Translated from the Russian, 3rd edition, Butterworth-Heinemann, Oxford, 1997.

[121] Leonas, V. B., Rodionov, I. D., *Perspectives of the microscopic approach in the investigation of extreme states of matter*, Preprint, Keldysh Institute of Applied Mathematics, Russian Academy of Sciences, Moscow, 1984, 21 p. (Russian)

[122] Latter, R, *Temperature behavior of the Thomas-Fermi statistical model for atoms*, Phys. Rev. **99** (1955), no. 6, 1854–1870.

[123] Lifshitz, E. M., Pitaevskii, L. P., *Relativistic Quantum Theory*, translated from the Russian, Pergamon Press, Oxford, 1974.

[124] Lotz, W., *Electron-impact ionization cross-sections for atoms up to* $Z = 108$, Z. Phys. **232** (1970), 101–107.

[125] Malkin, O. A., *Relaxation Processes in a Gas*, Atomizdat, Moscow, 1971. (Russian)

[126] Mancini, R. C., Hooper Jr., C. F, Coldwell, R. L., *Modeling of absorption spectra in dense argon plasma*, J. Quantit. Spectr. Radiat. Transfer **51** (1994), no. 1/2, 201–210.

[127] March, N. H., *Self-consistent Fields in Atoms (Hartree and Thomas–Fermi Atoms)*, Pergamon Press, Oxford, 1975.

[128] Mathews, J., Walker R. L., *Mathematical Methods of Physics*, 2nd edition, W. A. Benjamin, Inc., New York–Amsterdam, 1964.

[129] Matzubara, T., *A new approach to quantum-statistical mechanics*, Progress Theor. Phys. **14** (1955), no. 4, 351–378.

[130] Mayer, H. L., *Methods of Opacity Calculations*, Los Alamos Scientific Lab. Rep. LA-647 (AECD-1870) LASL, New Mexico, 1947.

[131] McQueen, R. G., Marsh, S. P., *Equation of state for nineteen metallic elements from shock wave measurements to two megabars*, J. Appl. Phys., **31**, (1960), 1253–1269.

[132] Merdji, H., Mißalla, T., Blenski, T., et al, *Absorption spectrocopy of a radiatively heated samarium plasma*, Phys. Rev. E **57** (1998), no. 1, 1042–1046.

[133] Mermin, N. D., *Thermal properties of the inhomogeneous electron gas*, Phys. Rev. A **137** (1965), no. 5, 1441–1443.

[134] Mihalas, D., *Stellar Atmospheres*, 2nd edition, W. H. Freeman, San Francisco, 1978.

[135] More, R. M., *Electronic energy levels in dense plasmas*, J. Quantit. Spectr. Radiat. Transfer **27** (1982), no. 3, 345–357.

[136] More, R. M., Warren, K. H., Young, D. A., Zimmerman, G. B., *A new quotidian equation of state* (QEOS) *for hot dense matter*, Phys. Fluids **31** (1988), no. 10, 3059–3078.

[137] Morse, P. M., Feshbach, H., *Methods of Theoretical Physics*, McGraw-Hill, New York–Toronto–London, 1953.

[138] Moszkowski, S. A., *On the energy distribution of terms and line arrays in atomic spectra*, Progress Theor. Phys. **28** (1962), no. 1, 1–23.

[139] Nardi, E., Zinamon, Z., *Radiative opacity of high-temperature and high-density gold*, Phys. Rev. A **20** (1979), no. 3, 1197–1200.

[140] Nguyen-Hoe, Drawin, H. W., Herman, L., *Effet d'un champ magnetique uniforme sur les profils des raies de l'hydrogene*, J. Quantit. Spectr. Radiat. Transfer **7** (1967), no. 3, 429–474.

[141] Nikiforov, A. F., *Numerical Methods for Solving some Problems of Quantum Mechanics*, Mosk. Gosud. Univ., Moscow, 1981. (Russian)

[142] Nikiforov, A. F., Novikov, V. G., Solomyannaya, A. D., *Analytical wave functions in self-consistent field models for high-temperature plasma*, Laser and Particle Beams **14** (1996), no. 4, 765–779.

[143] Nikiforov, A. F., Novikov, V. G., Solomyannaya, A. D., *Self-consistent hydrogen-like model of the average atom for matter with given temperature and density*, Teplofizika Vysokikh Temperatur **34** (1996), no. 2, 220–233 (Russian); English transl. High Temp. **34** (1996), no. 2, 214–227.

[144] Nikiforov, A. F., Novikov, V. G., Trukhanov, S. K., Uvarov, V. B., *Computations of the equation of state of aluminum in the high-temepratures region based on the Hartree-Fock-Slater model*, in: *Problems of Atomic Science and Technology. Methods and Codes for the Numerical Solution of Problems of Mathematical Physics*, **3** (1990), 62–74. (Russian).

[145] Nikiforov, A. F., Novikov, V. G., Uvarov, V. B., *A modified Hartree-Fock-Slater model and its application in deriving equations of state of matter in the high-temepratures region*, in: *Mathematical Modeling. Physico-Chemical Properties of Matter*, A. A. Samarskii and N. N. Kalitkin, eds., Nauka, Moscow, 1989. (Russian)

[146] Nikiforov, A. F., Suslov, S. K., Uvarov, V. B., *Classical Orthogonal Polynomials of a Discrete Variable*, Translated from the Russian, Springer Verlag, Berlin, 1991.

[147] Nikiforov, A. F., Uvarov, V. B., *Heat conduction of heavy substances in the high-temperature range*, Reports of the Department of Applied Mathematics of the V. A. Steklov Institute, Moscow, 1965. (Russian)

[148] Nikiforov, A. F., Uvarov, V. B., *Calculation of the Thomas-Fermi potential for a mixture of elements*, Zh. Vychisl. Mat. i Matem. Fiz. **9** (1969), no. 3, 715–720. (Russian)

[149] Nikiforov, A. F., Uvarov, V. B., *Light absorption coefficients in plasma*, Preprint, Keldysh Institute of Applied Mathematics, Russian Academy of Sciences, Moscow, 1969, 26 p. (Russian)

[150] Nikiforov, A. F., Uvarov, V. B., *Computation of opacities of stars with absorption of light in spectral lines accounted for*, Dokl. Akad. Nauk SSSR **191** (1970), no. 1, 47–49. (Russian)

[151] Nikiforov, A. F., Uvarov, V. B., *Description of the state of matter in the high-temperatures region based on the self-consistent field equations*, in: *Numerical Methods of the Mechanics of Continua*, Vol. 4, Moscow, 1973. (Russian)

[152] Nikiforov, A. F., Novikov, V. G., Uvarov, V. B., *A modified Hartree-Fock-Slater model for matter with given temperature and density*, in: *Problems of Atomic Science and Technology. Methods and Codes for the Numerical Solution of Problems of Mathematical Physics*, **4**(**6**) (1979), 16–26. (Russian).

[153] Nikiforov, A. F., Novikov, V. G., Uvarov, V. B., *Solving the Thomas-Fermi equations for a mixture of subtances by the double-sweep method*, in: *Problems of Atomic Science and Technology. Methods and Codes for the Numerical Solution of Problems of Mathematical Physics*, **1**(**9**) (1982), 12–17. (Russian).

[154] Nikiforov, A. F., Uvarov, V. B., *Special Functions of Mathematical Physics. A Unified Introduction with Applications*, Translated from the Russian, Birkhäuser Verlag, Basel, 1988.

[155] Nikiforov, A. F., Novikov, V. G., *Application of the phase method to the calculation of the energy eigenvalues*, Matemat. Modelirovanie **10** (1998), no. 10, 64–78. (Russian)

[156] Nikitin, A. A., Rudzikas, E. B., *Principles of the Theory of Spectra of Atoms and ions*, Nauka, Moscow, 1983. (Russian)

[157] Novikov, V. G., *Self-consistent field models with semiclassical approximation*, Preprint, Keldysh Institute of Applied Mathematics, Russian Academy of Sciences, Moscow, 1984, 19 p. (Russian)

[158] Novikov, V. G., *Shock compression of lithium, aluminum and iron according to the MFHS model*, Preprint, Keldysh Institute of Applied Mathematics, Russian Academy of Sciences, Moscow, 1985, 27 p. (Russian)

[159] Novikov, V. G., *Inclusion of individual states of ions in the modified Hartree-Fock-Slater model*, Teplofizika Vysokikh Temperatur **30** (1993), no. 4, 701–708. (Russian)

[160] Novikov, V. G., *A method for computing self-consistent atomic potentials for a mixture of substances in the relativistic Hartree-Fock-Slater approximation*, Zh. Vychisl. Mat. Matem. Fiz. **37** (1997), no.1, 107–116 (Russian); English transl. Comput. Math. Math. Phys. **37** (1997), no. 1, 104–112.

[161] Novikov, V. G., Vorob′ev, V. S., D′yachkov, L. G., Nikiforov, A. F., *Effect of a magnetic field on the radiation emitted by a nonequilibrium hydrogen and deuterium plasma*, Zh. Ekp. Teor. Fiz **119** (2001), no. 3, 509–523 (Russian); English transl. J. Exp. Theor Phys. **92** (2001), 441.

[162] Novikov, V. G., Nikiforov, A. F., Uvarov, V. B., Dragalov, V. V., *Photon absorption in hot plasma*, Preprint, Keldysh Institute of Applied Mathematics, Russian Academy of Sciences, Moscow, 1992, 22 p. (Russian)

[163] Novikov, V. G., Solomyannaya, A. D., *Spectral characteristics of plasma consistent with radiation*, Teplofizika Vysokikh Temperatur **36** (1998), no. 6, 858–864. (Russian)

[164] Novikov, V. G, Zakharov, S. V., *Modeling of non-equilibrium radiating tungsten liners*, J. Quantit. Spectr. Radiat. Transfer **81** (2003), nos. 1–4, 339–354.

[165] Olver, F. W. J., *Asymptotics and Special Functions*, Reprint of the 1974 original, A. K. Peters, Wellesley, MA, 1997.

[166] O'Sullivan G., Carroll P. K., 4d-4f *emission resonances in laser-produced plasmas*, J. Opt. Soc. Amer. **71** (1981), no. 3, 227–230.

[167] Palmer, R. G., Weeks, J. D., *Exact solution of the mean spherical model for charged hard spheres in a uniform neutralizing background*, J. Chem. Phys. **58** (1973), no. 10, 4171–4174.

[168] Parr, R. G., Yang, W., *Density-Functional Theory of Atoms and Molecules*, Oxford University Press, New York; Clarendon Press, Oxford, 1989.

[169] Partlo, W. N., Fomenkov, I. V., Ness, R., Oliver, R., *Progress toward use of a dense plamsa focus device as a light source for production* EUV *lithography*, Proceedings of SPIE, Vol. 4343–25, 2001.

[170] Perrot, F., *Hartree-Fock study of the ground-state energy and band structure of metallic copper*, Phys. Rev. B **11** (1975), no. 12, 4872–4884.

[171] Perrot, F., Dharma-wardana, M. W. C., *Hydrogen plasmas beyond density-functional theory: dynamic correlations and onset of localization*, Phys. Rev. A **29** (1984), no. 3, 1378–1390.

[172] Perry, T. S., Springer, P. T., Fields, D. F., et al, *Absorption experiments on X-ray-heated mid-Z constrained samples*, Phys. Rev. E **54** (1996), no. 5, 5617–5631.

[173] Pigarov, A. Yu., Terry, J. L., Lipschultz, B., *Study of the discrete-to-continuum transition in a Balmer spectrum from Alcator* C-Mod *divertor plasmas*, Plasma Phys. Control. Fusion **40** (1998), 1–18.

[174] Pomraning, G. C., *The Equations of Radiation Hydrodynamics*, Pergamon Press, Oxford, 1973.

[175] Post, D. E., Jensen, R. V., Tarter, C. B., et al., *Steady state radiative cooling rates for low density high temperature plasmas*, Atomic Data Nucl. Data Tables **20** (1977), 397–439.

[176] Rajagopal, A. K., *Theory of inhomogeneous electron systems: spin-density-functional formalism*, in: *Advances in Chemical Physics*, Vol. 41, pp. 76–136, G. I. Prigogine and S. A. Rice, eds., J. Willey, New York, 1980.

[177] Ralchenko, Yu. V., Griem, H. R., Bray, I., Fursa, D. V., *Quantum-mechanical calculation of Stark widths of* Ne VII $n = 3$, $\Delta n = 0$ *transitions*, Phys. Rev. A **59** (1999), no. 3, 1890–1895.

[178] Ralchenko, Yu.V., Griem, H.R., Bray I., Fursa, D.V. *Electron collisional broadening of* $2s3s - 2s3p$ *lines in* Be-*like ions*, J. Quantit. Spectr. Radiat. Transfer **71** (2001), nos. 2-6, 595–607.

[179] Rickert, A., *Review of the Third International Opacity Workshop and Code Comparison Study*, J. Quantit. Spectr. Radiat. Transfer, **54** (1995), nos. 1–2, 325–332.

[180] Rogers, F. J., Wilson, B. G., Iglesias, C. A., *Parametric potential method for generating atomic data*, Phys. Rev. A **38** (1988), no. 10, 5007–5020.

[181] Rogers, F. J., Iglesias, C. A., *Rosseland mean opacities for variable compositions*, Astrophys. J. **401** (1992), no. 1, 361–366.

[182] Rogers, F. J., Iglesias, C. A., *The* OPAL *opacity code: new results*, UCRL-JC-119032, in: Proceedings of the Astron. Soc. of the Pacific Conf. Ser., Hague, The Netherlands, 1994.

[183] Romanov, G. S., Stanchits, L. K., *On the computation of thermodynamic parameters using the full system of Saha equations*, Doklady Akad. Nauk. Biel. SSR **15** (1971), no. 3. (Russian)

[184] Romanov, G. S., Stankevich, Yu. A., Stanchits, L. K., Stepanov, K. L., *Thermodynamic and optical properties of gases in a wide range of parameters*, Int. J. Heat Mass Transfer **38** (1995), no. 3, 545–556.

[185] Rose, S. J., *Calculations of the radiative opacity of laser-produced plasmas*, J. Phys. B **25** (1992), no. 7, 1667–1681.

[186] Ross, M., *Matter under extreme conditions of temperature and pressure*, Rep. Prog. Phys. **48** (1985), 1–52.

[187] Ross, M., Young, D. A., *Theory of the equation of state at high pressure*, Ann. Rev. Phys. Chem. **44** (1993), 61–87.

[188] Rozsnyai, B. F., *Relativistic Hartree–Fock–Slater calculations for arbitrary temperature and matter density*, Phys. Rev. A **5** (1972), no. 3, 1137–1149.

[189] Rozsnyai, B. F., *Spectral lines in hot dense matter*, J. Quantit. Spectr. Radiat. Transfer **17** (1977), 77–88.

[190] Rozsnyai, B. F., *Quantum-statistical models for multicomponent plasmas. II.*, Phys. Rev. A **16** (1977), no. 4, 1687–1691.

[191] Rozsnyai, B. F., *An overview of the problems connected with theoretical calculations for hot plasmas*, J. Quantit. Spectr. Radiat. Transfer **27** (1982), no. 3, 211–217.

[192] Rozsnyai, B. F., *Bracketing the astrophysical opacities for the King* IVa *mixture*, Astrophys. J. **341** (1989), no. 1, 414–420.

[193] Rozsnyai, B. F., *Collisional-radiative average-atom model for hot plasmas*, Phys. Rev. E **55** (1997), no. 6, 7507–7521.

[194] Rozsnyai, B., Einwohner, T., *Theoretical analysis of spectral lines of one and two electron ions in hot plasma*, in: *Spectral Line Shapes* II, pp. 315–327, Walter de Gruyter, Berlin-New York, 1983.

[195] Rudzikas, Z. B., Nikitin, A. A., Kholtygin, A. F., *Theoretical Atomic Spectroscopy*, LGU, Leningrad, 1990. (Russian); see also Rudzikas, Z. B., *Theoretical Atomic Spectroscopy*, Cambridge University Press, Cambridge, New York, 1997.

[196] Salzmann, D., *Atomic Physics in Hot Plasmas*, Oxford University Press, New York–Oxford, 1998.

[197] Samarskii, A. A., *Introduction to Numerical Methods*, Nauka, Moscow, 1987. (Russian)

[198] Samarskii, A. A., Gulin, A. V., *Numerical Methods*, Nauka, Moscow, 1989. (Russian)

[199] Sampson, D. H., Parks, A. D., *Electron-impact cross sections for complex ions. II. Application to the isoelectronic series of helium and other light elements*, Astrophys. J. Suppl. **28** (1974), no. 263, 323–334.

[200] Sapozhnikov, A. T., Pershina, A. V., *Semi-empirical equation of state of metals in a wide range of densities and temperatures*, in: Voprosy Atom. Nauki Tekhn. *Methods and Numerical Codes for Solving Problems of Mathematical Physics*, Vol. 4 (1975), 47–56.

[201] Seaton, M. J., *Atomic data for opacity calculations*: XIII. *Line profiles for transitions in hydrogenic ions*, J. Phys. B: Atom. Mol. Opt. Phys., **23** (1990), 3256–3295.

[202] Sin′ko, G. V., *Utilization of the self-consistent field method for the calculation of electron thermodynamic functions in simple substances*, Teplofizika Vysokikh Temperatur **2** (1983), no 6, 1041–1052 (Russian); English transl. High Temp. **21** (1983), no. 6, 783–793.

[203] Slater, J. C., *A simplification of the Hartree-Fock method*, Phys. Rev. A **81** (1951), no. 3, 385–390.

[204] Slater, J. C., *Quantum Theory of Molecules and Solids, Vol. 4: The Self-Consistent Field of Molecules and Solids*, McGraw-Hill, New York, 1963–74.

[205] Sobelman, I. I., *Introduction to the Theory of Atomic Spectra*, Translated from the Russian, Pergamon Press, Oxford, New York, 1972.

[206] Spielman, R. B., Deeney, C., Chandler, G. A., et al., *Tungsten wire-array Z-pinch experiments at* 200 TW *and* 2 MJ, Phys. Plasmas **5** (1998), no. 5, 2105–2111.

[207] Springer, P. T., Fields, D. J., Wilson, B. G., et al, *Spectroscopic absorption measurements of an iron plasma*, Phys. Rev. Lett. **69** (1992), no. 26, 3735–3738.

[208] Springer, P. T., Fields, D. F., Wilson, B. G. et al, *Spectroscopic measurements of Rosseland mean opacity*, J. Quantit. Spectr. Radiat. Transfer, **52** (1994), nos. 1–2, 371–377.

[209] Springer, P. T., Wong, K. L., Iglesias, C. A., et al, *Laboratory measurements of opacity for stellar envelopes*, J. Quantit. Spectr. Radiat. Transfer, **58**, (1997), nos. 4–6, 927–935.

[210] Stein, J., Shalitin, D., Ron, A., *Average-atom models of line broadening in hot dense plasmas*, Phys. Rev. A **31** (1985), no. 1, 446–450.

[211] Steklov, V. A., *Asymptotic Behavior of Solutions of Linear Differential Equations*, Izd. Khar′kov. Gosud. Univ., Khar′kov, 1956. (Russian)

[212] Sviridov, D. T., Sviridova, R. K., Smirnov, Yu. F., *Theory of Optical Spectra of Transition Metal Ions in Crystals*, Nauka, Moscow, 1976. (Russian)

[213] Thomas, L. H., *The calculation of atomic fields*, Proc. Cambridge Phil. Soc., **23** (1926), p. 542–548.

[214] Trainor, K. S., *Construction of a wide-range tabular equation of state for copper*, J. Appl. Phys. **54** (1983), no. 5, 2372–2379.

[215] Tsakiris, G. D., Eidmann, K., *An approximate method for calculating Planck and Rosseland mean opacities in hot dense plasmas*, J. Quantit. Spectr. Radiat. Transfer, **38** (1987), no. 5, 353–368.

[216] Uvarov, V. B., Aldonyasov, V. I., *The phase method for the calculation of energy eigenvalues of the Schrödinger equation*, Zh. Vychisl. Mat. i Mat. Fiz., **7** (1967), no. 2, 436–439. (Russian)

[217] Uvarov, V. B., Nikiforov, A. F., *An approximation method for solving the Schrödinger equation*, Zh. Vychisl. Mat. i Mat. Fiz., **1** (1961), no. 1, 177–179. (Russian)

[218] Vainshtein, L. A., Sobel′man, I. I., Yukov, E. A. *Excitation of Atoms and Broadening of Spectral Lines*, translated from the Russian, Springer-Verlag, New York, 1995.

[219] Van Regemorter, H., *Rate of collisional excitation in stellar atmospheres*, Astrophys. J. **136** (1962), p. 906.

[220] Varshalovich, D. A., Moskalev, A. N., Khersonskii, V. K., *Quantum Theory of Angular Momentum: Irreducible Tensors, Spherical Harmonics, Coupling Coefficients, 3nj Symbols*, World Scientific, Teaneck, NJ, 1988.

[221] Veselov, M. G., Labzovskii, L. N., *Theory of Atoms. Structure of Electron Shells*, Nauka, Moscow, 1986. (Russian)

[222] Volkov, E. A., *Numerical methods*, translated from the Russian, "Mir", Moscow, 1986; Hemisphere Publishing Corp., New York, 1990.

[223] Voropinov, A. I., Gandel′man, G. M., Podval′nyi, V. G., *Electronic energy spectra and the equation of state of solids at high pressures and temperatures*, Uspekhi Fiz. Nauk **100** (1970), no. 2, 193–224 (Russian); English transl. Soviet Physics – Uspekhi **13** (1970), no. 1, 56–72.

[224] Voropinov, A. I., Gandel′man, G. M., Dmitriev, N. A., Podval′nyi, V. G., *Pressure in metals in the Hartree-Fock approximation*, Fizika Tverdogo Tela **19** (1977), no. 11, 3332–3338 (Russian); English transl. Soviet Physics - Solid State **19** (1977), no. 11, 1945–1949.

[225] Weizsäcker, C. F., Z. Phys., **96** (1935), p. 431.

[226] Wilkinson, J. H., *The Algebraic Eigenvalue Problem*, Oxford Science Publications. The Clarendon Press, Oxford University Press, New York, 1970.

[227] Winhart, G., Eidmann, K., Iglesias, C. A., et al, XUV *Opacity measurements and comparison with models*, J. Quantit. Spectr. Radiat. Transfer, **54** (1995), nos. 1-2, 437-446.

[228] Winhart, G., Eidmann, K., Iglesias, C. A., Bar-Shalom, A., *Measurements of extreme* uv *opacities in hot dense* Al, Fe *and* Ho, Phys. Rev. E **53** (1996), no. 2, R1332-R1335.

[229] Yan, Yu., Seaton, M. J., *Atomic data for opacity calculations*: IV. *Photoionization cross sections for* C II, J. Phys. B **20** (1987), 6409–6429.

[230] Yutsis, P. P., Levinson, I. B., Vanagas, V. V., *Mathematical Apparatus of the Theory of Angular Momentum*, translated from the Russian, Israel Program for Scientific Translations, Jerusalem, 1962; New York, Gordon and Breach, 1964.

[231] Zagonov, V. P., Novikov, V. G., *Local approximation of two-dimensional tables of thermophysical data*, Preprint, Keldysh Institute of Applied Mathematics, Russian Academy of Sciences, Moscow, 1996, 18 p. (Russian)

[232] Zagonov, V. P., Novikov, V. G., *Local approximation of two-dimensional tables on nonuniform meshes*, Preprint, Keldsyh Institue of Applied Mathematics, Russian Academy of Sciences, Moscow, 1997, 21 p. (Russian)

[233] Zav′yalov, Yu. S., Kvasov, B. I., Miroshnichenko, V. L., *Method of Spline Functions*, Nauka, Moscow, 1980. (Russian)

[234] Zeldovich, Ya. B., Raizer, Yu. P., *Physics of Shock Waves and High-Temperature Hydrodynamic Phenomena*, Translated from the Russian, Academic Press, New York, 1966–67.

[235] Zemtsov, Yu. K., Starostin, A. N., *Does the probability for spontaneous emission depend on the density and the temperature*? Zh. Eksp. Teor. Fiz. **103** (1993), no. 2, 345–373 (Russian); English transl. J. Exp. Theor. Phys. **76** (1993), no. 2, 186-199.

[236] Zhang, H. L., Sampson, D. H., Mohanty, A. K., *Fully relativistic and quasirelativistic distorted-wave methods for calculating collision strengths for highly charged ions*, Phys. Rev. A. **40** (1989), no. 2, 616–632.

[237] Zhdanov, V. P., *Dielectronic recombination*, Voprosy Teorii Plazmy (1982), no. 12, 79–93. (Russian)

[238] Zhernokletov, M. V., Zubarev, V. N., Trunin, M. F., Fortov, V. E., et al *Experimental data on shock compressibility and adiabatic expansion of condensed media for high energy densities*, VNIIEF, Chernogolovka, 1996. (Russian).

[239] Ziman, J. M., *The Calculation of Bloch Functions*, in: Solid State Physics, Vol. 26, Academic Press, New York, 1971.

[240] Zink, J. W., *Shell structure and the Thomas-Fermi equation of state*, Phys. Rev. **176** (1968), nos. 1–5, 279–284.

[241] Final Report of the Third International Opacity Workshop & Code Comparison Study. — Garshing: Max-Planck Institut für Quantenoptik. — 1995.

[242] Final Report of the Fourth International Opacity Workshop & Code Comparison Study. — Madrid: Institute of Nuclear Fusion, 1998 (in press).

[243] *Handbook of Optical Constants of Solids*, edited by E. D. Palik, Academic Press, Orlando, 1985.

[244] LASL *Shock Hugoniot Data*, S. P. Marsh, ed., University of California Press, Berkeley, 1980.

[245] Los Alamos Nat. Lab. Report SESAME'83: Report on the Los Alamos equation-of-state Library.T4-Group, Los Alamos $LANL - 83 - 4$ / New Mexico: LANL, 1983.

[246] Los Alamos Nat. Lab. Report SESAME: The Los Alamos National Laboratory Equation of State Database LA-UR-92-3407 New Mexico: LANL, 1992.

[247] *Proceedings of the 3rd International Conference on Radiative Properties of Hot Dense Matter*, B. Rozsnyai, ed., Williamsburg, Virginia, Oct. 14–19, 1985.

Index